Springer Water

The book series Springer Water comprises a broad portfolio of multi- and interdisciplinary scientific books, aiming at researchers, students, and everyone interested in water-related science. The series includes peer-reviewed monographs, edited volumes, textbooks, and conference proceedings. Its volumes combine all kinds of water-related research areas, such as: the movement, distribution and quality of freshwater; water resources; the quality and pollution of water and its influence on health; the water industry including drinking water, wastewater, and desalination services and technologies; water history; as well as water management and the governmental, political, developmental, and ethical aspects of water.

More information about this series at http://www.springer.com/series/13419

Ayad M. Fadhil Al-Quraishi ·
Abdelazim M. Negm
Editors

Environmental Remote Sensing and GIS in Iraq

 Springer

Editors
Ayad M. Fadhil Al-Quraishi 🆔
College of Engineering
Knowledge University
Erbil, Kurdistan Region, Iraq

Abdelazim M. Negm
Faculty of Engineering
Zagazig University
Zagazig, Egypt

ISSN 2364-6934   ISSN 2364-8198 (electronic)
Springer Water
ISBN 978-3-030-21346-6   ISBN 978-3-030-21344-2 (eBook)
https://doi.org/10.1007/978-3-030-21344-2

This Springer imprint is published by the registered company Springer Nature Switzerland AG
The registered company address is: Gewerbestrasse 11, 6330 Cham, Switzerland

Preface

This volume came into conception to highlight the use of remote sensing (RS) and geographic information system (GIS) and their applications in Iraq. This unique volume is authored by experts in the topic from Iraq and other countries too to present the results and findings of their research work and the related state of the art connected to the book title. The volume consists of five parts excluding the introduction and the conclusions parts. The book is comprised of 21 chapters written by more than 40 authors. **Part I** is an introduction to the environmental remote sensing and GIS in Iraq, where the editors present a general overview and highlight the technical elements of each chapter.

Part II of the volume is titled "**Soil Characterization, Modelling, and Mapping**" and contains three Chaps. 2, 3, and 4. Chapter 2 is titled "**Using Radar and Optical Data for Soil Salinity Modeling and Mapping in Central Iraq**". The chapter is aiming to ascertain the possibility to propose a simple and operational approach for soil salinity assessment by combining both radar and optical data as a complement to the available optical ones. The specific objectives were (1) to develop combined soil salinity model(s) by incorporating radar backscatter coefficient with biophysical indicators from optical data, and (2) to explore the potential to develop a radar-based model(s) for the same purpose. While Chap. 3 is titled "**Using Remote Sensing to Predict Soil Properties in Iraq**" and is devoted to demonstrating the development of some statistical models to predict some components of Iraqi soils using remote sensing techniques including spectral indices and electromagnet induction. The last chapter in Part II is titled "**Characterization and Classification of Soil Map Units by Using Remote Sensing and GIS in Bahar Al-Najaf, Iraq**". The chapter aims to find out the possibility of using remote sensing (RS) and geographic information system (GIS) techniques in contributing to soil surveys by selecting soil map units drawing and calculating spectral reflectance by satellite image of Landsat 8 provided with two sensors: Operational Land Imager (OLI) and Thermal Infrared Sensor (TIRS).

On the other hand, **Part III** is titled "**Proximal Soil Sensing**" and contains two Chaps. 5 and 6. Chapter 5 entitled "**Proximal Soil Sensing for Soil Monitoring**". The chapter brings together ideas and examples from developing and using proximal soil sensing (PSS) and proximal sensors for applications, such as precision agriculture and soil contamination monitoring with specific attention to Iraq soils. While Chap. 6 titled "**Proximal Soil Sensing Applications in Soil Fertility**" is presented to show how to predict soil total nitrogen (N) and available phosphor (P) by using Vis-NIR Spectroscopy technique. This is essential to help in fertilization management to avoid under- or over-fertilization, to reduce agricultural input costs, and to provide sustainability strategy.

Part IV is titled " **RS and GIS for Land Cover/Land Use Change Monitoring**" written in three Chaps. 7, 8, and 9. In Chap. 7 with the title "**Multi-temporal Satellite Data for Land Use/Cover (LULC) Change Detection in Zakho, Kurdistan Region-Iraq**," the author aims to provide information on LULC using multi-temporal Landsat images in Zakho district, Duhok, Kurdistan region, Iraq, for which detailed thematic maps are currently lacking. In the chapter, three main tasks are identified and achieved including: (a) identifying and defining types of LULC for the study area, (b) examining the variation in the distribution of LULC, and (c) providing an up-to-date database and produce accurate maps. While the authors of Chap. 8 under the title "**Monitoring of the Land Cover Changes in Iraq**" aim to evaluate the nature, and rate of climate and vegetation change from 2002 to 2016 using Tropical Rainfall Measuring Mission (TRMM) and Moderate Resolution Imaging Spectroradiometer-Normalized Difference Vegetation Index (MODIS-NDVI) time series by considering the variations in April. Moreover, the chapter deals with finding a statistical relationship between the density of vegetation and elevation in Iraq territory. Also, the authors monitored the change of some climate variables such us precipitation and temperature of Iraq using multi-in-situ and satellite data. Additionally, Chap. 9 entitled "**Effects of Land Cover Change on Surface Runoff Using GIS and Remote Sensing: A Case Study Duhok Sub-basin**" endeavors to examine the impact of the urban landscape pattern changes at a local level on the volume of runoff, which depends on a satellite image time series from 1990 to 2016 and using the city of Duhok, Kurdistan region, Iraq as a case study.

Part V has the theme "**Land Degradation, Drought, and Dust Storms**" and is covered in six chapters from 10 to 15. In Chap. 10 titled "**Monitoring and Mapping of Land Threats in Iraq Using Remote Sensing**," the author tries to demonstrate the output for some works, which have been done on monitoring and mapping spatial and temporal changes of Iraqi resources using remote sensing and GIS techniques. While in Chap. 11 under the title "**Agricultural Drought Monitoring Over Iraq Utilizing MODIS Products**," the authors use the remote sensing techniques for mapping agricultural drought maps from 2003 to 2015 in Iraq. They indicated that most of Iraqi agricultural lands are highly affected by one or more of the desertification processes due to poor management practices, dry climatic conditions, and effects of socio-economic factors.

On the other hand, Chap. 12 is titled "**The Aeolian Sand Dunes in Iraq: A New Insight**" and is devoted to detecting the aeolian sand dunes' changes between 2000 and 2016 using Landsat imagery for entire Iraq. Moreover, the authors determine the activity of the aeolian sand dunes' movement in an area in the central part of the

Mesopotamia by applying DInSAR technique for the period between March 2015 and August 2016 using Sentinel 1A. While Chap. 13 under the title "**Drought Monitoring for Northern Part of Iraq Using Temporal NDVI and Rainfall Indices**" combines both meteorological and remote-sensed indices to map drought conditions in the northern part of Iraq. The authors show that desertification has been increased in the study area. The authors aim to use the NDVI and SPI indices to detect the appearance and severity of the drought event for the northern part of Iraq.

Moreover, Chap. 14 is written under the title "**Remote Sensing and GIS for Dust Storm Studies in Iraq**" to demonstrate the spatial–temporal distributions of sources, causes, and atmospheric and wind patterns of dust storms as well as their environmental circumstances (air–soil–vegetation–water) in Iraq. Additionally, Chap. 15 is titled "**Drought Monitoring Using Spectral and Meteorological Based Indices Combination: A Case Study in Sulaimaniyah, Kurdistan Region of Iraq**". The authors attempt to highlight the benefits of the combination of the spectral and meteorological based indices for drought monitoring and mapping in Sulaimaniyah, Kurdistan region, Iraq.

Part VI contains five chapters, from 16 to 20 chapters. They are presented under the theme "**RS and GIS for Natural Resources**." Chapter 16 is titled "**Geo-Morphometric Analysis and Flood Simulation of the Tigris River Due to a Predicted Failure of the Mosul Dam, Mosul, Iraq**". The chapter covers two main objectives: (a) performing geo-morphometric analysis for a specific river basin in the northern part of Iraq that includes Mosul Dam, and (b) delineating flooded zone due to a predicted collapse of the Mosul Dam. While Chap. 17 under the title "**Hydrologic and Hydraulic Modelling of the Greater Zab River-Basin for an Effective Management of Water Resources in the Kurdistan Region of Iraq Using DEM and Raster Images**" presents the computational design of the hydrologic and hydraulic modeling system for the Greater Zab River Basin in the Kurdistan region of Iraq.

On the other hand, Chap. 18 titled "**Spatial Assessment of Drought Conditions Over Iraq Using the Standardized Precipitation Index (SPI) and GIS Techniques**". The chapter aimed to assessing, monitoring, and mapping of long-term drought and wet condition by using Standardized Precipitation Index (SPI), which is the best way to put efficient water policy and management in Iraq.

Additionally, Chap. 19 is presented under the title "**Assessing the Impacts of Climate Change on Natural Resources in Erbil Area, the Iraqi Kurdistan Using Geo-Information and Landsat Data**". It aims to quantitatively study the large-scale semi-aridization of the climate in the Kurdistan region of Iraq, revealed by the rise of temperatures and the decline of the amount of precipitations, and aims to quantify spatial and temporal dynamics of LULC, in particular, the changes in vegetation, surface water and urban and built-up areas in the study area. The last chapter in this part is titled "**Mapping Forest-Fire Potentiality Using Remote Sensing and GIS, Case Study: Kurdistan Region-Iraq.**" It is aiming to map the areas of most potential of firing to help the managers to follow preventive actions because a great number of forest fires have been recorded in the north of Iraq where it is almost the only area in Iraq forests are remaining.

The last chapter in this book (**Part VII**) is the conclusions and recommendations and is written by the editors. The chapter presents an update of the most recent findings, the most significant conclusions, and recommendations of the chapters contained in the volume.

Special thanks are due to all authors who contributed to this volume; without their efforts and patience, it would not have been possible to produce this unique volume on Environmental RS and GIS in Iraq. Also, thanks should be extended to include the Springer team who largely supported the authors and editors during the production of this volume.

Zagazig, Egypt   Abdelazim M. Negm
Erbil, Kurdistan Region, Iraq   Ayad M. Fadhil Al-Quraishi
April 2019

Contents

Part I   Introduction

1   Introduction to "Environmental Remote Sensing and GIS
    in Iraq"   3
    Ayad M. Fadhil Al-Quraishi and Abdelazim M. Negm

Part II   Soil Characterization, Modelling, and Mapping

2   Using Radar and Optical Data for Soil Salinity Modeling
    and Mapping in Central Iraq   19
    Weicheng Wu, Ahmad S. Muhaimeed, Waleed M. Al-Shafie
    and Ayad M. Fadhil Al-Quraishi

3   Using Remote Sensing to Predict Soil Properties in Iraq   41
    Ahmad Salih Muhaimeed

4   Characterization and Classification of Soil Map Units
    by Using Remote Sensing and GIS in Bahar Al-Najaf, Iraq   61
    Abdulameer S. Al-Hamdani and Hussein M. Al-Shimmary

Part III   Proximal Soil Sensing

5   Proximal Soil Sensing for Soil Monitoring   95
    Banaz M. Mustafa, Ayad M. Fadhil Al-Quraishi, Asa Gholizadeh
    and Mohammadmehdi Saberioon

6   Proximal Soil Sensing Applications in Soil Fertility   119
    Qassim A. Talib Alshujairy and Nooruldeen Shawqi Ali

Part IV   RS and GIS for Land Cover/Land Use Change Monitoring

7   Multi-temporal Satellite Data for Land Use/Cover (LULC)
    Change Detection in Zakho, Kurdistan Region-Iraq   161
    Yaseen T. Mustafa

8   Monitoring of the Land Cover Changes in Iraq   181
Arsalan Ahmed Othman, Ahmed T. Shihab, Ahmed F. Al-Maamar
and Younus I. Al-Saady

**9   Effects of Land Cover Change on Surface Runoff Using GIS
and Remote Sensing: A Case Study Duhok Sub-basin**   205
Hasan Mohammed Hameed, Gaylan Rasul Faqe and Azad Rasul

Part V   Land Degradation, Drought, and Dust Storms

**10   Monitoring and Mapping of Land Threats in Iraq
Using Remote Sensing**   227
Ahamd Salih Muhaimeed

**11   Agricultural Drought Monitoring Over Iraq Utilizing MODIS
Products**   253
Yousif S. Almamalachy, Ayad M. Fadhil Al-Quraishi
and Hamid Moradkhani

12   The Aeolian Sand Dunes in Iraq: A New Insight   279
Arsalan Ahmed Othman, Younus I. Al-Saady, Ahmed T. Shihab
and Ahmed F. Al-Maamar

**13   Drought Monitoring for Northern Part of Iraq Using Temporal
NDVI and Rainfall Indices**   301
Suhad M. Al-Hedny and Ahmad S. Muhaimeed

14   Remote Sensing and GIS for Dust Storm Studies in Iraq   333
Ali Darvishi Boloorani, Najmeh Neysani Samany, Saham Mirzaei,
Hossein Ali Bahrami and Seyed Kazem Alavipanah

**15   Drought Monitoring Using Spectral and Meteorological Based
Indices Combination: A Case Study in Sulaimaniyah, Kurdistan
Region of Iraq**   377
Ayad M. Fadhil Al-Quraishi, Sarchil H. Qader and Weicheng Wu

Part VI   RS and GIS for Natural Resources

**16   Geo-Morphometric Analysis and Flood Simulation
of the Tigris River Due to a Predicted Failure of the Mosul Dam,
Mosul, Iraq**   397
Younis Saida Saeedrashed and Ali C. Benim

**17   Hydrologic and Hydraulic Modelling of the Greater Zab
River-Basin for an Effective Management of Water
Resources in the Kurdistan Region of Iraq Using DEM
and Raster Images**   415
Younis Saida Saeedrashed

**18 Spatial Assessment of Drought Conditions Over Iraq Using
     the Standardized Precipitation Index (SPI)
     and GIS Techniques** 447
     Ayad Ali Faris Beg and Ahmed Hashem Al-Sulttani

**19 Assessing the Impacts of Climate Change on Natural Resources
     in Erbil Area, the Iraqi Kurdistan Using Geo-Information
     and Landsat Data** 463
     Huner Abdulla Kak Ahmed Khayyat, Azad Jalal Mohammed Sharif
     and Mattia Crespi

**20 Mapping Forest-Fire Potentiality Using Remote Sensing and GIS,
     Case Study: Kurdistan Region-Iraq** 499
     Iraj Rahimi, Salim N. Azeez and Imran H. Ahmed

Part VII Conclusions

**21 Updates, Conclusions, and Recommendations for Environmental
     Remote Sensing and GIS in Iraq** 517
     Ayad M. Fadhil Al-Quraishi and Abdelazim M. Negm

Part I
Introduction

Chapter 1
Introduction to "Environmental Remote Sensing and GIS in Iraq"

Ayad M. Fadhil Al-Quraishi and Abdelazim M. Negm

Abstract This chapter presents the main technical features of the book titled "Environmental Remote Sensing and GIS in Iraq". The book consisted of seven parts including the introduction (this chapter) and the conclusions (the closing chapter). The main body of the book consisted of five themes to cover Soil Characterization, Modelling, and Mapping in three chapters, Proximal Soil Sensing in two chapters, remote sensing (RS) and Geographical Information Systems (GIS) for Land Cover/Land Use Change Monitoring in three chapters, Land Degradation, Drought, and Dust Storms in six chapters, and RS and GIS for Natural Resources. The main technical elements of each chapter are presented under its relevant theme.

Keywords Remote sensing · GIS · Land degradation · Soil · Modeling · Mapping · Drought · Proximal · Natural resources · Iraq

1.1 Iraq: A Brief Background

Iraq, officially the Republic of Iraq (Fig. 1.1), is a country in Western Asia, bordered by Turkey to the north, Iran to the east, Kuwait to the southeast, Saudi Arabia to the south, Jordan to the southwest and Syria to the west. Iraq is located between latitudes $29° 5'$–$37° 22'$N and longitudes $38° 45'$–$48° 45'$E, with a total area of 438,317 km^2. Its capital is Baghdad, which is the largest city in the country.

A. M. F. Al-Quraishi
Environmental Engineering Department, College of Engineering, Knowledge University, Erbil 44001, Kurdistan Region, Iraq
e-mail: ayad.alquraishi@gmail.com; ayad.alquraishi@knowledge.edu.krd

A. M. Negm (✉)
Water and Water Structures Engineering Department, Faculty of Engineering, Zagazig University, Zagazig 44519, Egypt
e-mail: amnegm@zu.edu.eg

© Springer Nature Switzerland AG 2020
A. M. F. Al-Quraishi and A. M. Negm (eds.),
Environmental Remote Sensing and GIS in Iraq, Springer Water,
https://doi.org/10.1007/978-3-030-21344-2_1

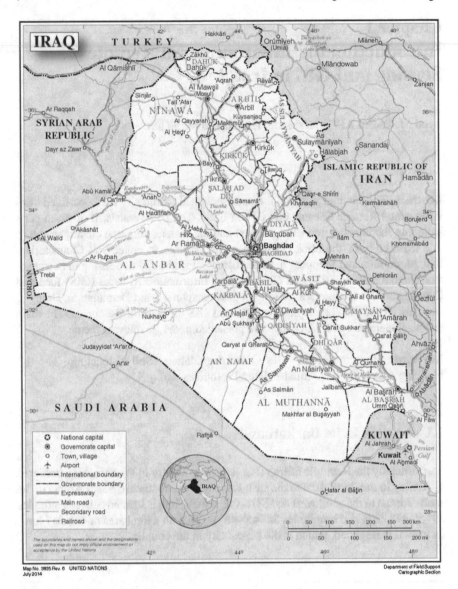

Fig. 1.1 The map of Iraq. *Source* Iraq, Map No. 3835 Rev. 6. July 2014. Courtesy of the United Nations, http://www.un.org/Depts/Cartographic/map/profile/iraq.pdf

During the ancient times, the lands that now constitute Iraq were known as *Mesopotamia*, which means "(Land Between the Rivers), a region whose extensive alluvial plains gave rise to some of the world's earliest civilizations, including those of Sumer, Akkad, Babylon, and Assyria. This wealthy region, comprising much of what is called the *Fertile Crescent*. Later became a valuable part of larger imperial polities, including sundry Persian, Greek, and Roman dynasties. Moreover after the 7th century, it became a central and integral part of the Islamic world. Iraq's capital, Baghdad, became the capital of the Abbāsid caliphate in the 8th century. The modern nation-state of Iraq was created following World War I (1914–18)" (https://www.britannica.com/place/Iraq).

Around 97% of the country is in arid lands with low and erratic rainfall. In most parts agriculture suffers from high rates of evapotranspiration that exceed rainfall. Temperature varies widely (10–40 °C) during the growing season, particularly in desert regions. Iraq is an agricultural country. Although a considerable portion of its agricultural lands is under irrigation, it still depends highly on the rainfed agriculture for grain and sheep production (http://www.fao.org/iraq/fao-in-iraq/iraq-at-a-glance/en/). "Rainfed agriculture is practiced in the northern parts where the mountains, foothills, and Jazeera desert are located" (https://documents.wfp.org/stellent/groups/public/documents/ena/wfp291289.pdf). These three regions depend mostly on rainfall for agricultural production and supply a substantial part of the grain (wheat and barley) consumed in the country. Of the 120,000 km^2 of cultivable land, which comprise 26.4% of the total area, there are 40,000 km^2 in the rainfed region.

The next section will present the technical aspects of the chapters under their related themes.

1.2 Soil Characterization, Modelling, and Mapping

This theme is covered in three chapters. Chapter 2 titled "Using Radar and Optical Data for Soil Salinity Modeling and Mapping in Central Iraq". In this chapter, Landsat 5 TM imagery and ALOS (Advanced Land Observing Satellite) L-band SAR data acquired at almost the same time were used for this study. Water Cloud Model (WCM) proposed by Attema and Ulaby (1978) was applied to minimize the impacts of vegetation cover and moisture on the soil components of the backscattering coefficients. After this processing, the soil components have gained an increase in correlation coefficients with the measured apparent soil salinity (ECa) by 7.5–25.6%.

A radar-optical combined dataset including all the biophysical indicators from Landsat TM images, L-band backscattering coefficients, and its soil components were produced. A multivariate linear regression (MLR) modeling was applied to couple the optical and radar indicators with the apparent soil salinity and to establish the combined and radar-based soil salinity models. Results revealed that the optical-radar combined or radar-based models can reliably predict soil salinity with an accuracy of 74.0–84.3% ($R^2 = 0.739$–0.843).

To sum up, removal of vegetation cover impacts can greatly improve the salinity prediction accuracy and reliability by radar data, but radar-optical combined dataset can deliver better soil salinity prediction and mapping results. Radar data have great potential for soil salinity mapping. Development of a radar-based approach for minimization of vegetation cover impacts on the backscattering coefficients without any dependence on the optical data is recommended in future work.

While Chap. 3 is focusing on "the use of Remote Sensing to Predict Soil Properties in Iraq". Therefore, the main objectives of this chapter are to demonstrate the development of some statistical models to predict some components of Iraqi soils using remote sensing techniques. The soil is one of the most important components of agricultural production and can have a dominant effect on crop yields and quality. "Different methodologies have been proposed for the estimation of soil parameters, based on different remote sensing sensors and techniques" (https://www.hindawi.com/journals/aess/2011/904561/). In-field soil information has been used for centuries by farmers to make decisions concerning crop management practices. Quantitative information and spatial distribution of soil properties are among the main prerequisites for achieving sustainable land management. The accuracy of soil information determines, to a large extent, the reliability of land resources management decisions. The results indicated that the remote sensing techniques are a very useful tool to predict soil organic matter (SOM), total Nitrogen, salinity level and some physical soil properties including the content of clay, silt, sand, and infiltration rate. The results revealed a highly significant relationship between the measured values for the selected soil properties and the predicted values. Finally, we can sum up that remote sensing multi-temporal satellite data has proven to be an important tool to predict some physical and chemical soil properties on a large scale using different indices, including NDVI, SAVI, and GDVI. Remote sensing techniques are very accurate, useful and helpful to predict the properties of the surface soil as they can consume efforts, time and money.

Moreover, Chap. 4 is dealing with the characterization and classification of soil map units by using remote sensing and GIS in Bahar Al-Najaf, Iraq. The chapter aims to find out the possibility of using remote sensing (RS) and GIS techniques to show the nature and interference effect of the relationship between some of the soil physical and chemical properties on its spectral reflectance values by choosing false-color composite RGB (753) and subsetting study area using ERDAS imagine 2013. Depending on the unsupervised classification, as well as soil indicators such as soil color, texture, natural plants and topography in determining the researcher movement paths to select 16 Pedons with 21 auger hole sites for surveying and isolating expected soil individuals which are identified using US modern classification. All the studied soils are within the Entisol order, which classified into two of Suborders. The first is Fluvents including great soil group Torri fluvents and subgroup Typic Torri fluvents, which includes 9 Soil series ME1, DW56, TW964, DM44, MW3, TE354, DW124, DE47, DE126. The second is Psamments including two great soil groups, The first Torri psamments containing subgroup Typic Torri psamments including DE33, DE34, DE74, TE334, TW446 soils series. The other quartzi psamments include subgroup Typic quartzi psamments containing two soil series ME1, DM14. The soil series are classified according to the proposed (Al-Agidi 1976) for alluvial Iraqi soils classification.

1.3   Proximal Soil Sensing

This theme is covered in two chapters. Chapter 5 is devoted to soil monitoring using proximal soil sensing technique. In a quest to derive information on potentials and limitations of soils and to increase the efficiency of soil survey and to make it cost-effective, PSS and soil spectroscopy have been used, and a perceptible improvement in the level of information that could be derived has been achieved. Finding suitable data preprocessing and calibration strategies for the application of Visible-Near Infrared-Short Wave Infrared (VIS-NIR-SWIR) spectroscopy is dramatically significant and influence the final accuracy. Moreover, inaccuracy and uncertainties occur at the different steps in the prediction procedure, and the accuracy of the resulting models can be insufficient for the planned purpose; therefore, it is important to assess the accuracy of the process. The chapter also provided a short study as a case study on employing soil spectroscopy in Sulaimani, Iraqi Kurdistan Region. The spectroscopy technique was used for monitoring and mapping total Fe, and Fe_2O_3 rich soils of some sites in the area. The statistical accuracy and distribution maps obtained using laboratory spectroscopy measurements indicated that, for the soil Fe and Fe_2O_3 under study, laboratory proximal level prediction could give a reasonable indicator based on spectra from soil samples. According to the study, the accuracy and distribution maps derived from laboratory spectroscopy measurements proved the capability of laboratory proximal level prediction for assessing the soil Fe and Fe_2O_3 under study area in Sulaimani, the Iraqi Kurdistan region.

Chapter 6 is dealing with the application of proximal soil sensing to monitor soil fertility. The chapter outlines a historical review of Proximal sensing along with its applications as a potential analytical technique for studying in different fields of study. To this end, the chapter contains background about the use of VIS-NIR-Spectroscopy in soil analysis such as soil minerals, texture, organic matter, water content and fertility with focusing on nutrients (N. P. and K.). The methodological part of this chapter overviews the sampling and preparation of soil for studying steps. The second section of the methodological part describes VIS-NIR Spectra analysis, calibration, spatial variability of total N, available P using GIS-Kriking Analysis and ended with the evaluation of prediction power. The chapter will finish with the discussion of the findings of this study.

The obtained prediction model quality parameter values were at best successfully model to predict soil total N and available P and well suited for a large variety of low to high concentrations under the condition of this study.

Finally, chapter puts its recommendations to pay attention toward strategic of the Iraqi Soil Spectral Library at the country to directly be used in soil monitoring and fertilization management.

1.4   RS and GIS for Land Cover/Land Use Change Monitoring

This theme is covered in Chaps. 7, 8 and 9. Chapter 7 is directed to use the multi-temporal satellite data for land use/cover (LULC) change detection in Zakho, Kurdistan region-Iraq. Land use/cover (LULC) affects local, global environment, climate and land degradation that reduces ecosystem services and functions. Monitoring of LULC is important to assess the change and manage the environment. Remote sensing is one of the effective techniques that plays an active role in accomplishing such tasks. Landsat data is widely used for change detection of LULC and mapping earth surface. Also, Landsat images used to map and extract several objects on the earth surface. The chapter aims to provide information on LULC using multi-temporal Landsat images in Zakho district, Duhok, Kurdistan region, Iraq for a period of 28 years (1989–2017). Such a task was performed by (1) identify and define types of LULC for the study area, (2) examining the variation in the distribution of LULC, and (3) provide an up-to-date database and produce accurate maps.

The authors indicated that the overall pattern of LULC change in the Zakho district over the past 28 years was one of the sprawl of crops and built-up lands. Adding to that, a substantial reduction of forest and grassland indicates an acceleration stage of agriculture and urbanization. Despite the pressing land requirements for urbanization, land development and consolidation in grass and forest areas, and the adjustment of the agricultural structure, the foundation was put for the transition to intensively use the land in the Zakho district. It is recommended to reduce urban expansion in agricultural areas, especially in the Sindi plain, moreover, to work on the vertical expansion of residential areas to accommodate population increment.

In Chap. 8, the authors used the RS to monitoring land cover changes in Iraq. The chapter describes the use of Moderate Resolution Imaging Spectroradiometer (MODIS) land cover maps—covering the period 2003 to 2013—for monitoring changes in the vegetation cover. They studied key factors controlling the spatial distribution of vegetation cover and types of vegetation and found a strong correlation between the NDVI values (using MODIS-NDVI scenes) and DEM within the Iraqi territory. There are 17 classes of land cover in Iraq, five of them are classified as non-vegetative, which include barren or sparsely vegetated; permanent wetlands; snow and ice; urban and built-up areas; and water. The rest are vegetation classes, which are closed shrublands, open shrublands, cropland/natural vegetation mosaic, croplands, deciduous broadleaf forest, deciduous needleleaf forest, evergreen broadleaf forest, evergreen needleleaf forest, mixed forest, grasslands, savannas, and woody savannas. Dominant vegetation classes in Iraq are shrublands (58.52%), croplands (27.29%) and grasslands (13.69%), which are located in Mesopotamia and the northern part. The western and southwestern parts of Iraq are mostly desert. The average area of shrublands, croplands, and grasslands estimated for the period, 2003–2013 was 66, 122, 30, 832, and 15,472 km^2, respectively.

The relationship between the digital elevation model and the NDVI values of the grasslands in Iraq shows good inverse correlation, where R^2 values were found to be greater than 0.7. There is no vegetation (NDVI value more than 0.2) above 2,600 m elevation. Moreover, no statistical relationships could be obtained between elevation and croplands or the shrublands. This is because croplands and shrublands are found at various elevations (i.e., lowlands and mountainous lands). On the contrary, grasslands occur in both the foothills and mountainous lands. The authors concluded that one of the major causes of eco-environmental degradation in Iraq is ineffective and wasteful utilization of land cover, particularly the vegetation cover that, to a large extent, contributes to climate change. Four main factors have resulted in the development of modern land cover in Iraq; these are climate, surface and subsurface water, lithology, and relief. Hence, a strong correlation was found between these factors, and the spatial distribution of the land cover classes is strong. It is important to reduce soil salinity and control sabkha development by preventing the use of Al-Tharthar Lake water for irrigation as it contributes to washing out of salts that ultimately enters the Euphrates River.

Moreover, Chap. 9 focus on the effects of land cover change on surface runoff using GIS and remote sensing with application to Duhok Sub-Basin, Kurdistan region, Iraq as a case study.

To quantify the impact of land cover changes on runoff, a simple hydrological model directly considering land use/land cover is used. The Soil Conservation Service Curve Number (SCS-CN) was used to assess surface water volume for a given rainfall event for the small watershed area. The integration of a GIS and remote sensing was used and "is recognized as a useful and effective tool in locating urban" growth (https://www.mdpi.com/2306-5338/4/1/12/pdf). The main objective of this study focuses on pattern changes of land cover due to urbanization in the Duhok sub-basin, Kurdistan region of Iraq, as well as their effect on rainfall surface runoff. The data used in this study were multispectral satellite imaginaries. Multitemporal and multispectral satellite images were used to generate historical land use/land cover and digital land use maps were generated to estimate the effect of land cover changes on the hydrologic evaluation. The ArcGIS and Idrisi were used to determine the spatial distribution of rainfall and land cover changes. Rainfall Grids were calculated and mapped for a selected rainfall depth per pixel. Inverse Distance Weighting (IDW) interpolation method was used to estimate the rainfall spatial distribution. The methodology used for assessing the spatio-temporal variations of runoff depth for all pixels of the study area using the Soil Conservation Service Curve Number model. To sump up, Duhok sub-basin was subjected to significant land use changes in the period from 1990 to 2016. The study indicates that the urban growth of the watershed increased from 10% in 1990 to 70% in 2016. Surface runoff volume increased from 12% in 1990 to 36% in 2016, while the vegetation land decreased from 47 to 14% in the same period.

1.5   Land Degradation, Drought, and Dust Storms

This theme is covered in six chapters from Chaps. 10 to 15. Chapter 10 is devoted to explaining the use of RS in monitoring and mapping of land threats in Iraq. There is a great need to monitor the natural resources and their properties in practical place and time. Remote sensing and GIS techniques are very useful for monitoring and mapping the main type of Iraq land degradation processes as reflected by high salt accumulation and sand dunes movements. More than 25% of the land area of Iraq has a serious erosion problem, while larger than 20% of the total area, mainly in southern Iraq was seriously affected by water lodging and more than 70% was affected by salinization. Lack of comprehensive information about national land resources increases the risk of releasing uninformed policy decisions, avoidable continued degradation of land, water resources and land cover. Remote sensing offers a unique opportunity in monitoring, assessing and empirically quantifying spatial and temporal changes in soil and vegetation taking place during a long period. The objective of this chapter is to demonstrate the important of using remote sensing and GIS techniques for monitoring and mapping spatial and temporal changes of Iraqi resources. The results for the application of remote sensing indicated that the main degradation processes are affecting the Iraq agricultural lands represented by salinity and sand dunes movement. More than 70% of agricultural lands are suffering from salt accumulation as reflected by the domination of bare land in Iraq.

While Chap. 11 is dealing with agricultural drought monitoring over Iraq utilizing MODIS. Remote sensing dataset and techniques were employed in this chapter. Four different spectral indices; Vegetation Health Index (VHI), Vegetation Drought Index (VDI), Visible and Shortwave infrared Drought Index (VSDI), Temperature–Vegetation Dryness Index (TVDI) were utilized, each of them is derived from the MODIS dataset of Terra satellite. The results revealed the year 2008 was the most severe drought year the period from 2003 to 2015 in Iraq, whereas the drought covered 37% of the vegetated lands, while 2009, 2011, and 2012 were the less-severe drought years dominated by mild or moderate drought with an areal coverage of 44, 50, and 48.5%, respectively. The Vegetation Drought Index (VDI) was found to be the more suitable drought index, which can be used for Iraq, because of the temperature integration in its structure in addition to meeting the assumptions it was based on.

In Chap. 12, authors present a new insight for the aeolian sand dunes in Iraq. As a matter of the fact that there are three major accumulations of aeolian sands existing in Iraq. The first accumulation occurs as a long strip along the Abu Jeer Fault Zone, where huge fields of sand dunes extend to near the Euphrates River, from north of Karbala to south of Al-Muthanna governorates. The second major accumulation is located in the foothills of Hemreen and Mak'hool mountains. These two dunes fields are relatively older and have existed for over 20 years. The third is a recent accumulation that occurs in the middle of the Mesopotamian Plain, lying between the two old accumulations. The last one includes three large dunes fields located to the east of

Baghdad, south of Al-Qadissiya, and south of Thi-Qar governorates. All three accumulations trend in a NW-SE direction. The three sand accumulations in Iraq include almost all of the aeolian sand dunes—i.e., sand accumulation related to topographic obstacles; sand accumulation related to vegetation; sand accumulation related to bed roughness; and those formed due to aerodynamic fluctuations. The accumulations range in length from 1 m to several 10's of km, and from few centimetres to about 150 m in height.

Arid to semi-arid climatic conditions, difficulty of access and high costs of ground surveys make it extremely difficult and expensive to study the migration patterns of the huge sand accumulations. Use of multitemporal remote sensing techniques offers an attractive and cost-effective option for mapping the desertification processes as it reduces the time and cost. Accordingly, Landsat data was utilized for spectral analyses to monitor spatiotemporal changes and activity of dunes, supported by ground check to determine the extent of aeolian sand accumulations. Results show that areas of sand dunes in Iraq have increased during the period 2000–2016 to about 4,528 km^2, as compared to about 9,891 km^2 in 2000, and about 14,419 km^2 in 2016. Additionally, the study applied DInSAR technique using Sentinel 1A data to a specific site in the central part of the Mesopotamian Plain (Hor Al-Dalmaj) to estimate the dunes displacement. Vertical displacement was found to range between ~33 and ~−40 cm and the volume of the sand dune ~5.9 times smaller than the volume of eroded sand dunes.

The authors conclude that the aeolian sediments in Iraq represent an active system whose on-going migration results in loss of agricultural land, highway obstruction, etc. Moreover, the desertification is increased in the Mesopotamian Plain for recent decades.

On the other hand, Chap. 13 investigate the drought monitoring for the Northern part of Iraq using temporal NDVI and rainfall indices. Therefore, the chapter contains background about: drought as a concept, drought impacts on soil properties, drought monitoring, remote sensing and GIS (how are they used in drought studies). In order to describe and quantify this disaster from different perspectives: meteorological, hydrological and agricultural, the chapter begins with a review of drought as a concept and definition. Drought phenomenon classified as a hazard natural disaster that depreciates the sustainable development of society. Due to its crawling nature, its effects may take weeks or months to appear in a reduction of surface/ground water to support crop growth and human activities. The severity of drought is often associated not only with the deficiency of precipitation, but also with other climatic factors such as high temperature, high wind, and low humidity. After giving this overview, the chapter will shift to the empirical part through analyzing the standardized precipitation index (SPI), which is used to classify the drought events according to meteorological data. To study the main features of intense drought events impacts on the vegetation cover, the Normalized Difference Vegetation Index (NDVI) has calculated as a most commonly used index to classify the drought severity based on NDVI anomalies. The chapter will finish with the discussion of the findings of this study.

Moreover, Chap. 14 explains the use of remote sensing and GIS for dust storm studies in Iraq to summarize the main characteristics of dust storms in Iraq. The method is based on the synthesis of RS and GIS knowledge and information to assess the situation of dust storms in Iraq. The author have mentioned some examples of case studies and projects in Iraq. Remote sensing for dust storm studies including dust sources mapping, detection of dust events using satellite imagery, wind speed, and direction, the trajectory of suspended dust particles, soil data and land surface coverage map/information. In addition to atmospheric patterns of dust storms dust sources modeling with GIS, knowledge-based approaches for dust studies with GIS, fuzzy inference systems, expert systems. The use of GIS for improving the monitoring and early warning system and spatiotemporal modeling of epidemic diseases are also presented. It is worth mentioning that dust storm studies require a comprehensive framework for implementing remote sensing and GIS in combination with other disciplines of the sciences and technologies.

Also, Chap. 15 is about drought monitoring using a combination of spectral and meteorological indices with a focus on Sulaimaniyah, Kurdistan region, Iraq as a case study. The drought has dramatically affected Iraq including the Kurdistan region throughout the last decades, which were characterized by a large drop in rainfall, and its main rivers discharge in general. This chapter aimed to investigate the role of the integration of NDVI and SPI for drought monitoring in the Kurdistan region of Iraq during the years of 1990, 2007, and 2008. The chapter utilized remote sensing and GIS techniques for monitoring and mapping the drought risk in Sulaimaniyah governorate of Kurdistan region, Iraq. The NDVI, LST, and TCW were derived from the Landsat images for 1990, 2007, and 2008, as well as to the SPI as spectral and meteorological based drought indices. The aforementioned indices were used for monitoring the droughts and their impacts in the study area. The combination of the NDVI-SPI indices was suggested in this chapter to assess and map the drought risk in Sulaimaniyah. The results revealed a significant increase in the total areas of extreme, severe, moderate drought classes in 2008 by a percentage of 81.2% more than in 2007. Moreover, Dukan Lake's surface area in Sulaimaniyah suffered a significantly shrunk by 16.5% and 32.5% in 2007 and 2008, respectively, compared with its total size in 1990.

1.6   Remote Sensing and GIS for Natural Resources

The technical aspects of this theme is covered in five chapters from Chaps. 16 to 20. Chapter 16 presents a geo-morphometric analysis and flood simulation of the Tigris River due to a predicted failure of the Mosul Dam. Since GIS, RS, hydrologic and hydraulic modeling systems are capable of integrating each other to perform required hydrologic and hydraulic investigation and analysis precisely. Therefore, in this chapter through a specific schematic workflow the methodology of this research is applied in an efficient and reliable way. In this study, remotely sensed images such as DEM, topographic maps, Landsat satellite image, and tabulated hydrograph data are used. Thus, obtained geo-morphometric parameters for the Mosul Dam River-Basin are tabulated. Also, the discharge flow based on unsteady flow analysis is simulated based on a computational-2D program which is so-called HEC-RAS in order to calculate related hydraulic parameters such as time of arrival, flood depth, water surface elevation, and stream power and to delineate flooded zone inside Mosul city, Iraq as well. The obtained result from this study compared to the previous studies SWISS 1984, IWTC 2009 and JRC 2016 show good agreement with those of authors, specifically with the results of the SWISS study which is more detailed comparatively.

The present predictions indicate that the initial flood wave will reach the Mosul city in around 2 h and the height of the flood wave will reach approx. 24 m within 8 h, while the average flood velocity is predicted to be 3.9 m/s.

On the other hand, Chap. 17 explain how to use the DEM and raster images to effectively managed the water resources in the Kurdistan region of Iraq for hydrologic and hydraulic modeling of the Greater Zab River-Basin. Digital elevation model (DEM) and raster images are used as essential raw data to build reliable modeling systems. The schematic workflow of this chapter shows the methodology that is applied in order to create computational hydrologic and hydraulic modeling systems. This methodology includes two main stages of processing; pre-processing of data and post-processing. In the first stage, both geo-morphological and hydrological feature classes are obtained and calculated by using ArcGIS tools such as Arc Hydro, HEC-GeoHMS, and HEC-GeoRAS. While, in the second stage, 2D-simulation programs such as HEC-HMS and HEC-RAS are used for creating both hydrologic and hydraulic modeling systems respectively. This chapter shows the results that have been obtained. First, the main hydrologic feature classes such as (basin, watersheds, sub-watersheds, catchments, streams, and rivers) are extracted from the digital elevation model. Thus, the main geo-morphometric parameters for the Greater Zab watershed are calculated. Second, the hydrologic model is designed for calculating Rainfall-Runoff and performing floodplain analysis.

In Chap. 18, authors present how to use the Standardized Precipitation Index (SPI) and GIS Techniques to assess spatially drought conditions over Iraq. Assessment of drought conditions in Iraq is a very important task because of its location in the arid and semi-arid region. To assess and monitor drought conditions in Iraq, monthly complete precipitation data from 18 weather stations for the period (1980–2010) have been used to calculate Standardized Precipitation Index (SPI) using SPI Generator software. Statistical analysis has been done for SPI data to calculate averages, counts, minimum, maximum, and frequency values of dry and wet conditions at each station. Statistical data have been joined with the location of weather stations in ArcGIS for mapping SPI. The IDW interpolation is used with specific parameters for mapping SPI average and SPI wet and dry frequencies data. The maps show that SPI frequency for wet and dry conditions give better and real spatial distribution and variation rather than average of SPI. Time series figures and summarize table for selected weather stations in Iraq were used to analyse the trend, duration, and frequency of wet and dry periods.

The SPI values indicate there are three main drought periods that have happened in Iraq in (1984, 1999, and 2008). The interval time between drought periods is decreased in recent years. Using the frequency of wet and dry SPI is better for mapping than average SPI.

While Chap. 19 is interested in assessing the impacts of climate change on natural resources with a focus on the Erbil area, in the Iraqi Kurdistan region using geo-information and Landsat data. It aims to quantitatively study the large-scale semi-aridization of the climate in the Kurdistan region of Iraq, revealed by the rise of temperatures and consequently a decline in the precipitations volume. It aims to quantify the spatial and temporal dynamics of LULCC, in particular, the changes in vegetation, surface water and urban and built-up areas in the study area.

Nowadays, the Kurdistan region of Iraq faces a large-scale semi-aridization of the climate, with negative effects visible, among others, in the desiccation of vegetation cover and surface water.

This critical situation was the main reason to develop this work, whose aim was to set up proper methods and perform a retrospective analysis about climate changes and LULCC that occurred in more than two decades (1992–2014) in Erbil Area. Attention was devoted to analyzing the role of climate change and urban and built-up areas expansion on the degradation of vegetation cover and surface water in this area.

The analysis was based on (i) the Modified Normalized Difference Water Index (MNDWI) to extract the water surface, and on (ii) the Modifies Soil Adjusted Vegetation Index 2 (MSAVI2) to extract the areas of vegetation cover. Then the authors measured the positive and negative and null changes between the two considered epochs; a segmentation classification was also used for extracting urban and built-up areas expansion and for quantifying their impacts on other land covers.

As regards climate data, spring and summer seasons were mainly affected by temperature increase and rainfall decrease; the vegetation cover was lost for more than 50%, mainly for both these climate change effects (>94%) and for the small remaining part (<6%) for urban and built-up areas expansion. Similarly, the surface water resources also suffered a strong reduction (>41%) due to the increase in temperatures and decrease of rainfall.

Due to the importance of forest as natural resources, Chap. 20 is devoted to mapping forest-fire potentiality using remote sensing and GIS with focus on Kurdistan region-Iraq as a case study. A forest fire is considered as an important issue for the environmental, economy, population, and human safety in many forested areas in the world. Satellite remote sensing (RS) is today regarded as the main source of data for mapping fire risk, assessing forest fuel, monitoring forest fires, as well as, estimating post-fire damages. This chapter, getting aid from the priceless ability of RS and GIS techniques, is aiming to provide a map to show the potentiality of fires among the Kurdistan region forest and rangeland areas. Two sets of data, fields data (the location and date of fires in 2014 and 2015), and satellite data (MODIS NDVI-product time-series) were used. It has been proposed in many studies to use the NDVI variations in order to estimate the proneness of vegetation to fire. By classifying the NDVI image by a supervised classification (the maximum likelihood method), using the ENVI software, a classified image produced based on the satellite images which was acquired in August 2010. This classified image was regarded as the fire potential map. Then, the location of the fires from 2014 to 2015 were added to the map. The result showed there is a high level of overlap between the fired locations recorded in 2014 and 2015, and the areas named as very high and high fire-potential areas on the developed map.

The book ends with the conclusions and recommendations Chap. 21.

Acknowledgements The editors who wrote this chapter would like to acknowledge the authors of the chapters for their efforts during the different phases of the book including their inputs in this chapter.

References

Al-Agidi WK (1976) Proposed soil classification at the series level for Iraqi soils. I. Alluvial Soil. Baghdad University, College of Agriculture. Tech. Bull, no. 2

Attema EPW, Ulaby T (1978) Vegetation modeled as a water cloud. Radio Sci 13(2):357–364. https://doi.org/10.1029/RS013i002p00357

Part II
Soil Characterization, Modelling, and Mapping

Chapter 2
Using Radar and Optical Data for Soil Salinity Modeling and Mapping in Central Iraq

Weicheng Wu, Ahmad S. Muhaimeed, Waleed M. Al-Shafie and Ayad M. Fadhil Al-Quraishi

Abstract As one of the environmental calamities, soil salinization has become a key concern in agricultural management, especially, in irrigated areas in dryland systems. How to quantify and map soil salinity in space and time to provide relevant advices for decision-makers and land managers for their agricultural development has become a pressing issue. Based on our previous works, this study was aimed to develop rather simple and operational approaches for such quantification and mapping taking the Mussaib site in Central Iraq as an example. In conjunction with the field samples, ALOS (Advanced Land Observing Satellite) PALSAR (Phased Array L-band Synthetic Aperture Radar) data and Landsat 5 TM (Thematic Mapper) imagery acquired at almost the same time were employed for this purpose. After derivation of different biophysical indicators from the TM images and removal of the impacts of vegetation water content (VWC) on the L-band radar backscattering coefficients, a multivariate linear regression (MLR) modeling was applied to establish the combined and radar-based soil salinity models. Results revealed that VWC removal procedure could significantly improve the correlation between the measured apparent soil salinity (ECa) and radar backscattering coefficients by 7.5–25.6%. The optical-radar combined models can reliably predict soil salinity with an accuracy of 77.0–83.7% (R^2 = 0.770–0.837). Merely, further improvement in reducing the impacts of vegetation cover and soil moisture by radar data themselves is still recommended. In conclusion, the optical-radar combined approaches and models developed in this chapter shall

W. Wu (✉)
Key Laboratory of Digital Land and Resources, East China University of Technology, Nanchang 330013, Jiangxi, China
e-mail: wuwc030903@sina.com; wuwch@ecit.cn

A. S. Muhaimeed
College of Agriculture, University of Baghdad, Baghdad, Iraq

W. M. Al-Shafie
GIS Division, Ministry of Agriculture, Baghdad, Iraq

A. M. F. Al-Quraishi
Department of Environmental Engineering, College of Engineering, Knowledge University, Erbil 44001, Kurdistan Region, Iraq

© Springer Nature Switzerland AG 2020
A. M. F. Al-Quraishi and A. M. Negm (eds.),
Environmental Remote Sensing and GIS in Iraq, Springer Water,
https://doi.org/10.1007/978-3-030-21344-2_2

be operational for soil salinity modeling and mapping; and radar-based approach has great potential for this purpose.

Keywords Soil salinity · ALOS L-band radar · Landsat TM · Modeling and mapping · Central Iraq

2.1 Introduction

Soil salinity has become one of the most active land degradation phenomena in Central and Southern Iraq since the Babylonian period (Jacobsen and Adams 1958; Buringh 1960; Wu et al. 2014a, b). It is estimated that approximately 60% of the cultivated land has been seriously affected by salinity, and 20–30% abandoned in Mesopotamia (Buringh 1960; Wu et al. 2014a, b). Even in the non-abandoned agricultural land, crop yield has declined by 30–60% compared to the normal ones as a consequence of salinization. This unfavorable biophysical process has led to a conversion of the most highly productive land into the salinized land and provoked negative impacts on crop production and food security in the country. It is hence of prime importance to investigate the distribution and intensity of soil salinity in space and time and analyze the causes of salinization to provide relevant advices and suggestions for decision-makers in agriculture management and development.

Regarding the soil salinization in Iraq, several authors have conducted analyses, and mapping works, for example, Jacobsen and Adams (1958), Buringh (1960), Dieleman (1963), Al-Layla (1978), Al-Mahawili (1983), and Abood et al. (2011). Buringh and the Ministry of Agriculture (MoA) in Iraq had jointly undertaken soil classification and produced a soil map of Iraq in 1960. The Food and Agriculture Organization of the United Nations (FAO) investigated soil salinization severity while assessing the irrigation condition in Western Asia including Iraq in 2008 (FAO 2008). However, outdating and low resolution of the available maps (e.g., soil map dated 1960 with a resolution of 5–10 km) cannot meet the requirement of land management and salinity control at the farm- and local-scale in the region. Recently, funded by the Australian Agency for International Development (AusAID), the International Center for Agricultural Research in the Dry Areas (ICARDA) has conducted an Integrated Soil Salinity Management Project in Central and Southern Iraq.[1] The project was implemented in cooperation with the MoA, Ministry of Water Resources (MoWR), Ministry of Science and Technology (MoST), and Ministry of Education (MoE) of the Iraqi Government in the period 2010–2014. Soil salinity modeling, quantification and mapping were one of the main tasks of the project. Based on field survey and multiyear optical remote sensing data, successful studies on salinity modeling and mapping with high reliability have been achieved in Mesopotamia (Wu et al. 2012, 2014a, b; Muhaimeed et al. 2013).

---

[1] https://www.icarda.org/iraq-salinity-project.

With application of optical remote sensing, several mature approaches for soil salinity mapping have been developed and discussed in the recent decades (Dwivedi and Rao 1992; Mougenot et al. 1993; Rao et al. 1995; Metternicht and Zinck 2003; Farifteh et al. 2006, 2007; Eldeiry and Garcia 2010; Wu et al. 2014a, b; Bannari et al. 2018). These approaches focused on best band combination, multiyear maxima-based multivariate modeling and quantification, etc. Some authors have even explored the possibility to detect soil salinity by microwave radar data as they are independent of weather condition (Sreenivas et al. 1995; Taylor et al. 1996; Shao et al. 2003; Aly et al. 2004; Lasne et al. 2008; Gong et al. 2013). The laboratory-based simulations by Shao et al. (2003), Lasne et al. (2008) and Gong et al. (2013) suggested that it is possible to use the microwave C-band, and especially L-band for detecting salinity in different settings. Because the radar signal can penetrate through the surface and reach subsoil at a depth of, e.g., >50–150 cm, depending on the wavelength or frequency of the emitted waves and soil moisture. They concluded that the real part of the soil dielectric permittivity is responsive to the moisture, and its imaginary part is associated with both soil moisture and salinity. However, successful and satisfactory radar-based salinity modeling and quantification have been rarely reported probably due to the difficulty to separate soil salinity from soil moisture, or rather, to remove the impacts of soil moisture and vegetation cover on the salinity part of the radar backscattering coefficient.

In view of this difficulty, our intention was to incorporate radar data, taking their advantages of independence of weather condition and penetration to subsoil, with optical imagery for salinity assessment. As a matter of fact, it has been reported that significant improvement in accuracy of land cover mapping had been achieved by combing radar with optical data (Li et al. 1980; Pereira et al. 2013), or a higher efficiency reached by adding radar data while assessing irrigation, soil moisture and crop performance (Clevers and van Leeuwen 1996; Moran et al. 1997; Fieuzal et al. 2011; He et al. 2014).

For this reason, the main objective of this study was to ascertain the possibility to propose a simple and operational approach for soil salinity assessment by combing both radar and optical data as a complement to the available optical ones. The specific objectives were (1) to develop combined soil salinity model(s) by incorporating radar backscattering coefficient (σ) with biophysical indicators from optical data, and (2) to explore the potential to develop radar-based model(s) for the same purpose. The research was implemented in the Mussaib site in Central Iraq where we had already conducted salinity modeling and mapping using multiyear optical remote sensing data (Muhaimeed et al. 2013; Wu et al. 2014a, 2018).

2.2 Methods and Materials

2.2.1 Study Area

Situated between the Tigris and the Euphrates Rivers in Central Iraq (Fig. 2.1), the study area has been a national agriculture development project site since 1950s. It has been developed for grain production including spring crops such as wheat and barley, and summer crops mainly corn, vegetables and fruits (e.g., tomato and watermelon), and locally cotton. Perennial alfalfa (*Medicago sativa*) and date palm as economic crop may be also locally cultivated. The total area of the project site is around 250,000 ha.

The dominant soil types are Aridisols and Entisols with texture class ranging from silt clay loam to silt loam with more than 20% of lime. Most of the soils are slightly saline to highly saline, e.g., from 4 to 30 dS/m (Muhaimeed et al. 2013; Wu et al. 2014a).

As for climate, the Mussaib site belongs to the subtropical region, characterized by short warm winter and long hot to extreme hot summer. Rainfall is concentrated in winter and spring from December to March with an annual average of about 82.5 mm in the past 60 years (recorded in the adjacent station, Hillah). The mean

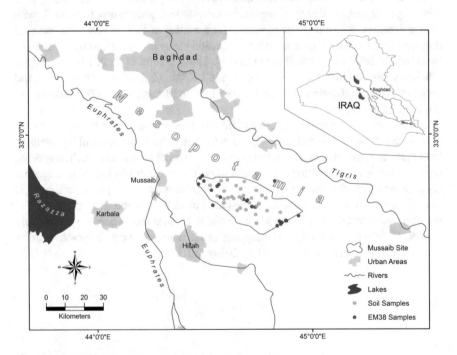

Fig. 2.1 Location of the study area and field sampling points

minimum temperature is 6.25 °C from December to February in winter, while the mean maximum temperature is 43.2 °C in July and August in summer.

2.2.2   Data

2.2.2.1   Field Data

Field surveys were conducted in the period from July 2011 to July 2012, including soil sampling, measurement of EM38-MK2 (EM38 hereafter, an electromagnetic instrument made by Geonics Ltd. to measure apparent electrical conductivity, ECa in mS/m), and land use/cover investigation.

Soil samples were taken in croplands or under halophyte from 13 pedons (the surface horizon part, 0–30 cm in depth) and 17 auger holes of 0–30 cm in the study area in July–Nov 2011, when EM38 instruments were not available. These samples were analyzed in laboratory to measure soil electrical conductivity (ECe) using 1:1 dilution method.

EM38 readings were conducted in two campaigns: one was in spring (March–April, $3 \times 15 = 45$ pairs of vertical and horizontal readings, respectively denoted as EM_V and EM_H) and the other was in early summer (June–July, when the dry season started after harvesting of barley and wheat, $3 \times 7 = 21$ pairs of readings) in 2012. As designed, both EM_H and EM_V readings were taken in the plots (1 m × 1 m) distributed at three corners of triangles. The averaged values of the three corners were regarded as the representatives of the triangle centers. The designed distance between each two corners of a triangle was about 15–20 m, so that the triangular area can represent more or less a Landsat TM pixel. The objective of such averaging was to have more comparability between the field sampling and satellite images, for example, Landsat TM data with a pixel size of 30 m. Also, two triangles of measurements situated nearby the Mussaib site conducted in the regional validation campaign in June 2013 (see Wu et al. 2014a) were also integrated into this study. In total, 24 averaged pairs of EM38 readings were made available.

EM38 readings were not measured at the same plots as soil samples due to accessibility problem (Fig. 2.1). Therefore, soil samples were used neither for calibrating the EM38 readings (ECa) nor for salinity modeling because of the difference in sampling locations between the two kinds of measurements and because of their poor representativeness for TM pixel-scale salinity due to the high spatial variability of salinity.

2.2.2.2   Satellite Data

As seen in Sect. 2.1, L-band radar images have great potential for salinity detection. Hence, the Japanese satellite ALOS (Advanced Land Observing Satellite) Level 1.5 PALSAR (Phased Array L-band Synthetic Aperture Radar) product, were obtained from the European Space Agency (ESA: https://alos-palsar-ds.eo.esa.int). The L-band images were produced by microwave radar with a wavelength of 23 cm and frequency of 1.27 GHz in Fine Beam Double (FBD) Polarization Mode (HH/HV). The images were acquired with off-nadir angle 34.3° and incidence angle of 7.5–60° on November 26, 2010. The Level 1.5 Product has a spatial resolution of 12.5 m.

Landsat 5 TM images (30 m in resolution) dated November 23, 2010, close to the acquisition date of ALOS images, were also obtained from ESA (https://landsat-ds.eo.esa.int).

When both satellite images were acquired, summer crops, e.g., mainly maize, started to become mature, and locally cultivated perennial forage crop, alfalfa, was still green. Winter crops such as barley and wheat were just sown, and the rainy season had not yet started in Mesopotamia. It is noted that in the surrounding weather stations of the study area, namely Baghdad, Karbala, and Hillah, etc., there was no rainfall recorded in September, October and November 2010. Thus, the rainfall-related moisture problem (Wu et al. 2014a, b) could be avoided.

2.2.3 Approaches and Procedures

As demonstrated in our previous works (Wu et al. 2014a, b), maximization of the multiyear biophysical indicators followed by multivariate regression modeling approach can minimize the problems related to crop rotation, fallow practice, soil moisture variation, and biotic stress, but need to process multiyear spring and summer images, which takes time. In this study, we made an effort on a single date of optical images incorporated with L-band radar data to reduce processing and save time but with high reliability and accuracy. The proposed approaches and procedures are briefly presented in Fig. 2.2.

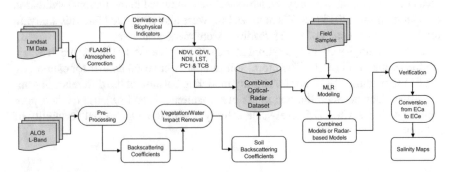

Fig. 2.2 Flowchart demonstrating the approaches and procedures in this study

2.2.3.1 TM Image Processing

The Landsat 5 TM images were radiometrically calibrated to convert Digital Number (DN) into radiance, and a FLAASH (Fast Line-of-sight Atmospheric Analysis of Spectral Hypercubes) model (Perkins et al. 2005) was applied for removing the additive atmospheric effects (Wu et al. 2014a, b).

As revealed earlier, the most relevant vegetation indices for salinization study, namely the Normalized Difference Vegetation Index (NDVI), the Normalized Difference Infrared Index from the difference between TM bands 4 (0.835 μm) and 5 (1.650 μm) or between bands 4 and 7 (2.215 μm) (NDII, Hardisky et al. 1983), which was renamed the Normalized Difference Water Index (NDWI) by Gao (1996), the Generalized Difference Vegetation Index (GDVI, Wu 2014) with power number of 2 and 3 (denoted respectively GDVI2 and GDVI3) were produced. Other biophysical indicators such as Land Surface Temperature (LST), Tasseled Cap Brightness (TCB, Crist and Cicone 1984) and the first Principal Component (PC1) were also respectively produced (Wu et al. 2014b).

2.2.3.2 L-Band Radar Processing

Level 1.5 radar product has been geometrically rectified, and its spatial deformation induced from landform and variation of incidence angle also corrected; pixels were resampled to 12.5 m in size by the provider. The DNs of the two polarization bands, HH and HV, were respectively calibrated and converted into backscattering coefficients (σ^0_{HH} and σ^0_{HV}) (in decibel, dB) in terms of Shimada et al. (2009):

$$\sigma^0[\text{dB}] = 10\log_{10}(DN)^2 + \text{CF} \tag{2.1}$$

where $\text{CF} = -83.0$ for Level 1.5 product.

Then, an Enhanced Lee filter (Lee 1980), 3×3 in size, was applied to the sigma nought (σ^0) to remove speckles and to reduce noise. σ^0_{HH} and σ^0_{HV} were hence derived and resampled to 30 m in pixel size to match the TM data.

2.2.3.3 Minimizing Impacts of Vegetation Cover

Water Cloud Model

As mentioned in Sect. 2.1, the difficulty to use radar backscattering coefficients to characterize soil salinity is related to soil moisture, especially, where there is vegetation cover. Attema and Ulaby (1978), and Ulaby et al. (1978) have proposed water cloud model for characterizing the effect of vegetation water content (VWC) on radar backscattering coefficient, which can be expressed as follows (Kumar et al. 2012):

$$\sigma^0 = \sigma_{veg}^0 + L^2 \sigma_{soil}^0 \qquad (2.2)$$

with

$$\sigma_{veg}^0 = A V_1 \cos(\theta_i)(1 - L^2) \qquad (2.3)$$

$$L^2 = \exp(-2B V_2 \sec(\theta_i)) \qquad (2.4)$$

where σ^0 is the total backscattering coefficient, σ_{veg}^0 is the backscattering contribution of the vegetation cover, σ_{soil}^0 is the backscattering contribution of the soil, and L^2 is the two-way vegetation attenuation; θ_i is the incidence angle; A and B are the vegetation parameters. V_1 represents a canopy descriptor, LAI (Leaf area index, m^2/m^2), and V_2 is the vegetation water content (VWC) (Kumar et al. 2012).

Determination of Canopy Descriptors and Vegetation Parameters

(1) V_1: LAI

Diker and Bausch (2003), Xavier and Vettorazzi (2004), Chaurasia et al. (2011), Wu (2014) and so on have investigated the relationships of LAI with vegetation indices such as NDVI, GDVI, EVI (Enhanced Vegetation Index, Huete et al. 1997), SAVI (Soil-Adjusted Vegetation Index, Huete 1988), SARVI (Soil Adjusted and Atmospherically Resistant Vegetation Index, Kaufman and Tanré 1992), WDRVI (Wide Dynamic Range Vegetation Index, Gitelson 2004), etc., for different vegetation types (Table 2.1). The works of Diker and Bausch (2003), Xavier and Vettorazzi (2004) look relevant to our study. However, as Wu (2014) has demonstrated, among the most frequently applied vegetation indices, GDVI2 has the best correlation with LAI across all land cover types including irrigated croplands (maize, cotton, fruits, vegetables), citrus, olive groves, rangelands, woodlands and forests in dryland ecosystems (Table 2.1).

(2) V_2: VWC

Jackson et al. (2004), Maggioni et al. (2006), Yi et al. (2007), Yilmaz et al. (2008), Huang et al. (2009), Gao et al. (2015) and so on, have conducted studies on quantification of VWC by spectral indices such as NDVI and NDWI for maize (corn), cereal grain (wheat and barley), legume (soybean, canola and lucerne/alfalfa) and grassland. Their results are summarized in Table 2.2.

(3) Vegetation Parameters A and B

Actually, several authors have undertaken the characterization of A and B (Prévot et al. 1993; Moran et al. 1997; Bindlish and Barros 2001; Dabrowska-Zielinska et al. 2007; Kumar et al. 2012; Leonard et al. 2013; Baghdadi et al. 2017), and their results are presented in Table 2.3.

Table 2.1 Relationships between LAI and vegetation indices (VIs)

Land cover and vegetation types	LAI-VIs equation (m^2/m^2)	Correlation coefficient	Sources
All land cover types including irrigated croplands (maize, cotton, vegetables and fruits), rangeland, olive groves, citrus, woodlands, forests and bare soil	$LAI = 0.1249$ $\exp(5.1005\ NDVI)$	$R^2 = 0.898$	Wu (2014)
	$LAI = 0.091$ $\exp(3.7579\ GDVI2)$	$R^2 = 0.932$	
	$LAI = 0.0631$ $\exp(3.646\ GDVI3)$	$R^2 = 0.925$	
	$LAI = 0.2625$ $\exp(5.0982\ SARVI)$	$R^2 = 0.866$	
	$LAI = 0.1602$ $\exp(4.8247\ EVI)$	$R^2 = 0.736$	
	$LAI = 0.1301$ $\exp(7.6536\ SAVI)$	$R^2 = 0.757$	
Maize cropland	$LAI = 4.59\ NDVI^{1.89}$	$R^2 = 0.990$	Diker and Bausch (2003)
Sugar cane, pasture, maize, eucalypt, and riparian forest	$LAI = 8.1954\ NDVI^{2.9154}$	$R^2 = 0.710$	Modified from Xavier and Vettorazzi (2004)
Wheat cropland	$LAI = 0.138$ $\exp(4.823\ NDVI)$	$R^2 = 0.840$	Chaurasia et al. (2011)

(4) Determination of the Relevant Parameters

Selection of the relevant vegetation parameters and descriptors is a critical procedure. To determine the best performed LAI (V_1), a number of tests were conducted on the LAI-VIs models mentioned above given the same VWC, A and B to relate the vegetation cover impact removed backscattering coefficient (σ^0_{soil}) with the measured apparent soil salinity. It was noted that the LAI-GDVI2 model of Wu (2014), i.e., $LAI = 0.091 \exp(3.7579\ GDVI2)$ ($R^2 = 0.932$) performed best followed by the LAI-NDVI model of Diker and Bausch (2003). Thus LAI-GDVI2 model was finally selected.

In the same way, the most relevant VWC model was determined given the same LAI, A, and B. After numerous tests and fittings, we found that the VWC-NDVI model for corn of Jackson et al. (2004) outperformed any other VWC model for corn and other crops and was hence chosen for our further vegetation impact removal processing.

Many tests were carried out for determining A and B under the same LAI and VWC condition. We noted that the 2nd case of Dabrowska-Zielinska et al. (2007) for ALOS L-band radar data, that is, $A = 0.0045$ (dB) and $B = 0.4179$, yielded the best fit between the vegetation removed backscattering coefficients (σ^0_{soil}) and our measured apparent soil salinity. Therefore, this pair of vegetation parameters was selected.

Table 2.2 Different spectral indices-based VWC models

	NDVI-based VWC		$NDWI_{1,640}$-based VWC		$NDWI_{2,150}$-based VWC		References
	Equation	R^2	Equation	R^2	Equation	R^2	
Maize	$y = 192.64x^5 - 417.46x^4 + 347.96x^3 - 138.93x^2 + 30.7x - 2.82$	0.990	$y = 9.82x + 0.05$	0.980			Jackson et al. (2004)
	$y = 0.098 \exp(4.225x)$	0.800	$y = 7.71x + 0.26$	0.620	$y = 10.51x - 4.11$	0.480	Huang et al. (2009)
			$y = 7.84x + 0.6$	0.870			Gao et al. (2015)
Cereal grains (wheat, barley)	$y = 4.81x - 0.55$	0.830	$y = 13.2x^2 + 1.62x$	0.790	$y = 10.99x - 3.07$	0.750	Maggioni et al. (2006)
	$y = 0.078 \exp(3.51x)$	0.590	$y = 2.45x + 0.57$	0.570	$y = 12.38x - 3.26$	0.840	Gao et al. (2015)
	$y = -51.73x^2 + 70.48x - 20.24$	0.690	$y = 12.5x - 0.44$	0.760	$y = 10.29x - 1.98$	0.840	Yi et al. (2007)
Legumes (soybean, canola, alfalfa)	$y = 7.63x^4 - 11.41x^3 + 6.87x^2 - 1.24x + 0.13$	0.990	$y = 1.44x^2 + 1.36x + 0.34$	0.970			Jackson et al. (2004)
			$y = 2.22x + 0.38$	0.870			Yilmaz et al. (2008)
			$y = 0.85x + 0.33$	0.310	$y = 0.97x - 0.01$	0.310	Huang et al. (2009)
	$y = 0.059 \exp(2.573x)$	0.510	$y = 1.74x + 0.34$	0.760			Gao et al. (2015)
Grassland	$y = 0.21x + 0.24$	0.920			$y = 0.78x + 0.01$	0.900	Maggioni et al. (2006)
	$y = 0.017 \exp(5.866x)$	0.520	$y = 1.16x + 0.45$	0.200	$y = 0.74x + 0.23$	0.310	Gao et al. (2015)

Note VGT—vegetation; 1.640 or 2.150 represent the wavelength (μm) of the shortwave infrared bands used to constitute NDWI or NDII

Table 2.3 Vegetation parameters A and B

Vegetation types	Radar band and polarization	A	B	Note	Source
Wheat, grass and rape	ERS C-band: $V_1 = V_2 = LAI$	0.0846	0.0615		Dabrowska-Zielinska et al. (2007)
	ERS C-band: $V_1 = V_2 = LWAI$	0.0957	0.0656		
	ERS C-band: $V_1 = V_2 = VMW$	0.0795	0.1464		
	ALOS L-band: $V_1 = V_2 = LAI$	0.0027	0.1957		
	ALOS L-band: $V_1 = V_2 = LWAI$	0.0045	0.4179		
	ALOS L-band: $V_1 = V_2 = VMW$	0.0110	0.0430		
Wheat	ASAR C-band VV	0.1836	0.0862	Genetic Algorithm	Kumar et al. (2012)
		0.1989	0.2075	SUMT	
Maize	RADARSAT-2 C-band VV	0.1200	0.8600		Leonard et al. (2013)
	RADARSAT-2 C-band HH	0.1800	0.6300		
	RADARSAT-2 C-band HV	0.0400	0.2300		
	RADARSAT-2 C-band VV	0.3900	0.3400		
	RADARSAT-2 C-band HH	1.2700	0.0500		
	RADARSAT-2 C-band HV	0.1300	0.1300		
Rangeland		0.0009	0.0320		Bindlish and Barros (2001)
Winter Wheat		0.0018	0.1380		
Pasture		0.0014	0.0840		
All land use		0.0012	0.0910		
Wheat	C HH	0.0000	0.0860		Prévot et al. (1993)
	X VV	0.0560	0.4230		
Wheat, Vineyard and grassland	Sentinel-1 C VV: $V_1 = V_2 = NDVI$	0.0950	0.5513		Baghdadi et al. (2017)
	Sentinel-1 C VH: $V_1 = V_2 = NDVI$	0.0413	1.1662		

Retrieval of the Soil-Related Backscattering Coefficient

After determination of the most relevant vegetation parameters and canopy descriptors, Eq. (2.2) can be transformed as follows:

$$\sigma_{soil}^0 = (\sigma^0 - \sigma_{veg}^0)/L^2 = \big[\sigma^0 - A * LAI * \cos(\theta_i)$$
$$* (1 - \exp(-2 * B * VWC * \sec(\theta_i)))\big]/\exp(-2 * B * VWC * \sec(\theta_i))$$

(2.5)

By inputting the selected A, B, LAI-GDVI2 and VWC-NDVI models, and 34.3° as the mean incidence angle (θ_i), vegetation backscattering coefficient (σ_{veg}^0) can be calculated for each pixel, and soil-related backscattering components (σ_{soil}^0) can be retrieved by removing the vegetation and moisture impacts on the total backscattering coefficient (σ^0, in our case, σ_{HH}^0 or σ_{HV}^0).

2.2.3.4   Soil Salinity Modeling

The produced biophysical indicator layers, namely NDVI, GDVI2, GDVI3, $NDII_{1.640}$, LST, TCB, PC1 and their natural logarithmic transformation (ln), together with the speckles removed L-band backscattering coefficients, σ_{HH}^0, σ_{HV}^0 and their sum, $\sigma_{SUM}^0 (= \sigma_{HH}^0 + \sigma_{HV}^0)$, were all imported into ArcGIS to extract their values corresponding to the field survey plots. The same were done for the radar backscattering coefficients whose vegetation and moisture influences were removed (Sect. 2.2.3.3), denoted respectively as $\sigma_{HH(soil)}^0$, $\sigma_{HV(soil)}^0$, and $\sigma_{SUM(soil)}^0$.

After that, all extracted biophysical and radar data together with the measured apparent soil salinity (EM_H and EM_V) were input into SYSTAT for correlation analysis and least-square multivariate linear regression (MLR) modeling by setting EM_V and EM_H as dependent variables (to be predicted and modeled) and all others as independent or predictive variables. The MLR modeling was conducted in a Stepwise (forward) manner with a confidence level of 95%, and the entrance threshold of the F-statistics was set 4.

2.2.3.5   Verification and Mapping

The combined and radar-based models derived from the above procedure were applied back to the combined indicators or radar backscatter coefficients to predict the apparent soil salinity (ECa). A verification against the field measured data was followed to check the reliability and accuracy of prediction. The most reliable models, either combined or radar-based, were used to produce the soil salinity maps of the study area.

2.2.3.6   Conversion from the Apparent Salinity (ECa) to Lab-Analyzed Salinity (ECe)

What MLR predicted was the apparent soil salinity (ECa in mS/m), and it had to be converted into the lab-measured ECe (dS/m) which would be more practical for land management. We applied hence our results obtained from the regional-scale sampling and lab-analysis in the whole Mesopotamia for this purpose. Regional sampling includes two transects and four pilot sites, where both soil and EM38 readings were sampled at the same plots. The ECe-EM38 readings (ECa) relationships were expressed as follows (Wu et al. 2014a, b):

$$ECe(\text{dS m}^{-1}) = 0.0005EM_V^2 - 0.0779EM_V + 12.655(\text{R}^2 = 0.850) \qquad (2.6)$$

$$ECe(\text{dS m}^{-1}) = 0.0002EM_H^2 + 0.0956EM_H + 0.0688(\text{R}^2 = 0.791) \qquad (2.7)$$

2.3   Results and Discussion

After the above processing and analysis, the obtained results are presented in Tables 2.4, 2.5, Figs. 2.3 and 2.4.

2.3.1   Correlation

As shown in Table 2.4, the measured salinity (EM_V and EM_H) are positively correlated with LST and its variant ln(LST), and PC1, and negatively correlated with the vegetation indices, such as NDVI, NDII, GDVI2 and GDVI3, especially, with the variant ln(GDVI3).

Interestingly, measured EM_V and EM_H show well negative correlation with radar backscattering coefficients (σ_{HH}^0 and σ_{HV}^0), in particular, with those whose influences of vegetation and moisture were removed ($\sigma_{HH(soil)}^0$ and $\sigma_{HV(soil)}^0$). It is seen that the removal procedure has gained an increase of 16.6–25.6% in correlation coefficients between the measured salinity and the HH polarization radar backscattering coefficient, σ_{HH}^0, and an increase of 11.5–21.4% in that between the measured salinity and σ_{HV}^0, while the sum of these two sigma nought, σ_{SUM}^0, has gained relatively less, 7.5–16.3%. This demonstrates that it is important and effective to conduct this removal procedure while employing radar data for salinity assessment.

Table 2.4 also revealed that for both EM_H and EM_V, HH and HV polarization backscattering coefficient (σ_{HH}^0 and σ_{HV}^0) are important salinity information carriers. Especially, after the removal of vegetation cover impact, both $\sigma_{HH(soil)}^0$ and $\sigma_{HV(soil)}^0$ show a strong negative correlation with the measured salinity (R^2 = 0.565–0.677); and σ_{HH}^0 and $\sigma_{HH(soil)}^0$ are better than σ_{HV}^0 and $\sigma_{HV(soil)}^0$.

Table 2.4 Pearson correlation coefficients between the measured soil salinity and biophysical indicators and backscattering coefficients

Field samples	LST	NDVI	NDII	GDVI3	GDVI2	TCB	ln(LST)	ln(NDVI)	ln(GDVI3)	ln(GDVI2)
EM$_V$	0.746	−0.416	−0.459	−0.475	−0.441	−0.056	0.746	−0.581	−0.612	−0.595
EM$_H$	0.739	−0.403	−0.428	−0.468	−0.430	−0.133	0.737	−0.586	−0.620	−0.601

Field samples	ln(TCB)	PCI	σ^0_{HH}	σ^0_{HV}	σ^0_{SUM}	$\sigma^0_{HH(soil)}$	$\sigma^0_{HV(soil)}$	$\sigma^0_{SUM(soil)}$
EM$_V$	−0.123	0.047	−0.684	−0.674	−0.721	−0.798	−0.752	−0.775
EM$_H$	−0.203	0.160	−0.655	−0.639	−0.687	−0.823	−0.776	−0.799

Table 2.5 Combined and radar-based soil salinity models

No	Model in ECa (mS/m)	Multiple R^2	RMS error	F-Ratio	p-Value
1	$EM_V = -45,016.864 + 7940.565 \ln(LST) + 204.956 \ln(GDVI2) - 4.066\sigma^0_{HH(soil)}$	0.821	±85.138	30.504	0.000
2	$EM_H = -8696.791 + 28.45\,LST + 420.642\,GDVI3 - 3.3\sigma^0_{HH(soil)}$	0.843	±79.036	35.899	0.000
3	$EM_V = 28.283 - 19.878\sigma^0_{HH(soil)} + 7.801\sigma^0_{SUM(soil)}$	0.739	±100.232	29.728	0.000
4	$EM_H = -22.971 - 12.286\sigma^0_{HH(soil)} + 7.65\sigma^0_{HV(soil)}$	0.783	±90.746	37.922	0.000

Note Observation sample number for modeling is 24

The sum of the two polarization backscattering coefficients (σ^0_{SUM} and $\sigma^0_{SUM(soil)}$) seems also a good indicator of salinity.

2.3.2 Soil Salinity Models and Maps

MLR modeling revealed very interesting soil salinity models (Table 2.5) linking the measured apparent soil salinity with radar backscattering coefficients and biophysical indicators from the optical data. Models 1 and 2 are the radar-optical combined models, i.e., EM_V is associated with LST, GDVI2 and $\sigma^0_{HH(soil)}$, while EM_H with LST, GDVI3 and $\sigma^0_{HH(soil)}$. Models 3 and 4 are radar-based salinity models, of which the impact of vegetation cover and moisture on backscattering coefficients has been removed.

The reliability of prediction of the combined Models 1 and 2 are presented in Fig. 2.3a, b (77.05 and 83.7% respectively). It is evident that both Models 1 and 2 seem reliable, and Model 2 (EM_H) can predict better than Model 1 (EM_V). Soil salinity maps derived from Model 2, which has been converted into ECe (dS/m), are presented in Fig. 2.4a, b.

Radar-based Models 3 and 4 also show high reliable soil salinity prediction. The agreement between the predicted salinity by Model 4 and field measured soil salinity reaches 79.2% (Fig. 2.3c). The map predicted by Model 4, which has been converted into ECe (dS/m), is presented in Fig. 2.4c.

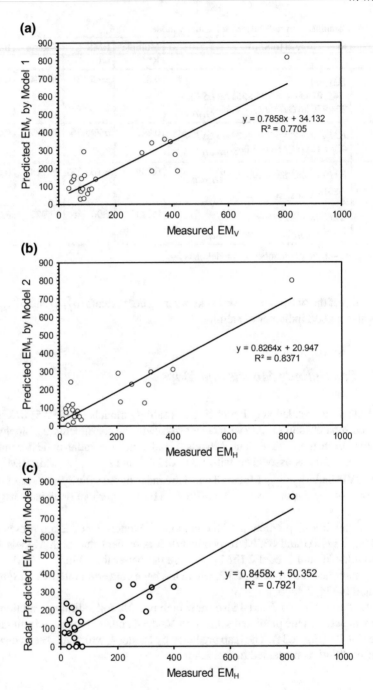

Fig. 2.3 Verification of the predicted soil salinity versus the field measured salinity. *Note* **a** Salinity predicted from the combined Model 1 (EM_V), **b** from the combined Model 2 (EM_H), and **c** from the radar-based Model 4

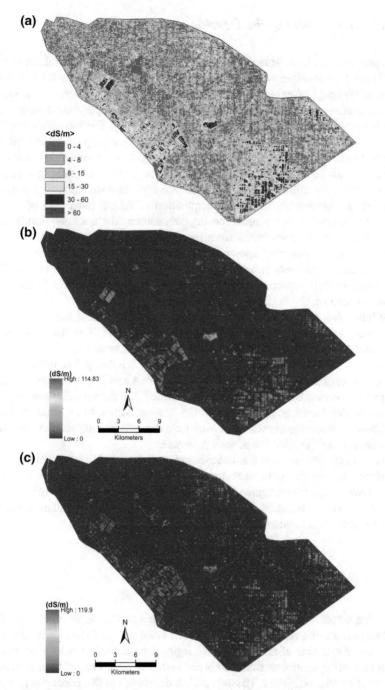

Fig. 2.4 Predicted soil salinity in ECe (dS/m) of the Mussaib site. *Note* **a** and **b**, predicted salinity from the combined Model 2 (EM$_H$), only (**a**) shown in classified grades of severity and (**b**) in continuity; **c** salinity from the radar-based Model 4 and shown in continuity

2.3.3   Assessment of the Developed Approaches

Our previous works in Mesopotamia (Wu et al. 2012, 2014a, b; Muhaimeed et al. 2013) involved multiyear spring and summer optical satellite data, and the developed approaches could overcome the problems related to crop rotation, fallow practice, soil moisture variation and biotic stress such as varietals or pest/insect outbreak, and provide good prediction of salinity ($R^2 = 0.830–0.869$). The inconvenience lies in that the approaches involve the production of multiyear biophysical indicators for vegetated and non-vegetated areas respectively, and their maximization followed by a multivariate linear regression modeling. Hence, a lot of processing is required. In this study, we utilized only one single late-autumn acquisition of Landsat TM imagery but incorporated with one acquisition of ALOS radar data of the same time as a complementary remote sensing data source. There was no discrimination of vegetated and non-vegetated areas for separate processing. Hence, the optical-radar combined approach proposed in this study is simpler and less time-consuming but able to achieve very reliable soil salinity prediction and mapping (83.7% of accuracy) as well. Hence, the newly developed approaches in this paper shall be more operational.

While using radar data for salinity assessment, it is evident that the removal of the impact of vegetation cover (e.g., water content (VWC)) on the backscattering coefficients is a critical procedure that leads to a substantial improvement in correlation with the measured salinity and in prediction reliability. The difficulty lies in the selection of the relevant vegetation parameters A and B, and canopy descriptors V_1 and V_2. As we have demonstrated in Sect. 2.2.3.3, the selection and evaluation should be conducted one by one given the same condition of other parameters for the observed vegetation types. Moreover, field data are the only standard to evaluate the relevance of the selection of each parameter.

This study demonstrated that radar-based models, e.g., Model 4, can also predict satisfying soil salinity with an accuracy of 79.2% (Fig. 2.3c) despite it is still a bit lower than that by the combined Model 2. We believe that it can be further improved, especially, when it becomes possible to remove vegetation cover and moisture based only on radar data themselves.

2.4   Conclusions

This chapter presents simple, fast and effective approaches for soil salinity modeling and mapping using ALOS L-band radar and Landsat 5 TM data. Since it involves only one single date of acquisition of images, it requires less processing time and shall be more operational. Both combined and radar-based models can predict well soil salinity with reliability. The approaches developed in this paper can provide an alternative or complementary method to the optical ones for salinity assessment in agricultural management. Through this study, we found that it is not a question of potential, but it is completely possible to apply radar data for such salinity mapping and assessment.

It is worthy of notice that while implementing either the combined or radar-based models to other similar environment for salinity assessment, it is recommended to select its appropriate LAI, VWC and vegetation parameters *A* and *B* in line with the vegetation types in the study areas. One of the future works should be focused on finding a way of minimization of the impact of vegetation cover and soil moisture by radar data themselves. Thus, we will be able to fulfil the mapping work by radar and field data only without any support from optical data.

2.5 Recommendations

A momentum has been gained in application of machine learning algorithms for salinity prediction and mapping (Farifteh et al. 2007; Taghizadeh-Mehrjardi et al. 2014; Wu et al. 2016, 2018). Especially, the work of Wu et al. (2018), using the same datasets as this study but employing Support Vector Regression (SVR) and Random Forest Regression (RFR) algorithms, achieved a milestone in salinity prediction and mapping with a high accuracy of 93–94%. However, MLR-based modeling as shown in this study, in spite of a bit lower accuracy (e.g., 78–84%), could provide intuitive salinity models. Therefore, using radar-optical combined dataset followed with a machine learning algorithm such as RFR and MLR modeling are approaches strongly recommended in future work.

Acknowledgements The authors would like to thank AusAID (Australian Agency for International Development) for their funding in our previous works in Mesopotamia (ICARDA Project No: LWR/2009/034, 2010–2014) while the first author was with ICARDA, and East China University of Technology for their financial support to Dr. Weicheng Wu for his research on assessment of sustainable use of environmental resources (Grant No: DHTP2018001, 2018–2021). Our sincere gratitude will go to ESA (https://alos-palsar-ds.eo.esa.int) for their provision of ALOS PALSAR and Landsat 5 TM data.

References

Abood S, Maclean A, Falkowski M (2011) Soil salinity detection in the Mesopotamian Agricultural Plain utilizing WorldView-2 imagery. Available online at: http://dgl.us.neolane.net/res/img/16bb89b080930a8ad1fdfc17665883e9.pdf. Accessed in Jan 2012

Al-Layla MA (1978) Effect of salinity on agriculture in Iraq. J Irrig Drain Div 104(2):195–207

Al-Mahawili SMH (1983) Satellite image interpretation and laboratory spectral reflectance measurement of saline and gypsiferous soils of west Baghdad, Iraq. M.Sc. thesis, Purdue University

Aly Z, Bonn FJ, Magagi R (2004) Modelling the backscattering coefficient of salt-affected soils: application to Wadi El-Natrun bottom, Egypt. EARSeL eProceedings, vol 3, 3/2004, pp 372–381

Attema EPW, Ulaby FT (1978) Vegetation modeled as a water cloud. Radio Sci J 13:357–364

Baghdadi N, El Hajj M, Zribi M, Bousbih S (2017) Calibration of the water cloud model at C-Band for winter crop fields and grasslands. Remote Sens 9(9):969. https://doi.org/10.3390/rs9090969

Bannari A, El-Battay A, Bannari R, Rhinane H (2018) Sentinel-MSI VNIR and SWIR bands sensitivity analysis for soil salinity discrimination in an arid landscape. Remote Sens 10:855. https://doi.org/10.3390/rs10060855

Bindlish R, Barros AP (2001) Parameterization of vegetation backscatter in radar-based, soil moisture estimation including vegetation scattering effects in a radar based soil moisture estimation model. Remote Sens Environ 76:130–137

Buringh P (1960) Soils and soil conditions in Iraq. MoA of Iraq, p 337

Chaurasia S, Nigam R, Bhattacharya BK, Sridhar VN, Mallick K, Vyas SP, Patel NK, Mukherjee J, Shekhar C, Kumar D, Singh KRP, Bairagi GD, Purohit NL, Parihar JS (2011) Development of regional wheat VI-LAI models using Resourcesat-1 AWiFS data. J Earth Syst Sci 120(6):1113–1125

Clevers JGPW, van Leeuwen HJC (1996) Combined use of optical and microwave remote sensing data for crop growth monitoring. Remote Sens Environ 56(1):42–51

Crist EP, Cicone RC (1984) Application of the tasseled cap concept to simulated thematic mapper data. Photogram Eng Remote Sens 50:343–352

Dabrowska-Zielinska K, Inoue Y, Kowalik W, Gruszczynska M (2007) Inferring the effect of plant and soil variables on C- and L-band SAR backscatter over agricultural fields, based on model analysis. Adv Space Res 39:139–148

Dieleman PJ (ed) (1963) Reclamation of salt affected soils in Iraq. International Institute for Land Reclamation and Improvement, Wageningen Publication, p 175

Diker K, Bausch WC (2003) Potential use of nitrogen reflectance index to estimate plant parameters and yield of maize. Biosys Eng 85(4):437–447. https://doi.org/10.1016/S1537-5110(03)00097-7

Dwivedi RS, Rao BRM (1992) The selection of the best possible Landsat TM band combination for delineating salt-affected soils. Int J Remote Sens 13(11):2051–2058

Eldeiry AA, Garcia LA (2010) Comparison of ordinary kriging, regression kriging, and cokriging techniques to estimate soil salinity using Landsat images. J Irrig Drain Eng 136:355–364

FAO (2008) Irrigation in the Middle East Region in figures. FAO Water Reports 34, AQUASTAT Survey, p 423

Farifteh J, Farshad A, George RJ (2006) Assessing salt-affected soils using remote sensing, solute modelling, and geophysics. Geoderma 130(3–4):191–206

Farifteh J, van der Meer F, Atzberger C, Carranza E (2007) Quantitative analysis of salt-affected soil reflectance spectra: a comparison of two adaptive methods (PLSR and ANN). Remote Sens Environ 110:59–78

Fieuzal R, Duchemin B, Jarlan L, Zribi M, Baup F, Merlin O, Hagolle O, Garatuza-Payan J (2011) Combined use of optical and radar satellite data for the monitoring of irrigation and soil moisture of wheat crops. Hydrol Earth Syst Sci 15:1117–1129

Gao B (1996) NDWI—a normalized difference water index for remote sensing of vegetation liquid water from space. Remote Sens Environ 58:257–266

Gao Y, Walker JP, Allahmoradi M, Monerris A, Ryu D, Jackson TJ (2015) Optical sensing of vegetation water content: a synthesis study. IEEE J Sel Topics Appl Earth Obs Remote Sens 8(4):1456–1464. https://doi.org/10.1109/JSTARS.2015.2398034

Gitelson AA (2004) Wide dynamic range vegetation index for remote quantification of crop biophysical characteristics. J Plant Physiol 161:165–173

Gong H, Shao Y, Brisco B, Hu Q, Tian W (2013) Modeling the dielectric behavior of saline soil at microwave frequencies. Can J Remote Sens 39(1):1–10

Hardisky MA, Klemas V, Smart RM (1983) The influences of soil salinity, growth form, and leaf moisture on the spectral reflectance of *Spartina alterniflora* canopies. Photogram Eng Remote Sens 49:77–83

He B, Xing M, Bai X (2014) A synergistic methodology for soil moisture estimation in an Alpine prairie using radar and optical satellite data. Remote Sens 6:10966–10985. https://doi.org/10.3390/rs61110966

Huang J, Chen DY, Cosh MH (2009) Sub-pixel reflectance unmixing in estimating vegetation water content and dry biomass of corn and soybeans cropland using normalized difference water index (NDWI) from satellites. Int J Remote Sens 30:2075–2104

Huete AR (1988) A soil adjusted vegetation index (SAVI). Remote Sens Environ 25:295–309

Huete AR, Liu HQ, Batchily K, van Leeuwen W (1997) A comparison of vegetation indices global set of TM images for EOS-MODIS. Remote Sens Environ 59:440–451

Jackson TJ, Chen D, Cosh M, Li F, Anderson M, Walthall C, Doriaswamy P, Hunt ER (2004) Vegetation water content mapping using Landsat data derived normalized difference water index for corn and soybeans. Remote Sens Environ 92:475–482

Jacobsen T, Adams RM (1958) Salt and silt in ancient Mesopotamian agriculture. Science 128:1251–1258

Kaufman YJ, Tanré D (1992) Atmospherically resistant vegetation index (ARVI) for EOS-MODIS. IEEE Trans Geosci Remote Sens 30:261–270

Kumar K, Hari Prasad KS, Arora MK (2012) Estimation of water cloud model vegetation parameters using a genetic algorithm. Hydrol Sci J 57(4):776–789

Lasne Y, Paillou P, Ruffie G, Serradilla C, Demontoux F, Freeman A, Farr T, McDonald K, Chapman B, Malezieux J-M (2008) Effect of salinity on the dielectric properties of geological materials: implication for soil moisture detection by means of remote sensing. IEEE Trans Geosci Remote Sens 46:3689–3693

Lee JS (1980) Digital image enhancement and noise filtering by use of local statistics. IEEE Trans Pattern Anal Mach Intell PAMI-2 2:165–168

Leonard A, Beriaux E, Defourny P (2013) Complementarity of linear polarizations in C-band SAR imagery to estimate leaf area index for maize and winter wheat. In: Proceedings of the ESA living planet symposium 2013, Edinburgh, UK, 9–13 Sept 2013 (ESA SP-722, Dec 2013)

Li RY, Ulaby FT, Eyton RJ (1980) Crop classification with a landsat/radar sensor combination. In: Proceedings of LARS symposia, Purdue University, West Lafayette, USA, pp 78–86

Maggioni V, Panciera R, Walker JP, Rinaldi M, Paruscia V (2006) A multi-sensor approach for high resolution airborne soil moisture mapping. In: Proceedings of the 30th hydrology and water resource symposium, 4–8 Dec 2006, Launceston, TAS, Australia

Metternicht GI, Zinck JA (2003) Remote sensing of soil salinity: potentials and constraints. Remote Sens Environ 85:1–20

Moran MS, Vidal A, Troufleau D, Qi J, Clarke TR, Pinter PJ Jr, Mitchell TA, Inoue Y, Neale CMU (1997) Combining multifrequency microwave and optical data for crop management. Remote Sens Environ 61:96–109

Mougenot B, Pouget M, Epema G (1993) Remote sensing of salt-affected soils. Remote Sens Rev 7:241–259

Muhaimeed AS, Wu W, AL-Shafie WM, Ziadat F, Al-Musawi HH, Saliem KA (2013) Use remote sensing to map soil salinity in the Musaib area in Central Iraq. Int J Geosci Geomatics 1(2):34–41. ISSN: 2052-5591

Pereira LDO, Freitas CDC, Anna SJSS, Lu D, Moran EF (2013) Optical and radar data integration for land use and land cover mapping in the Brazilian Amazon. GISci Remote Sens 50(3):301–321. https://doi.org/10.1080/15481603.2013.805589

Perkins T, Adler-Golden S, Matthew M, Berk A, Anderson G, Gardner J, Felde G (2005) Retrieval of atmospheric properties from hyper and multispectral imagery with the FLAASH atmospheric correction algorithm. In: Schäfer K, Comerón AT, Slusser JR, Picard RH, Carleer MR, Sifakis N (eds) Remote sensing of clouds and the atmosphere X. Proceedings of SPIE, vol 5979

Prévot L, Dechambre M, Taconet O, Vidal-Madjar D, Normand M, Galle S (1993) Estimating the characteristics of vegetation canopies with airborne radar measurements. Int J Remote Sens 14(15):2803–2818

Rao B, Sankar T, Dwivedi R, Thammappa S, Venkataratnam L, Sharma R, Das S (1995) Spectral behaviour of salt-affected soils. Int J Remote Sens 16:2125–2136

Shao Y, Hu Q, Guo H, Lu Y, Dong Q, Han C (2003) Effect of dielectric properties of moist salinized soils on backscattering coefficients extracted from RADARSAT image. IEEE Trans Geosci Remote Sens 41(8):1879–1888. https://doi.org/10.1109/TGRS.2003.813499

Shimada M, Isoguchi O, Tadono T, Isono K (2009) PALSAR radiometric and geometric calibration. IEEE Trans Geosci Remote Sens 47(12):3915–3931

Sreenivas K, Venkataratnam L, Rao PVN (1995) Dielectric properties of salt affected soils. Int J Remote Sens 16:641–649

Taghizadeh-Mehrjardi R, Minasny B, Sarmadian F, Malone BP (2014) Digital mapping of soil salinity in Ardakan region, central Iran. Geoderma 213:15–28. https://doi.org/10.1016/j.geoderma. 2013.07.020

Taylor GR, Mah AH, Kruse FA, Kierein-Young KS, Hewson RD, Bennett BA (1996) Characterization of saline soils using airborne radar imagery. Remote Sens Environ 57:127–142

Ulaby FT, Batlivala PP, Dobson MC (1978) Microwave backscatter dependence on surface roughness, soil moisture and soil texture—part I: bare soil. IEEE Trans Geosci Electron 16:286–295

Wu W (2014) The generalized difference vegetation index (GDVI) for dryland characterization. Remote Sens 6:1211–1233

Wu W, Muhaimeed AS, Platonov A, Al-Shafie WM, Abbas AH, Al-Musawi HH, Khalaf AJ, Salim KA, Christen E, De Pauw E, Ziadat F (2012) Salinity modeling by remote sensing in central and southern Iraq. American Geophysical Union (AGU) fall meeting, 03–07 Dec 2012, San Francisco, USA (extended abstract)

Wu W, Al-Shafie WM, Muhaimeed AS, Ziadat F, Nangia V, Payne W (2014a) Soil salinity mapping by multiscale remote sensing in Mesopotamia, Iraq. IEEE J Sel Topics Appl Earth Obs Remote Sens 7(11):4442–4452. https://doi.org/10.1109/JSTARS.2014.2360411

Wu W, Muhaimeed AS, Al-Shafie WM, Ziadat F, Nangia V, De Pauw E (2014b) Mapping soil salinity changes using remote sensing in Central Iraq. Geoderma Reg 2–3:21–31. https://doi.org/ 10.1016/j.geodrs.2014.09.002

Wu W, Zucca C, Karam F, Liu G (2016) Enhancing the performance of regional land cover mapping. Int J Earth Obs Geoinf 52:422–432. https://doi.org/10.1016/j.jag.2016.07.014

Wu W, Zucca C, Muhaimeed AS, Al-Shafie WM, Al-Quraishi AMF, Nangia V, Zhu M, Liu G (2018) Soil salinity prediction and mapping by machine learning regression in Central Mesopotamia. Land Degrad Dev 29(11):4005–4014. https://doi.org/10.1002/ldr.3148

Xavier AC, Vettorazzi CA (2004) Monitoring leaf area index at watershed level through NDVI from Landsat-7/ETM+ data. Sci Agricola 61(3):243–252 (Piracicaba, Braz.)

Yi YH, Yang DW, Chen DY, Huang JF (2007) Retrieving crop physiological parameters and assessing water deficiency using MODIS data during the winter wheat growing period. Can J Remote Sens 33:189–202

Yilmaz MT, Hunt ER, Jackson TJ (2008) Remote sensing of vegetation water content from equivalent water thickness using satellite Imagery. Remote Sens Environ 112:2514–2522

Chapter 3
Using Remote Sensing to Predict Soil Properties in Iraq

Ahmad Salih Muhaimeed

Abstract Soil composed of different percentages of solid, liquid and gasses materials. In reality, the percentages of these components vary tremendously, and the determination the precise amount for each fraction on world wide scale is expensive and time-consuming. The success of the precision agriculture depends strongly upon an efficient and accurate method for in-field soil property determination. For that reason, some helpful techniques were developed in order to predict and mapping soil components including remote sensing (RS) and Geographical Information Systems (GIS). Different methodologies have been used for the estimation of soil parameters, based on different remote sensing sensors and techniques. Many studies in soil science have used both RS bare-soil images and spectroscopic reflectance of soil samples for soil survey, mapping, and quantitative soil-property characterization. This chapter aims to demonstrate the development of some statistical models to predict some components of Iraqi soils using Remote sensing techniques including spectral indices and electromagnet induction.

Keywords Soil properties · Remote sensing · GIS · Soil composition

3.1 Introduction

Soil is one of the most important components of agricultural production and can have a dominant effect on crop yields and quality. Infield soil information has been used for centuries by farmers to make decisions concerning crop management practices. Traditionally, soil property information gathering is done by grid sampling. The USDA's Natural Resource Conservation Service (NRCS) implements a nationwide soil survey (Soil Survey Staff 2017).

A. S. Muhaimeed (✉)
College of Agriculture, Baghdad University, Baghdad, Iraq
e-mail: profahmad1958@yahoo.com

ACSAD, Damascus, Syria

© Springer Nature Switzerland AG 2020
A. M. F. Al-Quraishi and A. M. Negm (eds.),
Environmental Remote Sensing and GIS in Iraq, Springer Water,
https://doi.org/10.1007/978-3-030-21344-2_3

Soil is a natural body comprised of solids (minerals and organic matter), liquid, and gases that occurs on the land surface, occupies space, and is characterized by one or both of the following: horizons, or layers, that are distinguishable from the initial material by one or more properties, as a result of additions, losses, transfers, and transformations of energy and matter or the ability to support rooted plants in a natural environment (Simonson 1968; Soil Survey Staff 2014). FAO (1987) indicated that soil is a complex body composed of five major components:

(a) Mineral matter obtained by the disintegration and decomposition of rocks;
(b) Organic matter, obtained by the decay of plant residues, animal remains and microbial tissues;
(c) Water, obtained from the atmosphere and the reactions in soil (chemical, physical and microbial);
(d) Air or gases, from atmosphere, reactions of roots, microbes and chemicals in the soil;
(e) Organisms, both big (worms, insects) and small (microbes).

In general, soil composed of different percentages of solid, liquid and gasses materials (Fig. 3.1). In reality, these percentages of the four components vary tremendously. Soil air and water are found in the pore spaces between the solid soil particles. The ratio of air-filled pore space to water-filled pore space often changes seasonally, weekly, and even daily, depending on water additions through precipitation, through flow, groundwater discharge, and flooding. The volume of the pore space itself can be altered, one way or the other, by several processes. Organic matter content is usually much lower than 5% in South Carolina (typically 1% or less). Some wetland soils, however, have considerably more organic matter in them (greater than 50% of the solid portion of the soil in some cases) (FAO 1987).

The mineral components composed of different types of primary and secondary reflecting the effects of the main soil forming factors. Also, the organic matter consists of many fractions as shown in the following (Chenu 2004):

Soil Organic Matter Fractions:

1. Biochemistry:
   – Liquids
   – Polysaccharides
   – Proteins

Fig. 3.1 Approximation composition of soil

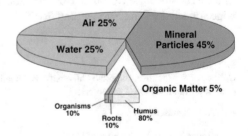

Air 25%

Water 25%

Mineral Particles 45%

Organic Matter 5%

Organisms 10%

Roots 10%

Humus 80%

- Lipids
- Non-Identified
2. Humus Chemistry:
   - Macro-organic matter
   - Fluvic Acids
   - Humic Acids
   - Humin
3 Physical fractionation:
   - POM > 2000 μm
   - POM 200–2000 μm
   - POM 50–200 μm
   - 20–50 μm
   - 2–20 μm
   - 0–2 μm

Determination of the precise amount for each fraction of world wide scale is expensive and time-consuming. The success of precision agriculture depends strongly upon an efficient and accurate method for in-field soil property determination. This information is critical for farmers to calculate the proper amount of inputs for best crop performance and least environmental effect (Omran 2012). Quantitative information and spatial distribution of soil properties are among the main prerequisites for achieving sustainable land management. The accuracy of soil information determines, to a large extent, the reliability of land resources management decisions (Mermut and Eswaran 2001; Salehi et al. 2003). Conventional soil surveys are usually used to derive information about soils and their distribution (Salehi et al. 1999). In soil science, the implementation of geomatics—geographical information systems (GIS), global positioning system (GPS), remote sensing (RS), and digital elevation models (DEMs)—is suggesting new alternatives (Mermut and Eswaran 2001; McBratney et al. 2003). Global coverage of high-resolution imagery and terrain data is increasing, and this enhances the popularity of digital soil mapping to generate accurate soil maps that capture many properties with reasonable effort (Browning and Duniway 2011).

For that reason, some helpful techniques were developed in order to predict and mapping soil components including RS and GIS. Remote sensing is expected to offer possibilities for improving incomplete spatial and thematic coverage of current regional and global soil databases. Soil properties that have been measured using remote or proximal sensing approaches include mineralogy, texture, soil iron, soil moisture, soil organic carbon, soil salinity and carbonate content (Mulder et al. 2011).

The great demand for information about properties of the surface and sub-surface soil which can be estimated by using spectral responses of the surface soil for different bands of radiometer and short time available for the work arises to employ remote sensing techniques in geotechnical engineering which reduce the time, cost, efforts, and staff. Yousif (2004), used RS techniques to classify the soil of Al-Najaf city by comparing the results obtained from remote sensing techniques with those obtained from traditional classification method. Remote sensing has shown high potential in soil characteristics retrieving in the last three decades (Zribi et al. 2011). Different methodologies have been proposed for the estimation of soil parameters, based on different RS sensors and techniques (Mulder et al. 2011). Many studies in soil science have used both RS bare-soil images and spectroscopic reflectance of soil samples for soil survey, mapping, and quantitative soil-property characterization (Dalal and Henry 1986; Agbu et al. 1990; Ben-Dor and Banin 1994, 1995). Compared to conventional methods (e.g., the pipette method for soil textures and the dry combustion method for soil organic carbon concentration), RS has proven to be cheaper, faster, and fairly accurate for certain applications.

The following sections will focus on the application of RS techniques to predict some soil properties. The main objectives of this chapter are to demonstrate the development of some statistical models to predict some components of Iraqi soils using RS techniques including spectral indices and electromagnet induction.

3.2 Case Studies

3.2.1 Prediction of Soil Organic Carbon and Nitrogen Forms

Organic material is a major part of the productive soils. Besides improving the physical and chemical properties of the soil, it is considered as a storehouse for many of the plant nutrients, especially carbon and nitrogen elements to a large degree, phosphorus, iron and sulfur to a lesser extent. The availability of many of these elements in organic form are affected by the existence of organic matter due to the many reactions between them. The measurement of soil total organic carbon and Nitrogen content in the fields is important to guide the reasonable application of nitrogenous fertilizer. Estimation of soil total nitrogen content with limited in situ data at an acceptable level of accuracy is important because laboratory measurement of nitrogen is a time- and lab or-consuming procedure. In order to build up a data base for soil organic carbon (SOC) and some Nitrogen forms on a large scale in Iraq, some attempts were done to predict these soil components using remote sensing. Muhaimeed and Taha (2014), Muhaimeed et al. (2017) used some spectral indices including NDVI, EVI and GDVI with soil laboratory data to produce the best fitting statistical model to predict SOC and Nitrogen forms.

3.2.1.1 The Study Area

The study area located in central of Iraq—Al-Kufa—Alnajaf province laying between 32° 10′ 04.76″N to 32° 10′ 04.76″N and 32° 10′ 04.76″N to 44° 34′ 49.28″E with a total area of 27,664 ha. Thirty-five sites were selected representing all variations within the study area, using a GPS device, and located on the landsat 8 image (Fig. 3.2). Soil samples were taken from each of the selected sites and analysed

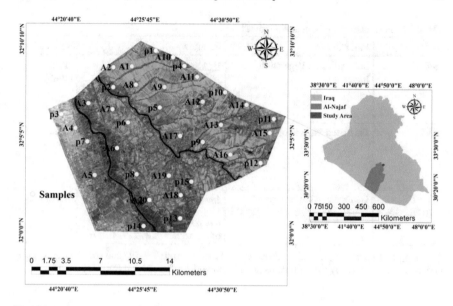

Fig. 3.2 Location of the study area and sampling sites (Muhaimeed et al. 2017)

in the laboratory to determine some physical and chemical properties including SOC and Nitrogen forms.

3.2.1.2 Satellite and Ancillary Data

The Landsat 8 OLI image (acquisition date: 1 Mar 2015) obtained from the USGS EROS Centre was used in this study. The OLI image consists of nine spectral bands with a spatial resolution of 30 m for Bands 1–7 and 9. The resolution for Band 8 (panchromatic) is 15 m. The time of acquisition corresponded closely to the field campaign, and occurred while the main cereal crop (wheat) was in the growing stages. Atmospheric correction for Landsat image was done using the FLASASH model (Perkins 2005) to correct both additive and multiplicative atmospheric effects. The corrected image was used to produce the following indices using ENVI 5 (Table 3.1).

3.2.1.3 Statistical Prediction Models

A Pearson correlation analysis was applied to compute the best statistical prediction models for SOC and Nitrogen forms. From multiple regression conducted using the SPSS, the coefficient of determination (R^2) between MNLI, SAVI, EVI and GDVI, with SOC and Nitrogen (NO_3, NH_4, TN) were derived to evaluate the usefulness of the various indices for prediction across the study area. The prediction accuracy was verified by comparing the predicted and measured values using 35

Table 3.1 Formulae of the vegetation indices

Index	Full name	Formula	References
SAVI	Soil-Adjusted Vegetation Index	$\frac{(1+L)(\rho_{NIR}-\rho_R)}{(\rho_{NIR}+\rho_R+L)}$ Low vegetation, $L = 1$, intermediate, 0.5, and high 0.25	Huete (1988)
EVI	Enhanced Vegetation Index	$G * \frac{(\rho_{NIR}-\rho_R)}{(\rho_{NIR}+C1*\rho_R-C2*\rho_B+L)}$ ρ_B = reflectance of blue band, $G = 2.5$, $C1 = 6$, $C2 = 7.5$ and $L = 1$	Huete et al. (1997)
GDVI	Generalized Difference Vegetation Index	$\frac{\rho_{NIR}^n-\rho_R^n}{\rho_{NIR}^n+\rho_R^n}$ n is power number, an integer of the values of 1, 2, 3, 4 ... n	Wu (2012)

Note ρ_{NIR} and ρ_R are respectively reflectance of the near infrared (NIR) and red (R) bands; ρ_B = and ρ_{MIR} are respectively that of blue band and the middle infrared band (like TM band 5)

randomly selected field observations. Then the best fitting models developed using the statistical correlation (Eqs. 3.1–3.4) were used to study the correlation between the predicted and the laboratory determined values of SOC, NO_3, NH_4 and TN (Muhaimeed et al. 2017). The best fitting models are represented by the following equations:

$$SOC = 32.588 + 8.495\ln(SAVI) - 57.965(SAVI) \quad R^2 = 0.924^{**} \quad (3.1)$$

$$NO_3 = 45.169 + 18.267\ln(SAVI) - 236.035(SAVI) + 35.34 \quad R^2 = 0.917^{**} \tag{3.2}$$

$$NH_4 = 8.974 + 10.307\ln(EVI) - 60.852(EVI) + 28.681e^{(EVI)} \quad R^2 = 0.993^{**} \tag{3.3}$$

$$Total.N = 104.356 + 22.723\ln(SAVI) - 258.974(SAVI) + 59.705(EVI)$$
$$R^2 = 0.991^{**} \tag{3.4}$$

**significant at the 0.01 probability level.

The results of the statistical correlation represented by the above equations indicate a strong correlation between the predicted value for all the studied soil properties and EVI and SAVI ($R^2 > 0.9$; $p < 0.01$) whereas the other indices did not perform as well.

The best developed statistical model (Eq. 3.1) was used to predict the content of SOC in the study area. The relationships between measured versus predicted content for SOC are shown in Fig. 3.3 (Muhaimeed et al. 2017). The results show a very strong correlation between the measured content of SOC and the predicted with $R^2 = 0.9531$ ($p < 0.01$).

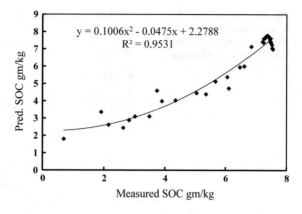

Fig. 3.3 Relationship between measured and predicted content for SOC (gm/kg) in the surface horizon of the study area

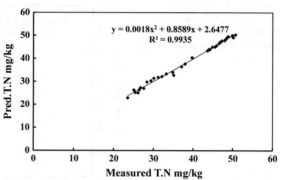

Fig. 3.4 The relationship between the measured and predicted total nitrogen content (mg/km) in the surface horizon of study area

Nitrogen exists in the soil system in many forms and changes (transforms) very easily form one form to another. Nitrogen is available to plants as either ammonium (NH_4^+–N) or nitrate (NO_3^-–N) (Lamb et al. 2014). Equations 3.2–3.4 represent the best developed statistical correlation prediction models used to compute the predicted values for Nitrogen forms content in the study area soils. Figure 3.4 shows a very strong correlation between measured and the predicted TN. Nitrogen content with $R^2 = 0.9934$.

The results of total content for Nitrogen forms indicated a very strong correlation between the predicted and the measured content for NH_4 and NO_3 with correlation coefficients $R^2 = 0.993$ and 0.988 respectively (Figs. 3.5 and 3.6). These results indicate that the developed statistical prediction models derived from EVI and SAVI were useful for spatial prediction of both NH_4 and NO_3 components.

The results of the prediction models for SOC and Nitrogen forms are in agreement with other researchers. They indicated that the correlation coefficients in the range of $0.87 < R^2 < 0.98$ between spectrally measured and chemically analysed samples have been obtained using mid infrared and combined diffuse reflectance spectroscopy (Barnes et al. 2003; Chang and Laird 2002; McCarty et al. 2002; Viscarra-Rossel et al. 2006).

Fig. 3.5 The relationship
between measured and
predicted content (mg/kg) of
NH4 in the surface horizon
of the study area

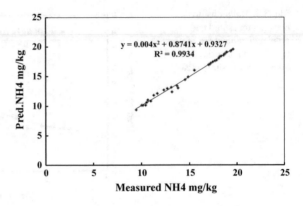

Fig. 3.6 The relationship
between the measured and
predicted content (mg/gm) of
NO3 in the surface of the
study area

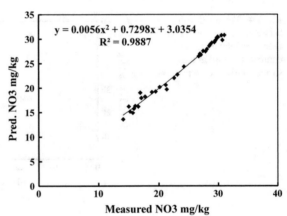

3.2.2 Prediction of Soil Salinity

3.2.2.1 Soil Salinity

The problem of soil salinity in Iraq is the main determinant of land degradation, and is the limiting factor for agricultural production, food security as well as for environment conditions. Salinization is a common problem for agriculture in dryland environments and it has greatly affected land productivity and even caused cropland abandonment in Central and Southern Iraq (Buringh 1960). Soil salinity assessment represents an important component in agriculture management and water allocation strategies. Excessive soil salinity can result in crop yield reduction, ground water contamination, and significant financial losses. The need for rapid, cost-effective appraisal techniques has become critical; as of approximately 60% of the cultivated land has been seriously salinized, of which 20–30% has been abandoned (FAO 2011; Farifteh et al. 2006) due to irrational land management (e.g., over irrigation and poor drainage) and other natural factors (e.g., flooding, drought, and impermeability of the underlying formation). More than 70% of the irrigated agriculture lands in the central and southern Iraq have been abandoned in the recent years and causing yield declined between 30 and 60% as a result, mainly, of salt accumulation by salinization process.

3.2.2.2 Measuring and Predicting of Soil Salinity

Traditionally, soil salinity has been measured via electrical conductivity (EC), by collecting in situ soil samples and analysing those samples in the laboratory to determine their solute concentrations or electrical conductivity. However, these methods are time-consuming and costly since dense sampling is required to adequately characterize the spatial variability of an area. A major constraint to using proximal and remote sensing data for mapping salinity is related to the fact that there is a strong vertical, spatial and temporal variability of salinity in the soil profile. Spectral data acquisition does not allow information to be extracted from the entire soil profile, since only the soil surface is observed. This can be overcome by integrating remote sensing data with simulation models and geophysical surveys (Farifteh et al. 2006; Mougenot et al. 1993). Direct and precise estimation of salt quantities is difficult using satellite data with a low spectral resolution because these fail to detect specific absorption bands of some salt types and the spectra interfere with other soil chromophores (Mougenot et al. 1993).

A different approach for rapidly characterizing soil salinity is offered by proximal sensing techniques using either visible near infrared diffuse reflectance spectroscopy (VisNIR DRS) or portable X-ray fluorescence (PXRF) spectrometry. Farifteh et al. (2006) studied diffuse reflectance spectroscopy (DRS) to determine its capability to identify different salt minerals in addition to quantifying soil salinity levels using samples artificially treated by different salt minerals in the laboratory, as well as those collected from a field experiment. Soil salinity is related to different spectral bands, ratios, and parameters extracted from satellite imagery using soil and vegetation based indices. The ability to diagnose and monitor field scale salinity conditions has been considerably refined and improved through the use of electromagnetic induction survey instruments (Rhoades 1992). Three types of portable instruments have been developed for measuring the apparent electrical conductivity (EC) of the soil: (1) four-electrode sensors, including either surface array or insertion probes, (2) remote electromagnetic (EM) induction sensors, such as the Geonics EM-31, EM-34, or EM-38, and 3 time domain reflectometry sensors (Rhoades 1992; Rhoades and Miyamoto 1990; McNeil 1980). The adaptation of electromagnetic induction sensors for soil electrical conductivity measurement greatly increases the speed with which reconnaissance surveys can be carried out. However, the conversion from EC, to soil salinity (EC) requires knowledge of soil properties, which are often too costly, difficult, or time consuming to measure during rapid survey work (Rhoades 1992).

Ghabour and Daels in (1993) agreed that detection soil salinity traditionally is time-consuming, but remote sensing data and techniques offer more efficiently and economically rapid tools and techniques for monitoring and mapping soil salinity.

Zhang et al. (2011) and Aldakheel (2011) indicated that soil salinity can be detected directly from remotely sensed data through salt features that are visible at the soil surface, such as bare soil with white salt crusts on the surface (Matinfar et al. 2013; Dematte et al. 2004) or indirectly from indicators such as the presence of halophytic plant, the performance level of salt-tolerant crops. Lhissou et al. (2014) showed the presence of a high correlation between soil reflectance and several soil properties such as organic matter content, soil moisture content, particle size distribution, iron content and surface condition. Remote sensing, GIS, modelling, geostatistical and advanced electromagnetic induction are the advanced technologies and tools for soil salinity assessment, mapping and monitoring.

3.2.2.3 Salinity Models Using Remote Sensing

Many spectral indices were used for monitoring and predicting soil salinity, which based on the visible spectral bands and found to be more sensitive to soil salinity in resent study (Wu et al. 2018, 2019; Mustafa et al. 2019; Sethi et al. 2010). Some of these indices included NDVI, SAVI and GDVI. The normalized difference vegetation index (NDVI) is considered as uncertain indicator for soil salinity assessment due to the possibility of growth of various plants in different salinity level (Allbed and Kumar 2013). The SAVI and other indices and enhanced models have helped the separation of soil from vegetation (Wu et al. 2014a). Furthermore, GDVI was developed in recent years has shown remarkable results for salinity assessment.

Matthew et al. (2000) and Wu et al. (2014b) build a prediction model for soil salinity level in the Mesopotamia plan using the electromagnetic induction instrument, field salinity reading and Multispectral satellite sensors data for Landsat ETM+, SPOT, Rapid Eye imagery and time-series of MODIS vegetation indices data (MOD13Q1), and land surface temperature (LST, MOD11A1 and A2). Atmospheric correction using FLAASH model (Rouse et al. 1973) for all Landsat ETM+, SPOT and Rapid Eye imagery. A set of mostly applied VIs such as NDVI (Normalized Difference Vegetation Index) (Kaufman and Tanre 1992), SAVI (Soil Adjusted Vegetation Index) (Huete 1988), SARVI (Soil Adjusted and Atmospherically Resistant Vegetation Index), EVI (Enhanced Vegetation Index) (Huete et al. 1997) were produced from the atmospherically-corrected and reflectance-based satellite imagery; a new vegetation index was used, the Generalized Difference Vegetation Index (GDVI) developed by Wu (2012) and in the form of:

$$\text{GDVI} = (\rho_{NIR}^n - \rho_R^n)/(\rho_{NIR}^n + \rho_R^n) \tag{3.5}$$

where ρ_{NIR} is the reflectance of the near-infrared band and ρ_R is that of the red band, and n is the power.

After separation of the vegetated and non-vegetated area, a multiple linear least-square regression analysis was undertaken to couple the EM38 measurements with

VIs for vegetated area and with NVIs for bare soils. The following remote sensing salinity models were obtained (Eqs. 3.6 and 3.7):

For vegetated area: $EMV = -824.134 + 918.536 * GDVI - 754.204 * \ln(GDVI)$

(Multiple $R^2 = 0.925$) $\hspace{9cm}$ (3.6)

For non-vegetated area: $EMV = 2{,}570{,}683.24 + 1821.24 * ST$

$$- 54{,}6476.07 * \ln(ST) \quad (\text{Multiple } R^2 = 0.829)$$

$\hspace{13cm}$ (3.7)

where EMV—vertical reading of EM38, ST—spring surface T in K. EMV can be converted into EC (electrical conductivity in dS/m) by the following relationships (Eq. 3.8):

$$EC = 0.0005 \, EMV^2 - 0.1007 \, EMV + 15.632 \quad (R^2 = 0.841) \qquad (3.8)$$

The study revealed that the was a high correlation between the ground value of soil salinity and the predicted EC using remote sensing techniques as shown in Fig. 3.7.

The spatial distribution of soil salinity in Musayb agricultural project in the middle of Iraq, was studied by Muhaimeed et al. (2013). They indicated that the development of the methodology for multitemporal salinity mapping by remote sensing and revealed the possibility to assess salinity by modelling approach coupling multiyear maximum remote sensing indicators with ground measurements. They indicated that

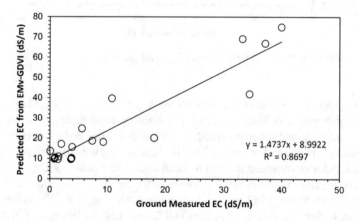

Fig. 3.7 The relationship between the ground EC and the predicted EC

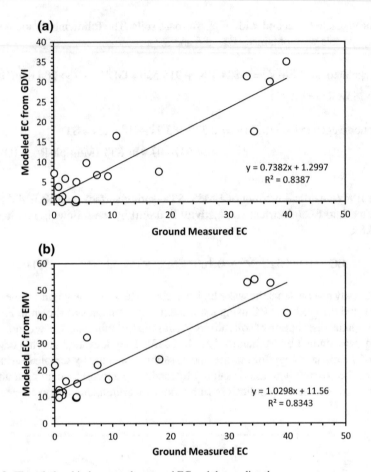

Fig. 3.8 The relationship between the ground EC and the predicted

EC-GDVI and EC-EMV-ST/NDII models are a good combination and operational for salinity mapping of both vegetated and non-vegetated areas. They added that salinity modelling and mapping could be successfully achieved by remote sensing and the GDVI and the spring surface T are the best salinity indicators respectively for vegetated and non-vegetated areas. The results showed a good correlation between the ground measured soil salinity and the predicted the value of EC from both EM38 and GDVI with R = 0.8343 and 0.8387, respectively (Fig. 3.8). These results are in agreement with the results of Allbed and Kumar (2013), Wu et al. (2014a) and Muhaimeed et al. (2017).

3.2.3 Physical Properties

3.2.3.1 Prediction of Soil Particles Size

According to Terra et al. (2015) soil properties directly detectable by VIS-NIR spectroscopy such as clay, iron and aluminium oxides, hydroxide contents and SOC can be modelled by first-order predictions. Viscarra-Rossel et al. (2006) reported an average R^2 for SOC prediction as 0.81 in the near infrared (NIR) region, 0.78 in the VIS region and 0.96 in the MIR region. Although MIR spectra generally produced more accurate results, especially for SOC, due to its stronger direct relation with MIR spectral data produced by fundamental vibrations. Clay content was predicted with the highest accuracy (R^2: 0.78), followed by soil organic carbon (R^2: 0.710), Sand, S/C, and CEC, were moderately well predicted from VIS-NIR spectra. Greater predictive performance for particle size is usually observed in the MIR compared to the VIS-NIR region. It can be explained by the stronger interaction between mid-infrared reflection spectra and soil particles by fundamental vibration processes when compared to the VIS-NIR (Viscarra-Rossel et al. 2006; Terra et al. 2015; Pirieet et al. 2005). They added that clay content showed a good fit (R^2 of 0.78) with MIR.

In Iraq, many studies have been done to predict soil components including soil particle size content using spectral indices. The studies of Abbas (2010) and Abbas and Muhaimeed (2012a, b) indicated that the result of stepwise analysis between remote sensing data and some soil properties, revealed the presence of significant correlations between spectral data and many physical soil properties for North Kut project—southern Iraq. They used images of Landsat 7 ETM+ acquired in 2009 with nine bands and some spectral indices, to build the best fitting statistical prediction models. They indicated the content of clay, silt and sand in the surface soil were highly correlated with the some spectral data as shown by the following Eqs. (3.9 and 3.10):

$$Clay\ \% = -2682 + 3.8\,B7 + 271RIN + 429VIN - 21.6\,B5 + 11.1\,B8 + 29.7\,B1$$
$$+ 0.119\,B62 \quad with\ R^2 = 0.996. \tag{3.9}$$

The results revealed a highly significant relationship between the measured clay content and the predicted clay content in the surface horizons as shown in Fig. 3.9.

They added that a high correlation between silt content in the surface soil and B2, B61 and B62 as shown in Eq. (3.10). Also, they found a strong correlation between the predicted silt content which was calculated by using the best fitting equation, and the measured silt content in the surface horizons as shown in Fig. 3.10.

$$Silt\ \% = 156 - 1.7\,B2 - 1.68\,B61 - 1.41\,B62 \quad with\ R^2 = 0.972 \tag{3.10}$$

Also, the results indicated a high correlation between sand content in the surface soil and some spectral data as shown in Eq. (3.11) and Fig. 3.11:

Fig. 3.9 The relationship
between the measured and
the predicted clay content in
the surface horizons

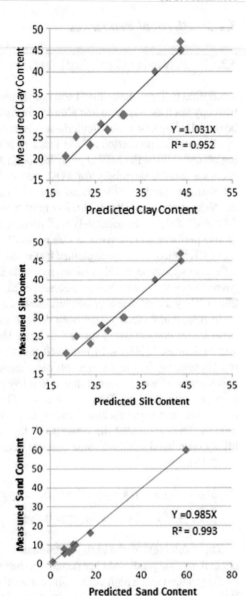

Fig. 3.10 The relationship
between the measured and
the predicted silt content

Fig. 3.11 The relationship
between the measured and
the predicted sand content

Sand % = −428−4.79 B4 + 9.26 B2 + 5.48 B62−5.72 B61 + 228RIN + 0.739 B7
with $R^2 = 0.998$ (3.11)

These results are in agreement with the result of Al-Shiakly (2001) and Muhaimeed and Al-Shiakly (2002) for some gypseferous soils in Al-Anbar, western Iraq. They used images of Landsat 5 TM sensor with six bands. The results for both simple linear and multi linear regression model show high correlation between clay content in the surface horizons and spectral data with $R^2 = 0.833$ and 0.97, respectively, as shown in the following Eqs. (3.12 and 3.13):

$$\% \, Clay = 108−212R4 \quad R^2 = 0.833 \tag{3.12}$$

$$\% \, Clay = 91.5−269TM1 + 1338TM2 + 814TM3 − 390TM4 + 190TM5$$
$$− 333TM7 \quad R^2 = 0.97 \tag{3.13}$$

Also, the results indicate a high correlation between the predicted and the measured values of clay content in the surface horizons with $R^2 = 0.829$ and 0.970, respectively for the simple and multi linear regressions, which are in agreement with the results of Colement et al. (1993).

3.2.4 Prediction of Some Soil Water Properties

Abbas and Muhaimeed (2012a) studied some water properties including infiltration rate, moisture content for some soils southern Iraq in a relationship with remote sensing data. They found the best fitting model to predict infiltration rat as shown in the following Eq. (3.14), which was used to calculate the predicted values for basic infiltration rate:

$$Basic \, Infiltration \, Rate = −40.0 + 0.398 \, B4 + 0.308 \, B2 + 0.0293 \, B62$$
$$With \, R^2 = 0.946 \tag{3.14}$$

There was a high correlation between the predicted and the measured values for the basic infiltration rate (Fig. 3.12).

Also, the results indicated that high correlation between the predicted and the measured values for Water Saturation Percentage (θs), and $R^2 = 0.998$ and available water (θv), with $R^2 = 0.998$ and 0.923 m, respectively.

Fig. 3.12 The relationship
between the measured and
the predicted infiltration rate

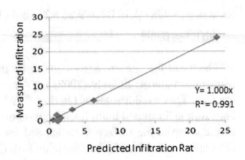

3.3 Conclusions

Remote sensing multi-temporal satellite data has proven to be an important tool to predict some physical and chemical soil properties on a large scale using different indices, including NDVI, SAVI and GDVI. The normalized difference vegetation index (NDVI) is considered as uncertain indicator for soil salinity assessment due to the possibility of growing of various plants in different salinity level multi-temporal satellite data help to delineate the various change of the earth surface. Remote sensing can be a useful tool to reduce efforts, time and coasts due to its varied applications and ability to allow users to collect, interpret, and manipulate data over dangerous areas.

References

Abbas AH (2010) Characterization and classification of soil units of North Kut project and prediction of some soil physical properties by using GIS and Remote Sensing. PhD thesis, College of Agriculture, Baghdad University, Iraq

Abbas AH, Muhaimeed AS (2012a) Prediction of infiltration and saturated hydraulic conductivity of the soils in the Iraqi central alluvial plain using remote sensing technique. J Iraqi Sci 53(4):971–976

Abbas AH, Muhaimeed AS (2012b) Prediction of some soil water content parameters using remote sensing & GIS techniques. Iraqi J Agric Sci 43(4):13–22

Agbu PA, Fehrenbacher DJ, Jansen IJ (1990) Soil property relationships with SPOT satellite digital data in east central Illinois. Soil Sci Soc Am J 54(3):807–812

Aldakheel YY (2011) Assessing NDVI spatial pattern as related to irrigation and salinity management in Al-Hassa Oasis, Saudi Arabia. J Indian Soc Remote Sens 39(2):171–180

Allbed A, Kumar L (2013) Soil salinity mapping and monitoring in arid and semi-arid regions using remote sensing technology: a review. ARS 2:373–385

Al-Shiakly FA (2001) Testing remote sensing techniques in determining soil units of different gypsum content and forecasting some of their features. Ms. thesis, Agriculture College, Baghdad University, Iraq

Barnes EM et al (2003) Remote- and ground-based sensor techniques to map soil properties. Photogramm Eng Remote Sens 69(6):3619–3630

Ben-Dor E, Banin A (1994) Visible and near-infrared (0.4–1.1 μm) analysis of arid and semiarid soils. Remote Sens Environ 48(3):261–274

Ben-Dor E, Banin A (1995) Near-infrared analysis as a rapid method to simultaneously evaluate several soil properties. Soil Sci Soc Am J 59(2):364–372

Browning DM, Duniway MC (2011) Digital soil mapping in the absence of field training data: a case study using terrain attributes and semi automated soil signature derivation to distinguish ecological potential. Appl Environ Soil Sci 42:1904–1910

Buringh P (1960) Soils and soil conditions in Iraq. Ministry of Agriculture of Iraq, 337 p

Chang CW, Laird DA (2002) Near-infrared reflectance spectroscopic analysis of soil C and N. Soil Sci 167(2):110–116

Chenu C (2004) Soil phases: the organic solid phase. In: Galcomo C, Scalenghe R (eds) Soils: basic concepts and future challenges. Cambridge University Press, Cambridge

Coleman TL, Agbu PA, Montgomery OL (1993) Spectral differentiation of surface soils and soil properties is it possible from space platform? Soil Sci 55:283–293

Dalal RC, Henry RJ (1986) Simultaneous determination of moisture, organic carbon, and total nitrogen by near infrared reflectance spectroscopy. Soil Sci Soc Am J 50(1):120–123

Dematte JAM, Gama MAP, Cooper M, Araui JC, Nanni MR, Fiorio PR (2004) Effect of fermentation residue on the spectral reflectance properties of soils. Geoderma 1203–4:187–200

FAO (2011) Country pasture/forage resource profiles: Iraq. FAO, Rome, Italy, p 34

FAO—UNDP (1987) Soil quality considerations in the selection of sites for aquaculture. Nigerian institute for oceanography and marine research. Project, RAT/82/009

Farifteh J, Farshad A, George RJ (2006) Assessing salt-affected soils using remote sensing, solute modelling, and geophysics. Geoderma 130(3–4):191–206

Ghabour T, Daels L (1993) Mapping and monitoring soil salinity of ISSN. Egypt J Soil Sci 33(4):355–370

Huete AR (1988) A soil adjusted vegetation index (SAVI). Remote Sens Environ 25:295–309

Huete AR, Liu HQ, Batchily K, van Leeuwen W (1997) A comparison of vegetation indices global set of TM images for EOS-MODIS. Remote Sens Environ 59:440–451

Kaufman YJ, Tanre D (1992) Atmospherically resistant vegetation index (ARVI) for EOS-MODIS. IEEE Trans Geosci Remote Sens 30:261–270

Lamb JA, Fernandez FG, Kaiser DE (2014) Understanding nitrogen in soil. University of Minnesota Extension

Lhissou R, El Harti A, Chokmani K (2014) Mapping soil salinity in irrigated land using optical remote sensing data. Aust J Soil Sci 3:82–88, 913–917

Matinfar HR et al (2013) Dedection of soil salinity changes and mapping land cover types based upon remotely sensed data. Arab J Geosci 6(3):6–12

Matthew MW, Adler-Golden SM, Berk A, Richtsmeier SC, Levine RY, Bernstein LS, Acharya PK, Anderson GP, Felde GW, Hoke MP, Ratkowski A, Burke H-H, Kaiser RD, Miller DP (2000) Status of atmospheric correction using a MODTRAN4-based algorithm. In: SPIE Proceedings of algorithms for multispectral, hyperspectral, and ultraspectral imagery VI, 4049, pp 199–207

McBratney AB, Mendonça Santos ML, Minasny B (2003) On digital soil mapping. Geoderma 117(1–2):3–52

McCarty GW, Reeves JB, Reeves VB, Follett RF, Kimble JM (2002) Mid-infrared and near-infrared diffuse reflectance spectroscopy for soil carbon measurement. Soil Sci Soc Am J 66(2):640–646

McNeil JD (1980) Electromagnetic terrain conductivity measurement at low induction numbers, Tech Note TN-6, Geonics Limited, Mississauga, Ont, Canada

Mermut AR, Eswaran H (2001) Some major developments in soil science since the mid-1960s. Geoderma 100(3–4):403–426

Mougenot B, Pouget M, Epema GF (1993) Remote sensing of salt affected soils. Remote Sens Rev 7(3–4):241–259

Muhaimeed AS, Al-Shiakly FF (2002) Forecasting some soil components using satellite images for TM sensor. In: 8th Scientific conference for foundation of technical education, pp 111–112

Muhaimeed AS, Taha AM (2014) Effect of land use and Irrigation water on amount and type of chemical compounds in some soils of Babylon's Governorate using remote sensing and GIS. Int J Environ Glob Clim Change 2(4):137–147

Muhaimeed AS, Wu W, AL-Shafi WM, Ziadat F, Katai HH, Saleim KA (2013) Use remote sensing to map soil salinity in Musaib area central Iraq. Int J Geosci Geomatics 2:34–41

Muhaimeed AS, Taha AM, Almashhadani HA (2017) Using remote sensing and GIS techniques for predicting soil organic carbon in Southern Iraq. In: Global symposium on soil organic carbon GSOC2017, 21–23 Mar. FAO, Rome, Italy

Mulder VL, Bruin S, de Schaepman ME, Mayr T (2011) The use of remote sensing in soil and terrain mapping: review. Geoderma 162:1–19

Mustafa BM, Al-Quraishi AMF, Gholizadeh A, Saberioon M (2019) Proximal soil sensing for soil monitoring. In: Al-Quraishi AMF, Negm AM (eds) Environmental remote sensing and GIS in Iraq. Springer Water

Omran EE (2012) On-the-go digital soil mapping for precision agriculture. Int J Remote Sens Appl 2(3):20–38

Perkins T, Adler-Golden S, Matthew M, Berk A, Anderson G, Gardner J, Felde G (2005) Retrieval of atmospheric properties from hyper and multispectral imagery with the FLAASH atmospheric correction algorithm. In: Schäfer K, Comerón AT, Slusser JR, Picard RH, Carleer MR, Sifakis N (eds) Remote sensing of clouds and the atmosphere X. Proceedings of SPIE, vol 5979

Pirieet A, Singh B, Islam K (2005) Ultra-violet, visible, near-infra-red, and mid-infra-red diffuse reflectance spectroscopic techniques to predict several soil properties. Aust J Soil Res 43:713–721

Rhoades JD (1992) Instrumental field methods of salinity appraisal. In: Topp GC, Reynolds WD, Green RE (eds) Advances in measurement of soil physical properties: bringing theory into practice. SSSA Spec. Publ., 30, pp 231–248

Rhoades JD, Miyamoto S (1990) Testing soils for salinity and sodicity. In: Westerman RL (ed) Soil testing and plant analysis. SSSA Book Ser, vol 3, 3rd edn, Soil Science Society of America, Madison, Wis

Rouse JW, Haas RH, Schell JA, Deering DW (1973) Monitoring vegetation systems in the Great plains with ERTS. In: Proceedings of the third ERTS-1 symposium, NASA, SP-351, vol 1, pp 309–317

Salehi CW, Baumgardner MF, Biehl LL (1999) Delineation of soil variability using geostatistics and fuzzy clustering analyses of hyperspectral data. Soil Sci Soc Am J 63(1):142–150

Salehi MH, Eghbal MK, Khademi H (2003) Comparison of soil variability in a detailed and a reconnaissance soil map in central Iran. Geoderma 111(1–2):45–56

Sethi M, Bundela DS, Lal K, Kamra SK (2010) Remote sensing and GIS for appraisal of salt affected soils in India. J Environ Qual 39(1):5–15

Simonson RW (1968) The concept of soil. Adv Agron 20:1–47

Soil Survey Division Staff (2017) Soil survey manual. USDA—SCS. In: Agriculture hand book, 18.3rd edn. U.S. Government printing Office, Washington, DC

Soil Survey Staff (2014) Key to soil taxonomy, 12th edn. USDA—NCRA. U.S. Government printing Office, Washington DC

Terra FS, Demattê JAM, Viscarra-Rossel RA (2015) Spectral libraries for quantitative analysis of tropical Brazilian soils: comparing VIS-NIR and MIR reflectance data. Geoderma 142:255–256, 81–93

Viscarra-Rossel RAV, Walvoort DJJ, McBratney AB, Janik LJ, Skjemstad JO (2006) Visible, near infrared, mid infrared or combined diffuse reflectance spectroscopy for simultaneous assessment of various soil properties. Geoderma 131:59–75

Wu W (2012) The generalized difference vegetation index (GDVI) for dryland characterization. Remote Sens 6:1211–1233

Wu W, Al-Shafie WM, Muhaimeed AS, Ziadat F, Nangia V, Payne WB (2014a) Soil salinity mapping by multiscale remote sensing in mesopotamia, Iraq. IEEE J Sel Topic Appl Earth Obs Remote Sens 7(11):4442–4449

Wu W, Muhaimeed AS, Al-Shafie WM, Ziadat F, Dhehibi B, Nangia V, De Pauw E (2014b) Mapping soil salinity changes using remote sensing in Central Iraq. Geoderma Reg 2–3:21–31

Wu W, Zucca C, Muhaimeed AS, Al-Shafie WM, Al-Quraishi AMF, Nangia V, Zhu M, Liu G (2018) Soil salinity prediction and mapping by machine learning regression in Central Mesopotamia, Iraq. Land Degrad Dev 29(11):4005–4014

Wu W, Muhaimeed AS, Al-Shafie AM, Al-Quraishi AMF (2019) Using radar and optical data for soil salinity modeling and mapping in Central Iraq. In: Al-Quraishi AMF, Negm AM (eds) Environmental remote sensing and GIS in Iraq. Springer Water

Yousif BF (2004) The use of remote sensing techniques in the classification of Al-Najaf Soil. M.Sc. thesis, Building and Construction Engineering Department, University of Technology

Zhang TT et al (2011) Using hyperspectral vegetation indices as a proxy to monitor soil salinity. Ecol Ind 11(6):1552–1562

Zribi M, Baghdadi N, Nolin M (2011) Remote sensing of soil. Appl Environ Soil Sci. Hindawi Publishing Corporation

Chapter 4
Characterization and Classification of Soil Map Units by Using Remote Sensing and GIS in Bahar Al-Najaf, Iraq

Abdulameer S. Al-Hamdani and Hussein M. Al-Shimmary

Abstract The Bahar Al-Najaf Lands is located in the west of the Al-Najaf Governorate. It is extended between 31° 39' 16"–32° 08' 08" N and 43° 47' 11"–44° 30' 15" E with a total area of 2,000.2 Km^2. This chapter aims to find out the possibility of using Remote sensing (RS) and Geographic Information Systems (GIS) techniques in contributing to soil surveys by selecting soil map units drawing and calculating spectral reflectance by satellite image of Landsat 8 provided with two Sensors; Operational Land Imager (OLI) and Thermal Infrared (TIRS). The image was acquired on 07/13/2014 from the USGS site and shows the nature and interference effect of the relationship between some of the soil physical and chemical properties on its spectral reflectance values by choosing false-color composite RGB (753), which is a specialist for studying soils and minerals and deducting study area using ERDAS imagine 2013. Some enhancements have been made (radiation, spectral and spatial) add to unsupervised classification, as well as using earth indicators (EIs). The used EIs include soil color, texture, natural plants and topography. These EIs are used in determining the researcher movement paths to select 16 Pedon's sites with 21 auger holes to a depth of 75 cm for surveying and isolating expected soil individuals within the study area which are identified using GPS. The unsupervised classification approach has shown considerable potential in the distribution of soil classes are close to supervised classification. The pedons revealed and described, according to the (Soil Survey Staff 1999). Disturbed soil samples were selected from each horizon, air-dried, crushed and passed through sieve openings of 2 mm diameter for execution physical and chemical analysis. The results of particle size analysis show that there is no specific pattern distribution of soil particles whether within the pedon itself or between the pedons. There was no observed variations in textures classes for the vertical direction within the same pedon as in (P9, P4A, P10) or got little changes as in P4B. Horizontally variations observed among sites which returns to the nature of sedimentary additives and helps to occur the Litho-

A. S. Al-Hamdani (✉)
Kufa Technical Institute/Al-Furat Al-Awsat Technical University, Najaf, Iraq
e-mail: abdulameer.1957@yahoo.com

H. M. Al-Shimmary
Remote Sensing Center/University of Kufa, Najaf, Iraq

© Springer Nature Switzerland AG 2020
A. M. F. Al-Quraishi and A. M. Negm (eds.),
Environmental Remote Sensing and GIS in Iraq, Springer Water,
https://doi.org/10.1007/978-3-030-21344-2_4

logical discontinuities in soils. All the pedon's horizons were strongly calcareous, while they varied in gypsum content in surface and subsurface horizons. Depending on the US modern classification, All the studied soils are within the Entisol order, which classified into two of Suborders. The first is Fluvents including great soil group Torrifluvents and subgroup Typic Torrifluvents which includes 9 Soil series ME1, DW56, TW964, DM44, MW3, TE354, DW124, DE47, DE126. The second is Psamments including two great soil groups, The first Torripsamments containing subgroup Typic Torripsamments including DE33, DE34, DE74, TE334, TW446 soils series. The other quartzipsamments include subgroup Typic quartzipsamments containing two soil series ME1, DM14. The soil series are classified according to the proposed (Al-Agidi 1976). for alluvial Iraqi soils classification, It was diagnosed nine families and 15 soil series in the study area.

Keywords Remote sensing · Landsate · Soil image · Classification · GIS

4.1 Introduction

Remote sensing (RS) is concerned with the use of devices to collect information about different objects and targets without direct physical contact with them by exploiting electromagnetic waves with their different wavelengths along the electromagnetic spectrum from short to high wavelengths and low frequencies (Elachi and Van 2006). There is an urgent demand for rapid information regarding the classification of soils and their properties. Since there are no adequate studies in this information, thus RS is considered as one of the rapid techniques that can provide these requires using digital classification according to their spectral characteristics producing thematic map used in various fields of sciences especially agriculture (Yousif 2004).

The Iraqi soils are different because of the variation in soil formation factors. Hence, they show different characteristics in both vertical and horizontal directions. Therefore, surveying and classification of these soils and isolation of their units are essential for the preparation of geomorphological and pedological maps. Soil mapping units, in fact, are modified units containing 85% of the characteristics of the nearest taxonomic unit. The spatial pattern and complexity of soil based on the interrelated effect of the soil formation factors at a given location and the extent of their variability within a given distance. Thus, the complexity becomes thornier as these variations become more substantial and on the basis of which the predictability of the pattern is determined at a particular classification level (Mutter 2008). Given the urgent need to increase agricultural production to avoid food shortages in the world, especially Iraq, as well as the severe climate variability facing humanity. There has been a need to use RS and GIS applications due to their economic feasibility in the field of soil survey and classification by speeding up the work, minimizing effort, cost as well as giving them continuous geospatial values of earth characteristics, natural and temporal improvements. The process of linking the soil's physiochemical and morphological characteristics with its spectral properties helps predict a number of factors affecting the soil conditions and its components. Therefore, it is necessary to conduct the spectral properties measurements of the soil units in each region, which are useful aids in soil surveying and classification (Abbas 2010).

The idea of the study is based on the use of RS techniques on the interactions between the fallen radiation from the electromagnetic spectrum with the soil surface. The reflected radiation depends on the soil's physical and chemical properties such as texture, structure, moisture, soil minerals, organic matter, gypsum, and carbonate which considered an essential feature in recording the spectral signature of the soil. As a result of these interactions, different types of soils can be characterized and separated (AL-Rajehy 2002; Zinck 2008).

For the purpose of codification, all the efforts, cost and time with the accuracy of the separation and classification of Bahar Al-Najaf soils are compared to the traditional methods, which includes the drilling of many profiles and Auger holes, The study is carried out to take advantage of these modern technologies (RS and GIS) to prepare soil map units and achieve the following objectives:-

1. The effect of the soil formation factors on the soil variations of Bahar Al-Najaf region.
2. Separating soil units depending on the variation of the earth's surface spectral reflections. Based on the soil morphological, physical, chemical and spectral characteristics.
3. Soil classification may contribute to the optimum use of good soil for the production of the ideal, and how to expand it within the vision of sustainable agricultural development.

4.2 Geology of the Study Area

Bahar Al-Najaf region is a transition zone between the sedimentary plain and the Western desert region, i.e. between the stable pavement and the unstable pier (Sallom and Segar 1994). The study area is located within the stable zone (Al-Rutbha-Al Jazeera) based on the geological divisions of Iraq (Barwari and Slewa 1995). The area is characterized by its simplicity and the lack of surface and sub-surface structures with the extension of the sediments of the Pleistocene and the discoveries of the Tertiary formation. In most parts of the region, they differ in their thickness and texture, where the deposit of valleys extends over areas adjacent to the seasonal valleys, The secondary gypsum deposits are spread over a very large area as a result of carbon rocks erosion belonging to the Tertiary period and the past (Barwari and Slewa 1995).

In some areas, the spread and extension of sand dunes and sediment deposits are noted. Also Bahar Al-Najaf contains natural plants, as well as Palm belt, is extended along Tar Al-Najaf, and it considered a separate limit between Tar Al-Najaf and Bahr Al-Najaf depression. Figure 4.1 shows the sequence of the geological strata of Al-Najaf Governorate.

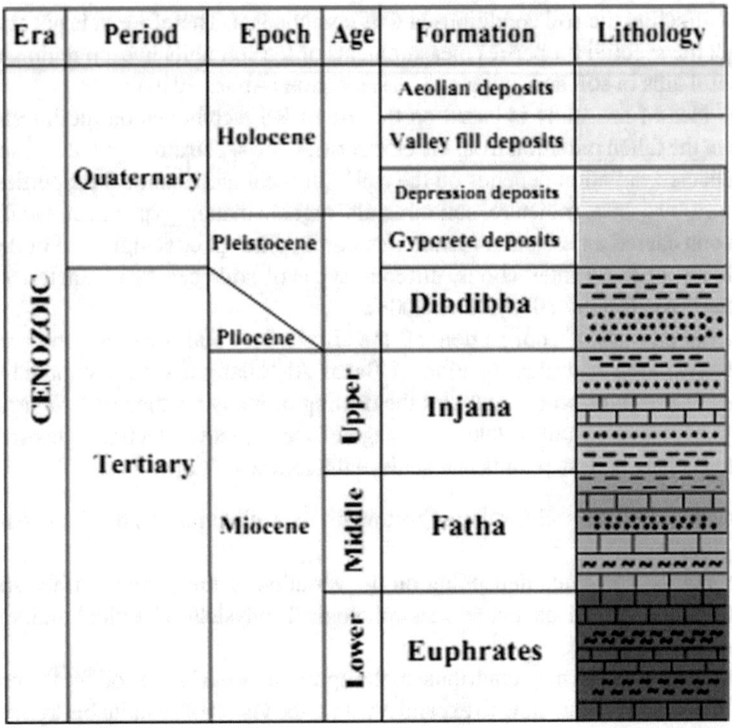

Era	Period	Epoch	Age	Formation	Lithology
CENOZOIC	Quaternary	Holocene		Aeolian deposits	
				Valley fill deposits	
				Depression deposits	
		Pleistocene		Gypcrete deposits	
	Tertiary	Pliocene		Dibdibba	
		Miocene	Upper	Injana	
			Middle	Fatha	
			Lower	Euphrates	

Fig. 4.1 Sequences of the geological formations of Al-Najaf Governorate from (Barwari and Slewa 1995)

4.2.1 Geomorphological and Hydrological Phenomena of the Study Area

4.2.1.1 Bahar Al-Najaf Depression

The Bahar Al-Najaf is a prominent hydrological phenomenon in Al-Najaf Governorate and the Middle Euphrates region. This is the precursor of Al-Najaf city since it was founded twelve centuries ago (Batatu 2002). It is located in the south and southwest of Al-Najaf holy city, and overlooking its plateau at a rate of about 130 m above sea level, within the range of the Iraqi sedimentary plain, The Bahar Al-Najaf lands covered, at the northwestern edge, with deposits of the Quaternary with thickness of (10–20 m) (Yacoub et al. 1981). The area of the Bahar Al-Najaf depression is about 360 km^2, and the height of its lowest point is 11 m above sea level. On the geological side, the rocky remains of the Anjana and the Dabdaba appear on its western edges (Al-Kadhimi et al. 1996).

The lands of Bahar Al-Najaf penetrated by the valleys, which descend from the western territory of Saudi Arabia Kingdom and have a length of hundreds of kilometers. Most notably the valleys of Shu'ayb Husub, Al-Kher, Al-Jal, Abu Talha, Al-Huwaimi. The waters of these valleys move in the winter and most of it flows into depression, a seasonal lake fed by water from the seasonal streams sloping from the high western parts as well as water from the Tar Al-Najaf after the rains and the surface drainage from all sides in the direction of Bahar Al-Najaf depression (Hassan 1983). The sediments of the Quaternary period, such as wind and mud, cover the slopes. The bottom of the depression is covered with clay, silt, sand and fine gravel with a quantity of salts in the upper layers. In addition, there is a large quantity of modern shells in the upper layers extending up to areas close to the foot of the Tar Al-Najaf indicating the maximum intensities of water level rise (Barwari and Slewa 1995).

4.2.1.2   Tar Al-Najaf

It is an insular line, about 65 km long, extended along the eastern and northern east edge of the study area and Al-Najaf plateau, which almost linearly cut. Its highest level is 188 m, but the point of contact with Tar Al-Sayyid, which lies within the borders of Karbala province is 177 m above sea level. One of the most important phenomena observed in Tar Al-Najaf is the depletion of water from some rock layers, Its groundwater stored in the Al-Najaf plateau (Al-Rawi 2007).

4.2.2   Classification of Iraqi Alluvial Soils

Buringh (1960) used the old American genetic classification system for classifying the Iraqi alluvial soils, identified as undeveloped soil of Azonal order and the Alluvial Soils for great group. Al-Agidi (1976) proposed a system for the classification of alluvial soils at the series level. It is the first Iraqi classification system to classify these soils which used two classes to diagnose and distinguish the series: Soil texture within the effective series for soil depth of the 120 cm thick and between (30–150 cm) depth as well as the internal drainage is expressed in mottling depth. These two characteristics are associated with many other soil properties.

Al-Ani (2006) studied the classification of agricultural land in the central region of the Iraqi alluvial plain, which represents the physiological units of the rivers and the irrigation channels prevailing in these areas, and based on the results of the morphological description as well as the physical, chemical and mineralogical characteristics of the studied soil pedons. It was found to be within the undeveloped soils of the Entisols order according to the modern American system, based on the presence of Ochric surface horizon and the absence of subsurface diagnostic horizons. The suborder were Fluvents, and Torrifluvents as great soil group and Typic Torrifluents as subgroup, These soils were classified into nine series based on the Al-Agidi (1976) proposal, named: ME9, DW116, DW25, DE97, DM97, DE44, MW9, DW95, MF8.

In a study conducted in the Iraqi alluvial regions, Hamad (2009) concluded that digital processing for satellite images show a new one that easily distinguishes soil types from other land uses and that the third, fifth and seventh spectral bands are suitable for determining soil units and water studies, With the eighth spectral package. Wahib (2012) showed that soil salinity was significantly correlated with spectral bands 5 and 7 for Landsat 7 satellite with partial correlation coefficients (−0.4229, −0.4030), respectively in the Al-Salman depression, southern Iraq, and recommended using the values of reflectivity as an alternative to digital numbers because they were more accurate in expressing the results.

4.2.3 Remote Sensing

The primary source of electromagnetic spectral radiation is the sun. When it reaches the earth's surface, it interacts with the atmosphere. Most of it is absorbed or dispersed by the atmosphere's components (water vapor and carbon dioxide and ozone), which prevents electromagnetic energy and specific wavelengths from reaching the earth's surface. The range of wavelengths in which the atmosphere is transmitted for energy, known as atmospheric windows, which also determines the type of sensors that can be used by the wavelength of these spectral bands.

4.2.3.1 Spectral Reflectivity of Soil

A significant advantage of multi-spectral imagery is the ability to detect important differences between surface materials by combination spectral bands. Different materials may appear virtually the same within a single band. Selecting a particular band combination, various materials can be contrasted against their background by using colors (Yao et al. 2011; Zhang et al. 2017; Mustafa et al. 2019; Al-Wassai 2003). The results showed a relationship between these characteristics and their reflectivity, depending on the degree of soil brightness (Katie and Mahimid 2002). Studied types of desert soils in the Razzaza area, in Karbala Governorate/Iraq, They found that the sandy and gypsious soil showed the highest values of spectral reflectivity comparing to the rest soils found there.

Ben-Dor et al. (2008) developed a method for characterizing soil profiles in the field by reading spectral reflectivity data for subsurface depths. It tested for moisture, organic matter, carbonate, free iron oxides, specific surface area for four sites in Palestine. This method is suitable for giving quantitative information to these pedons without digging trenches or sending soil samples to the laboratory. The results are for the studied sites only and need to be checked through the applications of the method in the future for other locations.

4.2.3.2   Digital Classification of Spectral Images

Classification using satellite imagery is one of the most essential techniques that contribute to accurate information about a given location in a short time frame. The classification aims to place all the pixels in groups according to their homogeneity and symmetry in the form of a classification map, through which the features, the types of the earth cover represented by these groups. Also, the accuracy in the production of such maps depends primarily on the accuracy of the process of classification of the study's images (Shuli 2008). Generally, the digital soil classification has two main types; unsupervised and supervised classification.

Unsupervised Classification

An implemented automatically without supervision, depending on the degree of similarity and convergence between the spectral signatures without prior knowledge of the identity of these items in the study area. The computer handles the automatic processing of the assemblies and does not require the selection of training areas and is distinguished by the researcher's knowledge of the region, the success of this method depends on two crucial points:

a. The number of spectral items assumed by the researcher must be sufficient to cover all visual parameters.
b. This classification is exploratory and requires a little intervention according to our knowledge of the region through a comparison with reference data such as maps and others to make the interpretation successful.

Supervised Classification

It is implemented based on preliminary samples of the spectral signatures of each class and their statistical measurements (i.e., the classification is under supervision). Each unit of images is classified on the basis of its affinity and matching with the training samples in terms of spectral response and some statistical calculations. One of its methods is the maximum likelihood classification. It calculates the amount of correlation and variation of the spectral response of the training samples to each other and on the basis of the distribution of unknown pixels to those known on the assumption that the sampling units are systematically distributed and all the spectral bands used. The second method is the classification of the minimum distance to mean classification, where the points of location are chosen accurately. In order to give us a clear reflection, and in this case behave other pixels which unknown reversal (as an unknown goal). It takes a reflectivity of the shortest distance between them and the objectives approaching it. In the case of the using colors, it takes the color of the nearest target, so that the unclassified pixel is similar to the classification closest to the center, so-called (Mean center).

4.2.4   Accuracy Assessment of Digital Image Classification

Numerous studies have carried out using both supervised and unsupervised classification methods. Some of which found that the unsupervised classification was more accurate (Borghuis et al. 2007). However the vast majority of studies suggest the opposite is true (Alrababah and Alhamad 2006; Bahadur 2009; Mukherjee and Mukherjee 2009). Accuracy assessment is a crucial part of the image classification process, as it evaluates the degree of acceptance between reference data and classified data (Tso and Olsen 2005). Alrababah and Alhamed (2006) achieved 69.1% overall accuracy of unsupervised classification without spatial improvements on satellite imagery and 73.7% accuracy with spatial improvements, while the overall accuracy of the supervised classification was (82.7, 78.8)% with and without spatial improvements, respectively.

AL-A'araage (2012) used the Landsat ETM + 7 classification technology for Badra City in Wasit Governorate/Iraq, verified the validity of the classified results of five ground covers using 100 control points, the researcher reached an overall accuracy of 92.00% and the statistical Kappa coefficient of 0.8996. Ahmed and Ahmed (2014) used supervised classification with maximum likelihood method of the Landsat TM satellite captured in 2007 for the land use-land cover map for an area of 5,700 Km^2 for Baghdad governorate, the data were classified into five categories. The classification accuracy was 93.7% and the statistical Kappa coefficient of 0.8833. Elkhrachy (2015) used the unsupervised classification method for an area in Najran, Saudi Arabia, to separate three categories of ground coverings (urban areas, barren land, agricultural land) using Erdas imagine and ArcGIS. The classification accuracy of all varieties varied between (61–87%). In Delhi, India, Nain and Kumar (2016) applied the unsupervised classification technology for Landsat ETM + 7 using Erdas Imagine to classify ground coverings and the ArcGIS, with an overall accuracy of 91.67% and a statistical Kappa coefficient of 0.896, which means actual acceptance and good performance of the work.

4.2.5   Applications of RS in the Field of Iraqi Environment and Soils

Qadir (2007) studied the reflectivity of land use varieties and their relationship to soil characteristics to determine their suitability for planting using B1, B2, B3 bands for Landsat TM. Twelve types of soils were identified in the study area depending on the state of variation in their reflectivity values and indicated a weak correlation between soil reflectivity and its characteristics. Sulaiman and Aboud (2012) conducted a semi-detailed survey in the center of the sedimentary plain south of Babylon province/Iraq to assess the land productivity. They dug 10 pedons representative of the soil series, morphological described and the soil texture estimated with some chemical characteristics. They reach that the soil of the study area is newly formed, characterized as stratified, medium to fine texture, and soils classified to the Entisols order and subgroups Typic Torrifluvent, Aquic Torrifluvent, Vertic Torrifluvent. The land assessment showed the presence of all land varieties in different proportions.

Azeez (2013) studied the use of Geo informatics techniques (GIS, RS and GPS) using Satellite images of Landsat 5 and 7 to identify and isolate soil map units in an area of 4,087.3 Km^2, Kurdistan Region, north of Iraq. Soil samples were taken from depth (0–30 cm) for 88 locations to determine physical and chemical properties. A total of 12 pedons were allocated to different map units in order to test the reliability of the soil map units. All pedons at the study area represent developed soils. The supervised classification was more harmony than unsupervised classification, as well as the soil map units in Arc GIS software package was more accordance with physical, chemical and morphological soil properties in comparison with supervised ERDAS classification. Several studies have been conducted on the environmental problems, such as drought events, land degradation, soil salinity, sand dunes encroachments in Iraq using satellite images, remote sensing techniques and GIS analysis in various areas of Iraq (Fadhil 2011; Wu et al. 2019; Fadhil 2009; Almamalachy et al. 2019; Wu et al. 2018; Al-Quraishi et al. 2019; Fadhil 2013).

4.2.6   Soil Surveying and Classification Studies of Bahar Al-Najaf Region

Yousif (2004) used Landsat 5 TM images to classify Al-Najaf governorate soil using three spectral bands. The results of the supervised classification using Maximum Likelihood Classifier method gave a good representation of the soil classes with a total accuracy of 92% while the unsupervised classification gave a good representation of some soil classes and merged them. The study found that the color composition 754 was optimal for soil moisture sensitivity and recommended that the thematic map is the best in soil classification and representation of the ground truth compared to the traditional method.

Mahmoud and Mahimid (2011) studied the inheritance and development of some gypsic soils in Iraq included Bahar Al-Najaf soils. They studied the morphological, physical and chemical properties of the pedons. They found that the Bahar Al-Najaf soil series KME 121 is undeveloped and with A-C horizons due to the dry climate, low vegetation cover, and relatively modern sedimentary parent material. All of which have helped to form undeveloped soil.

4.3 Methodology

4.3.1 The Geographical Location of Bahar Al-Najaf Region

The Bahar Al-Najaf Region Located in the west to southwestern part of Al-Najaf Governorate/Iraq, limited to longitude (43° 47′11″–44° 30′ 15″ E) and latitudes (31° 39′ 16″–32° 08′ 08″ N), and its area is estimated at 2,000.2 km^2 (Fig. 4.2).

The study area is in the hot and desert climate. The summer is characterized by its length and dryness, winter is relatively short and cold. The rainfall rate in winter is low, with an average of 7.2 mm per year, and the annual accumulated rainfall depth of the 86.2 mm distributed irregularly. The precipitation is interrupted in June, July, August, and September, which is accompanied by a small quantity and irregularity of rainfall in the fall with the rise in annual temperature rates at a value of 25.5 °C with a sharp increase in the summer months to reach the rate of 37.7 °C in July. So the soil of study area classified as hyperthermic type, because the annual average temperature is greater than 22 °C and the differences between the average temperature in summer and winter months are more than 5 °C. Its moisture system is of a Torric (Aridic) type due to the keeping of study area in dry conditions for more than 90 days on a continuous basis Where the amount of rainfall is less than double of the average monthly temperature, as well as the existence of a moisture deficit for an extended period of the year (Soil Survey Staff 1999).

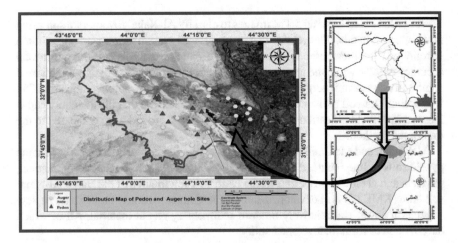

Fig. 4.2 Maps of location of Bahar Al-Najaf region, Al-Najaf governorate, and Iraq

4.3.2 FieldWork

Traffic and mobility routes were identified to determine and dug 16 pedons, as well as 21 auger holes, were drilled within the studied area. Their locations were recorded using the GPS type Garmin: GPSMAP76CSx via (Free Lance Soil Survey), while relying on the unsupervised classification for satellite image with some enhancements, as well as the use of land guidelines and directories such as variations in soil texture, natural plants in addition to topography. We relied on region's altitude levels based on Digital Elevation Model (DEM) files, which downloaded from the Indian Institute of Remote Sensing website (IIRS) (Fig. 4.3) (Table 4.1).

For surveying and isolating the map units representing the soil individuals expected to be present in the study area. The GPS dataset was projected to the geodetic system of UTM_WGS84_38N and saved into a text file. This file was imported into Arc GIS10.3 desktop and converted to shapefile layer to be overlapped with the Landsat 8 images. The pedons are morphologically described according to (Soil Survey Staff 1999). Soil samples from each horizon and auger holes surface depths were collected, air-dried, crushed to pass through a 2 mm sieve and analyzed for their physical and chemical properties, the results are shown in (Tables 4.2 and 4.3).

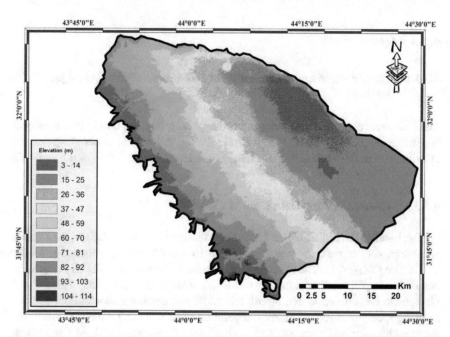

Fig. 4.3 Digital Elevation Model (DEM) of the study area

Table 4.1 Wavelengths and spatial accuracy of sensors bands (OLI, TIRS) for Landsat 8

Spectral band	Wavelength (μm)	Resolution (m)
Band 1—Coastal/aerosol	0.433–0.453	30
Band 2—Blue	0.450–0.515	30
Band 3—Green	0.525–0.600	30
Band 4—Red	0.630–0.680	30
Band 5—NIR	0.845–0.885	30
Band 6—SWIR1	1.560–1.660	30
Band 7—SWIR2	2.100–2.300	30
Band 8—Panchromatic	0.500–0.680	15
Band 9—Cirrus	1.630–1.390	30
Band 10—LIR10	10.30–11.30	100
Band 11—LIR11	11.50–12.50	100

NIR Near Infrared; *SWIR*$_1$ Short Wavelength Infrared; *SWIR*$_2$ Short Wavelength Infrared$_2$ *LIR*$_{10}$ Long Wavelength Infrared$_{10}$; *LIR*$_{11}$ Long Wavelength Infrared$_{11}$ http://fas.org/irp/imint/docs/rst/Front/overview.html

4.3.3 Office Work

4.3.3.1 Classification and Distribution of Soil Units in Bahar Al-Najaf Region

The soils were classified into order, suborder, great group and family levels according to the (Soil Survey Staff 1999). While for the series level under the proposed classification of Al-Agidi (1976) for Iraqi alluvial soils (Table 4.4; Figs. 4.4 and 4.7).

4.3.3.2 Remote Sensing Data Used in the Study

The Landsat 8 satellite, which carries two sensors (OLI and TIRS) with 11 bands was adopted to capture study area image on 13-7-2014 was downloaded from the USGS site (Table 4.1). The study area was deducted using the ERDAS Imagine to ensure a complete and accurate form for the study area. RGB (753) False Color Composite was chosen to distinguish terrestrial and geomorphological phenomena as being specialized for soil studies, separation of its units and minerals (Martinez-Rios and Monger 2002; Yacoub et al. 2008; Faleh and Shawan 2012; Al-Issawi et al. 2012).

Table 4.2 Some physical and chemical properties for soil surface layers of auger holes in the study area

Auger Points	Sand gm kg⁻¹	Silt	Clay	Texture	B. D Mg m⁻³	PH	EC ds/m	O.M %	Gyps. g kg⁻¹	Lime g kg⁻¹	Coordinates X	Y
L1	300	532	168	Si L	1.55	7.6	74.3	0.83	23.97	235	44.453	31.866
L2	780	92	328	S C L	1.23	7.55	177	0.51	52.83	230	44.392	31.772
L3	512	380	108	S L	1.62	7.52	4.07	0.32	14.4	620	44.371	31.76
L4	360	372	268	L	1.49	7.95	49.5	0.44	22.32	220	44.29	31.81
L5	760	72	168	S L	1.52	7.56	5.66	0.27	11.16	290	44.31	31.834
L6	674	198	128	S L	1.48	7.56	3.03	0.18	10.59	245	44.278	31.769
L7	662	78	260	S C L	1.45	7.38	8.77	0.23	13.46	590	44.187	31.713
L8	564	598	328	S C L	1.53	7.36	2.8	0.21	12.36	370	44.229	31.754
L9	552	160	288	S C L	1.09	7.39	2.12	0.11	11.02	230	44.278	31.724
L10	114	718	168	S L	1.46	8.26	41.2	3.22	18.17	315	44.293	31.968
L11	732	240	28	L S	1.57	8.33	3.63	3.15	13.83	120	44.297	31.998
L12	512	360	128	L	1.53	8.37	80.2	2.64	15.76	150	44.426	31.938
L13	422	368	208	L	1.52	7.86	39.5	2.48	24.21	185	44.396	31.923
L14	252	480	268	L	1.51	7.34	9.68	1.53	12.18	45	44.399	31.882
L15	124	648	228	Si L	1.66	7.15	3.04	2.36	13.58	345	44.452	31.913
L16	232	500	268	C	1.63	7.35	65.5	3.37	23.73	475	44.442	31.892
L17	624	248	128	S L	1.37	7.48	1.97	0.48	11.5	505	44.027	31.861
L18	632	340	28	S L	1.66	7.46	1.91	0.62	8.73	235	44.023	31.896
L19	532	400	68	S L	1.44	7.51	28.5	0.54	16.06	225	44.042	31.985
L20	732	198	70	S L	1.25	7.49	1.83	0.17	12.05	505	44.117	31.833
L21	632	200	168	S L	1.46	7.61	35.1	0.68	15.75	180	44.172	31.873

Table 4.3 Some physical and chemical properties for soil horizons of the pedons in the study area

Pedons	Horizons	Depth	Sand	Silt	Clay	Texture	B.D	PH	Ece 1:1	O.M	Gyp.	Lime	Coordinates	
		cm	gm kg^{-1}				Mg m^{-3}		dsm^{-1}	g kg^{-1}	g kg^{-1}	g kg^{-1}	X	Y
2	A	0–15	298.4	391.6	310	C L	1.44	7.82	2.99	4.7	10.2	445	43.96	31.95
	C$_1$	15–35	244	376	380	C L	1.49	7.71	3.79	4.6	4.23	445		
	C$_2$	35–68	242	348.8	407.2	C	1.48	7.75	3.44	3.2	4.34	450		
	C$_3$	68–110	569.6	163.2	267.2	SCL	1.5	7.64	2.77	1.8	5.76	500		
	C$_4$	110–150	578.7	165.3	256	SCL	1.64	7.68	2.82	0	6.12	535		
3	A	0–10	776	136	88	S L	1.7	7.72	2.5	6.6	10.2	225	44.12	31.89
	C$_1$	10–25	796	156	48	L S	1.73	7.68	2.5	4.5	10.4	425		
	C$_2$	25–50	684	148	164	S L	1.61	7.48	6.68	2.5	10.9	215		
	C$_3$	50–85	824	116	60	L S	1.64	7.59	3.02	1.1	9.95	405		
	C$_4$	85+	Rock											
4A	Ap	0–10	485.6	447.1	67.2	S L	1.21	7.28	121	4	26.2	205	44.09	31.97
	C$_1$	10–23	635.3	306	58.5	S L	1.12	7.49	47.4	3.5	14.9	163		
	C$_2$	23–46	585.6	374.4	40	S L	1.03	7.58	28	2.9	13.6	185		
	C$_3$	46–85	545.6	414.4	40	S L	1.11	7.72	11.8	2.7	11.7	180		
	C$_4$	85–160	605.6	344.4	50	S L	1.18	7.7	14.9	2.4	10.8	175		
4B	A	0–10	445.6	307.2	247.2	L	1.43	7.46	40	4.1	13.4	525	44.1	31.96
	C$_1$	10–38	465.6	494.4	40	S L	1.55	7.74	15	2.4	12.2	575		
	C$_2$	38–95	685.6	114.4	200	SL , SCL	1.65	7.63	9.41	3.3	11.3	514		

(continued)

Table 4.3 (continued)

Pedons	Horizons	Depth	Sand	Silt	Clay	Texture	B.D	PH	Ece 1:1	O.M	Gyp.	Lime	Coordinates	
		cm	gm kg^{-1}				Mg m^{-3}		dsm^{-1}	g kg^{-1}	g kg^{-1}	g kg^{-1}	X	Y
	C$_3$	95–150	705.6	234.4	60	S L	1.44	7.59	4.7	1.7	11.2	483		
	C$_4$	150+	665.6	294.4	40	S L	1.36	7.51	3.87	0	11.2	455		
5	A	0–10	376.8	301.6	321.6	C L	1.33	7.91	2.46	2.3	10	245	43.96	31.95
	C$_1$	10–30	439	496	65	L	1.41	7.7	8.69	4.2	10.7	255		
	C$_2$	30–70	302	288	424	C	1.34	7.64	3.91	3	10.8	300		
	C$_3$	70–130	211.2	348.8	440	C	1.45	7.34	2.64	1.5	10.9	355		
	C$_4$	130+	Rock											
6	A	0–5	785.6	154.4	60	L S	1.58	7.72	3	2.9	11.8	250	44.1	31.97
	C$_1$	5–15	931.2	31.6	37.2	S	1.62	7.8	2.54	2.4	11.1	220		
	C$_2$	15–47	941.2	21.6	37.2	S	1.65	7.79	2.88	2.5	10.8	230		
	C$_3$	47–60	931.2	31.6	37.2	S	1.68	7.79	2.7	1.3	10.7	220		
	C$_4$	60–103	758	46	196	S L	1.64	7.67	3.11	0.6	7.57	220		
	C$_5$	103+	Rock											
7	A	0–12	776	40	184	S L	1.69	7.59	9.73	5.1	10.8	570	44.14	31.86
	C$_1$	12–26	858.4	111.6	30	L S	1.65	7.71	3.43	4.8	10.7	475		
	C$_2$	26–46	566	390	44	S L	1.53	7.67	7.88	4	10.5	435		
	C$_3$	46–78	731.2	228	40	S L	1.59	7.68	8.45	4.9	10.7	450		
	C$_4$	78–110	876	70	54	L S	1.51	7.61	7.91	2.6	11.4	480		
	C$_5$	110–150	574	378	48	S L	1.54	7.53	7.23	0	10.1	490		

(continued)

Table 4.3 (continued)

Pedons	Horizons	Depth	Sand	Silt	Clay	Texture	B.D	PH	Ece 1:1	O.M	Gyp.	Lime	Coordinates	
		cm	gm kg⁻¹				Mg m⁻³		dsm⁻¹	g kg⁻¹	g kg⁻¹	g kg⁻¹	X	Y
8	A	0–15	262	288	464	C	1.39	7.41	48.1	3.7	16.4	390	44.23	31.92
	C_1	15–37	249.6	239.6	514.4	C	1.52	7.68	14.7	5.2	11.9	370		
	C_2	37–107	312	268	404	C	1.59	7.78	4.6	3.5	11.4	295		
	C_3	107–180	768	118	114	S L	1.65	7.65	5.17	3.1	10.9	270		
9	A	0–10	549.8	402	48.2	S L	1.33	7.34	88.9	1.5	22.2	245	44.22	31.86
	C_1	10–38	560	373.5	66.5	S L	1.16	7.44	30.3	1.2	14.2	235		
	C_2	38–106	568	38.48	47.2	S L	1.13	7.77	5.53	1.6	11.1	260		
	C_3	106–123	448	495.2	47.2	Si L	1.38	7.74	5.32	2.6	10.8	305		
	C_4	123–140	608	344.8	47.2	S L	1.47	7.79	4.63	2.2	10.9	345		
	C_5	140–185	651.2	271.6	77.2	S L	1.43	7.76	5.47	2.2	10.4	350		
10	Ap	0–10	584	358	58	S L	1.41	7.71	3.08	2	11.2	300	44.25	31.8
	C_1	10–38	568	364	68	S L	1.39	7.62	3.29	1.5	11.1	300		
	C_2	38–50	528	318	154	S L	1.33	7.58	4.5	0	6.44	280		
	R	50+	Rock											
11	A	0–15	122.4	577.4	300.2	SiCL	1.47	7.42	90	8.5	30.5	168	44.29	31.87
	C_1	15–37	433	506.3	60.7	Si L	1.53	7.58	45	4.3	12.5	340		
	C_2	37–60	451.2	307.4	241.4	L	1.49	7.5	18.6	2.5	10.6	472		
	C_3	60–88	378.1	467.3	154.6	L	1.55	7.8	6.3	1	8.56	426		
	C_4	88–115	707.1	232.6	60.3	S L	1.53	7.8	5.7	0	7.4	205		
	C_5	115–145	768	118	114	S L	1.48	7.85	4.5	0	8.3	218		

(continued)

Table 4.3 (continued)

Pedons	Horizons	Depth	Sand	Silt	Clay	Texture	B.D	PH	Ece 1:1	O.M	Gyp.	Lime	Coordinates	
		cm	gm kg^{-1}				Mg m^{-3}		dsm^{-1}	g kg^{-1}	g kg^{-1}	g kg^{-1}	X	Y
12	A	0–20	863	56.4	80.6	L S	1.52	7.59	54.3	2.5	10.4	357	44.27	31.9
	C$_1$	20–37	707.1	232.6	60.3	S L	1.48	7.46	45.9	1.7	8.8	448		
	C$_2$	37–47	439	496	65	L	1.51	7.32	41.2	0.4	5.41	402		
	C$_3$	47–60	732.5	205	62.5	S L	1.49	7.58	42.4	0	10.4	205		
	C$_4$	60–88	401.8	291.6	306.6	C L	1.41	7.47	38.2	0	29.6	263		
	C$_5$	88–120	145.6	507	347.4	SiCL	1.45	7.33	34.4	0	14.9	452		
13	A	0–25	217.3	561.6	221.1	Si L	1.51	7.44	62.4	6.5	14.4	297	44.33	31.86
	C$_1$	25–50	81.6	562.1	356.3	SiCL	1.5	7.54	14.8	3.7	7.81	248		
	C$_2$	50–60	378.1	467.3	154.6	L	1.54	7.39	16.2	1.4	7.94	262		
	C$_3$	60–80	18.9	588.8	392.3	SiCL	1.48	7.54	32.4	1.1	10.8	275		
	C$_4$	80–110	32.5	528.2	439.3	SiC	1.46	7.47	38.2	0	12.5	263		
	C$_5$	110–140	622	282.7	95.3	S L	1.53	7.25	36.4	0	9.57	227		
14	A	0–12	406.2	496.1	97.7	L	1.54	7.41	104	6.5	49.1	204	44.23	31.96
	C$_1$	12–40	368.7	471.2	160.1	L	1.53	7.62	36.5	3.1	16.5	386		
	C$_2$	40–70	438	506.6	55.4	Si L	1.56	7.57	11	1	19.2	216		
	C$_3$	70–115	145.4	507.3	347.3	SiCL	1.54	7.58	8.65	0	10.1	361		
	C$_4$	115–145	75.5	642.4	282.1	SiCL	1.52	7.64	14.2	0	12.1	376		
15	A	0–10	912	60	28	S	1.66	7.76	2.63	2.6	11.6	510	44.06	31.91
	C$_1$	10–60	826	138	36	S L	1.57	7.84	1.89	0.4	2.97	575		
	C$_2$	60–90	874	84	42	L S	1.61	7.88	1.45	Nil	2.24	590		

(continued)

Table 4.3 (continued)

Pedons	Horizons	Depth	Sand	Silt	Clay	Texture	B.D	PH	Ece 1:1	O.M	Gyp.	Lime	Coordinates	
		cm	gm kg⁻¹				Mg m⁻³		dsm⁻¹	g kg⁻¹	g kg⁻¹	g kg⁻¹	X	Y
16	A	0–12	372	200	428	L	1.52	7.74	1.73	1.4	2.67	210	44.27	31.75
	C_1	12–33	572	148	280	SCL	1.63	7.78	1.16	0.3	2.47	365		
	C_2	33–85	812	100	88	L S	1.61	7.85	0.43	Nil	Nil	450		
	C_3	85–150	592	120	288	S C L	1.55	7.61	1.58	Nil	Nil	485		

CL Clayey Loam; C Clayey; SCL Sand Clay Loam; SL Sandy Loam; LS Loamy Sand; L Loamy; S Sandy; SiCL Silty Clay Loam; SiL Silty Loam

Table 4.4 Classification of soils according to Soil Survey Staff (1999) and the proposal of Al-Agidi (1976) for Iraqi alluvial soil series

Family	Sub Great Group	Great Group	Sub order	Order	Series	Area km^2	Pedon No.
Fine silty,mixed, active, calcareous, hyper thermic, Typic torrifluvents	Typic torrifluvents	Torrifluvents	Fluvents	Entisols	DE126	10.31	P2
Sandy, mixed,active, gypsic, hyperthermic, Typic Haplocalcids	Typic torripsamments	Torripsamments	Psamments	Entisols	DE33	102.83	P3
Sandy, mixed, active, gypsic, hyperthermic, Typic Haplocalcids	Typic torri psamments	Torripsamments	Psamments	Entisols	DE34	44.97	P4A
Sandy, mixed,active, calcareous, hyperthermic, Typic torripsamments	Typic torripsamments	Torripsamments	Psamments	Entisols	DE74	31.73	P4B
Clayey, mixed, active, gypsic, hyperthermic , Typic torrifluvents	Typic Torrifluvents	Torrifluvents	Fluvents	Entisols	DE47	707.9	P5
Sandy , mixed ,active , gypsic , hyperthermic , Typic Haplocalcids	Typic quartzipsamments	Quartzipsamments	Psamments	Entisols	DM14	34.37	P6
Sandy, mixed ,active , calcareous , hyperthermic , Typic torripsamments	Typic torripsamments	Torripsamments	Psamments	Entisols	TE334	47.49	P7
Clayey, mixed,active, calcareous, hyperthermic, Typic Haplocalcids	Typic torrifluvents	Torrifluvents	Fluvents	Entisols	DW124	118.65	P8

(continued)

Table 4.4 (continued)

Family	Sub Great Group	Great Group	Sub order	Order	Series	Area km^2	Pedon No.
Loamy, mixed,active, gypsic, hyperthermic, Typic Haplosalids	Typic torrifluvents	Torrifluvents	Fluvents	Entisols	TE354	22.89	P9
Loamy, mixed, active, calcareous, hyperthermic, Typic Torripsamments	Typic torrifluvents	Torrifluvents	Fluvents	Entisols	MW3	168.83	P10
Loamy, mixed,active, gypsic, hyperthermic, Typic torrifluvents	Typic torrifluvents	Torrifluvents	Fluvents	Entisols	DM44	15.94	P11
Sandy, mixed,active, calcareous, hyperthermic, Typic torripsamments	Typic torripsamments	Torripsamments	Psamments	Entisols	TW446	113.58	P12
Loamy, mixed,active, gypsic, hyperthermic, Typic torrifluvents	Typic torrifluvents	Torrifluvents	Fluvents	Entisols	TW964	185.51	P13
Loamy, mixed,active, gypsic, hyperthermic, Typic Haplocalcids	Typic torrifluvents	Torrifluvents	Fluvents	Entisols	DW56	118.42	P14
Sandy, mixed,active, calcareous, hyperthermic, Typic Torripsamments	Typic quartzipsamments	Quartzipsamments	Psamments	Entisols	ME1	62.35	P15
Loamy, mixed,active, calcareous, hyperthermic, Typic Torripsamments	Typic Torrifluvents	Torripsamments	Fluvents	Entisols	ME1	48.19	P16

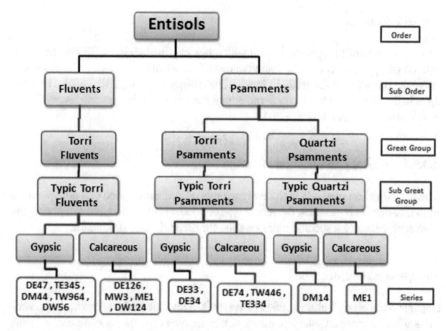

Fig. 4.4 Classification of the study soil according to the modern American system Soil Survey Staff (1999) and Al-Agidi's (1976) proposal for sedimentary soil series

4.3.3.3   Satellite Image Enhancement Processing

A series of enhancement processing was applied to the satellite image, which aimed to make the image's information more visible for purposes of visual interpretation. They included:-

Radiometric Enhancement

Means the removal of noise or interference of image elements because it affects the values of their spectral reflectivity, especially for red and near-infrared spectral bands.

Spectral Correction

Reflectivity is expressed by its true radiation by reversing the numerical numbers of the reflectivity (Digital Number, DN) to the actual radiation values, Top Of Atmosphere radiance (TOA). It considered as a radiometric treatment method which was important in the development of mathematical models that physically attach images data to ground truth (Lillesand and Kiefer 2000).

Spatial Enhancement

The integration of the spectral band 8 with other satellite bands, facilitated the acceleration of application and the maintenance of information accuracy and results to serve the purposes of the study in terms of the ability to distinguish well the types of soils in the image and to clearly demonstrate the process of integrating layers (Layer Stack) using ERDAS Imagine 2013.

4.3.3.4 Digital Classification of Image

The classification methods include algorithms of both supervised and unsupervised classification. In both cases, the image was divided and classified into categories that give each category a specific color or code for thematic map production.

Unsupervised Classification

Unsupervised classification applied using the K-Mean method, to determine the varieties without external interference to obtain an overview of the differences in the reflectivity of the study area soils. The calculator performs the classification process with higher accuracy and less distortion in the isolation of ground parameters (Fig. 4.5), because it works on the basis of pixels values represented by terrestrial

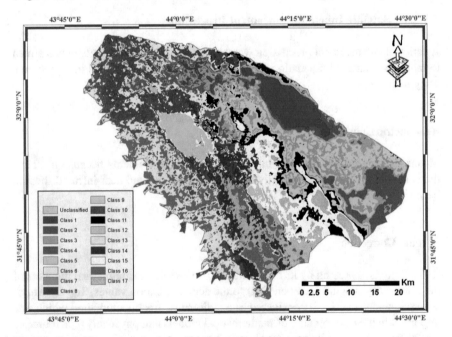

Fig. 4.5 Shows the unsupervised classification results of the study area

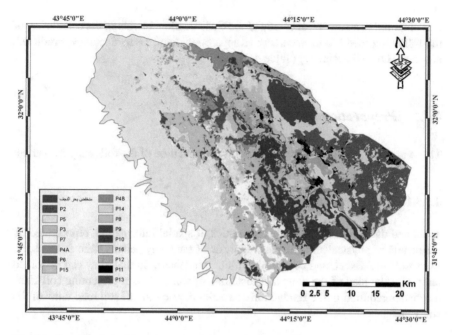

Fig. 4.6 Supervised classification to the study area soils based on pedons sites

features, which are separated and isolated by ERDAS Imagine 2013. This process was repeated several times to obtain satisfactory results (Lillesand and Kiefer 2000).

Supervised Classification

Was carried out using the Maximum Likelihood method, after obtaining the results of samples analysis taken for the pedons sites as in (Fig. 4.2). Training samples were then identified, which included all variances in the study area. The spectral signatures were taken for each training sample. This type of classification includes two stages (training and classification) (Fig. 4.6).

4.3.4 Accuracy Assessment for Digitalmap Classification

The accuracy of the classification was calculated by choosing 256 basis points, and the points outside the study area were excluded. Therefore, 140 points were adopted within the boundaries of the study area which compared with the points and locations of the classes values of the study area image which classified as supervised. The soil category code for each random point was fixed in the reference column of the accuracy assessment window. After the completion of the points, the final report was prepared to assess the accuracy of the classification, including the calculation of the following methods of accuracy: (Kappa coefficient, Users accuracy, Producers accuracy, Overall accuracy) (Table 4.7).

4.3.5   Preparation of GIS Database

This was done through the creation and implementation of the following measures.

4.3.5.1   Entering Quantity Data

Tabulated data containing the characteristics of spatial information, represented by the results of physical and chemical analysis of the study pedons were entered into Microsoft-MS Excel, and convert them into a database. So that, they can be joined later with the plan of the units and locations of soil pedons representing soil characteristics required in the production of spatial distribution of soil map units of the study area.

4.3.5.2   Map Digitizing

Performed by using the Arc GIS 10.3 software and implementing the screen digitization method for mapping soil units and pedons locations, and convert them into digital maps (shp.file) with two types of features: points representing the locations of pedons while polygons for soil individuals.

4.4   Results and Discussion

4.4.1   Classification and Distribution of Soil Units in the Study Area

The soils of the study area were classified at the levels of order, suborder, great soil group, subgroup and family according to (Soil Survey Staff 1999). Moreover at a series level according to the proposed classification of Iraqi alluvial soils by Al-Agidi (1976) (Figs. 4.4 and 4.7; Table 4.4), which depends on the presence of three different layers or textures except for the upper surface which goes to soil type. The thickness requirements of the first layer about 35 + 5 cm and 40 + 5 cm for the second and third layers (Table 4.3). The soil series was named with the following symbols:

a.   Number of layers: symbolized as (M, D, T) for (1, 2, 3) layers respectively.

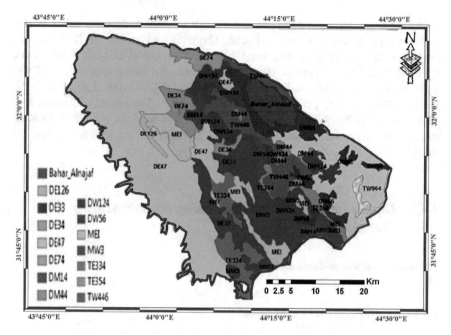

Fig. 4.7 Distribution of the soil series of the study area by adopting the supervised classification

b. Natural drainage class: This indicates the movement of water in the soil depending on the depth at which the mottling appear via 50% or more of the test exposed surface.

The results of the physical and chemical properties in (Table 4.3) indicate that The soils of the study area were undeveloped, classified as Entisols order. This is due to the nature of the environmental conditions prevailing in the study area as a result of the nature of the dry climate and the non-dense vegetation cover. These factors do not help the activity of the pedogenic processes responsible for the development of soil at an adequate level as well as the shorten chronological age of the parent material of these soils. The results were consistent with Mahmoud and Mahimid (2011) when studying the evolutionary status of Bahar Al-Najaf soils. Since all these soils are located in the alluvial plain, which receives new deposits of sedimentary material at frequent intervals of time. This has helped to weaken the pedogenic activity and characterized as A-C horizons alternation (Buringh 1960).

These soils were recognized by the presence of the surface horizon Ochric. It characterized by light color with low organic matter content and color value 4 or more in the moist state and 6 or more in the dry state, with the dominant of dry, moist system add to the Control section is not moist for most days in the year as well as the soil is unsaturated within the first meter of soil body. The soils order was classified into two suborders. The first Fluvents, which their parent materials are the result of added sediments, while the second Psamments which is located within the desert region and the parent materials are ancient deposits rich in coarse particles represented by gravel and sand. The main soil groups are, torrifluvents, torripsamments, quartzipsamments and three sub great groups are typic torrifluvents, typic torripsamments, Typic quartzipsamments and nine families, as well as 15 soil series, were named: DE126, DE33, DE34, DE74, DE47, DM14, TE334, DW124, TE354, MW3, DM44, TW446, TW964, DW56, ME1. The largest soil series area DE47 Occupied 707.9 Km^2 from the study area, The smallest area of 10.31 Km^2 was for the DE126 series. The illustrations of these details are shown in (Figs. 4.4 and 4.7; Table 4.4).

4.4.2 Digital Image Classification Using Remote Sensing Techniques

The most common methods are:

4.4.2.1 Unsupervised Classification

Figure 4.5 shows the results of the unsupervised classification of the study area, comprising 17 soil classes as well as unclassified sites located outside the boundaries of the study area.

4.4.2.2 Supervised Classification

Figures 4.6 and 4.7 illustrate the classification method for the study area soils in addition to the final map sorted to the level of the series with their respective area in Km^2 units.

4.4.3 Accuracy Assessment of Digital Classification

The classification accuracy technique is considered to be a quantitative analysis of the results obtained, It is necessary and important to assess the degree of error at the end of the work and to indicate the compatibility of the items titles in the thematic (classified) map with ground truths or reference data for the same sites (Jensen 2007; Taruvinga 2008).

The primary objective of assessing accuracy is to estimate the elements and components of the error or confusion matrix and different descriptive measurements of terrestrial data within the field of study and gives a decision on the accuracy of the spectral data and the chosen method (Nain and kumar 2016; Stehman 2013).

Table 4.5 The error matrix for the soil series units of the study area from Landsat 8 image

Classified data	Reference Data																Row total	Mapping accuracy (%)
	P2	P5	P3	P7	P4A	P6	P4B	P14	P8	P9	P10	P16	P12	P11	P13	P15		
P2	1	0	0	0	0	0	0	0	0	0	0	0	0	0	0	0	1	50
P5	0	38	0	1	0	0	0	0	0	0	0	0	0	0	0	0	39	95
P3	0	0	9	0	1	0	0	0	0	0	0	0	0	0	0	0	10	81.8
P7	0	1	0	10	0	1	0	0	0	0	0	0	0	0	0	0	12	77
P4A	0	0	1	0	3	0	0	0	0	0	0	0	0	0	1	0	5	43
P6	0	0	0	0	0	7	0	0	0	0	0	0	0	0	0	0	7	87.5
P4B	0	0	0	0	0	0	7	0	0	0	0	0	0	0	0	0	7	87.5
P14	0	0	0	0	1	0	1	9	0	0	0	0	0	0	0	0	11	75
P8	0	0	0	0	0	0	0	0	5	0	0	0	0	0	0	0	5	83.3
P9	0	0	0	0	0	0	0	0	0	1	0	0	0	0	0	0	1	100
P10	1	0	0	0	0	0	0	0	1	0	7	0	0	0	0	0	9	63.3
P16	0	0	0	0	0	0	0	0	0	0	1	0	0	0	0	0	1	0
P12	0	0	0	0	0	0	0	0	0	0	0	0	1	0	0	0	1	100
P11	0	0	0	0	0	0	0	0	0	0	0	0	0	2	0	0	2	66.7
P13	0	0	0	0	0	0	0	0	0	0	1	0	0	1	20	0	22	87
P15	0	0	0	0	0	0	0	1	0	0	0	0	0	0	0	5	6	83.3
Col. Totals	2	39	10	11	5	8	8	10	6	1	9	0	1	3	21	5	140	-

Table 4.5 shows the error matrix for fifteen soil units classified at the series level as suggested by Al-agidi (1976) resulting from a supervised classification using maximum likelihoods classifier. The validity of this classification was based on 140 reference points selected within the study area out of a total of 256 random points using the ERDAS 2013. The error matrix provides information on the correct and incorrect prediction of the image classification by comparing the classified map with the ground information. It works on the basis of class-by-item comparison between source data known as ground truth and the corresponding results on the classified image (Martine—Rios and Monger 2002; Proklamasi et al. 2015; Nash 2016).

The following methods are calculated for accuracy: Users accuracy, Producers accuracy, Overall accuracy, Kappa coefficient. Kappa coefficients are the most commonly used to measure the accuracy of the thematic map, as it shows the difference between actual agreement for reference data and the expected agreement of classified data by chance (Nain and kumar 2016). Koutroumpas et al. (2010) divided Kappa coefficient values into five categories of agreement classes (Table 4.6).

Table 4.7 shows that the producer accuracy of ME1, TW446, and TE354 soil series were 100%, this means that all the reference data pixels classified and located in the thematic map within those soil series. High values of producer accuracy (omission errors) as in the soil series DE47, DE33, TE334, TW964, DW56 are indicated that most of the terrestrial reference data pixels were classified in their corresponding positions in the classified thematic map. According to Koutroumpas et al. (2010) classification. The accuracy of the soil series units map of the study area were classified as very good for soil series ME1, TW964, TW446, TE354, DW124, DE74,

Table 4.6 Accuracy ranges of Kappa K_{hat} coefficients and agreement classes by Koutroumpas et al. (2010)

Agreement classes	Kappa K_{hat} coefficient range
Poor	$<= 0.2$
Fair	0.21–0.40
Moderate	0.41–0.60
Good	0.61–0.80
Very good	0.81–1.00

Table 4.7 Accuracy evaluation of the error matrix data for soil series of the study area

Class name	Soil series	Reference totals	Classified totals	Number correct	Producers accuracy (%)	Users accuracy (%)	Kappa coeffi- cient
P2	DE126	2	1	1	50.00	100.00	1
P5	DE47	39	39	38	97.44	97.44	0.9645
P3	DE33	10	10	9	90.00	90.00	0.8923
P7	TE334	11	12	10	90.91	83.33	0.8191
P4A	DE34	5	5	3	60.00	60.00	0.5852
P6	DM14	8	7	7	87.50	100.00	1
P4B	DE74	8	7	7	87.50	100.00	1
P14	DW56	10	11	9	90.00	81.82	0.8042
P8	DW124	6	5	5	83.33	100.00	1
P9	TE354	1	1	1	100.00	100.00	1
P10	MW3	9	9	7	77.78	77.78	0.7625
P16	ME1	0	1	0	–	–	0
P12	TW446	1	1	1	100.00	100.00	1
P11	DM44	3	2	2	66.67	100.00	1
P13	TW964	21	22	20	95.24	90.91	0.893
P15	ME1	5	6	5	100.00	83.33	0.8272
Totals		140	140	126			

Overall Classification Accuracy = 90.00%
Overall Kappa Statistics = 0.8846

DM14, DE47, DE33 and good for DM44, MW3, DW56, TE334 as well as Moderate for soil series DE34, DE126 (Tables 4.6 and 4.7).

Generally, it is noted that the overall accuracy of the soil units classification 90%, This means that the probability of 90% of the reference pixels is correctly classified within the classified map (Jensen 2007). The value of the Kappa coefficient is 0.8846. This means that there is a statistical acceptance of 88.46% between the pixels or reference points and their corresponding on the classified map. According to the Kappa equivalent value and the classification proposed by Koutroumpas et al. (2010). The overall accuracy of supervised classification of all sites in the study area falls within the category very good. It means, that is a very good agreement between ground truths and classified data.

4.5   Conclusions

1. There is a weak activity of the pedogenic process affecting the soil formation of the study area. Within the full dominion of the undeveloped soils (Entisols) order, as well as the influence of geomorphic processes. The nature of sedimentation sources from the highlands in the western and southwestern parts of Bahar Al-Najaf Region affects the quantity and nature of the distribution of soil components vertically and horizontally between pedons.
2. The remote sensing and GIS techniques contributed effectively and distinctly the in identification and separation of soil units. The overall accuracy of the matrix error obtained from the supervised classification using was 90% as well as 0.8846 for Kappa statistical coefficient. This means a very good acceptance of the thematic map to classify studied soils.
3. The study demonstrated the importance of remote sensing and GIS techniques for studying natural resources, including soil, its characteristics, depending on the analysis of the satellite images digitally and visually. As well as economic feasibility in the field of survey and classification of soils through the reduction of effort, cost, speed up the completion of work and reducing the difficulties compared to traditional surveys.

4.6   Recommendations

1. Emphasize the importance of using RS and GIS techniques in soil survey and classification studies because of the speed and accuracy of the work, preparation of soil survey and classification maps, add to reducing the effort and cost compared to field surveys.
2. Take advantage of remote sensing technologies and digital processing processes by selecting suitable spectral bands and integrating them into combinations that contribute to the study, detection and identification of desertification and land degradation areas.
3. In order to improve the results and to obtain the more specialized spectral signature curves as well as more details of the characteristics studied, especially in the field of surveying and classification of soils. We recommend using data from more than one sensor at the same time as their basis, or using images of sensors with high spectral and spatial discrimination capabilities that have multiple spectral bands with a narrow width (Hyper spectral imagery).

4. The need to use active remote sensing techniques to sense the internal soil characteristics, as well as the possibility of pairing with the passive technology to increase the efficiency of using remote sensing in soil studies.
5. The field measurements of the reflectivity are more accurate than the calculations of the satellite image because of many overlaps between the conditions and time of the images capture and the surface components of the earth coverings, which greatly affect the real values of their reflectivity. This requires calibration of the spectral reflectivity values calculated from the satellite images with the field reading using the spectroradiometer device.

References

Abbas AH (2010) Characterization and classification of soil units of North Kut Project and prediction of some soil physical properties by using GIS and Remote Sensing. Ph.D. thesis, College of Agriculture, University of Baghdad, Iraq

Ahmed MA, Ahmed WA (2014) Integration remote sensing and GIS techniques to evaluate land use/land cover of baghdad region and nearby areas. Iraqi J Sci 55(1):184–192

AL-A'araage AA (2012) Monitoring desertification in Badra Area Eastern Iraq by using landsat image data. M. Sc. thesis, Geology Science, University of Baghdad, Iraq

Al-Agidi W (1976) Proposed soil classification at series level for Iraqi soils: I-Alluvial soils. Faculty of Agriculture, University of Baghdad, Technical bulletin No. 1

Al-Ani AM (2006) Applications of numerical taxonomy to classify some soil series of river levees in Iraqi alluvial plain. Ph.D. thesis, Faculty of Agriculture, University of Baghdad, Iraqi

Al-Issawi DF, Ibrahim MK, Sayel KN (2012) The use of heterogeneity of the Values of spectral reflectivity to separate the soil units of Al-Hawija area south of Al-Fallujah city/Iraq. Iraqi J Desert Studies 4(1):26–35

Al-Kadhimi JM, Sissakian VK, Sattar AF, Deikran DB (1996) Tectonic Map of Iraq, 2nd edn. scale 1: 1000 000. GEOSURV., Baghdad, Iraq

Al-Quraishi AMF, Qader SH, Wu W (2019) Drought monitoring using spectral and meteorological based indices combination: a case study in Sulaimaniyah, Kurdistan Region of Iraq. In: Al-Quraishi AMF, Negm AM (eds) Environmental remote sensing and GIS in Iraq. Springer Water

Almamalachy YS, Al-Quraishi AMF, Moradkhani H (2019) Agricultural drought monitoring over Iraq utilizing MODIS products. In: Al-Quraishi AMF, Negm AM (eds) Environmental remote sensing and GIS in Iraq. Springer Water

Alrababah MK, Alhamad MN (2006) Land use/cover classification of arid and semi-arid Mediterranean landscapes using Landsat ETM. Int J Remote Sensing 27(13):2703–2718

AL-Rajehy AM (2002) Relationships between soil reflectance and soil physical and chemical properties, M.Sc. Thesis, Mississippi State University, Mississippi, USA, p 75

Al-Rawi MB (2007) studying the Hydrologic and morphometric properties of Bahr Al-Najaf Basin using Geographic Information system. M.Sc. thesis, College of Education (Abn-Rushd), University of Baghdad, Iraq

Al-Wassai FA (2003) Comparison between different method of satellite merge resolution. M.Sc. thesis, University of Baghdad, Iraq

Azeez SN (2013) Used Geo informatics techniques (GIS, RS and GPS) for identification and isolation of soil map units. In Garmyan district, Kurdistan Region—north of Iraq, Ph.D. Thesis, University of Sulaimania, Iraq

Bahadur KC (2009) Improving Landsat and IRS image classification: Evaluation of unsupervised and supervised classification through band ratios and DEM in a mountainous landscape in Nepal, Remote Sensing 1(4):1257–1272

Barwari AM, Slewa NA (1995) Geology of the Najaf plate, general company for geological survey and mining. Ministry of Industry and Minerals, Report No 20

Batatu HT (2002) The old social classes and revolutionary movement of Iraq. A study in Iraqs old landed and communist pp 88–91

Ben-Dor E, Heller D, Chudnovsky A (2008) A novel method of classifying soil Profiles in the field using optical means. Soil Sci Soc Am J 72:1113–1123

Borghuis AM, Chang K, Lee HY (2007) Comparison of automated and manual mapping of typhoon-triggered landslides from SPOT-5 imagery. Int J Remote Sens 28(8):1843–1856

Buringh P (1960) Soils and soil conditions in Iraq, soil survey and classification specialist, Ministry of Agriculture, Baghdad, Iraq, p 322

Elachi C, Van J (2006) Introduction to the physics and of Remote sensing (Wiley series in Remote sensing and Image Processing). Wiley-Inter science, p 18

Elkhrachy I (2015) Land use change detection using satellite images for Najran City, Kingdom of Saudi Arabia (KSA), WCS-CE-The World Cadastre Summit, Congress & Exhibition Istanbul, Turkey, 20–25 April

Fadhil AM (2009) Land degradation detection using geo-information technology for some sites in Iraq. J Al-Nahrain Univ Sci 12(3):94–108

Fadhil AM (2011) Drought mapping using Geoinformation technology for some sites in the Iraqi Kurdistan Region. Int J Digit Earth 4(3):239–257

Fadhil AM (2013) Sand dunes monitoring using remote sensing and GIS techniques for some sites in Iraq. In: Proceedings SPIE 8762, PIAGENG 2013: intelligent information, control, and communication technology for agricultural engineering, p 876206. http://dx.doi.org/10.1117/12.2019735

Faleh A, Shawan J (2012) GIS and remote sensing: principles and applications. Anvo Brandt Press. Fas. The Kingdom of Morocco

Hamad AI (2009) The use of remote sensing and geographic information systems of land evaluation in the center of the Iraqi alluvial plain. MSc. Thesis, Faculty of Agriculture, University of Baghdad, Iraq

Hassan AH (1983) Hydrological, hydrogeological and hydro chemical investigation of Bahar AL-Najaf area. MSc. Thesis, College of Science, University of Baghdad, Iraq, p (91). http://www.redalyc.org/articulo.oa?id=57320201

Jensen JR (2007) Introductory to digital image processing: a remote sensing perspective. Prentice Hall Series in Geographic Information Science

Katie HH, Mahimid AS (2002) The relationship between the geo morphological units and the distribution of soil units in the area west of Al-Razazah lake. Iraqi J Soil Sci 1(2):129–140

Koutroumpas A, Alexiou I, Vlychou M, Sakkas L (2010) Comparison between clinical and ultra-sonographic assessment in patients with erosive osteoarthritis of the hands. Clin Rheumatol 29:511–516

Lillesand TM, Kiefer RW (2000) Remote Sensing and Image Interpretation, 4th edn. Wiley, New York, USA

Mahmoud RA, Mahimid AS (2011) Inheritance and evolution of some gypsum soils in Iraq. Iraqi J Desert Res 24(5):88–99

Martinez-Rios JJ, Monger HC (2002) Soil classification in arid lands with thematic mapper data, TERRA, 20(2):89–100. Available in: http://www.redalyc.org/articulo.oa?id=57320201

Mukherjee S, Mukherjee P (2009) Assessment and comparison of classification techniques for forest inventory estimation: A case study using IRS-ID imagery. Int J Geo Info 5(2):63–73

Mustafa BM, Al-Quraishi AMF, Gholizadeh A, Saberioon M (2019) Proximal soil sensing for soil monitoring. In: Al-Quraishi AMF, Negm AM (eds) Environmental remote sensing and GIS in Iraq. Springer Water

Mutter AN (2008) Efficiency methods calculating plain soil mapping units of soil mid mesopotamian plain. MSc. Thesis, College of Agriculture, University of Baghdad, Iraq

Nain P, Kumar K (2016) Study for accuracy assessment of land use and land cover classification of New-Delhi, North India. Int J Comput Sci Trends Technol (I JCS T) 4(3):137–143

Nash N (2016) Detection and accuracy assessment of mountain pine beetle infestations using landsat 8 OLI and WorldView02 satellite imagery Lake Tahoe Basin-Nevada and California. Msc. Thesis, Geographic Information Science and Technology, Faculty of the USC Graduate School, University of Southern California, USA

Proklamasi SA, Indonesia G, Myint M (2015) Conducting initial vegetation classification through image analysis. In: Chapter 3, Version 1.0 March, Natural Resources Information Integration; Ihwan Rafina, TFT; and Tri A. Sugiyanto, PT SMART/TFT

Qadir MH (2007) Study of land cover—land use and its reflectivity in Shahrazur plain by using remote sensing techniques. MSc. Thesis, Collage of Agriculture, University of Sulaimania, Iraq

Sallom AJ, Segar RH (1994) Semi-detail of soil survey and hydrological investigations in the Najaf Sea Project. Department of Investigation and Soil, Ministry of Water Resources

Shuli MA (2008) Study of land coverings in the Nablus area using remote sensing technology. An-Najah National University in Nablus, Palestine

Soil Survey Staff (1999) Soil Taxonomy, A basic system of soil classification for making and interpreting soil survey, 2nd edn. Agriculture Handbook No. 436, USDA

Stehman SV (2013) Estimating area from an accuracy assessment error matrix. Remote Sens Environ 132:202–211

Sulaiman AA, Abboud NF (2012) Classification and evaluation some of alluvial Soils in the center of the sedimentary plain. Tikrit Univ J Agric Sci 12(3):155–162

Taruvinga K (2008) Gully mapping using remote sensing: case study in KwaZulu-Natal, South Africa. M.Sc. thesis, Geography Science, University of Waterloo, Ontario, Canada

Tso B, Olsen R (2005) Combining spectral and special information into hidden Marcov odels for unsupervised image classification. Int J Remote Sens 26:211–2133

Wahib QA (2012) Characteristics of spectral reflectivity of surface soil and ground coverings of Al-Salman depression in southern Iraq. J Iraqi Agric Sci 43(4)(special number):129–140

Wu W, Zucca C, Muhaimeed AS, Al-Shafie WM, Al-Quraishi AMF, Nangia V, Zhu M, Liu G (2018) Soil salinity prediction and mapping by machine learning regression in central mesopotamia, Iraq. Land Degrad Dev 29(11):4005–4014

Wu W, Muhaimeed AS, Al-Shafie AM, Al-Quraishi AMF (2019) Using radar and optical data for soil salinity modeling and mapping in central Iraq. In: Al-Quraishi AMF, Negm AM (eds) Environmental remote sensing and GIS in Iraq. Springer Water

Yacoub SY, Purser BH, Al-Hassani NH, Al-Azzawi M, Orzag-Sperber F, Hassan KM, Plaziat JC, Younis WR (1981) Preliminary study of the Quaternary Sediments of SE Iraq. Joint project between the geological survey of Iraq and University of Paris XI, Orsay, GEOSURV, int. rep. no. 1078

Yacoub SY, Abdul Jabbar MF, Shehab AT (2008) Study of determination of ferrous rocks explored using remote sensing techniques in Al-Kaara Area. Western Iraq, J Iraqi Geol Mining 4(1):43–52

Yao Y, Qin Q, Fadhil AM, Li Y, Zhao S, Liu S, Sui X, Dong H (2011) Evaluation of EDI derived from the exponential evapotranspiration model for monitoring China's surface drought. Environ Earth Sci 63(2):425–436

Yousif BF (2004) The use of remote sensing techniques in the classification of Al-Najaf Soil. M.Sc. Thesis, Building and Construction Engineering, University of Technology, Baghdad, Iraq

Zhang Z, Kang H, Yao Y, Fadhil AM, Zhang Y, Jia K (2017) Spatial and decadal variations in satellite-based terrestrial evapotranspiration and drought over Inner Mongolia Autonomous Region of China during 1982–2009. J Earth Syst Sci 126(8)

Zinck J (2008) Remote Sensing of soil salinization: impact on Land management, CRC Press, Technology & Engineering, pp 374

Part III
Proximal Soil Sensing

Chapter 5
Proximal Soil Sensing for Soil Monitoring

Banaz M. Mustafa, Ayad M. Fadhil Al-Quraishi, Asa Gholizadeh
and Mohammadmehdi Saberioon

Abstract The need for soil information is higher now than ever before. Agriculture and the way in which we use and manage our soils are being changed with the concerns over food security and global climate change. Mainly, soil data is necessary to be used in soil and natural resource management, e.g. for environmental modeling for a better understanding of soil processes and reducing risks in decision-making. Conventional soil survey cannot efficiently offer these data because the techniques are time-consuming and expensive. Proximal Soil Sensing (PSS), which has become a multidisciplinary area, aims to develop field-based techniques for acquiring information on the soil from close by, or within the soil. It can be used to monitor soil both surface and subsurface spatial and temporal information rapidly, cheaply and with less labor. This chapter reports on developments in PSS and its application in soil science and environmental assessment with specific attention to Iraq soils. It will

B. M. Mustafa
Department of Petroleum and Energy Engineering, Technical College of Engineering, Sulaimani Polytechnic University, Sulaimani, Kurdistan Region, Iraq
e-mail: banaz.muhammed@spu.edu.iq

Ministry of Agriculture and Water Resources. General Directorate of Agricultural Researches and Extension, Directorate of Agricultural Researches, Sulaimani, Kurdistan Region, Iraq

A. M. F. Al-Quraishi
Department of Environmental Engineering, College of Engineering, Knowledge University, Erbil 44001, Kurdistan Region, Iraq
e-mail: ayad.alquraishi@gmail.com; ayad.alquraishi@knowledge.edu.krd

A. Gholizadeh (✉) · M. Saberioon
Department of Soil Science and Soil Protection, Faculty of Agrobiology, Food and Natural Resources, Czech University of Life Sciences Prague, Prague, Czech Republic
e-mail: gholizadeh@af.czu.cz

M. Saberioon
e-mail: msaberioon@frov.jcu.cz

M. Saberioon
Laboratory of Signal and Image Processing, Institute of Complex Systems, South Bohemia Research Centre of Aquaculture and Biodiversity of Hydrocenoses, Faculty of Fisheries and Protection of Waters, University of South Bohemia in České Budějovice, Nové Hrady, Czech Republic

© Springer Nature Switzerland AG 2020
A. M. F. Al-Quraishi and A. M. Negm (eds.),
Environmental Remote Sensing and GIS in Iraq, Springer Water,
https://doi.org/10.1007/978-3-030-21344-2_5

review some of the technologies that may be used for PSS and proposes a framework for their use in Iraq soil and environmental monitoring. The chapter brings together ideas and examples from developing and using proximal sensors for applications such as precision agriculture and soil contamination monitoring, where there is a particular need for high spatial resolution information.

Keywords Proximal soil sensing · Diffuse reflectance spectroscopy · Soil spectroscopy · Environmental monitoring · Iraq

5.1 Introduction

The soil is the main natural resource for the production of energy and food. It controls the water movement in the landscape, working as a natural filter for contaminants that may leak into sensitive spheres of the environment (Stenberg et al. 2010). The soil is also recognized as a potential sink for carbon to mitigate climate change. Given the value of soil, there is a need for regular observing to detect variations in quantity and quality of parameters, which are temporally and spatially variable (Bouma 1997; Karlen et al. 1997). However, conventional soil survey cannot effectively offer the amount of data that is required. The major reason is well-known in case of requirement to spatial soil information (especially the functional parameters such as nutrient supply and water balance), conventional soil sampling and laboratory analyses are expensive and time-consuming (Viscarra Rossel and McBratney 1998a).

Proximal Soil Sensing (PSS) can obtain soil sample information cheaply and rapidly despite conventional methods and can be employed to assess both surface and subsurface soil properties, unlike Remote Sensing (RS) techniques (Viscarra Rossel and McBratney 1998b). Among others, PSS involves the use of optical methods including Diffuse Reflectance Spectroscopy (DRS). Visible–Near Infrared–Short Wave Infrared (VIS–NIR–SWIR) DRS is fast, cost-effective, nondestructive and reproducible analytical way (Reeves 2010). The use of VIS–NIR–SWIR DRS in soil science has attracted much attention over the last 28 years (1990–2018), with a considerable increase in the number of papers published in the literature (Ben-Dor and Banin 1995a; Stenberg et al. 1995; Viscarra Rossel and McBratney 1998a, b; Shepherd and Walsh 2002; Brown et al. 2006; Wetterlind et al. 2008; Gholizadeh et al. 2013a, 2016; Adeline et al. 2017). According to Viscarra Rossel et al. (2006), there are many reasons for the interest in VIS–NIR–SWIR reflectance spectroscopy. For instance, sample preparation contains only drying and crushing, the sample is not influenced by the analysis in any way, no chemical is needed, measurements take a few seconds, numerous soil attributes can be determined from a single scan and the technique is usable both in the laboratory and in situ. Therefore, the approach application requires to be expanded in order to prove its capability for soil attributes monitoring and mapping worldwide. To this end, the main aim of the chapter is to provide some information on the current state-of-the-art of the VIS–NIR–SWIR in soil science in general and soil spectroscopy for Iron Oxide (Fe_2O_3) retrieving and mapping Iraq soils in particular. We begin with a description of fundamentals in PSS and DRS. We then review the application of soil reflectance spectroscopy for prediction of different soil properties including soil contamination. We afterwards continue with the effects of external factors such as preprocessing and calibration algorithms on the final obtained accuracy. There then follows an example of a case study in Sulaimani, Iraqi Kurdistan region, and we finish with a discussion of the future aspects and recommendations.

5.1.1 Proximal Soil Sensing (PSS)

PSS refers to the quantitative information on soil attributes. It is the use of field-based sensors to collect soil information from close by (within 2 m) or within the soil body (Viscarra Rossel et al. 2010a). It is the field-based sensors that are sensitive to reflectance and emittance of radiation across the VIS–NIR–SWIR region. Various proximal sensors and techniques may be used for PSS purpose. According to Viscarra Rossel et al. (2010b), the technique may include Ground Penetrating Radar (GPR), magnetic susceptibility, Electromagnetic (EM) induction and electrical resistivity, γ-radio metrics, soil color, mechanical-draft systems, ion-selective electrodes (ISE), ion-selective field effect transistors (ISFET), DRS and Imaging Spectroscopy (IS), which two last cases are the most common techniques. The PSS sensors are mostly multispectral, hyperspectral and ultra-spectral spectrometers and cameras that attached to different platforms such as tripod stand, towers or crane, which are adequate for small-scale irregular monitoring and calibration. The purpose of the platform is to locate the sensor over an area of interest. The platform's type, therefore, is chosen by the needs of the measurements to be made (Dwivedi 2017).

Proximal soil sensors can be employed as alternatives to conventional soil sampling and laboratory analyses, which as previously mentioned, are slower, more labor and more expensive than PSS. However, the produced results by PSS may not be as accurate as conventional laboratory analysis; it facilitates the acquisition of larger amounts of quantitative spatial data applying simpler, cheaper and less laborious methods (Viscarra Rossel and McBratney 1998b). Furthermore, the assessment can be made in situ, presenting information of the soil at field conditions and promptly. Soil spectroscopy using laboratory and field spectroradiometers in VIS–NIR–SWIR range, which is sensitive to both organic and inorganic soil composition, is the most routine technique for soil monitoring that has been extensively studied and introduced as a potentially suitable and influential mean for PSS.

5.1.2 Soil Spectroscopy

Several types of research have suggested that the spectra from proximal VIS–NIR–SWIR DRS could provide inexpensive prediction of soil physical, chemical and biological properties. For generating a soil spectrum, radiation enclosing all applicable frequencies in the specific range is directed to the sample. Depending on the soil elements, the radiation will cause individual molecular bonds to vibrate and they will absorb light to different degrees with a certain energy quantum consistent to the variance between two energy levels (Stenberg et al. 2010). Due to the direct relation of energy quantum to frequency and its inverse relation to wavelength, the deriving absorption spectrum provides a typical shape that can be utilized for analytical reasons (Miller 2001). The technology is rapid, simple to use, fast and economical, which measurements need a small amount of sample preparation and a single spectrum can be used to predict multiple soil properties when joint to multivariate calibrations. These potentials simplify the acquisition of high-resolution soil data from different ranges over the spectrum. However, according to (Stenberg et al. 2010), the choice of spectral region to be used will highly depend on (i) the accuracy of the predictions, (ii) the cost of the technology and (iii) the amount of sample preparation needed.

5.1.3 Acquiring Reflectance Information from Soil

The spectral signature of soils alters with soil composition, environmental conditions as well as viewing geometry of the sensor. According to Dwivedi (2017), in many cases, the spectral response shape linked to a particular factor overlap with response shape of other's and thus prevent the evaluation of the effect of a particular factor. Therefore, it is essential to understand the physical activity and origin of soil's contents and environmental conditions. The soil is formed of minerals mainly clay and Fe_2O_3, Organic Matter (OM) both living and decomposed, as well as water in all the three phases (solid, liquid and gas). These elements have direct or indirect effects on soil spectra (Dwivedi 2017).

Soil spectra across the VIS–NIR–SWIR spectral range are specified by significant spectral features that allow quantitative analysis of various soil attributes. According to Viscarra Rossel et al. (2016), the information on soil composition, which contains minerals, organic compounds and water, is encoded by the obtained spectra. These encodings are shown in the spectra as absorptions at specific wavelengths of electromagnetic radiation that their measurements can define soil either qualitatively or quantitatively (Viscarra Rossel et al. 2016). On the other words, spectral signatures of soils are described by their reflectance or absorbance, as a function of wavelength. The signatures, under controlled conditions, are caused by electronic transitions of atoms and vibrational stretching and bending of structural groups of atoms that form molecules and crystals (Brown 2007). Certainly, soil spectra are sensitive to both organic and inorganic soil composition, considered by detecting the positive and negative peaks, which appear at specific wavelengths (Viscarra Rossel et al. 2006). Positive peaks are due to the component of interest, though negative peaks related to interfering components (Haaland and Thomas 1988). Absorptions in the VIS range (400–700 nm) primarily correspond to minerals that include Iron (Fe) (e.g. hematite, goethite) (Sherman and Waite 1985). Soil Organic Matter (OM) can also have extensive absorptions in the VIS, which are associated with chromophores and the darkness of humic acid and absorptions in the NIR (700–2500 nm) from the overtones and combination absorptions of O–H, C–H, and N–H (Clark et al. 1990; Clark 1999). To be more accurate, the regions around 1400 and 1900 nm are linked to vibrational frequencies of O–H groups in the water and hydroxyl absorption, and the features around 2000–2500 nm are related to the characteristics of SOM and clay minerals (Viscarra Rossel and Behrens 2010; Gholizadeh et al. 2017).

5.2 Background of Soil Spectroscopy

Soil spectroscopy is a developed discipline that had come a long way fairly since the mid-1960s when Bowers and Hanks (1965) published their paper on the correlation between soil reflectance and Soil Moisture Content (SMC). That pioneering study, followed by a series of papers by Hunt (1982), confirmed that water and minerals in the soil environment have unique spectral signatures that can be utilized for particular detection. Learning from several sectors' successes (e.g., food science, tobacco and textiles), Dalal and Henry (1986) used the PPS approach to soils for prediction of different properties. After presenting laboratory and field spectrometers to the market (around 1993), more scientists understood the potential of soil spectroscopy and accordingly, more Soil Spectral Libraries (SSLs) were assembled (Shepherd and Walsh 2002; Bellinaso et al. 2010; Viscarra Rossel et al. 2016). PPS has been studied on different soil attributes including spectral-active features as well as those are not spectrally-active parameters. The following sections present background materials on both categories.

5.2.1 Soil Spectroscopy for Soil Properties Monitoring

Over the last few decades, a large number of attempts have been made to predict soil properties with VIS–NIR–SWIR spectroscopy, which some has been mentioned below. Total and Organic Carbon (OC) determination are perhaps the most common, followed by Fe and clay content. This is logical as these attributes are the fundamental constituents of the soil and have well-known absorption parameters in the VIS–NIR–SWIR range (Viscarra Rossel et al. 2006). Some other studies frequently reported other spectrally-active properties, which their results were usually moderate and often very variable. According to Stenberg et al. (2010), this also makes sense as the covariations to constituents that are spectrally-active cannot be expected to be globally stable. Ben-Dor and Banin (1995a), as one of the pioneers in this field, demonstrated the power of reflectance spectroscopy in accounting for Carbonate Calcium ($CaCO_3$) content in the soil, and later in monitoring the structural composition of smectite soil minerals in the laboratory (Ben-Dor and Banin 1995b). The content of soil OM, which is one of the most critical indicators of soil quality, has also been determined using soil spectroscopy in several studies. Henderson et al. (1992) and Cozzolino and Moron (2006) found that VIS wavelengths had a strong correlation with soil OM for soils with the same parent material. Christy (2008) and Cecillon et al. (2009) reported very good assessments of soil OM using NIR spectroscopy in the laboratory and even under field conditions using on-the-go sensors. The performance of soil OM determination is highly variable. There are several probable descriptions for this. Despite the several absorption bands of soil OM over the VIS–NIR–SWIR range, it is frequently told that its signals in this region are weak (Viscarra Rossel and McBratney 1998b; Stenberg et al. 2010), mainly in soils with low OM contents. Later, various studies of VIS–NIR–SWIR spectroscopy were carried out in different configurations, in the laboratory, in the field using sampling or using on-the-go sensors embedded on a tractor, to determine the different spectrally-active soil parameters (Ben-Dor and Banin 1995a; Shepherd and Walsh 2002; Nanni and Dematte 2006; Gholizadeh et al. 2013a; Adeline et al. 2017).

Bellon-Maurel and McBratney (2011) reported that research using VIS–NIR–SWIR spectroscopy for soil analysis has been experiencing a boom over the last 30 years. However, most studies are associated with more soil quantity, and those linked to qualitative data are infrequent. The application of laboratory spectrometry to the specific prediction of soil quality began at the turn of the 21st century. Cohen et al. (2005) presented the first application to the rough assessment of a specific soil threat. They highlighted that VIS–NIR–SWIR spectroscopy exceeded a frequently used realistic model for sites classification based on the status of soil erosion. Cohen et al. (2006) employed another NIR-based method to assess global soil quality, including assessment of soil OM. They combined ordinal logistic regression and classification trees of soil NIR spectra to distinguish among ecological condition categories. Using classification trees, they recognized important spectral regions for ecological condition classification: 2200–2300, 1100–1200, and 500–600 nm. They concluded that using soil reflectance information for site classification, mainly for the discrimination of degraded sites, was more promising than using biogeochemical data. Consequently, soil NIR spectra offer an efficient mean for the fast analysis of the condition and ecosystems of the soil. Reeves et al. (2010) also proposed that VIS–NIR–SWIR could be used for predictions of soil OM quality as well as quantity.

5.2.2 Soil Spectroscopy for Soil Contamination Monitoring

Due to urbanization and fast economic progress in the last few decades, environmental pollution has become a progressively thoughtful issue. Extreme contaminants in the environment pose a fatal threat to human health by entering the food chain and migrating into drinking water sources (Shi et al. 2016). Research on soil contamination and soil alterations linked to contamination is getting more attention, especially in the soil ecosystems reclamation and sustainable use. As previously mentioned, conventional methods for determining the soil condition in vast areas need field sampling, chemical analyses in a laboratory and geostatistical interpolation, which are time-consuming and expensive (Ren et al. 2009). Also, these approaches are ineffective at sensing small soil changes and hence benefits of small changes detection, their controlling before they become large and able to bring greater effects, will be ignored (Sanches et al. 2013). However, any major change of soil condition requires being sensibly assessed using existing high-tech sensors including proximal soil spectrometers for early detection of soil status due to soil contamination (Gholizadeh et al. 2018a).

Soil contamination has been measured using spectroscopy method, even though they do not have spectral features within the VIS–NIR–SWIR range at low concentration levels, via inter-correlation with soil parameters that are spectrally-active in this range (Ben-Dor et al. 1997; Kooistra et al. 2001; Wu et al. 2005; Song et al. 2012; Gholizadeh et al. 2015b). Therefore, through the correlation between spectrally-active soil attributes and soil contaminants concentrations, reflectance spectra can be employed for the indirect assessment of contamination in soil samples from polluted areas. For example, Kooistra et al. (2001) discovered that there was a positive correlation between the soil OM content and the contents of Zn and Cd in floodplains along the river Rhine in the Netherlands, and accordingly the Zn and Cd contamination levels were predicted. Kemper and Sommer (2002) successfully used DRS to estimate As, Pb, Fe, Hg, S and Sb contents in the Aznalcollar mine area in Spain. Ren et al. (2009) have also determined Cu and as concentrations using reflectance spectroscopy of areas near mining activities. Pandit et al. (2010) found very high correlation coefficients between laboratory-determined and predicted values of several toxic parameters including Pb. In a study by Gholizadeh et al. (2015b), VIS–NIR–SWIR spectroscopy was utilized to assess potentially toxic elements in dumpsites of the Czech Republic. They mentioned although intense bands in the VIS–NIR–SWIR spectra are not directly connected to the presence of contaminants, it is obvious that these parameters can interact with the main spectrally-active components of soil.

5.3 Spectroscopic Preprocessing and Calibration

According to Ge et al. (2007), spectroscopy modeling involves the following four key steps: (i) measurement in which contents of selected soil parameters are assessed and soil spectroscopic reflectance is achieved; (ii) preprocessing in which spectroscopic reflectance spectra are preprocessed; (iii) calibration in which a subset of samples is utilized to develop regression models and (iv) validation in which the residual samples are used to determine the validity of the regression models for estimating parameters contents. The principal challenge preventing the application of spectroscopy for the soil properties determination is finding appropriate data preprocessing and calibration strategies to gain acceptable accuracy. So far, different preprocessing and calibration algorithms have been recommended to link spectra to the measured attributes of materials, which will be discussed in the following sections.

5.3.1 Preprocessing

The most limiting factor in VIS–NIR–SWIR spectroscopy modeling is the spectra quality (Couteaux et al. 2003). Even the most dominant software for calibration is not able to provide consistent predictions from poor spectra. Spectrum preprocessing may improve the calibrations quality when powerful software is utilized, but the quality of the spectra and the accuracy of the reference data are the key elements of reliable models. Couteaux et al. (2003) stated that the preprocessing of soil spectra before model calibration is required to decrease interference. Several studies showed an improvement in accuracy of VIS–NIR–SWIR spectroscopy predictions after various preprocessing algorithms to minimize noise effects (Van Waes et al. 2005; Brunet et al. 2007; Gholizadeh et al. 2015a). Spectral preprocessing methods are used to eliminate any unsuitable information that cannot be controlled properly by the modeling techniques. This step may take more time than the analysis itself. Spectral preprocessing methods include a range of mathematical methods, which correct light scattering in reflectance measurements before using the data in the calibration process. Preprocessing techniques are divided into four categories namely (i) smoothing, (ii) baseline removal, (iii) scaling and (iv) normalization (Xie et al. 2012; Gholizadeh et al. 2015a).

Smoothing is usually used for noise reduction. Savitzky-Golay smoothing as one of the most common approaches is an averaging algorithm that fits a least squares polynomial to the data points, and then the value to be averaged is predicted from the polynomial (Savitzky and Golay 1964). In some forms of spectroscopy, a baseline or background signal can be encountered that is far away from the zero level. As this affects some measurements such as peak area and peak height, it is of the highest significance to correct for such phenomena; hence, the second category of preprocessing is the baseline removal. There are some baseline removal approaches existing namely spectral derivative transformation that according to Duckworth (2004), is one of the best approaches for removing baseline effects. The First Derivative (FD) is very efficient for removing baseline offset; although the Second Derivative (SD) is efficient enough for both the baseline offset and linear trend from a spectrum (Duckworth 2004; Rinnan et al. 2009). An alternative way to eliminate scatter effects in spectroscopy is Multiplicative Scatter Correction (MSC), which is known as a transformation method employed to compensate for multiplicative and/or additive scatter effects in the data (Chu et al. 2004). Another preprocessing technique is range scaling that is appropriate in situations that the total intensity in the spectra is sample-dependent and samples require to be scaled in a way that intensities can be matched. Standard Normal Variety (SNV) is one of the range scaling methods, which is widely used in NIR applications. It can correct the multiplicative interferences of light scatter and particle size by centering and scaling each spectrum (Duckworth 2004). After scaling, every spectrum will have a mean of zero and a standard deviation of 1. Another method for preprocessing is normalization, which is normally utilized for absorption feature improvement. Continuum Removal (CR) is one of the common normalization methods, which produces new spectral data by dividing the envelope curve of a continuum on raw reflectance spectra (Clark and Roush 1984). CR is efficient at isolating particular absorption features and eliminating the effects of changing slopes and overall reflectance levels (Kokaly et al. 2003).

5.3.2 Calibration

Due to the overlapping absorption of soil elements, soil DRS in the VIS–NIR–SWIR region are mostly nonspecific. This characteristic is compounded by scattering effects produced by soil structure or specific elements such as quartz (Dwivedi 2017). All of these factors yield complex absorption trends, which require be mathematically extracting from the spectra and associating to soil attributes. Therefore, the analyses of soil DRS need the application of multivariate calibrations. Calibration is connecting a set of spectral parameters that are resulting from the spectral information to the materials in question (Gholizadeh et al. 2013b). Application of a set of recognized calibration techniques makes this procedure achievable. Selecting the most robust calibration method can support to attain a more consistent prediction model. Robustness means that the modeling performance does not analytically alter in case of using new calibration samples (Geladi and Kowalski 1986). Moreover, the performance of a model is connected to its predictive accuracy and robustness. A good model presents a higher coefficient of determinations (R^2) and lower Root Mean Square Error (RMSE) value (Aichi et al. 2009; Chen et al. 2012).

The most frequent calibration algorithms for soil applications are based on linear regressions such as Stepwise Multiple Linear Regressions (SMLR), Principal Component Regression (PCR) and Partial Least Squares Regression (PLSR). The key sense for employing SMLR is the insufficiency of more conventional regression methods such as Multiple Linear Regression (MLR) and absence of users' awareness about the availability of full spectrum information compression methods such as PCR and PLSR, which both can cope with data containing large numbers of predictor that are very collinear (Dwivedi 2017). Nevertheless, MLR, PCR and PLS are linear models, the data-mining techniques can manage nonlinear data. Viscarra Rossel and Lark (2009) used wavelets joint with polynomial regressions to decrease the spectral data, account for nonlinearity and provide accurate calibrations according to the chosen wavelet coefficients. Mouazen et al. (2010) compared Artificial Neural Networks (ANN) with PCR and PLS for the prediction of various soil properties. They showed combined PLSR-ANN models could develop better predictions as compared to PLSR and PCR. Viscarra Rossel and Behrens (2010) compared the use of PLSR to some data-mining approaches as well as various feature selection methods namely Multivariate Adaptive Regression Splines (MARS), Random Forests (RF), Boosted Trees Regressions (BTR), Support Vector Machine Regressions (SVMR), ANN and wavelets, for clay, OC and pH prediction. Their results highlighted that data-mining algorithms yield more accurate performances than PLSR.

Furthermore, they showed that some of the techniques offer information on the particular wavelength importance of the models. Recently, Gholizadeh et al. (2016)

explored whether the new Memory-Based Learning (MBL) method performs better than the other methods namely PLSR, BTR and SVMR for prediction of soil texture. They concluded that MBL improved the model accuracy, decreased the number of samples to be examined for precision management applications in the field and could be used as additional materials in combination with spatial statistical methods to assess soil conditions.

5.4 Accuracy and Uncertainty in Soil Spectroscopy

Uncertainty is an expression of confidence in our knowledge and is thus subjective (Heuvelink et al. 2007). Uncertainty evaluation is needed for the data value establishment as inputs for decision-making (Brown 2004) and for the consistency judgment of decisions that are learned from the data (Beven 2000). It is also essential for determining the uncertainty reasons in environmental research and for pointing resources to data quality improvement (Brown and Heuvelink 2005).

According to Brodsky et al. (2013), as the VIS–NIR–SWIR reflectance spectroscopy steps are often expanded, there is necessary to quantify and assess the linked uncertainties in the context of the application examples. There are some parameters, which negatively affect the accuracy of the soil spectroscopy predictions as changing soil structure, texture, moisture and mineralogy (Stenberg et al. 2010). The predictive capability of spectroscopy is limited regarding some parameters adversely influencing the expected prediction that needs more attention and work. These factors rise from an absence of standardized methods in relation to various protocols, sampling techniques, sample preparation, instrument specifications, spectral acquisition and analytical algorithms (Whiting et al. 2004; Gholizadeh et al. 2013b). A number of strategies have been suggested to overcome these issues and similar limitations. Some studies recommended that spiking local samples into general calibration models could improve predictive accuracy under laboratory or in-situ measurement conditions (Wetterlind and Stenberg 2010; Kuang and Mouazen 2013). The spiking of general or global calibrations has been suggested and has formed progressive performance for moving a calibration from one geographical area to another on the country level (Viscarra Rossel et al. 2008) as well as field level (Reeves et al. 1999) and has improved the results for a small catchment using a global calibration (Brown 2007). Standardization of the rules for determining the accuracy of VIS–NIR–SWIR spectroscopy predictions is another essential way and a necessity for the comparison of the VIS–NIR–SWIR spectroscopy presentations across various studies. In other words, the differences in condition explain the significance to establish a simple procedure for standardization that diminishes the systematic and random effects and permits unification of SSLs (Ben-Dor et al. 2015; Viscarra Rossel and Bouma 2016).

5.5   Soil Spectroscopy for Iron Oxide Prediction: A Case Study in Sulaimani, the Iraqi Kurdistan Region

Iron (Fe) is one of the most important elements on the Earth crust and is the fourth most abundant element in the lithosphere after oxygen, silicate and aluminium, which composed 5% of the earth crust and its average content in soils, is estimated at 3.8% (Lindsay 1979). The deposits and occurrences of Fe ores in Iraq are located in the mountains folded zone of the country in Asnawa, Mishau, and Benavi in the platform area western desert of Gaara and Hussainiyat (Etabi 1982). Fe and Fe_2O_3 have significant influences on the spectral reflectance characteristics of soil, and many of the absorption features in soil reflectance spectra are due to the presence of Fe in various forms, the increase in Fe and Fe_2O_3 contents can cause a significant decrease in reflectance, at least in the VIS wavelengths (Matthews 1972). Removal of the Fe_2O_3 from the soil will cause a marked increase in reflectance throughout the 500 to 1100 nm region, but the reflectance above 1100 nm is not particularly affected (Mustafa 2015). Therefore, in an experiment, spectroscopy technique was used for detecting and mapping Fe and Fe_2O_3 rich soils of some sites in Sulaimani, Iraqi Kurdistan Region because no such a research has been done so far in the proposed study area, and there are no up-to-date maps that present and show soil Fe and Fe_2O_3 in Sulaimani (Mustafa 2015).

5.5.1   Materials and Methods

5.5.1.1   Study Area

The study area (Fig. 5.1) was a part of Sulaimani Governorate (locates in the Northeast of Iraq) in the Iraqi Kurdistan Region, which covers an area of 9,829 Km^2. The geographical location extends from latitude 35° 11′ 59″ to 36° 14′ 20″ N and from longitude 44° 48′ 50″ to 46° 01′ 9″ E. The climate of the study area shows typical characteristics of the Mediterranean environments. The lowest average temperature of the area recorded in 2013 was 12.6°C and the highest was 22.7°C. Moreover, the lowest average rainfall recorded was 46.85 mm and the highest was 131.83 mm.

5.5.1.2   Soil Sampling and Analysis

Thirty-nine soil samples were collected from the soil surface layer depth (0–15 cm), air-dried, ground and sieved (≤2 mm) and thoroughly mixed before analyzing. The total Fe was digested by the Kjeldahl (Bremner and Mulvaney 1982), while Fe_2O_3 removed from soils by a dithionite-citrate system buffered with sodium bicarbonate and then measured by atomic absorption spectrometer (AA700) with wavelength of 248.3 nm according to Mehra and Jackson (1953).

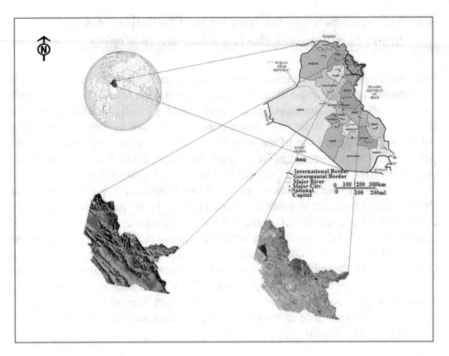

Fig. 5.1 The location map of the study area (Mustafa 2015)

5.5.1.3 Spectral Reflectance Measurement

Spectral reflectance was calculated across the 350–2500 nm range using an ASD FieldSpec III Pro FR spectroradiometer (ASD Inc., Denver, Colorado, USA) with a high-intensity contact probe. The spectral resolution of the spectroradiometer was 2 nm for the region of 350–1050 nm and 10 nm for the region of 1050–2500 nm. Moreover, the radiometer bandwidth from 350–1000 nm was 1.4 nm while it was 2 nm from 1000–2500 nm. The instrument ran for about 30 min to allow the spectrometer and lamp to warm. Soil samples were placed in 9 cm diameter petri dishes and formed 2 cm layers of soil. This was done to avoid beam reflectance from the bottom of the dish due to down-welling solar and sky radiation penetrating into the soil at approximately 1/2 the wavelength (Jensen 2007), which could have an unwanted effect for modifying the soil spectra. Samples were leveled off using a stainless steel blade to guarantee a flat surface flush with the top of the petri dish as a smooth soil surface ensures maximum light reflection and a high Signal to Noise Ratio (SNR) (Mouazen et al. 2005). We measured all spectral readings in the center of the samples (3 replications each) in a dark room to avoid interference from stray light. The spectroradiometer was optimized using a white Spectralon™ (Lab-sphere, North Sutton, New Hampshire, USA) before the first scan and after every six measurements

Fig. 5.2 Schematic diagram
of methodology steps
(Mustafa 2015)

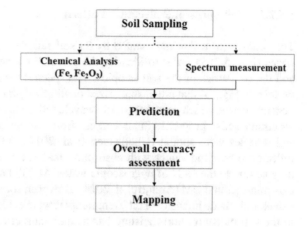

(Shi et al. 2016). For each soil measurement, 30 spectra were averaged to improve
SNR.

5.5.1.4   Prediction Assessment and Mapping

Mustafa (2015) developed simple regression models using regression procedure and
considered as independent variables of Fe and its spectra reflectance that allowed an
assessment of the percentage of dependent variables total variance, to be expressed
from the independent variable in the model. While regression models were derived
to predict Fe and Fe_2O_3 using SPSS software, transformed, linear, logarithmic,
quadratic and cubic models were evaluated to find the best-fitting model. Calibration
and validation of regression models were evaluated by calculated R^2 and RMSE using
statistical software, and the maps of soil properties were generated using ArcGIS software at the same scale and an equal number of classes to allow easier comparison.
The interpolated maps of Fe and Fe_2O_3 were generated. The schematic diagram of
general methodology steps can be seen in Fig. 5.2.

5.5.2   Results

5.5.2.1   Fe Laboratory Assessment

The lowest and highest values of Fe were 1.89 and 49.40 gm kg^{-1} and the lowest
and highest data for the Fe_2O_3 were 0.22 and 18 gm kg^{-1}, respectively. Generally,
Fe was found in high concentrations at most of the soil samples. Abdul Hameed and
Athmar (2012) pointed out that there is no limit to the concentration of Fe in soil
because it is abundant in soil.

5.5.2.2 Soil Spectral Reflectance Pattern

The study area, which represented by 39 soil samples, was categorized into three zones (Fig. 5.3) in order to identify the correlations between the spectral curves and the properties of the soil samples. The spectral responses of soil samples can be seen in Fig. 5.4 that are characterized qualitatively by observing the positive and negative peaks, which occur at specific wavelengths. Due to the presence of the same spectrally-active properties in all samples from various locations, the spectra of all soil samples were similar (Gholizadeh et al. 2015a, 2018b). All samples showed reflectance baseline values with absorption features at 1400 and 1900 nm, which may be due to the O–H of hygroscopic water. At 2300 nm, this can be assigned to clay minerals and OM (Araujo et al. 2014). However, some differences in reflectance values and albedo intensity of different samples were evident, which can be attributed to the soil moisture. Soil moisture has a significant effect on the composition and amount of reflected and emitted energy from a soil surface that reduces the reflectance over the entire spectrum (Nocita et al. 2013).

The obtained results of three zones showed that the zone 3 had the highest reflectance values (0–0.6), in the zone 1 the reflectance range was between 0 and 0.35 (Fig. 5.4). The absorption between 450 and 850 nm referred to increase in Fe_2O_3; soils with higher contents of free Fe showed more pronounced concavity. In general, the concavity centered at 850 nm. This spectral effect of Fe is observed because of the reduced content of OM in the samples. The presence of OM in soil causes

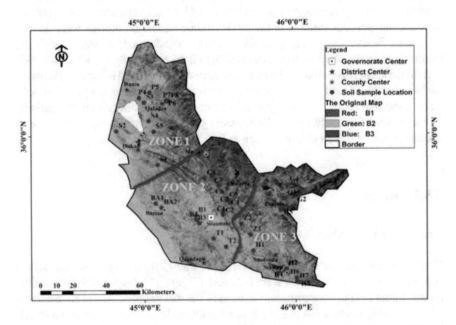

Fig. 5.3 Map of the three zones of the study area (Mustafa 2015)

Fig. 5.4 Soil samples spectra based on different zones (Mustafa 2015)

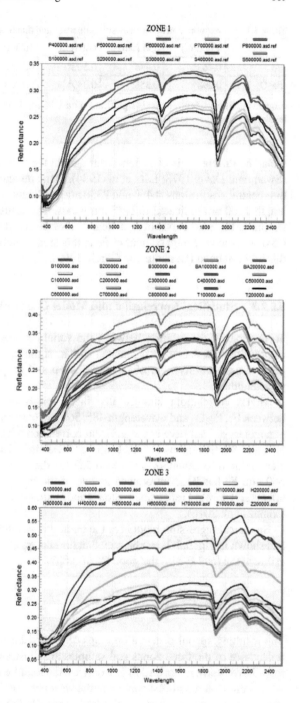

Table 5.1 Correlation matrix between soil properties and their spectral reflectance (Mustafa 2015)

	485 (nm)	560 (nm)	660 (nm)	1660 (nm)	1665 (nm)	2215 (nm)
Fe	0.463[a]	0.445[b]	0.562[b]	0.221 n.s.	0.220 n.s.	0.283 n.s.
Fe_2O_3	0.298[a]	0.353[a]	0.349[a]	0.145 n.s.	0.144 n.s.	0.203 n.s.

n.s. non-significant, [a]correlation is significant at the 0.05 level, [b]correlation is significant at the 0.01 level

a marked difference in reflectance throughout the 500–1100 nm wavelength region (Swain and Davis 1978; Latz et al. 1984). Dematte and Nanni (2003) Showed that the spectral absorptions at 460 and 2330 nm are diagnostic of soil Fe_2O_3 while wavelength 1100 nm indicated to Fe^{2+} ion bands. The results of zone 2 were relatively similar to the zone 1; however, a clear difference could be seen in spectra of sample C5 due to the very high content of Fe in this area, which influences the concavity on this spectral band (Dematte and Garcia 1999).

5.5.2.3 Statistical Correlation and Model Calibration

To understand the relation between the variables, the Pearson correlation coefficients of the raw data were calculated with its significance values (Table 5.1). Strong positive correlations were observed between soil properties and spectra (selected wavelengths).

Matrix correlations showed that there was a significant positive correlation between Fe, Fe_2O_3 and wavelength 485, 560 and 660 nm (VIS range); however, these parameters did not have any significant correlation with the selected wavelengths in NIR and SWIR regions (1660, 1665 and 2215 nm). The highest correlation coefficient value of parameters was r = 0.562 for the correlation of Fe and wavelength 660 nm. Fe has been introduced as one of the most important parameters, which affects soil spectral reflectance (Stoner and Baumgardner 1981; Dematte et al. 2007; Gomez et al. 2016).

Statistical regression model for the Fe at the wavelength 660 nm was developed and shown in Fig. 5.5. It can be seen that the spectra at the wavelength 660 nm could estimate the total F with the good R^2 of 0.837.

5.5.2.4 Prediction Maps

The resulting spatial distribution maps of Fe and Fe_2O_3 derived from the spectral reflectance of the three zones soil samples are illustrated in Figs. 5.6 and 5.7. The maps have been categorized in five classes, the total Fe map (Fig. 5.6) shows that Fe was recorded at the lowest content at the west parts of the study area but the highest content was at the east parts of the study area. On the other hand, the lowest value of Fe_2O_3 was recorded at the west parts of the study area and highest was at the

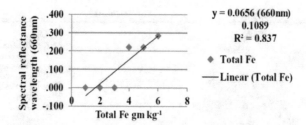

Fig. 5.5 The relation between total Fe and spectral reflectance at the wavelength 660 nm (Mustafa 2015)

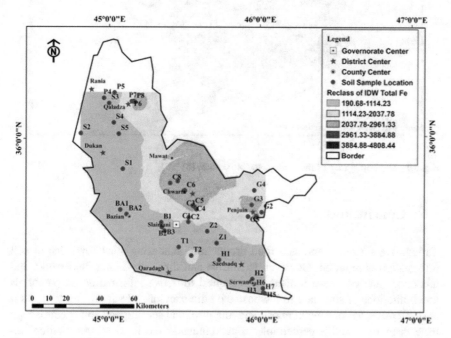

Fig. 5.6 The total Fe map of the study area (Mustafa 2015)

south parts as shown in Fig. 5.7. The results highlighted that the predicted and the measured results defined the reality of nature, the spatial distribution maps had the highest value of total Fe and Fe_2O_3 in the sites Dere and Kani Sard3 (P_8 and C_5), respectively.

Overall, the statistical accuracy and distribution maps obtained using laboratory spectroscopy measurements indicated that in the current study area, laboratory PSS could be considered as a reliable technique for the prediction of soil Fe and Fe_2O_3 using spectra from soil samples (Mustafa 2015).

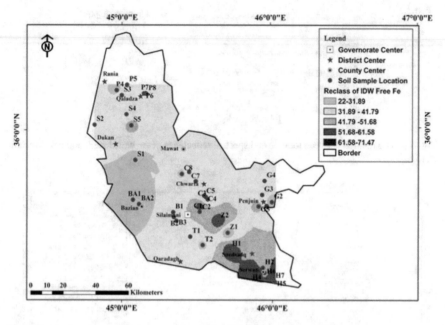

Fig. 5.7 The Fe$_2$O$_3$ map of the study area (Mustafa 2015)

5.6 Conclusions

To improve sustained and safe food production, increasing our knowledge of soil information is required. However, expensive and time-consuming field works and laboratory analysis have limited it. In a quest to derive information on potentials and limitations of soils and to increase the efficiency of soil survey and to make it cost-effective, various techniques including proximal sensing and soil spectroscopy have been used and a perceptible improvement in the level of information has been achieved. Extensive research has demonstrated that obtained spectra from VIS–NIR–SWIR spectroscopy carry unique and important information related to many of soil attributes. The most obvious ones are clay content, Fe, soil OC, soil OM and soil water. Even soil contaminants, which are not spectrally-active, have been predicted using their inter-correlation with soil properties that are spectrally-active in this region.

Finding suitable data preprocessing and calibration strategies for the application of VIS–NIR–SWIR spectroscopy are dramatically significant. This chapter discussed developments in preprocessing and calibration algorithms and concluded that the selection of the preprocessing and calibration strategies influence the final accuracy and therefore should be highly considered. Inaccuracy and uncertainties occur at the different steps in the prediction procedure: soil sampling, spectral collection, predictive model building (choosing the number of training samples and model parameters) and final spatial prediction. The accuracy of the resulting models can be insufficient for the planned purpose, therefore it is important to be able to assess the accuracy of the process. This will be a significant step toward helping users of soil data and models, address and manage uncertainty instead of ignoring it. The chapter also provided a short study as a case study on employing soil spectroscopy in Sulaimani, Iraqi Kurdistan Region. The spectroscopy technique was used for monitoring and mapping total Fe and Fe_2O_3 rich soils of some sites in Sulaimani governorate, Iraqi Kurdistan Region as any such a research had not been conducted in the region, and there was no up-to-date map that present and show soil Fe and Fe_2O_3 in Sulaimani governorate. According to the study, the accuracy and distribution maps derived from laboratory spectroscopy measurements proved the capability of laboratory proximal sensing prediction for assessing the soil Fe and Fe_2O_3 under the study area.

5.7 Recommendations and Future Aspects

There are many questions that remain to be answered before we can move this technology from local research to global use and from samples in a single study to broad, practical calibrations. It should be noted that there are still some limitations in using proximal spectroscopy data, which are mainly associated with the lack of standardization and an agreed-upon protocol. Therefore, adherence to a consistent protocol is needed, which will improve the accuracy and the comparability of results. Internal standards, a standard protocol and controlled conditions are a few of the factors that can assist in researchers' sharing and compare soil spectra worldwide. A better understanding of the complexity of soil and the physical basis for soil reflection as well as applications of spectra for making inferences about soil fertility and function, and prediction of soil fertility are the other recommended and required steps. More importantly, to provide more effective use of spectroscopic data, development of SSLs with additional details about the sampling area is required. Also, detail chemical and physical data need to be recorded. Therefore, further work is needed to deal with the abovementioned limitations and issues.

References

Abdul Hameed MJ, Athmar AM (2012) Heavy metal contaminations in urban soil within Baghdad city, Iraq. J Environ Prot 4:72–82
Adeline K, Gomez C, Gorretta N, Roger JM (2017) Predictive ability of soil properties to spectral degradation from laboratory Vis-NIR spectroscopy data. Geoderma 288:143–153
Aichi H, Fouad Y, Walter C, Viscarra Rossel RA, Zohra Lili C, Mustafa S (2009) Regional predictions of soil organic carbon content from spectral reflectance measurements. Biosystem Eng 104(3):442–446
Araujo SR, Wetterlind J, Dematte JAM, Stenberg B (2014) Improving the prediction performance of a large tropical Vis-NIR spectroscopic soil library from Brazil by clustering into smaller subsets or use of data mining calibration techniques. Eur J Soil Sci 65:718–729

Bellinaso H, Dematte JAM, Araujo SR (2010) Spectral library and its use in soil classification. The Revista Brasileira de Ciencia Solo 34:861–870

Bellon-Maurel V, McBratney AB (2011) Near infrared (NIR) and mid infrared (MIR) spectroscopic techniques for assessing the amount of carbon stock in soils-critical review and research perspectives. Soil Biol Biochem 43(7):1398–1410

Ben-Dor E, Banin A (1995a) Near-infrared analysis as a rapid method to simultaneously evaluate several soil properties. Soil Sci Soc Am J 59:364–372

Ben-Dor E, Banin A (1995b) Near infrared analysis (NIRA) as a method to simultaneously evaluate spectral featureless constituents in soils. Soil Sci 159:259–270

Ben-Dor E, Inbar Y, Chen Y (1997) The reflectance spectra of organic matter in the visible near-infrared and short wave infrared region (400-2500 nm) during a controlled decomposition process. Remote Sens Environ 61:1–15

Ben-Dor E, Ong C, Lau IC (2015) Reflectance measurements of soils in the laboratory: standards and protocols. Geoderma 245–246:112–124

Beven KJ (2000) On model uncertainty, risk and decision making. Hydrol Process 14(14):2605–2606

Bouma J (1997) Soil environmental quality: a European perspective. J Environ Qual 26(1):26–31

Bowers S, Hanks RJ (1965) Reflectance of radiant energy from soils. Soil Sci 100:130–138

Bremner JM, Mulvaney CS (1982) Nitrogen-total. In: Page AL (ed) Methods of soil analysis, Part 2, Chemical and microbiological properties. Agron Monogr 9. ASA, Madison, WI, pp 595–622

Brodsky L, Vasat R, Klement A, Zadorova T, Jaksık O (2013) Uncertainty propagation in VNIR reflectance spectroscopy soil organic carbon mapping. Geoderma 199:54–63

Brown DJ (2004) Knowledge, uncertainty and physical geography: towards the development of methodologies for questioning belief. Trans Inst British Geogr 29(3):367–381

Brown DJ (2007) Using a global vis–NIR soil-spectral library for local soil characterization and landscape modeling in a 2nd-order Uganda watershed. Geoderma 140:444–453

Brown DJ, Heuvelink GBM (2005) Assessing uncertainty propagation through physically based models of soil water flow and solute transport. In: Anderson MG (ed) Encyclopedia of hydrological sciences. Chichester, UK, John Wiley, 2(79), 1181–1196

Brown DJ, Shepherd KD, Walsh MG, Mays MD, Reinsch TG (2006) Global soil characterization with VNIR diffuse reflectance spectroscopy. Geoderma 132:273–290

Brunet D, Barthes BG, Chotte JL, Feller C (2007) Determination of carbon and nitrogen contents in Alfisols, Oxisols and Ultisols from Africa and Brazil using NIRS analysis: effects of sample grinding and set heterogeneity. Geoderma 139(1–2):106–117

Cecillon L, Barthes BG, Gomez C, Ertlen D, Genot V, Hedde M, Stevens A, Brun JJ (2009) Assessment and monitoring of soil quality using near-infrared reflectance spectroscopy (NIRS). Eur J Soil Sci 60(5):770–784

Chen Q, Guo Z, Zhao J, Ouyang Q (2012) Comparisons of different regressions tools in measurement of antioxidant activity in green tea using near infrared spectroscopy. J Pharm Biomed Anal 60:92–97

Christy CD (2008) Real-time measurement of soil attributes using on-the-go near infrared reflectance spectroscopy. Comput Electron Agriculture 61(1):10–19

Chu XL, Yuan HF, Lu WZ (2004) Progress and application of spectral data pretreatment and wavelength selection methods in NIR analytical technique. Progress Chemistry 16:528–542

Clark RN (1999) Spectroscopy of rocks and minerals and principles of spectroscopy. In: Rencz N (ed) Remote sensing for the earth sciences: manual of remote sensing. Wiley, New York, pp 3–52

Clark RN, King TVV, Klejwa M, Swayze G, Vergo N (1990) High spectral resolution reflectance spectroscopy of minerals. J Geophys Res 95:12653–12680

Clark RN, Roush TL (1984) Reflectance spectroscopy: quantitative analysis techniques for remote sensing application. J Geophys Res 89:6329–6340

Cohen MJ, Shepherd KD, Walsh MG (2005) Empirical reformulation of the universal soil loss equation for erosion risk assessment in a tropical watershed. Geoderma 124(3–4):235–252

Cohen MJ, Dabral S, Graham WD, Prenger JP, DeBusk WF (2006) Evaluating ecological condition using soil biogeochemical parameters and near infrared reflectance spectra. Environ Monit Assess 116(1–3):427–457

Couteaux MM, Berg B, Rovira P (2003) Near infrared reflectance spectroscopy for determination of organic matter fractions including microbial biomass in coniferous forest soils. Soil Biol Biochem 35(12):1587–1600

Cozzolino D, Moron A (2006) Potential of near-infrared reflectance spectroscopy and chemometrics to predict soil organic carbon fractions. Soil Tillage Res 85(1–2):78–85

Dalal RC, Henry RJ (1986) Simultaneous determination of moisture, organic carbon and total nitrogen by near-infrared reflectance spectroscopy. Soil Sci Soc Am J 50:120–123

Dematte JAM, Garcia GJ (1999) Alteration of soil properties through a weathering sequence as evaluated by spectral reflectance. Soil Sci Soc Am J 63:327–342

Dematte JAM, Nanni MR (2003) Weathering sequence of soils developed from basalt as evaluated by laboratory (IRIS), airborne (AVIRIS) and orbital (TM) sensors. Int J Remote Sens 24:4715–4738

Dematte JAM, Nanni MR, Formaggio AR, Epiphanio JCN (2007) Spectral reflectance for the mineralogical evaluation of Brazilian low clay activity soils. Int J Remote Sens 28(20):4537–4559

Duckworth J (2004) Mathematical data preprocessing. In: Roberts CA, Workman J, Reeves JB (eds) Near-infrared spectroscopy in agriculture. Madison, ASA-CSSA-SSSA, pp 115–132

Dwivedi RS (2017) Remote sensing of soils. Springer

Etabi W (1982) Iron ore deposits and occurrences of Iraq. *Geosurvey* International Reports, 1324

Ge Y, Morgan CLS, Thomasson JA, Waiser T (2007) A new perspective to near infrared reflectance spectroscopy: a wavelet approach. Trans ASABE 50(1):303–311

Geladi P, Kowalski B (1986) Partial least squares regression: a tutorial. Anal Chim Acta 185:1–17

Gholizadeh A, Amin MSM, Saberioon MM, Boruvka L (2013a) Visible and near-infrared reflectance spectroscopy to determine chemical properties of paddy soils. J Food Agric Environ 11(2):859–866

Gholizadeh A, Boruvka L, Saberioon MM, Vasat R (2013b) Visible, near-infrared, and mid-infrared spectroscopy applications for soil assessment with emphasis on soil organic matter content and quality: State-of-the-art and key issues. Appl Spectrosc 67:1349–1362

Gholizadeh A, Boruvka L, Vasat R, Saberioon MM (2015a) Comparing different data preprocessing methods for monitoring soil heavy metals based on soil spectral features. Soil Water Res 10(4):218–227

Gholizadeh A, Boruvka L, Vasat R, Saberioon MM, Klement A, Kratina J, Tejnecky V, Drabek O (2015b) Estimation of potentially toxic elements contamination in anthropogenic soils on a brown coal mining dumpsite by reflectance spectroscopy: a case study. PLoS ONE 10(2):e0117457

Gholizadeh A, Carmon N, Ben-Dor E, Boruvka L (2017) Agricultural soil spectral response and properties assessment: effects of measurement protocol and data mining technique. Remote Sensing 9:1078

Gholizadeh A, Saberioon MM, Ben-Dor E, Boruvka L (2018a) Monitoring of selected soil contaminants using proximal and remote sensing techniques: background, state-of-the-art and future perspective. Critical Rev Environ Sci Technol 48(3):243–278

Gholizadeh A, Zizala D, Saberioon MM, Boruvka L (2018b) Soil organic carbon and texture retrieving and mapping using proximal, airborne and Sentinel-2 spectral imaging. Remote Sens Environ 218:89–103

Gholizadeh A, Saberioon MM, Boruvka L, Vasat R (2016) A memory-based learning approach as compared to other data mining algorithms for the prediction of soil texture using diffuse reflectance spectra. Remote Sensing 8:341

Gomez C, Gholizadeh A, Boruvka L, Lagacherie P (2016) Using legacy data for correction of soil surface clay content predicted from VNIR/SWIR hyperspectral airborne images. Geoderma 276:84–92

Haaland DM, Thomas EV (1988) Partial least-squares methods for spectral analyses: 1. Relation to other quantitative calibration methods and the extraction of qualitative information. Anal Chem 60:1193–1202

Henderson TL, Baumgardner MF, Franzmeier DP, Stott DE, Coster DC (1992) High dimensional reflectance analysis of soil organic matter. Soil Sci Soc Am J 56(3):865–872

Heuvelink GBM, Brown JD, Van Loon EE (2007) A probabilistic framework for representing and simulating uncertain environmental variables. Int J Geogr Inf Sci 21(5):497–513

Hunt GR (1982) Spectroscopic properties of rock and minerals. In: Stewart CR (ed) Handbook of physical properties rocks. CRC Press, pp. 295

Jensen JR (2007) Remote sensing of the environment: an Earth resource perspective. Prentice Hall, Upper Saddle River, New Jersey

Karlen DL, Mausbach MJ, Doran JW, Cline RG, Harris RF, Schuman GE (1997) Soil quality: a concept, definition, and framework for evaluation. Soil Sci Soc Am J 61(1):4–10

Kemper T, Sommer S (2002) Estimate of heavy metal contamination in soils after a mining accident using reflectance spectroscopy. Environ Sci Technol 36:2742–2747

Kokaly RF, Despain DG, Clark RN, Livo KE (2003) Mapping vegetation in Yellowstone National Park using spectral feature analysis of AVIRIS data. Remote Sens Environ 84:437–456

Kooistra L, Wehren R, Leuven RSE, Buydens LMC (2001) Possibilities of visible-near-infrared spectroscopy for the assessment of soil contamination in river flood plains. Anal Chim Acta 446:97–105

Kuang B, Mouazen AM (2013) Effect of spiking strategy and ratio on calibration of on-line visible and near infrared soil sensor for measurement in European farms. Soil & Tillage Research 128:125–136

Latz K, Weismiller RA, Van Scoyoc GE, Baumgardner MF (1984) Characteristic variations in spectral reflectance of selected eroded Alfisols. Soil Sci Soc Am J 48:1130–1134

Lindsay WL (1979) Chemical equilibria in soils. Wiley Inter Science, New York

Matthews HL (1972) Application of multispectral remote sensing and spectral reflectance patterns to soil survey research. Ph.D. Dissertation, Pennsylvania State University, College Station, PA, USA

Mehra OP, Jackson ML (1953) Iron oxide removal from soils and clays by a dithionite citrate system buffered with sodium bicarbonate. Madison, WI, USA

Miller CE (2001) Chemical principles of near-infrared technology. In: Williams P, Norris K (eds) Near-infrared technology in the agricultural and food industries. The American Association of Cereal Chemists Inc, St Paul, MN, USA, pp 19–37

Mouazen AM, De Baerdemaeker J, Ramon H (2005) Towards development of on-line soil moisture content sensor using a fibre-type NIR spectrophotometer. Soil Tillage Research 80:171–183

Mouazen AM, Kuang B, De Baerdemaeker J, Ramon H (2010) Comparison among principal component, partial least squares and back propagation neural network analyses for accuracy of measurement of selected soil properties with visible and near infrared spectroscopy. Geoderma 158:23–31

Mustafa BM (2015) Mapping and detection of iron oxides using Geoinformatics in Sulaimani Governorate, Kurdistan region, Iraq. Unpublished Master of Science (M.Sc.) thesis, College of Agriculture, Salahaddin University, Erbil, Kurdistan Region, Iraq

Nanni MR, Dematte JAM (2006) Spectral reflectance methodology in comparison to traditional soil analysis. Soil Sci Soc Am J 70(2):393–407

Nocita M, Stevens A, Noon C, Van Wesemael B (2013) Prediction of soil organic carbon for different levels of soil moisture using Vis-NIR spectroscopy. Geoderma 199:37–42

Pandit CM, Filippelli GM, Li L (2010) Estimation of heavy-metal contamination in soil using reflectance spectroscopy and partial least-squares regression. Int J Remote Sens 31(15):4111–4123

Reeves JB (2010) Near- versus mid-infrared diffuse reflectance spectroscopy for soil analysis emphasizing carbon and laboratory versus on-site analysis: where are we and what needs to be done? Geoderma 158(1–2):3–14

Reeves JB, McCarty GW, Meisinger JJ (1999) Near infrared reflectance spectroscopy for the analysis of agricultural soils. J Near Infrared Spectrosc 7:179–193

Ren HY, Zhuang DF, Singh AN, Pan JJ, Qid DS, Shi RH (2009) Estimation of As and Cu contamination in agricultural soils around a mining area by reflectance spectroscopy: a case study. Pedosphere 19:719–726

Rinnan A, Van den Berg F, Engelsen SB (2009) Review of the most common pre-processing techniques for near-infrared spectra. Trends Anal Chem 28:1201–1222

Sanches ID, Souza Filho CR, Magalhaes LA, Quiterio GCM, Alves MN, Oliveira WJ (2013) Unravelling remote sensing signatures of plants contaminated with gasoline and diesel: an approach using the red edge spectral feature. Environ Pollut 174:16–27

Savitzky A, Golay MJE (1964) Smoothing and differentiation of data by simplified least squares procedures. Anal Chem 36:1627–1639

Shepherd KD, Walsh MG (2002) Development of reflectance spectral libraries for characterization of soil properties. Soil Sci Soc Am J 66:988–998

Sherman DM, Waite TD (1985) Electronic spectra of Fe^{3+} oxides and oxyhydroxides in the near infrared to ultraviolet. Am Miner 70:1262–1269

Shi T, Wang J, Chen Y, Wu G (2016) Improving the prediction of arsenic contents in agricultural soils bycombining the reflectance spectroscopy of soils and rice plants. Int J Appl Earth Obs Geoinf 52:95–103

Song Y, Li F, Yang Z, Ayoko GA, Frost RL, Ji J (2012) Diffuse reflectance spectroscopy for monitoring potentially toxic elements in the agricultural soils of Changjiang river delta, China. Applid Clay Science 64:75–83

Stenberg B, Nordkvist E, Salomonsson L (1995) Use of near infrared reflectance spectra of soils for objective selection of samples. Soil Sci 159:109–114

Stenberg B, Viscarra Rossel RA, Mouazen AM, Wetterlind J (2010) Visible and near infrared spectroscopy in soil science. In: Sparks DL (ed) Advances in agronomy. Burlington, VT, Elsevier, 107, 163–215

Stoner ER, Baumgardner MF (1981) Characteristic variations in reflectance of surface soils. Soil Sci Soc Am J 45:1161–1165

Swain PH, Davis SM (1978) Remote sensing: the quantitative approach. McGraw-Hill, New York

Van Waes C, Mestdagh I, Lootens P, Carlier L (2005) Possibilities of near infrared reflectance spectroscopy for the prediction of organic carbon concentrations in grassland soils. J Agric Sci 143(6):487–492

Viscarra Rossel RA, Behrens T (2010) Using data mining to model and interpret soil diffuse reflectance spectra. Geoderma 158:46–54

Viscarra Rossel RA, Behrens T, Ben-Dor E, Brown DJ, Dematte JAM, Shepherd KD, Shi Z, Stenberg B, Stevens A, Adamchuk V, Aichi H, Barthes BG, Bartholomeus HM, Bayer AD, Bernoux M, Bottcher K, Brodsky L, Du CW, Chappell A, Fouad Y, Genot V, Gomez C, Grunwald S, Gubler A, Guerrero C, Hedley CB, Knadel M, Morras HJM, Nocita M, Ramirez-Lopez L, Roudier P, Rufasto Campos EM, Sanborn P, Sellitto VM, Sudduth KA, Rawlins BG, Walter C, Winowiecki LA, Hong SY, Ji W (2016) A global spectral library to characterize the world's soil. Earth Sci Rev 155:198–230

Viscarra Rossel RA, Bouma J (2016) Soil sensing: a new paradigm for agriculture. Agric Syst 148:71–74

Viscarra Rossel RA, Jeon YS, Odeh IOA, McBratney AB (2008) Using a legacy soil sample to develop mid-IR spectral library. Aust J Soil Res 46(1):1–16

Viscarra Rossel RA, Lark RM (2009) Improved analysis and modelling of soil diffuse reflectance spectra using wavelets. Eur J Soil Sci 60:453–464

Viscarra Rossel RA, McBratney AB (1998a) Soil chemical analytical accuracy and costs: Implications from precision agriculture. Aust J Exp Agric 38:765–775

Viscarra Rossel RA, McBratney AB (1998b) Laboratory evaluation of a proximal sensing technique for simultaneous measurement of clay and water content. Geoderma 85:19–39

Viscarra Rossel RA, McBratney AB, Minasny B (2010a) Proximal soil sensing, Springer

Viscarra Rossel RA, McKenzie NJ, Grundy MJ (2010b) Using proximal soil sensors for digital soil mapping. In: Boettinger JL, Howell DW, Moore AC, Hartemink AE, Kienast-Brown S (eds) Digital soil mapping, progress in soil science. Springer, Dordrecht

Viscarra Rossel RA, Walvoort DJJ, McBratney AB, Janik LJ, Skjemstad JO (2006) Visible, near infrared, mid infrared or combined diffuse reflectance spectroscopy for simultaneous assessment of various soil properties. Geoderma 131:59–75

Wetterlind J, Stenberg B (2010) Near-infrared spectroscopy for within field soil characterization: Small local calibrations compared with national libraries spiked with local samples. Eur J Soil Sci 61(6):823–843

Wetterlind J, Stenberg B, Soderstrom M (2008) The use of near infrared (NIR) spectroscopy to improve soil mapping at the farm scale. Precision Agric 9:57–69

Whiting ML, Li L, Ustin SL (2004) Predicting water content using Gaussian model on soil spectra. Remote Sens Environ 89:535–552

Wu Y, Chen J, Wu X, Tian Q, Ji J, Qin Z (2005) Possibilities of reflectance spectroscopy for the assessment of contaminant elements in suburban soils. Appl Geochem 20:1051–1059

Xie X, Pan XZ, Sun B (2012) Visible and near-infrared diffuse reflectance spectroscopy for prediction of soil properties near a Copper smelter. Pedosphere 22:351–366

Chapter 6
Proximal Soil Sensing Applications in Soil Fertility

Qassim A. Talib Alshujairy and Nooruldeen Shawqi Ali

Abstract Soil samples were collected from two locations: Samawa and Rumetha from Al-Muthana governorate in the south of Iraq. The samples were divided into two sets to be used as a calibration and validation set. Vis-NIR reflectance (350–2500 nm) and GIS-Kriging were used in combination with Partial Least Square (PLS) to predict soil available Phosphorus and total N. For total N, only two regions reported higher determination coefficient R^2 and lower Root Mean Square Error (RMSE) than the other wavelength regions. PLS calibration models yielded an R^2 of 0.96 and 0.97 for Rumetha and 0.87 and 0.94 for Samawa location in bands at 500–600 and 800–1000 nm, respectively. The validation dataset from both study locations were used to predict the new unknown soil samples based on NIRS and GIS-Kriging models. The GIS-Kriging models were unfavorable predicted with an Q^2 of 0.28 between laboratory-measured and predicted total N values for Rumetha and 0.43 for Samawa location. While NIRS- based validation models achieved highly predictive power with an R^2v of 0.84 between laboratory-measured and predicted total N values for Rumetha and 0.85 for Samawa location. These results reveal an extreme decrease in model predictive ability when shifting from NIR Spectroscopy method to GIS-Kriging. According to the results of this study, three wavelength regions were reported as the main sensitive bands for soil available P. The best prediction ability was achieved for Rumetha location at 1400–1600 nm with an R^2 of 0.85, lowest RMSE of 1.405, and lowest standard deviation of 1.577 and for Samawa location at 900–1000 nm with an R^2 of 0.81, RMSE of 2.666 and lowest standard deviation of 2.879. At wavelength region 2100–2200 nm, both studied locations showed lowest R^2, highest RMSE, and highest standard deviation. The capability of the NIRS-based and GIS-Kriging prediction models were evaluated by using cross-validation values Q^2 and R^2 between measured and predicted soil available P of each model. The selection principle parameters showed the best prediction by NIRS-models with an R^2 of 0.79 for Rumetha and 0.75 for Samawa location. While the prediction abil-

Q. A. T. Alshujairy (✉)
College of Agriculture, University of Muthana, Samawah, Iraq
e-mail: qassimtalib@mu.edu.iq

N. S. Ali
College of Agriculture, University of Baghdad, Baghdad, Iraq

© Springer Nature Switzerland AG 2020
A. M. F. Al-Quraishi and A. M. Negm (eds.),
Environmental Remote Sensing and GIS in Iraq, Springer Water,
https://doi.org/10.1007/978-3-030-21344-2_6

119

ity of GIS-Kriging models were in worst with an Q^2 of 0.17 for Samawa location and reasonable with an Q^2 of 0.58 for Rumetha location. This empirical result is an evidence of the superiority of NIRS-based models for prediction soil available P over the GIS-Kriging models. To simulate soil fertilization conditions, 15 soil samples from each location have mixed with Urea fertilizer in the rates ranged from 0.1 to 0.8 g kg^{-1} soil (90–720 kg N ha^{-1}) and Monopotassium Phosphate (MPP) fertilizer in the rates ranged from 160 to 700 kg P_2O_5 ha^{-1} (70–300 kg P ha^{-1}). Based upon values of coefficients between lab-measured and absorbance spectra R^2, four wavelength regions were highly correlated with total N. Among the four selected sensitive wavelengths for total N, the highest Q^2 were observed at 2100–2200 nm with a Q^2 of 0.97 for Rumetha and 0.95 for Samawa location. Three main bands for prediction soil available P are described in this study. The capability of NIRS-based models was best at 900–1000 nm with an Q^2 of 0.75 for Samawa and 0.91 for the Rumetha location. The yielded quality parameter values were at best successfully models to predict total N and soil available P and well suited for a large variety of low to high concentrations. These results indicate that, NIR Spectroscopy is an effective tool for a rapid assessment of soil information under field conditions, through which decision on fertilizer requirements can be based (under the conditions of this study).

Keywords Proximal soil sensing · Spectroradiometer

6.1   Introduction

Up-to-date and accurate information is needed to improve our knowledge of local field conditions. A site-specific management is of high interest to avoid under or over-fertilization, reduce agricultural input costs and to provide sustainability strategy. Most of the soil properties are both spatially and temporally variable over a short distance (Harris et al. 1996; Karlen et al. 1997) all of these variations need to be detected in its status at farm or field scales to apply the Precision Fertilization concept. This option called as fertilization management, which is not taken into account in conventional agriculture practices. In recent years, there is much focus on developing accurate inexpensive soil analysis methods (Bricklemyer et al. 2005).

Vis-NIR Spectroscopy is a viable alternative for rapidly analyzing soil properties in comparison with conventional methods. There are many reasons for this attention, for example: several properties can be sensed in a single scan, no chemical is required, easily use in both laboratories and situ and keeping sample pretreatment to a minimum (Wetterlind et al. 2008).

In addition to the application of Vis-NIR spectroscopy in different fields of study, the technique is considered as a useful tool for soil analysis. Soil Nitrogen, Phosphorus and Potassium are the main required nutrients to derive soil fertility or health index (Idowu et al. 2008). Due to this fact, methods for determining their levels attract significant interest in developing the simplicity and rapidity of determination techniques (Stenberg et al. 2010). Vis-NIR spectroscopy method is based on light absorption by different materials in a soil sample. The main complexity of the method lies in the interpretation of absorption/reflectance information (Tiruneh 2014). The process of extracting appropriate information from such wide and complex spectra is called as Chemometrics (Cao 2013). Partial Least Square (PLS) is the most commonly used tools in Vis-NIR Spectroscopy to estimate the correlation between spectral wavelength and chemically measurement of the studied property (Burns and Ciurczak 1992). Calibration model performance can be evaluated by using randomly with held dataset as a validation set (Brown et al. 2006). The validation step is crucial to evaluate the potential of the calibrated model to predict future unknown samples. A number of different statistically measurements such as R^2, Q2, RMSE and Standard deviation are used as a main parameters to assess the quality of the prediction model (Dunn et al. 2002).

Geostatistics is another approach of describing soil variability based on the regionalized variable theory. GIS-Kriging as a one of Geostatistics method makes it possible to determine values at the un-sampled location by using the spatial correlation between estimated points to predict un-estimated samples. Dense data locations with a uniformly distribution will get good predictive model. While as data with large gabs between each other will get an unreliable estimate. However, Kriging technique is widely used in soil nutrients studies to estimate levels in difficult to measure samples based on more easily determination primary soil samples (Qu et al. 2006).

6.2 Background: Near-Infrared Spectroscopy Historical Review

The Near Infrared Reflection was discovered by Frederick William Herschel in 1800. The German-born British scientist referred to his discovery as "calorific rays" (Hindle 1997; Pasquini 2003). Sir Herschel found that the temperature increased clearly beyond the red end region, which is now identified as near-infrared (Reich 2005). Later this region described with Greek prefix "infra" to name the first non-visible portion of the electromagnetic spectrum. In 1900, Coblentz validated the value of using infrared to identify the organic functional groups, this finding was the first absorbance spectra of pure organic substances. In the 1960s, Karl Norris searched for new methods to determine moisture in agricultural products (Hart et al. 1961). Norris, who is named "NIR father", published attractive information for both theoretical and instrumental progress in 1965 (Pasquini et al. 2000). After this work, NIR spectroscopy was ignored for a long time. This period was described in a paper "Near-Infrared Analysis-Sleeper among Spectroscopic Techniques", published by Wetzel in 1983 (Wetzel 1983). Since 1980s NIR spectral region had been developed, where greater than 1000 papers were published with titles transferred NIR spectral from sleeping techniques to the sunrise of spectroscopy (Blanco and Villarroya 2002). The potential of using NIR spectroscopy is mainly associated with instrumental (spectrophotometer) developments, data acquisition and their handling (computer), which provided the tools for obtaining information and facilitated its use (Pasquini 2003). In recent years, NIR spectroscopic technology gained wide acceptance because of its ability to provide a chemical free, rapid analysis, real-time measurement at field site as well as its cost-effectiveness. On the other hand, the advanced mathematical and statistical analysis applied to develop NIR spectroscopy into a useful technique to produce an ideal analytical method.

6.2.1 The Potential of Vis-NIR Spectroscopy and Its Applications to Soil Analysis

Over the last 30 years, there has been a growing interest in VIS-NIR-Spectroscopy as a potential analytical technique for studying materials in many different fields such as Agriculture, food, textiles, polymers, wool, biomedical and pharmaceutical (Blanco and Villarroya 2002; Niemoller and Behmer 2008; Guerrero et al. 2010). Vis-NIR-Spectroscopy has attracted much attention from soil scientist as a promising technique to determine a lot of soil properties from a single scan, reduce sample preparation, diminish the hazardous of using chemicals, and measurement can be taken both in laboratory and field with a few seconds (Ben-Dor and Banin 1995a, b; Chang et al. 2001; Viscarra Rossel et al. 2006). The various soil spectrum could be obtained according to the constituents of soils, this could be expressed as a spectral characteristic shape. Absorption in the visible region is mainly caused by electrons transition between molecular orbits. While the vibration in molecular bonds is mainly occurs in near-infrared region (Wetterlind et al. 2013). The frequencies at which light are absorbed match the difference between two energy levels and are displayed in %R for analytical purposes (Miller 2001).

Soil reflectance is mostly influenced by sample chemical constituent, functional groups and bonds are the main reasons for broad and overlapping absorption features of soil spectra (Miller 2001). Soil reflectance is mostly influenced by Soil properties such as particle size, surface structure, and water films on the soil surface (Twomey et al. 1986).

Generally, this review intends to provide important information on the use of Vis-NIR-Spectroscopy in soil analysis. To this end, a review on issues related to the essential soil attributes such as soil texture, soil organic matter, minerals, water, and fertility with focusing on nutrients (N, P, and K) will be provided in the following sections.

6.2.1.1 Soil Minerals

Soil minerals represent half of the soil volume and determine the range of soil properties such as texture, structure, and CEC which affecting other soil properties (Schulze 2002). Valuable information on soil minerals can be detected in the vis-near infrared region. Minerals absorb photons through electronic and vibration processes, and it is this variety of processes and wavelength at which light is absorbed that are utilized to determine the chemistry of a mineral under study (Clark 1999). In the spectra of minerals, most of the electronic process is due to unfilled electrons shells of transition elements (Ni, Cr, Co, Fe, etc.). The splitting of the d orbital energy state when the atom is located in a crystal field, enables an electron to move from the lower level into a higher level by absorption matching the difference between two levels (Burn 1993). The variation of crystal field with crystal structure enables the same ion to absorb differently from mineral to mineral (Burns 1993). Clay minerals have distinct spectral fingerprints in the NIR region due to the strong absorption of the SO_4^{-2} and CO_3^{-2} and OH^- groups (Clark 1999; Burns and Ciurczak 1992). A significant absorption takes place in the vis-near infrared region due to O–H stretch combinations and metal–OH (Viscarra Rossel et al. 2006). While absorption in the visible region (400–780 nm) is mainly correlated with iron-bearing minerals (Sherman and Waite 1985; Stenberg et al. 2010). Iron-bearing minerals such as haematite (α-Fe_2O_3) and goethite (α-FeOOH) show strong absorption between 400 and 660 nm (Sherman and Waite 1985).

Goethite shows three distinct absorptions in the visible region, one near 660 nm, another near 480 nm and one near 420 nm (Morris et al. 1985). A study by Sherman and Waite (Sherman and Waite 1985) attributed goethite weak absorption near 1700 nm to the first overtone of a stretching vibration of OH. In both goethite and haematite, the absorption band that occurs near 930 and 880 nm, respectively, are generally attributed to excitation from a ground energy state to the first higher energy state. The vivid red haematite and yellow goethite colors are caused by the absorption in the visible range, while the weak absorption near 1700 nm is due to the stretching vibration of OH in the crystal structure of goethite (Morris et al. 1985). In the 2200–2500 nm region most of clay minerals absorption are due to O–H stretch and metal–OH bend, where the absorption near 2200 nm is due to Al–OH, this absorption as in kaolinite, montmorillonite, and illite. The absorption near 2290 nm is due to Fe–OH, while the absorption near 2300 nm is mainly attributed to Mg–OH as in illite and montmorillonite (Clark et al. 1990; Post and Noble 1993). Clay minerals absorptions near 1400, 1900, and 2200 nm as in smectite can be attributed to the first overtone of structural O–H stretching vibration mode in octahedral layer as one part and to combination vibrations of water bound in the interlayer as hydrated cations plus adsorbed water on particle surfaces as second part of absorption (Bishop et al. 1994). Such water may diagnostically distinguish between clay minerals for example dry kaolinitic soils show a very weak absorption near 1900 nm due to the absence of water films and water in interlayer (Post and Noble 1993). According to Campbell et al. (2013), the most common soil minerals can successfully be predicted based on the creation of soil minerals models with multivariate calibrations using NIR spectroscopy. Brown et al. (Burns and Ciurczak 1992) evaluated the accuracy of Vis-NIR models to predict Kaolinite and montmorillonite as well as clay content, CEC, soil organic carbon and extractable Fe for 4100 soil samples and reported strong predictability in the model.

Carbonates have the strongest absorption near 2,335 nm with some weaker absorption near 2,160 nm, which are due to overtones and combination bands of the CO_3 fundamentals, but these bands vary with composition (Hunt and Salisbury 1970; Ben-Dor and Banin 1994; Clark 1999). Vis-NIR region is insensitive to several minerals; quartz is one of these minerals that diagnostically has no absorption features in this wavelength region (Clark 1999).

Due to the mentioned above facts, many studies quantitatively characterized clay minerals by using Vis-NIR spectroscopy, but only some of these studies quantified their concentration in soil (Farmer 1974; Hunt 1977; Clark et al. 1990; Bishop et al. 1994). Recently, several authors (Burns and Ciurczak 1992; Viscarra Rossel et al. 2006; Viscarra 2009; Curcio et al. 2013) have used diffuse reflectance spectroscopy to make accurate to moderate estimations of both soil minerals compositions and contents.

6.2.1.2 Soil Texture

Soil texture is the main variable with the large effect on several key soil properties such as structure, CEC, water dynamics, aeration, nutrient cycling processes, and many other important environmental aspects (Stenberg 1999; Jarvis 2007). Both direct and indirect effects make textural fractions an important feature for understanding and managing soils (Conforti et al. 2015). Labor intensive and time-consuming methods for determining soil texture drive the scientific community to develop indirect estimation methods based on proximal or remote sensors, including diffuse reflectance spectroscopy (Ben-Dor et al. 2009). Several studies described this technique as an efficient tool to estimate textural fractions (sand, silt and clay). Ben-Dor and Banin (1995a), reported that bands for clay content, specific surface area (SSA) and CEC are mostly related to O–H in water and Mg–, Al–, and Fe–OH in the crystal minerals. The results obtained by several studies confirmed that the clay usually predicted well with r^2 values between 0.72 and 0.95 (Malley et al. 2000; Shepherd and Walsh 2002; Chang et al. 2005; Sørensen and Dalsgaard 2005).

Conversely, in term of reflectance the results obtained by Conforti et al. (2015) showed that the soils with sand over 70% have relatively high reflectance due to quartz content in sand fractions. Other studies considered clay and organic carbon contents as important factors that are affecting soil spectral behavior and decreasing reflectance (Schwanghart and Jarmer 2011; Conforti et al. 2013). Curcio et al. (2013) registered high coefficient of determination (R^2) for clay and silt fractions $R^2 = 0.87, 0.80$, respectively and satisfactory for sand with $R^2 = 0.60$ for 100 soil samples collected to estimate soil texture from visible, near-infrared and shortwave infrared (400–1200 and 1200–2500 nm) reflectance. Silt is the hardest to predict by VNIR-spectroscopy in comparison with clay which is best predicted followed by sand content prediction based on results obtained by a number of studies (Shepherd and Walsh 2002; Islam et al. 2003; McCarty and Reeves 2006).

The wavelengths at 1400, 1900 and 2200 nm are highlighted as the most significant bands to successively predicting of soil texture. These bands considered as essential for the estimation of soil textures due to clay minerals spectral features, while quartz and feldspar as the main minerals in sand fractions are spectrally featureless (Hunt and Salisbury 1970; Hunt 1982; Ben-Dor 2002).

6.2.1.3 Soil Organic Matter (SOM)

Soil organic matter consists of different stages of plant and animal residual decomposition (Schnitzer et al. 1991). This matter plays an important role in the physical, chemical and biological soil properties in addition to the global C cycle (Schlesinger 1990). In the Vis-NIR region, overtones and combination bands are usually related with the stretching and bending of NH, CH, and CO groups (Goddu and Delker 1960; Ben-Dor et al. 1999; Bokobza 1998).

Bands in 1100, 1600, 1700–1800, 2000 and 2200–2400 have been recognized as important bands for detection of soil organic matter and total nitrogen (Krishnan et al. 1980; Dalal and Henry 1986; Henderson et al. 1992; Martin et al. 2002; Stenberg 2010). Clark (1999) identified a band in 2,300 nm as a combination a band and first and second overtones in 1700 and 1100 nm. C–H stretch fundamentals near 3400 nm that assigned by Clark (1999) has been confirmed with humic acids peaks at 3,413 nm from Chinese soil (Xing et al. 2005) and 3,509 nm from Massachusetts soils (Kang and Xing 2005), where both studies ascribed these peaks to aliphatic CH_2 stretching.

At wavelengths beyond 2000 nm, the absorption by organic matter is difficult to interpret, as several organic and inorganic molecules may absorb in these regions (Goddu and Delker 1960; Clark et al. 1990). Viscarra Rossel et al (2006) suggested that Vis-NIR recognizes soil organic matter better than NIR region alone. This suggestion is also reported by other studies for example, Fystro (2002) and Islam et al. (2003) achieved better results in the visible region (350–700 nm) for Australian and Norwegian soils, respectively. While, Stenberg (2010) achieved only a small improvement in visible region calibration for Swedish soils.

The organic matter signals are often weak for soils with low content of organic matter, particularly for soil with highly variable minerals (Viscarra Rossel and McBratney 1998). According to Ben-Dor and Banin (1995a) related the spectra patterns with the stage of organic matter decomposition. Their study suggested that the prediction changes with quality and quantity of soil organic matter for two sets of soils, one set with up to 4% SOM mostly consists of humus and second set with more than 4% SOM mostly consists of decomposed litter.

More recently, a study by Cozzolino and Morón (2006), reported that the absorption spectrum in the region between 700 and 800 nm is a function of the composition of compounds during decomposition of SOM and plant residues. Several studies (Chang et al. 2001; Fidencio et al. 2002; Fystro 2002; Mouazen et al. 2007) reported strong predictability for SOM and or SOC with R^2 ranging from 0.80 to 0.92. Contradictory, other studies reported weak predictability with R^2 ranging from 0.46 to 0.66 (Ben-Dor and Banin 1995a; Stenberg et al. 2002; Sørensen and Dalsgaard 2005)

They ascribed the low prediction ability to the sand content, where the sandiest soils will influence the accuracy of calibration due to the role of Quartz in the scattering of light. Another likely clarification is the same amount of organic matter will be distributed in thicker coatings in soils with the smaller surface area (sandy soils).

Stenberg (2010) observed that the prediction had higher errors with increased content of sand fraction and he improved the calibration by removing the sandiest soils. The C/N ratio is one of the soils biological activity indicators, and authors suggested successfully using Vis-NIR spectroscopy to predict the spatial distribution of C/N ratio (Chang and Laird 2002; Fystro 2002; Shaddad et al. 2014). Their results reported that both C and N have similar features in general, but their peaks diverge in heights. In an attempt to test the possible independence between predictions of C and N, Martin et al. (2002) found that N is best predicted when it well correlated to C and the correlation between predicted C and N was 0.96.

6.2.1.4   Soil Nutrients and Fertility Assessment

The assessment of soil fertility is normally associated with soil nutrients levels and organic matter content as main soil properties for either Precision agriculture or to maintain soil from degradation (Gobeille et al. 2006). This techniques are needed to acquire data and on line soil analysis, since the soil analysis in terms of sampling and laboratory are costly and time-consuming (Chang et al. 2001; Pirie et al. 2005; Burns and Ciurczak 1992; Nanni and Dematte 2006). As described by Desbiez et al. (2004), soil fertility is a function of soil properties therefore, it is a comprehensive concept more than to measure directly.

Many studies have been devoted to determining soil nutrient and micronutrient concentration, as a crucial factor in soil fertility assessment. In particular, very acceptable estimations have been reported by using Vis-NIR spectroscopy under laboratory condition and in situ field measurements.

Soil nitrogen and organic carbon were generally yielded the most optimistic correlation coefficients (higher than ($R^2 > 0.90$)) between the actual and estimated concentrations. Dalal and Henry (1986) studied the prediction of total soil nitrogen based upon selection of the combination of three wavelengths 1702, 1870 and 2052 nm. Soil samples with the low contents of organic matter and total nitrogen showed higher standard errors of the prediction and correlation coefficient with value of $R^2 = 0.92$ and the standard errors for the prediction were much larger for soil samples with low contents of organic matter. The best wavelengths to predict total soil nitrogen with correlation coefficient value of $R^2 = 0.94$ were reported in the range of 1100–2300 nm (Reeves and McCarty 2001).

The very successful calibration coefficient range of 0.80–0.98 of N and C to infrared radiation values pushed researchers toward soil N and organic C analyses (Reeves and McCarty 2001; Cozzolino and Moron 2006; Stevens et al. 2008). Several parameters were evaluated in addition to total N and C, microbial C and N with reported R^2 in the range 0.60 to over 0.90 (Chang et al. 2001; Ludwig et al. 2002; Chodak et al. 2002; Cécillon et al. 2008). The accumulated mineral N under incubation conditions have been deeply studied with correlation coefficient values between 0.70 and 0.80 (Palmborg and Nordgren 1993; Fystro 2002; Schimann et al. 2007). Others studies have reported fewer correlation values ($R^2 < 0.5$). The disappointing results have been attributed to a small subset of samples, diverse locations and/or predicting one experiment with calibration on the other (Reeves et al. 1999; Chang et al. 2005, 50). The importance of the sample sets predict biological properties with NIR spectroscopy were underlined by Terhoeveb-Urselmans et al. (2006). The temporal changes in soil solution chemistry are usually related with a prediction with very low correlation coefficient with values of $R^2 < 0.50$ (Janik et al. 1998).

Several factors influence soil N content. Soil particle-size fractions are one of these factors which has been studied by indirect measurement of available soil N by using NIR spectroscopy (Barthès et al. 2008). The infrared absorption bands are often overlapped, some of the interference is associated with soil carbonate contents which is considered another factor influencing the accuracy of soil N content (Linker et al. 2005; Jahn et al. 2006). Borenstein et al. (2006) considered carbonate overlapping as the largest source of error for nitrate determination in calcareous soils, where the nitrate band is disturbed by the absorbance band of carbonate.

When connected to soil P forms, P in soil has been well studied by soil scientists as the second nutrient after N in terms of its importance for plants. Soil P is not easy to determine from spectral data like SOM and total N. Some authors reported Vis-NIR as a good way to predict soil P with very successful correlations ($R^2 > 0.90$) and others reported the absolute failure of prediction with very low correlations ($R^2 < 0.50$). Udelhoven et al. (2003) attributed poor prediction of extractable P to the size of the yield as P and other nutrients are removed with the yield. Other authors attributed low correlations to the measurement methods (Mamo et al. 1996; Zbiral and Nemec 2002). Moron and Cozzolino (2007) identified poor accessibility prediction with R^2 of 0.61, they suggested that the extractable and available P measurement methods are not at consistently well correlated with absorbance.

Bogrekci and Lee (2005) investigated the effects of the four most common P compounds on reflectance spectra using UV-VIS-NIR. Their results recognized the highest absorbance for calcium phosphate at 2516 nm, while the Aluminum phosphate showed lower absorbance between 1374 and 2550 nm and highest absorbance at 228 nm in the UV range. The same study observed four small distinguishable peaks at 871, 1413, 1934, and 2213 nm for Fe-associated P and very significant peak at 281 nm in the UV region. According to Daniel et al. (2003) high absorbance peaks at 400–1100 nm for available P were recognized with high correlation ($R^2 = 0.81$) between actual concentration and absorbance. Factors affecting P prediction by Vis-NIR spectroscopy are also used in literature to improve the prediction of soil Bogrekci and Lee (2005) obtained very strong correlation R^2 of 0.92 between total P and absorbance of dried and sieved soil samples. For the same aim, Bogrekci and Lee (2006) obtained more accurate prediction of soil P from spectra of moisture-free soil than those of moist soil.

NIR spectroscopy is a well-established technique in the evaluation of soil nutrients (Daniel et al. 2003). In particular, most of the soil properties are interrelated, and thus make the predictions of these properties achievable (Cohen et al. 2005).

These promising attempts suggest hope for further application such as soil Potassium. Absorption depth, area, and width are the main absorption features that are typically used to identify the wavelength ranges sensitive to soil K. Based on results of sensitivity analyses, the wavelength ranges that relate to K concentration in soils were found to be between 2450 and 2470 nm (Ben-Dor and Banin 1995a; Udelhoven et al. 2003; Luleva et al. 2011).

Generally, the instability of K prediction and the poor results with R^2 0.47 and 0.68 depend on the variability and the capacity of the set calibration (Ludwig et al. 2002; Burns and Ciurczak 1992; He et al. 2007). Several authors reported different coefficients (R^2) for available K (0.56–0.83), exchangeable K (0.11–0.55), Ca (0.75–0.89), Fe (0.64–0.91), Na (0.09–0.44), and Mg (0.53–0.82). The local conditions, the form of measured element (extractable, available, or total), and the reference methods available for the assessment of these parameters could be the explanation for such variation in calibration coefficients (Ehsani et al. 1999; Shibusawa et al. 2001; Groenigen et al. 2003; Pereira et al. 2004).

Despite the high variation in finding relationships between measured nutrients and reflectance at the specific absorption band discussed above, the Vis-NIR Spectroscopy technique has proven as very successful in many soil parameters studies.

6.3 Materials and Methods

6.3.1 Study Site Description and Soil Sampling

The study area (Al-Muthana province) is located in Southern Iraq (Fig. 6.1). The second location is located in the Al-Rumetha at (45°15′07.32″–45°15′18.10″ E, 031′29′49.70″–031′29′56.04″ N) the northern border of the first study location Al-Samawa (45°29′08.24″–45°29′13.63″ E, 31°17′57.16″–31°18′03.69″ N), about 25 km from the center of Al-Samawa.

The case study regions receive an average annual rainfall of 89.5 mm. The temperature is moderate with highs of 32 and lows of 17 °C.

The main land use is dominated by a croplands "wheat area", at each site. The soil is considered to be fertile clay loam and sandy clay loam for Rumetha and Samawa location respectively with a potential for moderate productivity. Forty (40) georeferenced soil samples were collected from each study site. A Garmin GPS receiver was used to reach the pre-defined geo-referenced sampling places. Soil samples were taken from the top 20 cm of the soil.

All collected soil samples were air dried and cleaned and ground to pass through a 2 mm sieve to prepare for organic matter content, pH, EC, CEC, texture analyze as shown in Table 6.1.

Following the common soil preparation steps, and read the reflectance spectra for all samples, a representative sample (is obtained by mixing all samples of each location) were mixed with urea and Monopotassium Phosphate (MPP) fertilizers to sim-

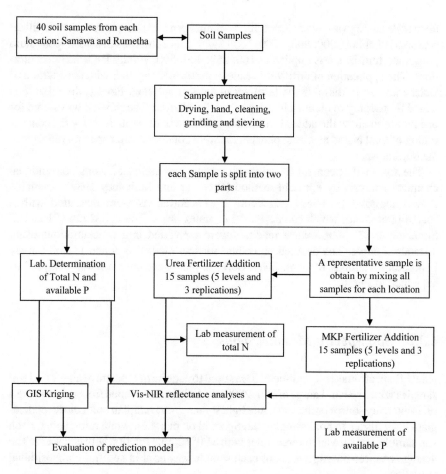

Fig. 6.1 Preparation steps for the NIR reflectance analyses and total N and P determinations

Table 6.1 Some soil properties of the studied locations

Property		Location		
		Rumetha	Samawah	
Soil classification		Typic torrifluvent	Typic torrifluvent	
Parent material		Alluvium	Alluvium	
Climate		Arid	Arid	
(EC$_e$)		8.7	23	dS m^{-1}
(pH)		7.5	7.7	
Soil Particles	Clay	395	325	g kg^{-1} soil
	Silt	235	175	
	Sand	370	500	
Texture		Clay loam	Sandy clay loam	

ulate soils having variable concentrations of nitrogen and phosphorus. Urea fertilizer was applied at 200, 500, 800, 1200, 1600 kg/ha concentrations and Monopotassium phosphate fertilizer was applied at 160, 250, 400, 500, 700 kg P_2O_5/ha concentrations. The application of fertilizer is usually performed together with the addition of water into soil to dissolve the fertilizer in the soil solution. Finally, the mixture is mixed thoroughly to obtain a homogeneous distribution. The mixture was stored for one month to allow the added fertilizers equilibrate in the soil. Tests for the concentration of total N and available phosphorus were conducted after spectral reflectance measurements.

The soil total N content for each soil sample was estimated using conventional chemical analysis by Kjeldahl method (Bremner and Mulvaney 1982). Available P was analyzed by Olsen extraction, P concentrations were measured with a spectrophotometer which can detect the complex absorption at 882 nm (Olsen and Sommers 1982). Absorbance readings were converted into concentrations using a standard curve. The detailed procedure for this method is found in Soil Survey Laboratory Methods Manual, 2004. Soil samples were prepared and analyzed as explained in Fig. 6.1.

6.3.2   Soil Spectra Measurements

An ASD spectrometer- FieldSpec® 3 was used to acquire the spectra of air—dried soil samples with a spectral range of 350–2500 nm. Soils were scanned from below using a high-intensity source probe and white light source with Duraplan borosilicate optical-glass Petri dishes to hold samples and Spectral on panel for white referencing. Each soil sample was measured three times with a 90° sample rotation between scans. The average of three measurements of each sample was used in spectral-compositional modeling.

The wavelength-dependent signal-to-noise ratio (S/N) for our instrument was estimated by taking repeated irradiance measurements of Spectral on the white-reference panel over a 5 min interval. Replicate soil scans were compared (reflectance and first derivatives) to check for errors, then averaged to produce a single reflectance scan for each sample. Using the weights and smoothing parameters described above.

6.3.3   Data Analysis: Calibration

6.3.3.1   Vis-NIR Spectra Analysis

The Log(1/Reflectance) function was used to transforme reflectance to absorbance, where the absorbance directly related to concentration of the studied properties according to Beer Lambert law. The soil samples were randomly divided into two sets: 32 samples for calibration and 28 samples for validation with 8 and 11 soil samples for Total N and available P respectively before model development to pre-dict the mentioned nutrients. Spectral absorbance was correlated to total N and soil available P concentration using statistical analysis software XLSTAT 2014 program to create predictor models. Models with the highest coefficient of determination (R^2) and lowest root mean square error (RMSE) were plotted to identify the significant portion of wavelength for total N and soil available p prediction.

6.3.3.2   GIS-Kriging Analysis

Soil spatial variability of total N and soil available P was described using geosta-tistical method (Oridinar Kriging). The position of 40 samples from both studied locations (Samawa and Rumetha) were geo-referenced using a hand-held global position system (GPS). Of 40 soil samples, only 32 sampling points were selected randomly for total N and 28 sampling points for soil available P prediction. The geostatistical analyst extension available in ESRI ArcMap v. 10 was used to perform the Ordinary Kriging. The program goes through two- analysis steps to estimate the model of autocorrelation as a first step and to predict the unmeasured sampling points as a second step. The rest 8 and 11 sampling points were used for cross-Validation. Finally, the spatial prediction distribution map of soil total N and soil available P in both studied locations were produced.

6.3.4   Evaluation of the Prediction Power

The model prediction capability is different from the fitness of model. The ability of the Partial Least Square (PLS) and GIS-Kriging calculated models were evaluated by comparing the predicted values with the measured values. Different assessing parameters were used to evaluate the accuracy of the predicted concentrations of total N and soil available P from soil samples that do not participate to the calculated models. The coefficient of determination (R^2v) for validation, the Corresponding coefficient of determination Q^2, RMSE, and Standard deviation were determined to obtain some measures of model ability. The cross-validation technique is the most simple way, which can easily determined and allows model comparison. This parameter is called corresponding coefficient Q^2 with a value of $Q^2 \leq 1$ and even negative for bad predictive models (Consonni et al. 2010). The best predictive models were chosen by simply looking at the highest value of the coefficient of determination for validation R^2v/or the highest corresponding coefficient Q^2 and the lowest value of RMSE and standard deviation (Fig. 6.2).

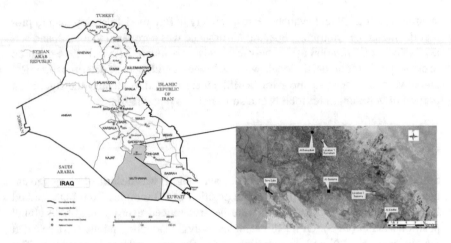

Fig. 6.2 Locations and extent of the study area

6.4   Results and Discussion

6.4.1   Vis-NIR Spectra of Soils

Spectra of the 40 soil samples from Rumetha and Samawa locations are plotted in
Figs. 6.3 and 6.4. All spectra showed typical soil absorption form at each region of
wavelength. As it can be seen from both figures, there are two distinguished peeks
at approximately 1400 and 1900 nm. The main absorptions near 1400 nm could
be related to the first overtone of an O–H stretch vibrat02+ion of water (H–O–H)
or metal–O–H vibration (Whiting et al. 2004). The peaks near 1900 nm could
be related to combination vibrations of water (H–O–H) and O–H stretch (13ywd-
pLJ9iEA93BtCwez2w8zXFPWxoDota). Soil samples from both studied locations,
showed small peaks at 2200 nm, and absorbance values continued to increase from
2200 to 2400 nm. The absorbance peaks near 2200 nm could be associated with
hygroscopic water and clay lattice OH, while the absorption features at approximately 1400 and 1900 nm, can be strongly related to free available water (Hunt
1982; Ben-Dor et al. 1999).

Generally, soil spectra fingerprints are comparable for both Rumetha and Samawa
locations, which are mainly attributed to several functional groups in soil samples.
Rumetha soil samples (Fig. 6.3) showed relatively higher absorbance values than
those of Samawa soil samples (Fig. 6.4). The differences in absorbance values
between the two locations can be attributed to the differences in soil texture (Table 6.1)
and in particular clay contents as Mouazen et al. (2005) and Conforti et al. (2015)
mentioned. In term of spectra basic shape, both locations of this study have soil with
distinct fingerprints similar to those observed by other studies (Shepherd and Walsh
2002; Xie et al. 2011; Xu et al. 2016).

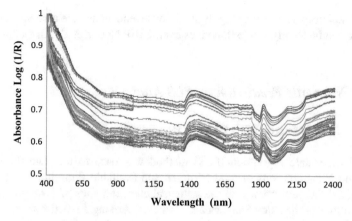

Fig. 6.3 Absorbance graph of Samawa location soil near-infrared spectroscopy

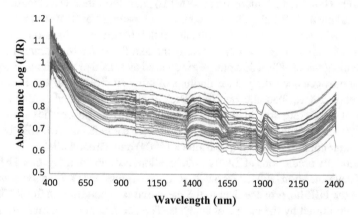

Fig. 6.4 Absorbance graph of Rumetha location soil near-infrared spectroscopy

The soil samples from both locations (Rumetha and Samawa) gave rise to different peaks intensities in wavelength despite the similarity in the absorption pattern across the wavelength. Factors that have the main influence on the shape of soil spectra are temperature, moisture, composition, and structural properties, which cause light scattering effects while absorbance decreases as the particle size increases (Clark 1999; Dahm and Dahm 2007). Despite these facts, other studies have considered the pretreatments of soil samples before NIR analyses as the main reasons behind the variations in the intensity of absorption peaks (Krishnan et al. 1980; Stenberg et al. 2000).

As a typical soil spectra signature, the spectra curves of both studied locations are consistent with other research findings (Johnston and Aochi 1996; Haberhauer and Gerzabek 2001; Burns and Ciurczak 1992). The relationship between spectral data and soil properties can be highly spatially dependent and the associated spectral variation can be locally stable (Savvides et al. 2010; Nocita et al. 2014).

6.4.2 Nutrients Prediction by PLS Analysis

6.4.2.1 Total Nitrogen (TN) Content Prediction

Partial least square regression (PLS) method was used to translate the spectral curve into concentrations of total Nitrogen and available Phosphorous. Table 6.2 summarized the performance of the statistical characteristics of Total Nitrogen for both locations of the study Samawa and Rumetha. Among the entire spectral region of 350–2500 nm only two regions reported higher determination coefficient R^2 and lower Root Mean Square Error (RMSE) than the other wavelength regions. The absorbance of Total Nitrogen contents in band at 500–600 nm were well correlated with $R^2 = 0.95$ for Rumetha and 0.869 for Samawa location and in band at 800–1000 nm with $R^2 = 0.97$ for Rumetha and 0.938 for the Samawa location. The wavelengths of 900–1000 nm were reported as sensitive wavelengths to detect soil total nitrogen according to the findings of other studies (Dalal and Henry 1986; Martin et al. 2002; Shao and He 2011; Dick et al. 2013; Vohland et al. 2014). The models of predicting soil total nitrogen were built from absorbance in region of 800–1000 nm and the band at 1400–2200 nm were excluded due to NIR absorption of soil water content According to Bullock (2004) and Tiruneh (2014).

Statistically predictions of TN for both locations are shown in Fig. 6.5. The coefficient of determination (R^2) for Partial least square regression (PLS) analysis was used to relate the NIR absorbance data to the measured concentration of Total Nitrogen, which determined by the chemical analysis. The coefficient of determination (R^2c) for calibration set, which obtained by PLS analysis of soil samples for both locations were 0.979 and 0.938 for Rumetha and Samawa, respectively. The determination coefficients (R^2v) for the external validation set (the remaining 8 soil samples) of both locations were calculated to prove the success of PLS model.

The results of Tables 6.3 and 6.4 prove the ability of the calibration models to obtain reliable quantitative prediction. The lower RMSE values of 0.016 for Rumetha

Table 6.2 Statistical characteristics (calibration) for total nitrogen in samples of Rumetha and Samawa locations

Location	Wavelength (nm)	Observations	R^2	RMSE	Std. deviation (Std.)
Rumetha	500–600	32	0.955	0.023	0.027
	800–1000	32	0.979	0.016	0.019
Samawa	500–600	31	0.869	0.045	0.050
	800–1000	31	0.938	0.039	0.042

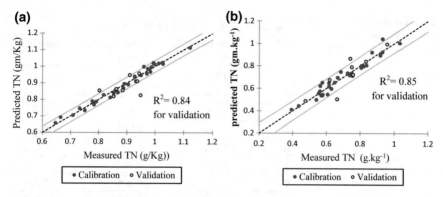

Fig. 6.5 Measured versus predicted values of total nitrogen obtained by using PLS models for both sets of calibration and validation sets at band 800–1000 nm for: **a** Rumetha location and **b** Samawa location

and of 0.039 for Samawa confirmed the high power of the predictive model. This Results are quite promising to make approximate quantitative predictions for new unknown samples.

In the statistical terms of R^2 and RMSE, the calibration of total Nitrogen was fairly accurate. Based upon literature, the prediction is considered reliable with $R^2 > 0.50$ and accurate with $R^2 > 0.80$ (Shepherd and Walsh 2002; Alrajehy 2002; Ludwig et al. 2002). The findings are in trend with other studies (Bilgili et al. 2010; Kleinebecker et al. 2013; Bansod and Thakare 2014) for achieving the highly predictive power of NIR spectroscopy model for Total Nitrogen in the soil.

6.4.2.2 Total Nitrogen Content Prediction Model Quality

To simulate soil fertilization conditions, 15 soil samples from each location have mixed with Urea fertilizer in the rates ranged from 0.1 to 0.8 g kg^{-1} soil (90–720 kg N ha^{-1}). The mean NIR spectra of treated soil samples from both locations have basic shapes as plotted in Fig. 6.6. The curves show very short peaks in the wavelength region 1000–1200 nm, which shifted in the broad peaks toward long waveband direction. All the treated samples exhibited a slightly rising peak at approximate 1400 nm and high peaks in the wavelength region 1900–2000 nm. In the bands 2000–2300 nm, the spectra are mostly broad with very close short absorption features. Soil spectra curve is a representation of the absorption pattern across the wavelength range. The wavelengths at which light is absorbed are highly related to the soil constituents. As can be seen from Fig. 6.6, the absorbance was higher for Rumetha location with Clay loam texture with (39.5%) clay content (Table 6.1), Compared with the Samawa location with Sand Clay loam texture with (50%) sand content (Table 6.1). Generally, the absorbance of soil samples from both locations

Table 6.3 Prediction results for total nitrogen by PLS model for Rumetha location

Sample No.	PLS model for (800–1000 nm)		
	Measured TN (g kg^{-1})	Predicted TN (g Kg^{-1})	Residual
1	0.812	0.841	−0.029
2	0.746	0.732	0.014
4	0.939	0.927	0.012
5	0.769	0.784	−0.015
6	0.790	0.777	0.012
7	0.967	0.983	−0.016
9	0.843	0.868	−0.026
10	0.827	0.818	0.010
11	0.946	0.950	−0.004
12	0.890	0.905	−0.014
13	1.018	0.994	0.024
15	0.998	0.960	0.038
16	0.840	0.854	−0.014
17	1.015	1.027	−0.012
19	0.986	0.974	0.012
20	0.895	0.891	0.004
21	0.734	0.751	−0.017
22	0.704	0.708	−0.004
23	0.787	0.791	−0.004
24	0.657	0.647	0.010
25	0.935	0.928	0.007
28	0.869	0.887	−0.018
29	1.018	1.012	0.006
30	0.956	0.971	−0.015
32	0.855	0.860	−0.005
33	1.113	1.121	−0.008
34	1.021	0.994	0.027
35	1.022	1.025	−0.003
37	0.692	0.665	0.027
38	0.910	0.931	−0.021
39	1.014	0.997	0.016
40	0.995	0.987	0.008

Sample No.	PLS model for (800–1000 nm)		
Table 6.4 Prediction results for total nitrogen by PLS model for Samawa location	Measured TN (g kg^{-1})	Predicted TN (g kg^{-1})	Residual
2	0.622	0.564	0.058
3	0.410	0.394	0.016
5	0.538	0.577	−0.040
6	0.557	0.554	0.003
7	0.545	0.578	−0.034
8	0.651	0.576	0.075
9	0.574	0.561	0.013
11	0.893	0.920	−0.026
12	0.536	0.570	−0.033
13	0.616	0.616	0.000
14	0.596	0.631	−0.034
15	0.735	0.725	0.010
16	0.784	0.807	−0.023
17	0.546	0.608	−0.062
18	0.647	0.616	0.031
19	0.735	0.750	−0.015
20	0.444	0.437	0.007
22	0.735	0.726	0.009
23	0.711	0.754	−0.043
25	0.920	0.871	0.049
26	0.805	0.838	−0.033
28	0.797	0.820	−0.024
29	0.717	0.737	−0.020
30	0.498	0.540	−0.042
31	0.652	0.649	0.003
32	0.747	0.686	0.062
35	0.923	0.932	−0.009
36	0.696	0.667	0.028
37	1.004	1.032	−0.028
38	1.042	0.932	0.109
39	0.801	0.808	−0.007

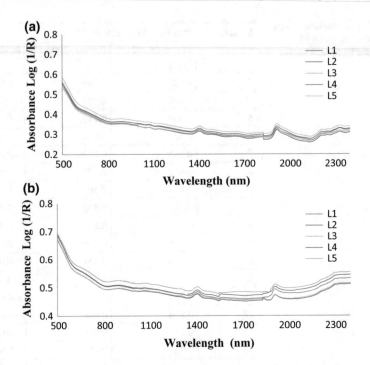

Fig. 6.6 Average absorbance spectra of treated soil samples at different Urea fertilizer rates: **a** Samawa location and **b** Rumetha location. Each spectrum is an average of three samples

was directly related to soil texture fractions as Islam et al. (2003) and Conforti et al. (2015) indicated.

In order to get a reasonable interpretation from soil spectra curves, the relationship between the absorption spectra and the treated soil sample TN contents are analyzed. Based upon PLS analysis, the absorption at bands 800–1000, 1100–1200, 2100–2200 and 2200–2300 nm acquired the highest coefficient of determination R^2 of 0.88, 0.99, 0.92 and 0.97, respectively for Rumetha location and R^2 of 0.99, 0.72, 0.99 and 0.92, respectively for Samawa location.

TN is recognition with R^2 of 0.92 in the wavelength range of 1100–2500 and with R^2 of 0.94 in the wavelength range of 1100–2300 nm as reported by many researches (Dalal and Henry 1986; Reeves and McCarty 2001; Bansod and Thakare 2014). At bands approximate 1400 and 1900 nm, the coefficient of determination between soil spectra and the measured concentration of TN content were very weak with R^2 of 0.14 and 0.07 for Rumetha location and R^2 of 0.46 and 0.50 for Samawa location. This means, that the PLS analysis helps to select the sensitive wavebands for predicting Total Nitrogen content under the interference of water absorption wavebands of 1400, 1900 nm. The absorbance features for the 4 wavelengths mentioned above were correlated with the higher Total Nitrogen content (Fig. 6.6).

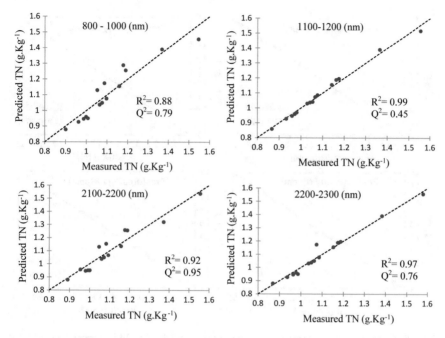

Fig. 6.7 Comparison of measured total N content versus predicted by PLS regression on NIR soil spectra for Rumetha location

Figures 6.7 and 6.8 show Total Nitrogen measured by chemical analysis versus predicted by PLS on NIR treated soil spectra across both locations of study. As can be seen from the figures, both the coefficient of determination R^2 and the corresponding coefficient of determination Q^2 are used to determine the predictive quality of the model. Among the four selected sensitive wavelengths for TN contents, the largest Q^2 were observed in the bands 2100–2200 nm with Q^2 of 0.95 and 0.97 for the Rumetha and Samawa locations, respectively. cross-validation is highly recommended to estimate the most favorable number of factors to avoid overfitting (Shenk and Westerhaus 1991; Freschet et al. 2011). On the whole, NIRS and PLS yielded the most accurate Q^2 and R^2 of Total Nitrogen contents at waveband 2100–2300 nm for both studied locations. Accurate prediction of TN using NIR spectroscopy has been widely reported (Brunet et al. 2007; Wang et al. 2008).

6.4.2.3   Available Phosphorous Prediction

Partial least squares (PLS) regression analysis was used to predict the concentration of available soil P in the soil samples collected from Rumetha and Samawa locations. As shown in Table 6.5, only wavelength regions with high determination coefficients R^2 are tabulated. The best predictive ability was achieved for Rumetha location

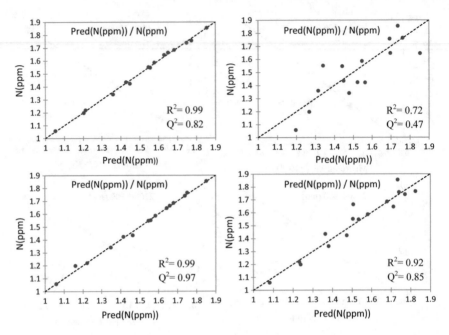

Fig. 6.8 Comparison of measured total N content versus predicted by PLS regression on NIR soil spectra for Samawa location

at 1400–1600 nm with R^2 of 0.85, lowest RMSE of 1.405, and lowest standard deviation of 1.577 and for the Samawa location at 900–1000 nm with highest R^2 of 0.81, lowest RMSE of 2.666, and lowest standard deviation of 2.879. At wavelength region 2100–2200 nm, both studied locations showed lowest R^2, highest RMSE, and highest standard deviation. The determination coefficient values for soil available P at the three reported wavelength regions were less than those have been obtained for T N. In literature, generally the mentioned above wavelength regions were well-known as P bands. According to this literature, the absorption wavelength regions are strongly related to the level of P in solution, the chemistry of soil, and the type of the target P compounds (Varvel et al. 1999; Bogrekci et al. 2003; Turner et al. 2003; Bogrekci and Lee 2005; Maleki et al. 2006). All primary soil properties like N, C, SOM, moisture content, and particle size distribution have possible absorption in NIR regions arise from bonds like O–H, C–H, and N–H (Malley et al. 2004; Niederberger et al. 2015). Generally, soil spectrum is obtained by passing radiation through a soil sample, which excites molecules with dipole moment to vibrational stretching of atoms form molecule. That means, the energy at which peak appears corresponding to the chemistry of the soil matrix. Unlike N, P is one of the secondary soil properties in responding to NIRS radiation due to the weak P–O dipole moment (Malley et al. 2004).

In spite of specific absorption peak is not being observed for available P, the prediction worked well for both studied locations. Each calibration equation was

Table 6.5 Statistical characteristics (calibration) for available phosphorous in samples of Rumetha and Samawa locations

Location	Wavelength (nm)	Observations	R^2	RMSE	Std. deviation
Rumetha	900–1000	28	0.81	1.958	2.109
	1400–1600	28	0.85	1.405	1.577
	2100–2200	28	0.65	2.455	2.698
Samawa	900–1000	28	0.81	2.666	2.879
	1400–1600	28	0.75	2.910	3.143
	2100–2200	28	0.75	3.192	3.522

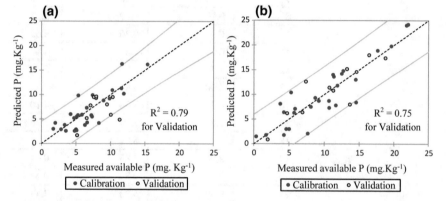

Fig. 6.9 Plots of measured versus predicted soil available P in the calibration and validation sets obtained via PLS models at 900–1000 nm for: **a** Rumetha location and **b** Samawa location

developed from 28 soil samples as a calibration set and validated with 11 soil samples as a validation set. For each, the NIR-predicted values were correlated with their measured values. The measured values versus predicted soil available P are plotted in Fig. 6.9. The parameter coefficient of determination (R^2v) of the cross-validation was used as selection principle to build a robust model. Based on R^2v values, the studied locations showed to be best predicted by a model built from measurements at 900–1000 nm with R^2v of 0.79 for Rumetha location and 0.75 for the Samawa location. The determination coefficient of validation represents empirical evidence of the capacity of the prediction model to make accurate predictions for new unknown data. Depending on the cross-validation results, the best wavelength was selected to make an accurate estimate of the prediction performance of calibration.

The main goal of using PLS analysis is to predict the value of a dependent variable for new samples from the same location. In this case, the results presented in Tables 6.6 and 6.7 show the potential of NIRS to predict soil available P concentration. The values of the lowest predicted soil available P were 1.74 and 0.59 mg kg^{-1} and the highest predicted soil available P was 15.39 and 22.00 mg kg^{-1} for Rumetha and Samawa locations, respectively. While the values of the lowest measured soil

Table 6.6 Prediction results for available soil P by PLS model for Rumetha location

PLS model for (900–1000 nm)			
Sample No.	Measured available P (mg kg^{-1})	Predicted available P (mg kg^{-1})	Residual
1	5.90	5.85	0.05
2	6.00	4.22	1.78
3	4.30	6.61	−2.31
4	10.20	11.91	−1.71
5	5.50	7.32	−1.82
7	5.80	5.28	0.52
8	9.90	7.39	2.51
10	10.90	10.26	0.65
11	16.30	11.66	4.64
12	11.30	11.57	−0.20
13	2.60	4.81	−2.21
16	4.20	8.30	−4.10
17	3.00	1.74	1.26
18	2.90	5.01	−2.11
19	2.90	2.86	0.04
20	9.20	9.84	−0.64
21	9.80	5.33	4.47
25	4.20	2.03	2.17
26	5.20	4.74	0.47
27	3.80	3.38	0.42
30	3.80	6.28	−2.48
31	5.50	5.11	0.39
33	16.20	15.39	0.82
34	9.50	7.88	1.62
35	8.80	8.96	−0.16
36	5.80	7.20	−1.40
37	2.70	5.09	−2.39
38	2.60	3.56	−0.96

available P were 2.60 and 1.50 mg kg^{-1} and the highest measured soil available P were 16.30 and 24.2 mg kg^{-1}. These results verify the potential of the prediction model to capture the variation of soil available P concentrations in the locations under study. Several studies reported that smaller or more similar areas have resulted in better prediction and are more applicable in practice for most of the soil properties (Viscarra Rossel et al. 2006; Sankey et al. 2008; Wetterlind et al. 2010).

Table 6.7 Prediction results for available soil P by PLS model for Samawa location

PLS Model for (900–1000 nm)

Sample No	Measured available P (mg kg^{-1})	Predicted available P (mg kg^{-1})	Residual
2	8.10	3.76	4.34
3	8.70	9.23	−0.52
4	9.30	8.72	0.58
6	8.40	14.56	−6.16
7	7.50	8.10	−0.60
8	6.40	4.28	2.12
11	7.10	5.86	1.24
12	2.10	7.66	−5.56
13	3.00	5.03	−2.03
14	3.00	4.62	−1.62
15	6.30	5.73	0.57
16	1.80	1.75	0.05
17	1.80	4.29	−2.49
18	10.40	5.30	5.10
19	1.50	0.36	1.14
21	14.80	12.72	2.08
23	14.00	11.25	2.75
25	13.00	14.50	−1.50
26	11.80	11.88	−0.08
27	14.20	12.50	1.70
28	13.60	11.37	2.23
30	8.70	10.67	−1.97
31	7.30	10.70	−3.40
32	7.80	11.65	−3.85
34	24.00	21.77	2.23
35	18.90	17.66	1.24
37	24.20	22.00	2.20
39	19.80	19.61	0.19

6.4.2.4   Available Phosphorous Content Prediction Model Quality

To cover the majority of a large variation in available soil P, soil samples from both studied locations were treated with different levels of Monopotassium Phosphate (MPP) fertilizer. Average absorbance spectra of the treated soils samples with different fertilizer levels ranged from 160 to 700 kg P_2O_5 ha^{-1} (70–300 kg P ha^{-1}) are given in Fig. 6.10. For both studied locations, each spectrum is an average of three

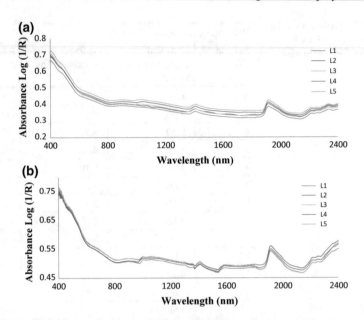

Fig. 6.10 Average absorbance spectra of treated soil samples at different Monopotassium Phosphate (MPP) fertilizer concentrations: **a** Samawa location and **b** Rumetha location. Each spectrum is an average of three samples

samples. As Beer-Lambert's law explains, soil absorbance of both studied locations increased with an increase in P addition levels. This relationship was observed clearly in the region 800–1,000 nm for Rumetha location and 800–1,200 nm for the Samawa location. Although the Rumetha absorbance curve was overlapped at all P levels, the mean soil absorbance values for the Rumetha location were slightly higher than the Samawa location. Different spectral behavior as shown in Fig. 6.10 could be related to their reflectance intensity, which can be influenced by texture, mineralogy, components and orientation of soil samples (Hummel et al. 2001; Madejova and Komadel 2001; Kusumo et al. 2010). In this study, the characteristics of absorption bands resulted in a large number of broad peaks and often masked due to the possibility of interference from other soil components. Since the O–H has a strong influence on the soil absorbance curve, water absorption was very clearly at 1400 and 1900 nm mutually in both studied locations. Thus, most of these broad, overlapped and poorly defined peaks could not be only identified by simple evaluation of soil spectra.

PLS analysis was used to produce a reliable NIRS model for predicting available soil P at the different levels of concentrations added. Soil available phosphorous were well predicted at regions 900–1000, 1400–1600 and 2100–2200 nm, which were spectrally active for prediction of available P with R^2 ranged from 0.93 to 0.96 for the Rumetha location and 0.84 to 0.93 for Samawa location (Fig. 6.11). It might be difficult to evaluate the most robust model among the three addressed wavelengths as good models to predict soil available P. The corresponding coefficient of determination Q^2 were calculated for measured and predicted values for soil samples contain different levels of P. The coefficients of determination R^2 and corresponding coefficient of determination Q^2 for the three sensitive wavelengths for available soil P are given in Fig. 6.11. This test showed a reduction in model quality at 2,100–2,200 nm with the lowest corresponding coefficient of determination Q^2 of 0.51 for Rumetha and 0.12 for Samawa location. The capability of soil available P prediction model was better at 900–1000 nm with the highest corresponding coefficient of determination Q^2 of 0.91 for the Rumetha and 0.75 for Samawa location. The obtained quality parameter values (Q^2) were at best successfully model to predict soil available P and well suited for a large variety of low to high concentrations. Similar successfully results were observed by other studies (Chang et al. 2001; Malley et al. 2004; Maleki et al. 2006; Genot et al. 2011), who tested the accuracy and capability of Vis- NIRS to predict soil P fractions.

6.4.3   Nutrients Prediction by GIS-Kriging

Laboratory-measured soil data set of 32 and 28 point samples from Rumetha and Samawa were used for statistical modeling and prediction process. The spatial concentration of each of total N and available P were generated with the use of the GIS-Kriging Ordinary Spherical technique for the (x, y) locations point samples from Rumetha and Samawa. The spatial characteristics of predicted total N and available P concentrations showed an increasing trend from center to southwest part of Samawa as shown in Fig. 6.12. In contrast, the spatial distribution of total N showed increasing trend from southwest to northeast part, while the highest values for available P are found in the north part of Rumetha location (Fig. 6.13). The spatial distribution trend of total N, soil available P status as shown in Figs. 6.11 and 6.12 could be significantly affected by slop gradient, land use, and soil type.

The quality of the predicted total N and available P values by GIS-Kriging models were generally assessed by using validation dataset with 8 soil points for total N and 11 soil points for available P. The same validation dataset as was used in the validation of NIRS model prediction was used to evaluate the Kriging model's prediction accuracy. Partial Least Square Regression analysis was done between total N and available P Kriging-predicted values and lab-measured values. The evaluation criteria for assessing the final prediction GIS-Kriging model are summarized in Table 6.8. The results as shown in Table 6.8, indicated that the coefficient of determination R^2 was in worst for available P with a value of 0.65 for Rumetha and 0.22 for Samawa reasonable value. While both studied locations showed convergent R^2 values for total N with values of 0.60 and 0.61 for Samawa and Rumetha, respectively. Compared with R^2 results, other statistical measurements values were unsatisfactorily predicted with low values of Q^2 and high values of standard deviation and Root Mean Square Error as presented in Table 6.8. Based upon these evaluation criteria, the smaller the RMSE and Standard deviation values, the better the prediction ability of the model (Hengl et al. 2007).

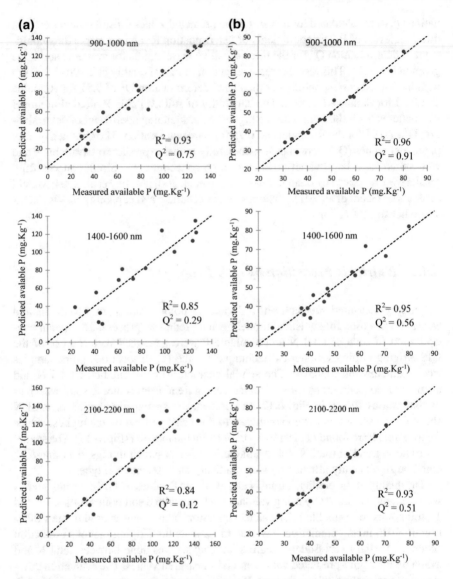

Fig. 6.11 Comparison of soil available P content as measured versus predicted by PLS regression on NIR soil spectra for: **a** Samawa location and **b** Rumetha location

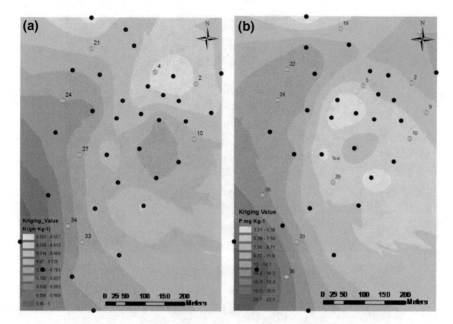

Fig. 6.12 Spatial distribution map of total N (**a**) and soil available P (**b**) in Samawa location

To define the potential of prediction models, it is always important not to look at a single measurement. In this study, prediction capability of the Ordinary Kriging models for total N and available P were generally poor in terms of all statistical evaluation parameters (Table 6.8). The total N and available P predicted with both the NIRS-based models and GIS-Kriging models were compared to the laboratory-measured in the validation dataset. In term of cross-validation, the coefficient of determination for validation was as much higher for NIRS-based models with R^2v values of 0.85 and 0.84 for total N and 0.79 and 0.75 for available P in Rumetha and Samawa, respectively. This statistical comparison is evident of the superiority of NIRS-based models over the GIS-Kriging models. At this point, the Vis-NIR method can be recommended for practical prediction of total N and available P with sufficient goodness of model fit. A number of limitations of using GIS-Kriging for practical prediction of soil parameters have been identified by several previous studies. The most reported limitations were: quality of input data, a large number of samples needed to build a reliable prediction model, the different assumption could lead to different prediction values, and the local neighborhood, where only points at a short distance are well predicted (Neter et al. 1996; Hengl et al. 2007; Webster and Oliver 2001; Kumar 2009; Gao et al. 2011).

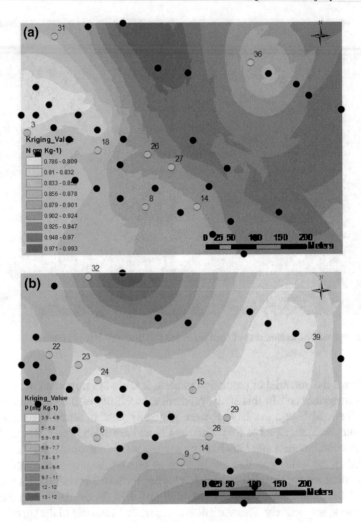

Fig. 6.13 Spatial distribution map of total N (**a**) and soil available P (**b**) in Rumetha location

Table 6.8 Summary of validation of total N and available P predicted by GIS-Ordinary Kriging for studied locations

Location	Target variables	Observations	R^2	Q^2	RMSE	Std. deviation (Std.)
Rumetha	Total N	8	0.60	0.28	0.023	0.027
	Available P	11	0.65	0.58	0.866	0.957
Samawa	Total N	8	0.61	0.43	0.059	0.068
	Available P	11	0.22	0.17	3.023	3.342

6.5 Conclusions

The overall goal of this study was to predict soil total N and available P by using Vis-NIR Spectroscopy technique. The main conclusion can be derived from the obtained results are:

1 Soil spectra fingerprints are comparable for both studied locations: Samawa and Rumetha, which are mainly attributed to several functional groups in soil samples. The obtained typical absorption showed two distinguished peeks at approximately 1,400 and 1,900 nm.

2 Due to the difference in clay contents, Rumetha location showed relatively higher absorbance values than those of Samawa location.

3 N in aid of qualitative interpretation (PLS), the spectral bands that showed good correlations with total N and available P have been identified. For total N, only two regions reported high R^2 and low RMSE: for bands at 500–600 nm with R^2 = 0.95 for Rumetha and 0.87 for Samawa and 800–1,000 nm with R^2 = 0.97 for Rumetha and 0.94 for Samawa location. While soil available P correlated differently with NIR Spectra, where the high R^2 for Rumetha location was achieved at 1,400–1,600 nm with R^2 = 0.85 and at 900–1,000 nm for Samawa with R^2 of 0.81.

4 The reported wavelength regions were well-known as N and P bands. At all reported bands, the R^2 for soil available P were less than those have been obtained for total N.

5 The performance of both prediction model: Kriging- based models and NIRS-based models have been assessed by using validation set. In terms of all statistically evaluation parameters, prediction capability of the GIS- Kriging models for total N and available P were as much lower and poor in comparison with NIRS-based model.

6 Fertilizers addition at different levels from Urea and Monopotassium Phosphorus were used to cover the majority of a large variation in soil total N and available P. Among four wavelength bands reported high determination coefficient R^2, only band at 2,100–2,200 nm reported highest R^2 for total N of 0.92 for Rumetha and 0.99 for Samawa and highest Q^2 of 0.95 for Rumetha and 0.97 for Samawa location. While the capability of soil available P prediction model showed reduction in model quality at 2,100–2,200 nm with lowest Q^2. The highest Q^2 of 0.75 for Samawa and 0.91 for Rumetha were reported at 900–1,000 nm.

7 The obtained prediction model quality parameter values were at best successfully model to predict soil total N and available P and well suited for a large variety of low to high concentrations under the condition of this study.

6.6 Recommendations

Despite the NIR soil spectroscopy has been well studied for more than decade, there are still needing to gain better understanding of variability and complexity of soil. The future studies could focus on the following recommendations:

1. A detailed study of NIR spectroscopy interaction with soil properties such as mineral composition, clay contents and water would be helpful for understanding the impact of soil components on absorbance spectra.
2. Developing cleverer uses of NIR spectroscopy for making presumption about soil quality by directly using in soil mapping and soil monitoring.
3. Attention should be paid toward strategic of the Iraqi Soil Spectral Library at country to encourage its adoption in soil science.
4. A complementary study could be also conducted with a field as a real-time onsite prediction.
5. The research should make an attempt to use the hyperspectral airborne and space-borne system to determine spatial variation of soil properties over wide area.
6. A final recommendation for further research is to produce an annual survey about soil fertility situation. The increased knowledge of how to develop fertilizer management would be beneficial for the farmers themselves, Ministry of Agricultural, Scientific Institutions at country level, and perhaps even for international bodies.

References

Alrajehy A (2002) Relationships between soil reflectance and soil physical and chemical properties. M.Sc. Thesis, Mississippi State University

Bansod SJ, Thakare SS (2014) Near infrared spectroscopy based a portable soil nitrogen detector design. Int J Comput Sci Inf Technol (IJCSIT) 5(3):3953–3956

Barthès BG, Brunet D, Hien E, Enjalric F, Conche S, Freschet GT, d'Annunzio R, Toucet-Louri J (2008) Determining the distributions of soil carbon and nitrogen in particle size fractions using near-infrared reflectance spectrum of bulk soil samples. Soil Biol Biochem 40:1533–1537

Ben-Dor E (2002) Quantitative remote sensing of soil properties. Adv Agron:172–243 (Academic press)

Ben-Dor E, Banin A (1994) Visible and near-infrared (0.4–1.1 μm) analysis of arid and semiarid soils. Remote Sens Environ 48:261–274

Ben-Dor E, Banin A (1995a) Near-Infrared analysis (NIRA) as a method to simultaneously evaluate spectral featureless constituents in soils. Soil Sci 159(4):259–270

Ben-Dor E, Banin A (1995b) Near-Infrared analysis as a rapid method to simultaneously evaluate several soil properties. Soil Sci Soc Am J 49:364–372

Ben-Dor E, Chabrillat S, Demattê JAM, Taylor GR, Hill J, Whiting ML, Sommer S (2009) Using imaging spectroscopy to study soil properties. Remote Sens Environ 113:38–55

Ben-Dor E, Irons JR, Epema GF (1999) Remote sensing for the earth sciences: manual of remote sensing. Wiley, New York, pp 111–188

Bilgili AV, van Es HM, Akbas F, Durak A, Hively WD (2010) Visible-near infrared reflectance spectroscopy for assessment of soil properties in a semi-arid area of Turkey. J Arid Environ 74:229–238

Bishop JL, Pieters CM, Edwards JO (1994) Infrared spectroscopic analyses on the nature of water in montmorillonite. Clays Clay Minerals 42:702–716

Blanco M, Villarroya I (2002) NIR spectroscopy: a rapid response analytical tool. Trac-Trends Anal Chem 21:240–250

Bogrekci I, Lee WS (2005) Spectral measurement of common soil phosphate. Am Soc Agric Eng 48(6):2371–2378

Bogrekci I, Lee WS (2006) Effect of soil moisture content on absorbance spectra of Sandy soil in sensing phosphorus concentrations using UV-VIS-NIR spectrascopy. Am Soc Agric Bio Eng 49(4):1175–1180

Bogrekci I, WS Lee, J Herrera (2003) Assessment of P concentrations in the Lake Okeechobee drainage basins with spectroscopic reflectance of VIS and NIR. ASAE Paper No. 031139. ASAE, St. Joseph, Michigan

Bokobza L (1998) Near infrared spectroscopy. J Near Infrared Spectrosc 6:3–17

Borenstein A, Linker R, Shmulevich I, Shaviv A (2006) Determination of soil nitrate and water content using attenuated total reflectance spectroscopy. Appl Spectrosc 60:1267–1272

Bremner JM, Mulvaney CS (1982) Methods of soil analysis. Part 2. Chem Microbiol Prop, ASA, SSSA 9.2:595–624

Bricklemyer RS, Miller PR, Paustian K, Keck T, Nielsen GA, Antle JM (2005) Soil organic carbon variability and sampling optimization in Montana dryland wheat fields. J Soil Water Conserv 60:42–51

Brown DJ, Shepherd KD, Walsh MG, Mays MD, Reinsch TG (2006) Global soil characterization with VNIR diffuse reflectance spectroscopy. Geoderma 132:273–290

Brunet D, Barthès BG, Chotte JL, Feller C (2007) Determination of carbon and nitrogen contents in Alfisols, Oxisols, and Ultisols from Africa and Brazil using NIRS analysis: effects of sample grinding and set heterogeneity. Geoderma 139:106–117

Bullock D (2004) Moving from theory to practice: an examination of the factors that pre-service teachers encounter as the attempt to gain experience teaching with technology during field placement experiences. J Technol Teach Educ 12(2):211–237

Burns PA, Ciurczak EW (1992) Handbook of near infrared analysis. Marcel Dekker, New York, USA

Burns RG (1993) Rates and mechanisms of chemical weathering of ferromagnesian silicate minerals on Mars. Geochim Cosmochim Acta 57:4555–4574

Campbell D, Shiley D, Curtiss B (2013) Measurement of soil mineralogy and CEC using near-infrared reflectance spectroscopy. ASD Inc., PANalytical company, Boulder, Colorado, 80301, USA

Cao N (2013) Calibration optimization and efficiency in near infrared spectroscopy. Graduate Theses and Dissertations. Paper 13199

Cécillon L, Cassagne N, Czarnes S, Gros R, Brun JJ (2008) Variable selection in near infrared spectra for the biological characterization of soil and earthworm casts. Soil Biol Biochem 40:1975–1979

Chang GW, Laird DA, Hurburgh GR (2005) Influence of soil moisture on nearinfrared reflectance spectroscopic measurement of soil properties. Soil Sci 170(4):244–255

Chang CW, Laird DA (2002) Near-infrared reflectance spectroscopic analysis of soil C and N. Soil Sci 167:110–116

Chang CW, Laird DA, Mausbach MJ, Jr CR (2001) Near- infrared reflectance spectroscopy- principle components regression analyses of soil properties. SSSAJ 65:480–490

Chodak M, Ludwig B, Partap K, Beese F, Khanna P (2002) Use of near infrared spectroscopy to determine biological and chemical characteristics of organic layers under spruce and beech stands. J Plant Nutr Soil Sci 165:27–33

Clark RN (1999) Spectroscopy of rocks and minerals and principles of spectroscopy. Wiley, Chichester, UK

Clark RN, King TVV, Klejwa M, Swayze GA, Vergo N (1990) High spectral resolution reflectance spectroscopy of minerals. J Geophys Res 95:12653–12680

Cohen MJ, Prenger JP, DeBusk WF (2005) Visible-Near infrared reflectance spectroscopy for rapid, non-destructive assessment of wetland soil quality. J Environ Qual 34:1422–1434

Conforti M, Buttafuoco G, Leone AP, Aucelli PPC, Robustelli G, Scarciglia F (2013) Studying the relationship between water-induced soil erosion and soil organic matter using Vis-NIR spectroscopy and geomorphological analysis: a case study in a southern Italy area. CATENA 110:44–58

Conforti M, Froio R, Matteucci G, Buttafuoco G (2015) Visible and near infrared spectroscopy for predicting texture in forest soil: an application in southern Italy. iForest 8:339–347

Consonni V, Ballabio D, Todeschini R (2010) Evaluation of model predictive ability by external validation techniques. J. Chemometrics 24:194–201

Cozzolino D, Moron A (2006) Potential of near-infrared reflectance spectroscopy and chemometrics to predict soil organic carbon fractions. Soil Tillage Res 85:78–85

Curcio D, Ciraolo G, Asaro FD, Minacapilli M (2013) Prediction of soil texture distributions using VNIR-SWIR reflectance spectroscopy. Procedia Environ Sci 19:494–503

Dahm DJ, Dahm KD (2007) Interpreting diffuse reflectance and transmittance—a theoretical introduction to absorption spectroscopy of scattering materials. NIR Publications, Chichester, UK

Dalal RC, Henry RJ (1986) Simultaneous determination of moisture, organic carbon and total nitrogen by near infrared reflectance spectrophotometry. Soil Sci Soc Am J 50:120–123

Daniel KW, Tripathi NK, Honda K (2003) Artificial neural network analysis of laboratory and in situ spectra for the estimation of macronutrients in soils of Lop Buri (Thailand). Aust J Soil Res 41:47–59

Desbiez A, Matthews R, Tripathi B, Joues J (2004) Perception and assessment of fertility by farmers in the mid-hills of Nepal. Ecosyst Environ 103:191–206

Dick WA, Thavamani B, Conley S, Blaisdell R, Sengupta A (2013) Prediction of beta-glucosidase and betaglucosaminidase activities, soil organic C, and amino sugar N in a diverse population of soils using near infrared reflectance spectroscopy. Soil Biol Biochem 56:99–104

Dunn BW, Beercher HG, Batten GD, Ciavarell S (2002) The potential of near-infrared reflectance spectroscopy for soil analysis—a case study from the Riverine Plain of south-eastern Australia. Aust J Exp Agric 42:607–614

Ehsani MR, Upadhyaya SK, Slaughter D, Shafii S, Pelletier M (1999) A NIR technique for rapid determination of soil mineral nitrogen. Precision Agric 1:217–234

Farmer VC (ed) (1974) The infra-red spectra of minerals. Mineralogical Society, London, UK, pp. 1–539

Fidencio PH, Poppi RJ, De Andrede JC (2002) Determination of organic matter in soils using radial basis function networks and near infrared spectroscopy. Anal Chim Acta 453:125–134

Freschet GT, Barthès BG, Brunet D, Hien E, Masse D (2011) Use of Near Infrared Reflectance Spectroscopy (NIRS) for predicting soil fertility and historical management. Commun Soil Sci Plant Anal 42:1692–1705

Fystro G (2002) The prediction of C and N content and their potential mineralisation in heterogeneous soil samples using Vis-NIR spectroscopy and comparative methods. Plant Soil 246(2):139–149

Gao Y, Gao J, Chen J (2011) Spatial variation of surface soil available phosphorous and its relation with environmental factors in the Chaohu Lake Watershed. Int J Environ 8(8):3299–3317

Genot V, Colinet G, Bock L, Vanvyve D, Reusen Y, Dardenne P (2011) Near infrared reflectance spectroscopy for estimating soil characteristics valuable in the diagnosis of soil fertility. J Near Infrared Spectrosc 119:117–138

Gobeille A, Yavitt J, Stalcup P, Valenzuela A (2006) Effects of soil management practices on soil fertility measurements on Agave tequilana plantations in Western Central Mexico. Soil Tillage Res 87:80–88

Goddu RF, Delker DA (1960) Spectra-structure correlations for the near- infrared region. Anal Chem 32:140–141

Groenigen JW, Mutters CS, Horwath WR, Kessel C (2003) NIR and DRIFT-MIR spectrometry of soils for predicting soil and crop parameters in a flooded field. Plant Soil 250:155–165

Guerrero C, Rossel RAV, Mouazen AM (2010) Special issue diffuse reflectance spectroscopy in soil science and land resource assessment. Geoderma 158:1–2

Haberhauer G, Gerzabek MH (2001) FTIR-spectroscopy of soils characterization of soil dynamic processes. Trends Appl Spectrosc 3:103–109

Harris RF, Karlen DL, Mulla DJ (1996) A conceptual framework for assessment and management of soil quality and health. In: Jones AJ, Doran JW (eds) Methods for assessing soil quality, vol 49. SSSA, Special Publication, Madison, WI, pp 61–82

Hart JR, Norris KH, Golumbic C (1961) Determination of the moisture content of seeds by near infrared spectrophotometry of their methanol extracts. Cereal Chem 39:94–99

He Y, Huang M, Garcia A, Hernandez A, Song H (2007) Prediction of soil macronutrients content using near-infrared spectroscopy. Comput Electron Agric 58:144–153

Henderson TL, Baumgardner MF, Franzmeier DP, Stott DE, Coster DC (1992) High dimensional reflectance analysis of soil organic-matter. Soil Sci Soc Am J 56:865–872

Hengl T, Heuvelink GBM, Rossiter DG (2007) About regression-Kriging: from equations to case studies. Comput Geosci 33(10):1301–1315

Hindle PH (1997) Towards 2000-the past, present and future of on-line NIR sensing. Process Control Qual 9(4):105–115

Hummel JW, Sudduth KA, Hollinger SE (2001) Soil moisture and organic matter prediction of surface and subsurface soils using an NIR soil sensor. Comput Electron Agric 32:149–165

Hunt GR (1977) Spectral signatures of particulate minerals in visible and near-infrared. Trans Am Geophys Union 58:553

Hunt GR (1982) Spectroscopic properties of rocks and minerals. In: Carmichael RS (ed) Handbook of physical properties of rocks. CRC Press, Boca Raton, pp 295–385

Hunt GR, Salisbury JW (1970) Visible and near-infrared spectra of minerals and rocks. I Silicate Minerals Mod Geol 1:283–300

Idowu OJ, van Es HM, Abawi GS, Wolfe DW, Ball JI, Gugino BK, Moebius BN, Schindelbeck RR, Bilgili AV (2008) Farmer-oriented assessment of soil quality using field, laboratory, and VNIR spectroscopy methods. Plant Soil 3071–2:243–253

Islam K, Singh B, McBratney A (2003) Simultaneous estimation of several soil properties by ultraviolet, visible, and near-infrared reflectance spectroscopy. Aust J Soil Res 41(6):1101–1114

Jahn BR, Linker R, Upadhyaya S, Shaviv A, Slaughter D, Shmulevich I (2006) Mid-infrared spectroscopic determination of soil nitrate content. Biosys Eng 94:505–515

Janik LJ, Merry RH, Skjemstad JO (1998) Can mid-infrared diffuse reflectance analysis replace soil extractions? Aust J Exp Agric 38:681–696. https://doi.org/10.1071/EA97144

Jarvis NJ (2007) A review of non-equilibrium water flow and solute transport in soil macropores: principles, controlling factors and consequences for water quality. Eur J Soil Sci 58:523–546

Johnston CT, Aochi YO (1996) Fourier transform infrared and Raman spectroscopy. In: Bartels JM, Bigham JM (eds) Methods of soil analysis, Part 3. SSSA Inc, American Society of Agronomy, Inc., Madison, pp 269–320

Kang SH, Xing BS (2005) Phenanthrene sorption to sequentially extracted soil humic acids and humins. Environ Sci Technol 39:134–140

Karlen DL, Mausbach MJ, Doran JW, Cline RG, Harris RF, Schuman GE (1997) Soil quality: a concept, definition, and framework for evaluation. Soil Sci Soc Am J 61:4–10

Kleinebecker T, Poelen MDM, Smolders AJP, Lamers LPM, Holzel N (2013) Fast and inexpensive detection of total and extractable element concentrations in aquatic sediments using Near-Infrared Reflectance Spectroscopy (NIRS), vol 8(7). PLoS ONE

Krishnan P, Alexander DJ, Butler B, Hummel JW (1980) Reflectance technique for predicting soil organic matter. Soil Sci Soc Am J 44:1282–1285

Kumar N (2009) Investigating the potentiality of regression Kriging in the Estimation of Soil Organic Carbon versus the Extracted result from the existing soil map. M.Sc. Thesis, ITC and IIRS, Enschede, Netherlands

Kusumo BH, Hedley MJ, Tuohy MP, Hedley CP, Arnold GC (2010) Prediction of soil carbon and nitrogen concentrations and pasture root densities from proximally sensed soil spectral reflectance. In: Viscarra Rosse RA, McBRatney AB, Minasny B (eds) Proximal soil sensing. Springer, pp. 177–190

Linker R, Shmulevich I, Kenny A, Shaviv A (2005) Soil identification and chemometrics for direct determination of nitrate in soils using FTIR-ATR mid-infrared spectroscopy. Chemosphere 61:652–658

Ludwig B, Khanna PK, Bauhus J, Hopmans P (2002) Near infrared spectroscopy of forest soils to determine chemical and biological properties related to soil sustainability. For Ecol Manage 171:121–132

Luleva MI, Werff HVD, Jetten V, Meer FVD (2011) Can infrared spectroscopy be used to measure change in potassium nitrate concentration as a proxy for soil particle movement. Sensors 11:4188–4206

Madejova J, Komadel P (2001) Baseline studies of the clay minerals society source clays: infrared methods. Clays Clay Minerals 49:410–432

Maleki MR, Van Holm L, Ramon H, Merckx R, De Baerdemaeker J, Mouazen AM (2006) Phosphorus sensing for fresh soils using visible and near infrared spectroscopy. Biosyst Eng 95:425–436

Malley DF, Martin PD, McClintock LM, Yesmin L, Eilers RG, Haluschak P (2000) Feasibility of analysing archived Canadian prairie agricultural soils by near infrared reflectance spectroscopy. In: AMC Davies, R Giangiacomo (eds) Near infrared spectroscopy: proceedings of the 9th international conference. NIR Publications, Chichester, UK, pp 579–585

Malley DF, Martin PD, Ben-Dor E (2004) Applications in analysis of soils. In: Roberts CA, Workman J Jr, Reeves JB III (eds) Near-infrared spectroscopy in agriculture. ASA, CSSA, SSSA, Madison, Wisconsin, USA, pp 729–784

Mamo T, Richter C, Heiligtag B (1996) Comparison of extractants for the determination of available phosphorus, potassium, calcium, magnesium and sodium in some Ethiopian and German soils. Commun Soil Sci Plant Anal 27:2197–2212

Martin PD, Malley DF, Manning G, Fuller L (2002) Determination of soil organic carbon and nitrogen at the field level using near-infrared spectroscopy. Can J Soil Sci 82(4):413–422

McCarty GW, Reeves JB (2006) Comparison of near infrared and mid infrared diffuse reflectance spectroscopy for field-scale measurement of soil fertility parameters. Soil Sci 171(2):94–102

Miller CE (2001) Chemical principles of near infrared technology. Near-Infrared Technol Agric Food Ind 2

Moron A, Cozzolino D (2007) Measurement of phosphorus in soils by near infrared reflectance spectroscopy: effect of reference method on calibration. Commun Soil Sci Plant Anal 38:1965–1974

Morris RV, Lauer HV, Lawson CA, Gibson EK, Nace GA, Stewart C (1985) Spectral and other physicochemical properties of submicron powders of hematite (Alpha-Fe2O3), maghemite (Gamma-Fe2O3), magnetite (Fe2O4), goethite (Alpha-FeOOH), and lepidocrocite (Gamma-FeOOH). J Geophys Res 90:3126–3144

Mouazen AM, Maleki MR, De Baerdemaeker J, Ramon H (2007) On-line measurement of some selected soil properties using a VIS-NIR sensor. Soil Tillage Res. 93:13–27

Mouazen AM, Karoui R, De Baerdemaeker J, Ramon H (2005) Classification of soil texture classes by using soil visual near infrared spectroscopy and factorial discriminant analysis techniques. J Near Infrared Spectrosc 13:231–240

Nanni MR, Dematte JAW (2006) Spectral reflectance methodology in comparison to traditional soil analysis. Soil Sci Soc Am J 70:393–407

Neter J, Kutner MH, Nachtsheim CJ, Wasserman W (1996) Applied linear statistical models, 4th edn, 67(3–4). McGraw- Hill, pp 215–226

Niederberger J, Todt B, Boča A, Nitschke R, Kohler M, Kühn P, Bauhus J (2015) Use of near—infrared spectroscopy to assess phosphorus fractions of different plant availability in forest soils. Biogeosciences 12:3415–3428

Niemoller A, Behmer D (2008) Use of near infrared spectroscopy in the food industry. Nondestr Test Food Qual:67–118

Nocita M, Stevens A, Toth G, Panagos P, van Wesemael B, Montanarella L (2014) Prediction of soil organic carbon content by diffuse reflectance spectroscopy using a local partial least square regression approach. Soil Biol Biochem 68:337–347

Olsen SR, Sommers LE (1982) Phosphorus. In: Page AL, et al (eds) Methods of soil analysis, Part 2, 2nd edn, Agron Monogr 9. ASA and ASSA, Madison WI, pp 403–430

Palmborg C, Nordgren A (1993) Modeling microbial activity and biomass in forest soil with substrate quality measured using near infrared reflectance spectroscopy. Soil Biol Biochem 25:1713–1718

Pasquini C (2003) Near infrared spectroscopy: fundamentals, practical aspects and analytical applications. J Braz Chem Soc 14:198–219

Pasquini C, da Silva HEB, Guchardi R (2000) In near infrared spectroscopy: proceedings of the 9th International conference. In: Davies AMC, Giangiacomo R (eds). NIR Publications, Chichester, p. 109

Pereira AG, Gomez AH, He Y (2004) Near-infrared spectroscopy potential to predict N, P, K and OM content in a loamy mixed soil and its combination with precision farming. In: CIGR international conference Beijing. Beijing, China

Pirie A, Singh B, Islam K (2005) Ultra-violet, visible, near-infrared and mid-infrared diffuse reflectance spectroscopies techniques to predict several soil properties. Aust J Soil Res 43:713–721

Post JL, Noble PN (1993) The near-infrared combination band frequencies of dioctahedral smectites, micas, and illites. Clays Clay Miner 41:639–644

Quyang Y, Zhang JE, Ou LT (2006) Temporal and spatial distributions of sediment total organic carbon in an estuary river. J Environ Qual 35:93–100

Reeves JB, McCarty GW, Meisinger JJ (1999) Near infrared reflectance spectroscopy for the analysis of agricultural soils. J Near Infrared Spectrosc 7(3):179–193

Reeves JB III, McCarty GW (2001) Quantitative analysis of agricultural soils using near infrared reflectance spectroscopy and a fiber-optic probe. J Near Infrared Spectrosc 9:25–43

Reich G (2005) Near-infrared spectroscopy and imaging: basic principles and pharmaceutical applications. Adv Drug Deliv Rev 57:1109–1143

Sankey JB, Brown DJ, Bernard ML, Lawrence RL (2008) Comparing local versus global visible and near-infrared (Vis-NIR) diffuse reflectance spectroscopy (DRS) calibrations for the prediction of soil clay, organic C and inorganic C. Geoderma 148(2):149–158

Savvides A, Corstanje R, Baxter SJ, Rawlins BG, Lark RM (2010) The relationship between diffuse spectral reflectance of the soil and its cation exchange capacity is scale-dependent. Geoderma 154:353–358

Schimann H, Joffre R, Roggy JC, Lensi R, Domenach AM (2007) Evaluation of the recovery of microbial functions during soil restoration using near-infrared spectroscopy. Appl Soil Ecol 37:223–232

Schlesinger WH (1990) Evidence from chronosequence studies for a low carbonstorage potential of soils. Nature (London) 348:232–234

Schnitzer M, Schulten HR, Schuppli P, Angers DA (1991) Extraction of organic matter from soils with water at high pressures and temperatures. Soil Sci Am J 55:102–108

Schulze DG (2002) An introduction to soil mineralogy. In: Dixon JB, Schulze DG (eds) Soil mineralogy with environmental applications. Soil Science Society of America Inc., Madison, WI, pp 1–35

Schwanghart W, Jarmer T (2011) Linking spatial patterns of soil organic carbon to topography—a case study from south-eastern Spain. Geomorphology 126:252–263

Shaddad SM, Madrau S, Mouazen AM (2014) Mapping of C/N ratio as an indicator of soil biological activity by on-line vis-NIR sensor in a field in UK. In: Proceedings international conference of Agriculture Engineering, Zurich, 6–10 July

Shao YN, He Y (2011) Nitrogen, phosphorus, and potassium prediction in soils, using infrared spectroscopy. Soil Res 49:166–172

Shenk JS, Westerhaus MO (1991) Population definition, sample selection, and calibration procedures for near infrared reflectance spectroscopy. Crop Sci 31:469–474

Shepherd KD, Walsh MG (2002) Development of reflectance spectral libraries for characterization of soil properties. Soil Sci Soc Am J 66(3):988–998

Sherman DM, Waite TD (1985) Electronic spectra of Fe^{3+} oxides and oxyhydroxides in the near infrared to ultraviolet. Am Mineral 70:1262–1269

Shibusawa S, Made Anom SW, Sato HP, Sasao A (2001) Soil mapping using the real-time soil spectrometer. In: Gerenier G, Blackmore S (eds) ECPA 2001, vol 2. Agro Montpellier, Montpellier, pp. 485–490

Sørensen LK, Dalsgaard S (2005) Determination of clay and other soil properties by near infrared spectroscopy. Soil Sci Soc Am J 69(1):159–167

Stenberg B (2010) Effects of soil sample pretreatments and standardised rewetting as interacted with sand classes on Vis-NIR predictions of clay and soil organic carbon. Geoderma 158(1–2):15–22

Stenberg B (1999) Monitoring soil quality of arable land: microbiological indicators. Acta Agric Scand, Sect B Soil Plant Sci 49:1–24

Stenberg B, Jonsson A, Börjesson T (2002) Near infrared technology for soil analysis with implications for precision agriculture. In: Davies A, Cho R (eds) Near infrared spectroscopy: proceedings of the 10th international conference. NIR Publications, Chichester, UK, pp 279–284

Stenberg B, Jonsson A, Bo rjesson T (2000) Snabbmetoder för bestämning av lerhalt och kalkbehov med hjälp av NIR-analys. (Rep. No. 2000–2. ODAL FoU, Lidköping)

Stenberg B, Rossel RAV, Mouazen AM, Wetterlind J (2010) Visible and near infrared spectroscopy in soil science. Adv Agron 107:163–215

Stevens A, Wesemael B, Bartholomeus H, Rosillon D, Tychon B, Ben-Dor E (2008) Laboratory, field and airborne spectroscopy for monitoring organic carbon content in agricultural soils. Geoderma 144:395–404

Terhoeven-Urselmans T, Michel K, Helfrich M, Flessa H, Ludwig B (2006) Near-infrared spectroscopy can predict the composition of organic matter in soil and litter. J Plant Nutr Soil Sci 169:168–174

Tiruneh GG (2014) Rapid soil quality assessment using portable visible near infrared (VNIR) spectroscopy. M.Sc. Thesis, Uppsala University, Sweden

Turner et al (2003) A framework for vulnerability analysis in sustainability science. Proc Nat Acad Sci 100(14):8074–8079

Twomey SA, Bohren CF, Mergenthaler JL (1986) Reflectance and albedo differences between wet and dry surfaces. Appl Opt 25(3):431–437

Udelhoven T, Emmerling C, Jarmer T (2003) Quantitative analysis of soil chemical properties with diffuse reflectance spectrometry and partial least-square regression: a feasibility study. Plant Soil 251:319–329

Varvel GE, Schlemmer MR, Schepers JS (1999) Relationship between spectral data from an aerial image and soil organic matter and phosphorus levels. Precision Agric 1(3):291–300

Viscarra Rossel RA, McBratney AB (1998) Soil chemical analytical accuracy and costs: implications from precision agriculture. Aust J Exp Agric 38:765–775

Viscarra Rossel RA (2009) The Soil Spectroscopy Group and the development of a global soil spectral library. NIR News 20:17–18

Viscarra Rossel RA, McGlynn RN, McBratney AB (2006). Determing the composition of mineral-organic mixes using UV-vis-NIR diffuse reflectance spectroscopy. Geoderma 137:70–82

Vohland M, Ludwig M, Thiele-Bruhn S, Ludwig B (2014) Determination of soil properties with visible to near- and mid-infrared spectroscopy: effects of spectral variable selection. Geoderma 223:88–96

Wang SQ, Shu N, Zhang HT (2008) In-site total N content prediction of soil with VIS/NIR spectroscopy. Spectrosc Spect Anal 28(4):802–812

Webster R, Oliver MA (2001) Geostatistics for environmental scientists. Wiley, Chichester

Wetterlind J, Stenberg B, Jonsson A (2008) Near infrared reflectance spectroscopy compared with soil clay and organic matter content for estimating within-field variation in N uptake in cereals. Plant Soil 302:317–327

Wetterlind J, Stenberg B, Soderstrom M (2010) Increased sample point density in farm soil mapping by local calibration of visible and near infrared prediction models. Geoderma 156(3–4):152–160

Johanna Wetterlind, Bo Stenberg, Rossel Viscarra, Raphael A (2013) Soil analysis using visible and near infrared spectroscopy. Methods Mol Biol 953:95–107

Wetzel DL (1983) Near-infrared reflectance analysis—sleeper among spectroscopic techniques. Anal Chem 55(12):1165–1171

Xie HT, Yang XM, Drury CF, Yang JY, Zhang XD (2011) Predicting soil organic carbon and total nitrogen using mid- and near- infrared spectra for Brookston clay loam soil in Southwestern Ontario, Canda. Can J Soil Sci 91:53–63

Xing BS, Liu JD, Liu XB, Han XZ (2005) Extraction and characterization of humic acids and humin fractions from a black soil of China. Pedosphere 14:1–8

Xu S, Shi X, Wang M, Zhao Y (2016) Effects of Subsetting parent materials on prediction of soil organic matter content in a Hillv Area using Vis- NIR Spectroscopy, vol 11(3). PLoS ONE

Zbiral J, Nemec P (2002) Comparison of Mehlich 2, Mehlich 3, CAL, Egner, Olsen, and 0.01 M CaCl2 extractant for determination of phosphorus in soils. Commun Soil Sci Plant Anal 33:3405–3417

Part IV
RS and GIS for Land Cover/Land Use Change Monitoring

Chapter 7
Multi-temporal Satellite Data for Land Use/Cover (LULC) Change Detection in Zakho, Kurdistan Region-Iraq

Yaseen T. Mustafa

Abstract Historical and current status of the land is essential for efficient environmental management. This can especially be noticed in regions that are vitally affected by climate variability and human activities such as Zakho district, Kurdistan Region-Iraq. The information and status of land use/cover (LULC) help to design an efficient and sustainable environmental management program. The present study illustrates the spatiotemporal dynamics of LULC in Zakho district, Iraq. Landsat satellite imageries of two different time periods, i.e., Landsat Thematic Mapper (TM) of 1989 and Landsat Operational Land Imager (OLI) of 2017 were acquired and the changes in Zakho over a period of 28 years were quantified. Supervised classification methodology has been employed using Maximum Likelihood Algorithm. The images were categorized into eight different classes namely dense forest, sparse forest, grass, rock, soil, crop, built-up and water body. The results showed that during the last 28 years, build-up land had been increased from 9 km^2 in 1989 to 49 km^2 in 2017. Crops and rocks lands have been increased as well by about 102.1 and 15.39 km^2, respectively. Moreover, a very slight increase has been observed in water body and soil by about 3.5 and 0.98 km^2, respectively. On the other hand, dense forest, spare forest, and grass lands have been decreased by 92.83, 14.26, and 53.68 km^2, respectively. This chapter concluded that a major change in Zakho district land happened in a negative trend regarding the natural environment.

Keywords Environmental management · Land use/cover · Remote sensing · Landsat data · Zakho

Y. T. Mustafa (✉)
Environmental Science Department, Faculty of Science, University of Zakho, Duhok, Kurdistan Region, Iraq
e-mail: yaseen.mustafa@uoz.edu.krd

© Springer Nature Switzerland AG 2020
A. M. F. Al-Quraishi and A. M. Negm (eds.),
Environmental Remote Sensing and GIS in Iraq, Springer Water,
https://doi.org/10.1007/978-3-030-21344-2_7

7.1   Introduction

In most developing countries, the environment and its preservation do not count within the priority and are almost neglected. The constant neglect of the environment could lead to irreparable damage and cannot be recovered (French 2003). On this basis, the role of researchers and environmental conservatives comes to warn persons in charge and provide scientific reports. One aspect that helps to assess the state of the environment is the estimated changes in the land use/cover.

Land use/cover (LULC) affects local, global environment, climate and land degradation that reduces ecosystem services and functions (Tolessa et al. 2017). Monitoring of LULC is important to assess the change and manage the environment (Jawarneh Rana and Biradar 2017; Butt et al. 2015; Xi et al. 2017). Such a task can be achieved by using modern techniques. Remote sensing is one of the effective techniques that play an active role in accomplishing such tasks. This is because the information can be obtained from remotely sensed data efficiently and cheaply and repetitive coverage at short intervals with consistent image quality (Mustafa and Habeeb 2016). Thus, change detection has become a major application of remotely sensed data.

One of the consistent and remote sensed data regarding spatial and temporal coverage is the Landsat satellite data. Landsat data is widely used for change detection of LULC and mapping earth surface. Also, Landsat images used to map and extract several objects on the earth surface. This is because it has been providing data since 1972 and it has a fixed acquisition schedule. On this basis, its use has improved knowledge of land-surface processes across spatio-temporal scales (Cohen and Goward 2004; Vogeler et al. 2018). Fadhil (2010) used Landsat images to report and show the spatial distribution of the drought states in a part of the north of Iraq. Different techniques with remotely sensed data were evaluated to map lakes and rivers as Elsahabi et al. (2016) used Landsat data to examine the performance of several techniques for extracting water surface of Lake Nubia, in Egypt. Moreover, Mustafa et al. (2012) used Landsat images to monitor and evaluate the change in urban areas, vegetation, soil and water bodies in Duhok city, Kurdistan Region of Iraq. They found that the urban area has expanded at the expense of green areas.

This chapter aims to provide information on LULC using multi-temporal Landsat images in Zakho district, Duhok, Kurdistan region-Iraq for which detailed thematic maps are currently lacking. This can be split into three subtasks: 1) Identify and define types of LULC for the study area, 2) examining the variation in the distribution of LULC, and 3) provide an up-to-date database and produce accurate maps.

7.2 Materials and Methods

7.2.1 Study Area

Zakho district is located between lat. 37° 22′–37° 00′ N and long. 43° 11′–42° 22′ E in the northwest of Duhok province, Kurdistan Region-Iraq. It covers an area of about 1,454 km^2 and includes three sub-districts: Rezgary, Batifa, and Darkar (Fig. 7.1). The elevation ranges from 350 to 2,250 m. It includes a series of mountains started from the southeast to the northwest of the study area. The terrain of the area is varied between mountainous areas (covering an area of about 559 km^2), rolling areas and hills (approximate area 425 km^2), and plains and valleys (with an area about 470 km^2). The latest area type is considered the most important areas of the study, as the human settlement are located there. Moreover, it is considered the largest production in terms of agricultural crops, given the fertility of soil, which was formed by the deposition of rivers and floods.

Zakho has a strategic location as it lies at the junction of the borders of Iraq, Turkey and Syria. There is one river passing the area with its tributaries and is known as the Khabour river. It is the main source of Zakho water supply for its various uses. The length of the Khabour river reaches 166.8 km, and it is expanded within the boundaries of Zakho district. The climate in winter is a rainy cold and in summer is sunny dry, where area shows a variety of land cover (Othman 2008).

According to Directorate of statistics in Duhok (Directorate of Statistics Duhok 2011), the population of Zakho has a remarkable change during the period (1987–2010). The population reached 93,377 people in the year 1987 and increased to become 115,303 in 1998.

7.2.2 Data and Pre-processing

7.2.2.1 Landsat Images

Landsat data of 30 m spatial resolution with Path/Row (170/34) were used in this study. Two Landsat products were downloaded from USGS (United States Geological Survey) website for the years 1989, and 2017. They are Landsat Thematic Mapper (TM) acquired on 28 June 1989 and Landsat Operational Land Imager (OLI) acquired on 17 June 2017. For each Landsat image, five bands were considered in this study: blue, green, red, near infrared and middle infrared.

The images were calibrated and corrected to remove the noise. This is performed first by calibrating digital number (DN) into radiance by using the information from the metadata files. Then, the radiance data were converted to surface reflectance. These processes were achieved using FLAASH (Fast Line-of-sight Atmospheric

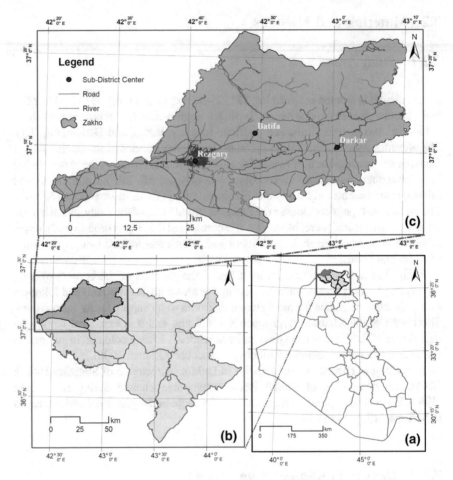

Fig. 7.1 Map of the study area; **a** Iraq, **b** Duhok Province, **c** Zakho district

Analysis of Spectral Hypercubes) module within ENVI 5 (Yuan and Niu 2008). Most of the required information by FLAASH were obtained from the metadata files, while others were already included in the module (Module 2009).

To get good alignment of pixels in the respective images, an image-to-image co-registration was performed with the root mean square error (RMSE) of half a pixel. Moreover, the images were georeferenced to the UTM (Universal Transverse Mercator) Zone 38 North with a WGS (World Geodetic System) 84 datum.

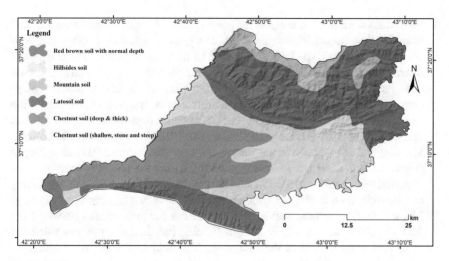

Fig. 7.2 Soil type in Zakho district

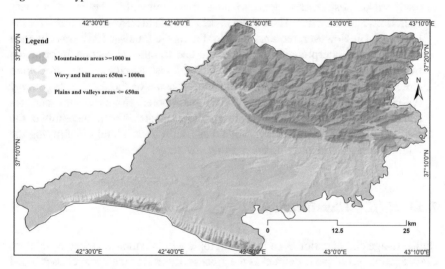

Fig. 7.3 The topography of Zakho district

7.2.2.2 Reference Data Sources

Reference data were also included in the LULC change detection. They are:

1 **Elevation and slope data** that were derived from the DEM (Digital Elevation Model) of ASTER with a spatial resolution of 30 m × 30 m.
2 **Soil and geomorphology** were derived and obtained from (Sissakian et al. 2011; Buringh 1960), and maps were created after rasterizing and resampling with 30 m using ArcGIS 10.3 (Figs. 7.2 and 7.3, respectively).

3 **Field data** were used as samples for training and validation of the created LULC
   from Landsat images. Before the field survey, the Landsat OLI image was pan-
   sharpened to have a spatial resolution of 15 m using Gramm-Schmidt Spectral
   Sharpening algorithm. This technique is well explained and presented in (Mustafa
   and Habeeb 2014). It is a process of combining the higher resolution image
   containing fine spatial information (panchromatic pixel) with the lower-resolution
   (color pixel) imagery containing spectral information to produce a high-resolution
   spectral image (Pu and Landry 2012; Yan et al. 2018). This image was interpreted
   visually, and the samples were distributed over the images. Data sampling in
   this study was designed following a stratified random sampling method. This
   is performed to assure a reliable representation of the LULC variability, taking
   into consideration the topographic diversity of the area when surveyed (Tso and
   Mather 2009). The field survey was carried out for three weeks starting from
   July 15, 2017, with a handheld GPS (Global Positioning Systems) using the
   same projection system of the study area. The field survey did not include the
   mountainous areas in the north and south of the study area due to the lack of
   roads. Also, most of these areas are unsafe; and fall within the line of military
   operations and activities. The collected data from field survey was useful for the
   Landsat OLI. However, required data for the image Landsat 1989 was obtained
   through visual interpretation of the image and through collecting information
   from the government offices as a municipality, forest directorate, and agriculture
   directorate … etc. The LULC classification scheme of the Geological Survey
   (USGS) (Mannion 2002) and FAO (2000) was adopted. However, some changes
   and modifications were made to represent these categories with the reality of the
   study area (Table 7.1). Moreover, that was useful to accurately identifying the
   training areas and identifying the existing coverings.

7.2.3   LULC Classification

Digital image classification is an important topic within remote sensing techniques.
Classification technique is defined as the process of sorting pixels into classes based
on their data file values in remotely sensed imagery (Butt et al. 2015). Maximum
Likelihood (ML) algorithm was used for Landsat images classification. ML technique is based on a statistical decision criterion, which assists in the classification
of overlapping signatures. In this way, pixels will be assigned to the class or area of
highest probability (Tso and Mather 2009; Xu et al. 2018). The training data were
used for the classification process, and classified maps were created. Polygons were
created from the training data for each LULC class and used as a vector layer of
training areas. This process was achieved using ArcGIS 10.3. Next, polygons were
converted to the region of interest (ROI) using ENVI 5, then the classification process
performed.

Table 7.1 Adapted LULC classification scheme from USGS and FAO (Mannion 2002; FAO 2000)

LULC classes	Description	Key
Dense forest	This area of LULC is covered with various types of trees, including natural and industrial, and thickly	DF
Sparse forest	This type of LULC is covered with grass, low-density trees and shrubs	SF
Grass	This area was characterized by land that covered the grass with scattered shrubs	G
Rock	This type of LULC is mainly covered by rocks (Conglomerate rocks, sand rocks, gypsum and calcareous rocks, and gravel)	R
Soil	This category includes areas covered with soils scattered in the study area and small spaces	S
Water	The water body in the study area includes rivers and dams	W
Cropland	This type of cover includes crop cultivation areas	C
Built-up area	This type of LULC includes areas that is covered by buildings	B

7.2.4   LULC Mapping: Post Classification Change Detection

Post classification process within this work consisted of two steps: smoothing classified map (as a raster) and change detection. Smoothing classified maps is an important step that needs to be performed after classification results. This is because the spectral overlap between the LULC patterns within the satellite image results in the appearance of some extreme pixel in the classification process. This may lead to mix and interference pixels with other classes, especially in the case of the small area of LULC patterns. Two processes were used for smoothing technique. They are sieve, and clump processes. Sieve is run first to remove the isolated pixels, and that is based on a size threshold. While clump is run second to add spatial coherency to existing classes (EXELIS 2013).

Post-classification change detection that is based on pixel-based was performed using the classified maps 1989 and 2017. This step includes the process of comparing each LULC maps from two different years (1989 and 2017) to determine change areas. It points out on "from to" change detection technique that provides a change matrix. Moreover, to determine the quantity of conversions from a particular LULC to another LULC category, cross-tabulation analysis on a pixel by pixel basis was performed. Thus, a new thematic map was produced containing different combinations of "from to" change classes.

7.2.5 Accuracy and Area Assessment

Accuracy assessment of 1989 and 2017 classification images was carried out to check the quality of the derived information from the data. It is defined as the process of comparing the classification results with the truth (validation) data that were collected from the field survey. Thus, to implement the classification accuracy process, a confusion matrix was calculated. The non-diagonal values of in the columns of confusion matrix refer to the omission error, while the non-diagonal values in the rows of confusion matrix refer to the commission error. The most important indicators that are resulted with the confusion matrix are overall accuracy (OA), and Kappa coefficient (KC) (Butt et al. 2015).

7.3 Results and Discussion

7.3.1 LULC Analysis

The classified LULC map is generated based on the defined classes (Table 7.1). Few classified pixels have been incorporated within neighboring classes following the techniques that were explained in the previous section (Sect. 2.4). Those are only the pixels that their proportion is too small and is not even reflected clearly. The ultimate classified LULC map of Zakho district of years 1989 and 2017 is given in Fig. 7.4a, b. The classified LULC were evaluated using confusion matrix. The resulted classification accuracy gained for 1989 and 2017 is reported in Tables 7.2 and 7.3, respectively. The number of columns and rows in these tables is equal to the LULC classes that are used in the study. The columns represent the LULC patterns in the classified image, and the rows represent the LULC patterns in reality. From Tables 7.2 and 7.3, we observed that the total number of pixels that are used for 1989 of all LULC is 1976 pixels, and for 2017 is 3968 pixels. While, the total number of the corrected classified pixel for the year 1989 is 1713, and 3740 for the year 2017.

The overall accuracy of the classified images (1989 and 2017) was found to be 89.67 and 94.62%, respectively with a Kappa coefficient of 0.87, and 0.94. This shows that the classified maps (1989 and 2017) were created successfully. This is reported in several publications (Tolessa et al. 2017; Tso and Mather 2009) as the acceptable overall accuracy for LULC classification exceed 85% with Kappa values greater than 0.80. The overall accuracy of the classification varies between the selected years (1989 and 2017). This is because the used data of assigning training areas is different in the classification process as the field data were used to classify the image of 2017. While, in the classification of the 1989 image, the historical data and topographic maps were used in the classification process and these data were less accurate than field data. In addition, the latter satellite image (Landsat-8) is more sharpen as it contains 11 spectral bands.

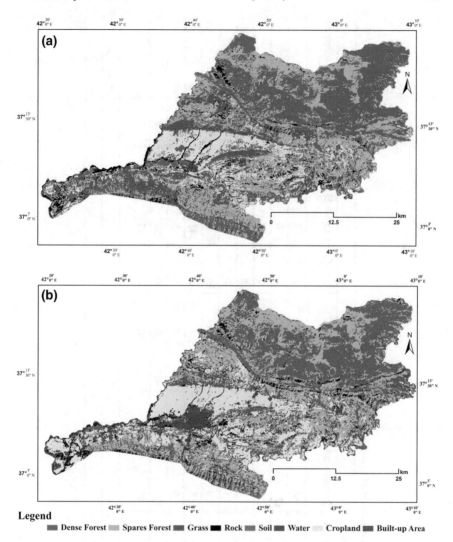

Fig. 7.4 Classified LULC map of Zakho district **a** 1989, **b** 2017

7.3.1.1 LULC of 1989

The classified LULC maps were converted from raster to vector using ENVI 5. This is performed to calculate the area of each class. Table 7.4 shows the area of each LULC class; for the year 1989, which is calculated using ArcGIS 10.3. It shows that the area of dense forest covers the major part of the study area as it counts 505 km^2 with a ratio of 34.68%. As shown in the Fig. 7.4a, most of this LULC type is located in north and south of Zakho. The next major cover type of the LULC is the sparse

Table 7.2 Accuracy (confusion matrix) of the classified LULC for the year 1989

LULC		Classified								Total	User's accuracy (%)
		Dense forest	Sparse forest	Grass	Rock	Soil	Water	Cropland	Built-up		
Referenced	Dense forest	**514**	14	14	18	0	0	2	5	567	91
	Sparse forest	20	**130**	2	8	0	0	0	0	160	81
	Grass	0	20	**185**	0	0	2	19	0	226	82
	Rock	0	1	2	**366**	0	5	7	25	406	89
	Soil	0	1	0	2	**20**	0	1	0	24	83
	Water	0	0	0	2	0	**56**	0	58	116	95
	Cropland	3	4	17	0	0	2	**347**	0	373	92
	Built-up	0	0	0	4	0	0	5	**95**	104	89
	Total	537	170	220	400	20	65	381	183	1976	
	Producer's accuracy (%)	95	77	81	90	93	88	95	80		
Overall accuracy											89.67
Kappa coefficient											0.87

Table 7.3 Accuracy (confusion matrix) of the classified LULC for the year 2017

LULC	Classified									
Referenced		Dense forest	Sparse forest	Grass	Rock	Soil	Water	Cropland	Built-up	Total	User's accuracy (%)
	Dense forest	**625**	22	0	1	0	0	2	0	650	96
	Sparse forest	22	**236**	2	4	0	0	5	0	269	83
	Grass	0	20	**385**	0	0	0	35	1	441	90
	Rock	0	0	1	**327**	0	0	0	33	361	91
	Soil	0	0	0	0	**64**	0	17	0	81	84
	Water	0	0	0	0	0	**43**	0	2	45	92
	Cropland	0	0	24	2	4	0	**1701**	2	1733	97
	Built-up	0	0	0	28	0	1	0	**359**	388	96
	Total	647	278	412	362	68	44	1760	397	3968	
	Producer's accuracy (%)	95	98	88	89	92	90	95	93		
Overall accuracy											94.62
Kappa coefficient											0.94

Table 7.4 LULC areas with their percentage of Zakho district for the years 1989 and 2017

LULC	1989		2017	
	Area (km^2)	%	Area (km^2)	%
Dense forest	504.71	34.72	411.88	28.34
Sparse forest	408.24	28.09	393.98	27.10
Grass	241.00	16.58	187.32	12.89
Rock	65.48	4.50	80.87	5.56
Soil	8.94	0.62	9.92	0.68
Water	2.04	0.14	5.54	0.38
Cropland	214.26	14.74	316.36	21.76
Built-up	8.87	0.61	47.67	3.29
Total	1453.54	100.00	1453.54	100.00

forest with an area of 408 km^2 and a ratio of 28.02%. This type of LULC is located in the north, northeast and southeast of the map (Fig. 7.4a).

The grass is ranked third based on the area that was occupied (239 km^2 and 16.41%). It is located in the hills and the mountains in the north, west and south of the plains area. The third major LULC type in the study is croplands. They cover an area of 219 km^2 with a ratio of 15.04% of the total area, and they are located in the centre and east of Zakho.

The rest of the LULC types counts the minority. Rock covers an area of 65 km^2 and with a ratio of 4.46% that is appeared in the high mountains, south-west and south-east of the study area and on the banks of the rivers. Whereas, soil LULC covers an area of 9 km^2 (0.62%) of the total area, which is particularly prominent in Mount Deira and Kira and at the foothills of the hills. The built-up area covered an area of about 9 km^2 (0.62%) of the total area of the study area and occupied at the plains areas. Finally, the lowest LULC type in the study area is water, which is estimated at 2 km^2 (0.15%) and mainly existed within major rivers.

7.3.1.2   LULC of 2017

Figure 7.4b shows the resulted LULC of the study for the year 2017. The area of each LULC class with their percentage is extracted from Fig. 7.4b and shown in Table 7.4.

Dense forest LULC occupied the largest area of the study area, around 401 km^2, with a ratio of 27.53%. It is concentrated in the northern parts of Zakho. While the sparse forests that are located in the northern, eastern and southern parts of the study area count 373 km^2 (25.62%). Croplands come third based on their occupied area. They covered an area of 351 km^2 with 24.11% of the total area of Zakho district and located in the southeast, southwest and middle of Zakho. The grass covers an area of 183 km^2 with 12.57% of the total area. From the Fig. 7.4b, grass LULC is concentrated in the hills and some eastern parts of Zakho. Regarding rock LULC, they occupied different parts of the region, particularly in the high mountains, south-east and south-west, and at the river banks, with an area of 85 km² and 5.84%. The built-up areas occupied a ratio of 3.37% with an area of 49 km² of the total area. While soil and water LULC cover an area of 7 km² (0.48%), and 7 km² (0.48%); respectively, of the total area of the study area.

7.3.2   LULC Patterns Change in the Study Area Between 1989 and 2017

The changed areas with their percentage within the Zakho district over 28 years are summarized in Table 7.5. The negative/positive sign refers to the decrease/increase of the LULC area. This is also shown in Fig. 7.5, as the percentage decline and improvement of the LULC areas are graphically presented. The comparison of each LULC class of 1989 and 2017 showed that there had been a marked LULC change during the study period. Dense and sparse forests land were the predominant LULC (34.72 and 28.09%) during 1989, although they declined with a rate of 6.38 and 0.99% as they reached to 28.34 and 27.10% respectively, in 2017. This is due to climate changes in general, overgrazing, logging and fire. Also, the region experienced a difficult economic situation especially during 11 years (1990–2000) because of the lack of fuel and electricity, which forced people to use trees as a main source of energy. All reasons above have led to a lack of forest area.

Moreover, a noticeable decrement was observed on grass LULC from 241 to 187.32 km², with a declined rate of 3.69%. This is due to the degradation of these lands and human activities of the infrastructure in the region. Also, agricultural land is expanded at the expense of grassland, and this is observed as croplands significantly increased with a rate of 7.02%. The reason behind that is the adoption of agricultural development projects at the regional level in general, and in the study

Table 7.5 Area and percentage change of LULC from 1989 to 2014

LULC	1989–2017	
	Change in area (km²)	Change in percentage (%)
Dense forest	−92.83	−6.38
Sparse forest	−14.26	−0.99
Grass	−53.68	−3.69
Rock	15.39	1.06
Soil	0.98	0.06
Water	3.5	0.24
Cropland	102.1	7.02
Built-up	38.8	2.68

The negative sign refers to the decrease in the area

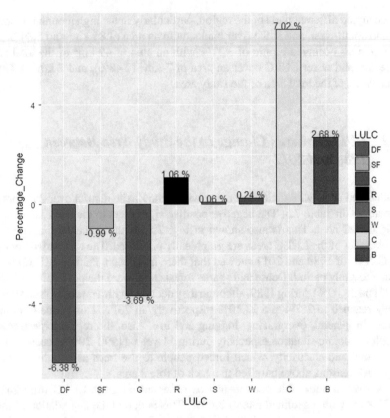

Fig. 7.5 Gain and loss percentage change of the 8 classes of LULC (DF: Dense Forest, SF: Sparse Forest, G: Grass, R: Rock, S: Soil, W: Water, C: Cropland, B: Built-up) during 28 years (1989–2017)

area in particular. Also, the study area enjoys wide areas of plains land. Adding to that, the agricultural craft is the main profession in the plains area, particularly growing of cereal crops. This reflects what has been obtained in the classification results as a major increment occurred in cropland.

The second class which faced significant increment (improvement) in the total area was a Built-up area with 2.68% (Fig. 7.5). It occupied an area of 9 km^2 in 1989 and increased to become 49 km^2. The region has witnessed a horizontal expansion in residential and commercial areas. It expanded along the main roads, especially south of the Khabur river, in the direction of the border crossing of Ibrahim al-Khalil. It was exploited for extensive commercial activity. It also expanded into the agricultural plain and farmlands at Al-Mahdaria north of the district centre and towards Talker, which was part of the district centre. This expansion was mainly at the expense of agricultural land, and this is essentially being due to migration from the countryside to the city and the increase in population. Migration from the countryside to the city was the most important form of internal migration in Zakho, which contributed to the increase in population and expansion of urban centres (Eklund 2015).

According to the information revealed by classification results, the rock area showed a minor increase (4.5–5.56%). This is because of the human activities for the provision of building materials, especially with regard to gravel quarries, which witnessed a significant expansion of the industrial and economic development in the study area. The need for building materials became an urgent necessity, adding to that the study area is rich in gravel in all its forms and sizes. The total area of quarries during 2007 was 3.4 km², compared to 0.35 km² in 1972 (Directorate of Statistics Duhok 2008).

An increment was observed in the water body, as the rate of increment reached 0.24%, and this is accounted for the establishment of several agricultural dams. A minor improvement was observed on soil LULC with a change percentage of 0.06%. The area was increased to 9.92 km² in 2017 from 8.94 km² in 1989. This is because in recent years the Ministry of Agriculture has implemented some plantation projects in the affected areas by erosion especially in sloping areas (Directorate of Forests and Range in Duhok 2013).

7.3.3 LULC Conversions for the Last 28 Years (1989–2017)

Change detection comparison was performed under the post-classification step. This step produced a change map for comprehending the spatial pattern of the change between 1989 and 2017 (Fig. 7.6). The two LULC classified maps were overlaid to

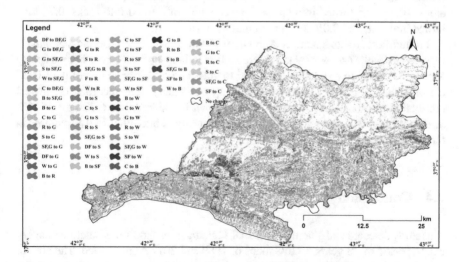

Fig. 7.6 LULC changes between 1989 and 2017; where DF: Dense Forest, SF: Sparse Forest, G: Grass, R: Rock, S: Soil, W: Water, C: Cropland, and B: Built-up

produce the LULC change map. Moreover, the cross-tabulation matrix between the two dates is also calculated. The conversions among LULC were calculated using ArcGIS 10.3.

The cross-tabulation matrices (Table 7.6) show the shift in the LULC classes and the areas of the converted LULC were calculated. Table 7.6 showed that 832.5 km² with 57.27% of the LULC area remained stable with no change in the Zakho district during the study period. This is shown in Fig. 7.6a white area with the map of Zakho district during the study period. Out of 504.71 km² that was dense forests in 1989, 344.49 km² was still dense forest in 2017, but 160.22 km² was converted to the other LULC classes. Out of the converted dense forests, 95.49 km² was turned into sparse forests, 21.91 km² converted to field crops, 17.27 km² to grass, 17.03 km² to the rocks, 6.63 km² to the built-up area and 1.89 km² to soil and water body.

The second highest shift occurred in grassland where it was of 241 km² in 1989, and 85.76 km² of the area was lost mainly to croplands, the sparse forests retained 51.07, 8.64 km² for dense forests, rock got 10.16, and 3.97 km² to built-up areas. While soil and water body retained the lowest area from grass with 2.5 and 0.08 km², respectively.

The sparse forests decreased and turned into other LULC types by 181.87 km². The main class that was taken from sparse forest was grass with an area of 54.51 km². While, 51.21 km² was converted to crops, 49.40 km² to dense forests, and to rocks by 14.82 km². Adding to that sparse forests area of 7.73 km² was converted to the built-up areas, 3.10 km² to the soil, and 1.10 km² to the water body.

From Table 7.5 and Fig. 7.5, we noticed that the major increased change occurred in croplands with an improvement rate of 7.02%. Although the area of cropland was increased in 2017, 66.71 km² of croplands were converted into the other LULC classes. A cropland area of 26 km² turned into grass, 14.53 km² turned into built-up areas, 10.78 km² turned into sparse forests, 9.65 km² turned into rocks, 0.76 km² turned into soil, and 0.05 km² turned into the water body.

Furthermore, rocks land turned about 39.14 km² into other LULC classes. This is reported in Table 7.6, as 9.09 km² were shifted to built-up areas, 8.71 km² to sparse forests, 8.05 km² to croplands, 7.78 km² to grass and 5.51 km² to rest of the LULC classes. The left of the LULC (soil, built-up and water) areas recorded the lowest conversion to other LULC types. This is shown in Table 7.6 as 6.73 km² of soil, 3.32 km² of built-up area and 0.87 km² of water turned into other LULC patterns in the study area.

7.4 Conclusion

This study detected and quantified the LULC changes by classifying satellite imagery over a period of 28 years. LULC maps of 1989 and 2017 were created. The area of each LULC class (1989 and 2017) was calculated with its percentage including the shifted areas as well.

Table 7.6 LULC patterns conversion (cross-tabulation) between 1989 and 2014 in km²

LULC		2017								
		Dense forest	Sparse forest	Grass	Rock	Soil	Water	Cropland	Built-up	Total (km²)
1989	Dense forest	**344.49**	95.49	17.27	17.03	1.22	0.67	21.91	6.63	504.71
	Sparse forest	49.4	**226.37**	54.51	14.82	3.1	1.1	51.21	7.73	408.24
	Grass	8.64	51.07	**78.82**	10.16	2.5	0.08	85.76	3.97	241
	Rock	3.46	8.71	7.78	**26.34**	0.04	2.01	8.05	9.09	65.48
	Soil	0.3	1.15	2.82	0.7	**2.21**	0.06	1.65	0.05	8.94
	Water	0.15	0.04	0	0.56	0	**1.17**	0	0.12	2.04
	Cropland	4.94	10.78	26	9.65	0.76	0.05	**147.55**	14.53	214.26
	Built-up	0.5	0.37	0.12	1.61	0.09	0.4	0.23	**5.55**	8.87
	Total (km²)	411.88	393.98	187.32	80.87	9.92	5.54	316.36	47.67	
	No change area									832.50

Observable changes were noticed and reported in this study. Remarkably, agricultural expansion, development activities as urbanization, and deforestation have caused changes in LULC status. This is the first study attempting to assess, evaluate and provide up-to-date information about the LULC changes in the region. This is performed by analysing the spatial patterns of change at a reasonable resolution. The LULC shift in the Zakho district was evident by the decline in the area of dense, sparse forests and grass class (6.38, 0.99 and 3.69% respectively) and augmentation of area covered by classes of rock (1.06%), soil (0.06%), water (0.24%), crops (7.02%) and built-up class (2.68%).

Although Directorate of Forests in Duhok did afforestation, the forests class decreased. This is mainly due to the political and economic situation that the region witnessed from 1990 to 1996. During that period there were violent clashes between the PKK and the Kurdistan Regional Government on the one hand and the PKK and the Turkish government on the other (Rogg and Rimscha 2007). This caused great damage to the region through the burning of large areas of forests. Moreover, the difficult economic conditions experienced by the people of the region characterised by the lack of oil and electricity led the people to rely on trees as a main source of energy in all uses.

The overall pattern of LULC change in the Zakho district over the past 28 years was one of the sprawl of crops and built-up lands. Adding to that, a substantial reduction of forest and grassland indicates an acceleration stage of agriculture and urbanization. Despite the pressing land requirements for urbanization, land development and consolidation in grass and forest areas, and the adjustment of the agricultural structure, the foundation was put for the transition to intensively use the land in the Zakho district.

7.5 Recommendations

This chapter recommend the followings:

1. Utilizing the provided data by remote sensing and GIS techniques for studying land cover changes. This is because an up-to-date information can be gained that help in monitoring changes occurring within the natural environment. Ultimately, they will help decision makers to take an appropriate decision in solving emergency problems
2. Establishing a remote sensing and GIS Center to take care of all domains in the region. This will help in creating databases to be part of a comprehensive national system that serves different purposes and would be available to specialists, researchers and planners.
3. Work on forming institutions that are concerned with environmental affairs at the regional, local and national levels, and provide them with accurate applied research to show other factors that have a significant impact on land cover and land use such as soil, rain, human and political variables etc. ....

4. Reducing urban expansion in agricultural areas, especially in the Sindi plain. Moreover, work on the vertical expansion of residential areas to accommodate population increment.

References

Buringh P (1960) Soils and soil conditions in Iraq. Director General of Agricultural Research, Baghdad

Butt A, Shabbir R, Ahmad SS, Aziz N (2015) Land use change mapping and analysis using Remote Sensing and GIS: a case study of Simly watershed, Islamabad, Pakistan. Egypt J Remote Sens Space Sci 18(2):251–259. https://doi.org/10.1016/j.ejrs.2015.07.003

Cohen WB, Goward SN (2004) Landsat's role in ecological applications of remote sensing. Bioscience 54(6):535–545

Directorate of Forests and Range in Duhok PD, Ministry of Agriculture and Water Resources, Kurdistan Region-Iraq (2013) Plantation Projects

Directorate of Statistics Duhok AD, High Commission for Statistics, Ministry of Planning, Kurdistan Regional Government of Iraq (2008) Area of gravel quarries

Directorate of Statistics Duhok PD, High Commission for Statistics, Ministry of Planning, Kurdistan Regional Government of Iraq (2011) Population data for years (1996–2010)

Eklund L (2015) "No friends but the mountains" Understanding population mobility and land dynamics in Iraqi Kurdistan. Department of Physical Geography and Ecosystem Science, Lund University

Elsahabi M, Negm AM, Ali K (2016) Performances evaluation of water body extraction techniques using Landsat ETM+ Data: case-study of Lake Nubia, Sudan. Egypt Int J Eng Sci Technol 19(2):7

EXELIS (2013) ENVI Classic Tutorial: classification methods. http://www.exelisvis.com/portals/0/pdfs/envi/Classification_Methods.pdf. Accessed 4 July 2013

Fadhil AM (2010) Drought mapping using Geoinformation technology for some sites in the Iraqi Kurdistan region. Int J Digital Earth 4(3):239–257. https://doi.org/10.1080/17538947.2010.489971

FAO (2000) Land cover classification system (LCCS): classification concepts and user manual. http://www.fao.org/docrep/003/x0596e/x0596e00.HTM. Accessed 10 July 2017

French DA (2003) Environmental damage in international and comparative law: problems of definition and valuation. JSTOR

Jawarneh Rana N, Biradar CM (2017) Decadal national land cover database for Jordan at 30 m resolution. Arab J Geosci 10(22)

Mannion AM (2002) Dynamic world: land-cover and land-use change. Arnold, Hodder Headline Group, London

Module F (2009) Atmospheric correction module: QUAC and FLAASH user's guide. Version 4:44

Mustafa YT, Ali RT, Saleh RM (2012) Monitoring and evaluating land cover change in the Duhok city, Kurdistan region-Iraq, by using remote sensing and GIS. Int J Eng Inv 1(11):28–33

Mustafa YT, Habeeb HN (2014) Object based technique for delineating and mapping 15 tree species using VHR WorldView-2 imagery. In: SPIE remote sensing. SPIE, p 13

Mustafa YT, Habeeb HN (2016) Landsat LDCM imagery for estimating and mapping burned forest areas caused by Jet Attacks in Duhok Governorate, Kurdistan Region—Iraq. Journal of University of Duhok 19(1):683–690

Othman HA (2008) Urban planning strategies, towards sustainable land use management in Duhok city Kurdistan region, Iraq. University of Duhok, Duhok

Pu R, Landry S (2012) A comparative analysis of high spatial resolution IKONOS and WorldView-2 imagery for mapping urban tree species. Remote Sens Environ 124:516–533. https://doi.org/10.1016/j.rse.2012.06.011

Rogg I, Rimscha H (2007) The Kurds as parties to and victims of conflicts in Iraq. Int Rev Red Cross 89(868):823–842

Sissakian V, Ahad A, Hamid A (2011) Geological hazards in Iraq, classification and geographical distribution. Iraqi Bull Geol Mining 7(1):1–28

Tolessa T, Senbeta F, Kidane M (2017) The impact of land use/land cover change on ecosystem services in the central highlands of Ethiopia. Ecosyst Serv 23 (Supplement C):47–54. https://doi.org/10.1016/j.ecoser.2016.11.010

Tso B, Mather PM (2009) Classification methods for remotely sensed data, 2nd edn. CRC Press, Boca Raton

Vogeler JC, Braaten JD, Slesak RA, Falkowski MJ (2018) Extracting the full value of the Landsat archive: inter-sensor harmonization for the mapping of Minnesota forest canopy cover (1973–2015). Remote Sens Environ 209:363–374. https://doi.org/10.1016/j.rse.2018.02.046

Xi J, Zhao X, Wang X, Zhang Z (2017) Assessing the impact of land use change on soil erosion on the Loess Plateau of China from the end of the 1980s to 2010. J Soil Water Conserv 72(5):452–462

Xu X, Li W, Ran Q, Du Q, Gao L, Zhang B (2018) Multisource remote sensing data classification based on convolutional neural network. IEEE Trans Geosci Remote Sens 56(2):937–949. https://doi.org/10.1109/TGRS.2017.2756851

Yan J, Zhou W, Han L, Qian Y (2018) Mapping vegetation functional types in urban areas with WorldView-2 imagery: integrating object-based classification with phenology. Urban For Urban Greening 31:230–240. https://doi.org/10.1016/j.ufug.2018.01.021

Yuan J, Niu Z (2008) Evaluation of atmospheric correction using FLAASH. In: Earth observation and remote sensing applications. EORSA 2008. International Workshop on 2008. IEEE, pp 1–6

Chapter 8
Monitoring of the Land Cover Changes in Iraq

Arsalan Ahmed Othman, Ahmed T. Shihab, Ahmed F. Al-Maamar and Younus I. Al-Saady

Abstract Moderate Resolution Imaging Spectroradiometer (MODIS) data is one of the longest records of global-scale medium spatial resolution earth observation data. The current methods for monitoring a large area of land cover change using spatial resolution imagery (100–1000 m) typically employs MODIS data. NDVI time series has proven to be a very active index for vegetation change dynamic monitoring over a long period. This chapter aims at capturing global vegetation change dynamics within 10-days MODIS-NDVI time series by considering the variations in April throughout the long-term observation period (2000–2017). Due to the scaling factor (10^{-4}), the results of the derived NDVI MODIS values range from $-10,000$ to $10,000$, and most of the pixels of vegetation cover have values greater than 2000. To calculate the change in the vegetation cover, we used eleven MODIS land cover maps for the period from 2003 to 2013 to monitor these changes. MODIS-NDVI scenes were used to detect the relationship between the NDVI values and both of the annual precipitation and the elevation. Croplands and annual precipitation seem to be correlated. It is a direct relationship between the accumulative annual precipitation amount and the croplands area, which attribute mainly to the irrigation type, where vast of croplands (wheat and barley in particular) are irrigated by rainfall. The NDVI values were quantified using the elevation values extracted from SRTM. An inverse relationship was observed for the NDVI values of the grasslands class and the elevation values, where the R^2 were more than 0.7. Also; the analysis of elevation values may give some information on the NDVI values and the vegetation density.

Keywords MODIS · NDVI · TRMM · Iraq · Change detection

A. A. Othman (✉)
Iraq Geological Survey, Sulaymaniyah Office, Sulaymaniyah, Iraq
e-mail: arsalan.aljaf@gmail.com

A. T. Shihab · A. F. Al-Maamar · Y. I. Al-Saady
Iraq Geological Survey, P.O. Box 986, Baghdad, Iraq

© Springer Nature Switzerland AG 2020
A. M. F. Al-Quraishi and A. M. Negm (eds.),
Environmental Remote Sensing and GIS in Iraq, Springer Water,
https://doi.org/10.1007/978-3-030-21344-2_8

8.1   Introduction

Iraq is the birthplace of irrigated agriculture since 5000 BC., which includes most
of "Mesopotamia". Mesopotamia is a historical region in the middle east located
between the Tigris–Euphrates river system and spread southwards from the Zagros
foothills (Bulliet et al. 2010). Tigris and Euphrates Rivers originates rise in the
highlands of eastern Turkey and join to form the Shatt al Arab waterway before
discharging into the Arabian Gulf south of Basra City. Consequently, the overall
impact of the aridity was exacerbated by the construction of large dams in Turkey,
Syria and Iran, which have resulted in a decrease of water quantity, deterioration of
the soil quality and land productivity. Environmental degradation has been causing
severe problems around the world, especially in the Mesopotamia region, which is
semi-arid area because of their ecosystems' vulnerability (Jabbar and Zhou 2011;
Hadi et al. 2014). One of the major causes of eco-environment degradation in Iraq is
an inefficient use of the land cover and especially vegetation cover that contributes
to climate change (Jabbar and Zhou 2011). The old irrigation habit's particularly
in the Mesopotamia zone result in deterioration of vast agriculture lands. This old
irrigation approach leads to wasting of a high amount of water and increases salt
content in the agriculture lands.

Land cover is the physical condition and biotic component of the earth's surface,
whereas land use is the modification of the human activities and climate change on the
land cover (Friedl et al. 2010; Liu et al. 2016; Usman et al. 2015). Determination of
modification in land cover over a certain time period is called change detection. The
change in the land cover is the most important aspect of global environment change
(Usman et al. 2015). Several processes of the earth's surface such as atmospheric
circulation, vegetation productivity, biogeochemical cycle, energy cycle, etc. are
affected by the change in land cover (Liu et al. 2016).

Change detection is the procedure of identifying differences in the state of an
object by monitoring it at different periods; it involves the ability to quantify temporal
effects using multitemporal datasets (Khiry 2007; Othman et al. 2014). The previous
studies that dealt with the change detection of vegetation in Iraq were focused on
the spatial distribution such as Othman et al. (2014), Hamad et al. (2017), Qader
et al. (2015, 2018), Hadi et al. (2014), Jabbar and Zhou (2011), Jasim et al. (2016).
However, these researches did not address any relationship between the amount of the
precipitation received and vegetation distribution in the area, which will be covered
in this chapter.

Geographic Information System (GIS) and remote sensing are suitable tools to
identify and monitor land use and land cover (Hashemi et al. 2015). Remote sensing
data can supply accurate and appropriate information about vegetation and crop distribution, which helps decision-makers in various fields. The modern remote sensing
data for agriculture purposes can overcome the deficiency in information concerning
land use land cover (LULC). Use of satellite remote sensing data in practice started
more than 40 years ago for monitoring vegetation and crop distribution (Shao et al.

2001). While, GIS can be used to compare thematic maps, which are extracted from satellite data to detect the changes (Hadi et al. 2014).

The main objective of this chapter is to evaluate the nature, and rate of climate and vegetation change from 2002 to 2016 using Tropical Rainfall Measuring Mission (TRMM) and Moderate Resolution Imaging Spectroradiometer-Normalized Difference Vegetation Index (MODIS-NDVI) time series by considering the variations in April. Moreover, this chapter deals with finding a statistical relation between the density of vegetation and elevation in Iraq territory. In addition to the vegetation, we also monitored the change of some climate variables such us precipitation and temperature of Iraq using multi in situ and satellite data.

8.2 Climate Changes in the World

The last decades have witnessed the terrible phenomenon of climate change, which has shaped challenges around the world. During the last years, there has been a phenomenal increase in natural disasters and hostile weather events (droughts, floods, tropical cyclones, hurricanes and tropical cyclones and so on) throughout the world (WMO 2012), which are mainly attributed to variation in climate (Alexander et al. 2013).

Clouds and water vapor are the main anthropogenic contributor to the Earth's greenhouse effect, but a new study of the atmosphere-ocean climate modeling proves that the planet's temperature ultimately depends on the atmospheric concentration of CO_2. However, clearing of plants releases large amounts of greenhouse gases (Fearnside et al. 2009). We can say that the atmospheric carbon dioxide controls the global temperature of Earth. There is a direct relationship between rising global temperature and rising atmospheric carbon dioxide (Lacis 2010). The Carbone dioxide natural emissions are ~20 times higher than its anthropogenic emissions (Dawson and Spannagle 2008). Since the Industrial Revolution in 1750, the carbon cycle has been changed due to the human activities. As a consequence, the current level of carbon dioxide in the atmosphere is higher compared to its earlier level at any point in the past. In 2017, the concentration reached 405 ppmv, with a growth of 2 ppmv per year in the last decade (Keeling et al. 2001).

For the last few years, the atmospheric concentration of carbon dioxide, methane, and nitrous oxide have increased to levels unmatched since the last 800,000 years (Alexander et al. 2013). Carbon dioxide concentration has increased by about 25% since 1958 and 40% since the 1750s (Fig. 8.1; Alexander et al. 2013).

Figure 8.2 shows the anomalies of the annual mean global surface temperature (°C) and the five-year mean (°C) for the period from 1956 to 2016. The temperature of the earth's climate system is on the rise that has led to substantial environmental changes such us warming of ocean and atmosphere, the retreat of glaciers snow, the rise in sea level. Global temperature in 2016 was about 1.19 °C warmer than 1956. Two-thirds of the warming have occurred since 1986.

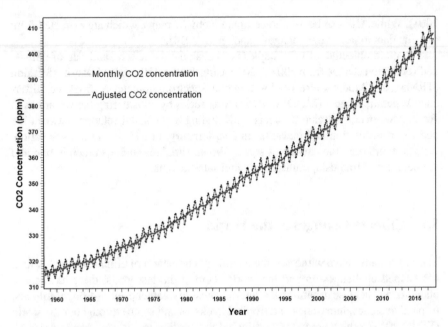

Fig. 8.1 The atmospheric concentration of carbon dioxide (CO_2) from Mauna Loa, Observatory, Hawaii since 1958. Reprinted and modified after Keeling et al. (2001)

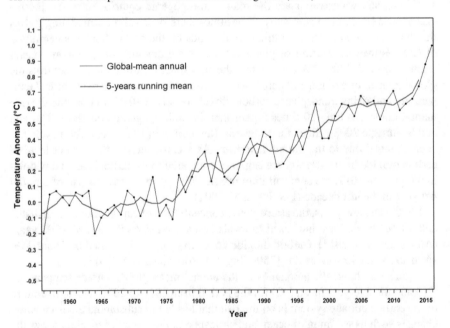

Fig. 8.2 Global surface temperature anomalies relative to 1956–2016. The dotted black line is the global annual mean; the bold red line is the five-year mean. Reprinted and modified after (GISTEMP 2017) update from Hansen et al. (2010)

8.3 Rainfall and Temperature

The in situ data of precipitation collected from the meteorological stations in Iraq was a great challenge. There are limited numbers of gauging stations that provides this information that are not free of charge. While satellite-based techniques provide us with potential continuous meteorological data, which are free of charge and can be used in many different models. In this study, we have used data based on ground and remote sensing measurements as inputs to detect the change in the climate of Iraq. We also used the Climate Change Knowledge Portal and the TRMM data to determine the climate situation of Iraq. This dataset of the Climate Change Knowledge Portal was derived from observational data that provides quality controlled temperature and rainfall values of thousands weather stations worldwide, as well as derivative products including monthly climatologies and long-term historical climatologies (The Climate Change Knowledge Portal 2017). The TRMM (GSFC DAAC 2017) was a joint space mission between NASA and the Japan Aerospace Exploration Agency (JAXA). It was designed to measure the rainfall of tropical and subtropical regions of the world. As the total territory of Iraq falls within the subtropical region, therefore, we use the TRMM rainfall acquired from September 2002–August 2017. The type of the data used is TRMM (3B43-V7), which combines precipitation with a spatial resolution of $0.25° \times 0.25°$ (Kummerow et al. 1998). The suitability of TRMM data in Iraq was evaluated by making a comparison with observed precipitation dataset of Erbil meteorological station. The data of Fifty-eight observed precipitation measurements were used for the period from October 2002 to March 2010. We find a good linear relationship between the monthly TRMM dataset and the observed precipitation, where the coefficient of determination (R^2) is of 0.603 (Fig. 8.3). The coefficient of determination (R^2) of 0.603, and the slope and intercept were 0.7915 and 20.18, respectively.

The climate of Iraq is characterized by greater changes throughout the year. The climate is described by long hot summer with occasional dust storms and short winter with limited and seasonal rainfall (Fig. 8.4). Most of the annual precipitation (165 mm) occurs between October to May. January shows the highest precipitation with an average value of 30.65 mm. Monthly temperatures vary between 9.5 °C (January) and 34.5 °C (August). The snowfall takes place few days per year on an average between November and April in the north of Iraq. Above 1500 m, heavy snowfall occurs in the winter (Fig. 8.4). According to the Köppen–Geiger climate classification system (Kottek et al. 2006), the climate of Iraq can be classified as warm temperate with dry and hot summer (Csa).

To fully understand the distribution of vegetation in Iraq, it is important to understand the spatial distribution of rainfall. There is a noticeable variation in the pattern of rainfall between different regions of Iraq. Precipitation in Iraq varies from 62.29 mm/year in the southwest (~70 km southwest of Al-Nukhayb Town) to 980.28 mm/year in the northeast of Iraq (~30 km east of Rawandoz Town). The precipitation in Al-Jazeera area (west-northwest of Iraq) and the Mesopotamian floodplain (central part of Iraq) ranges between 100 and 400 mm/year. The foothills

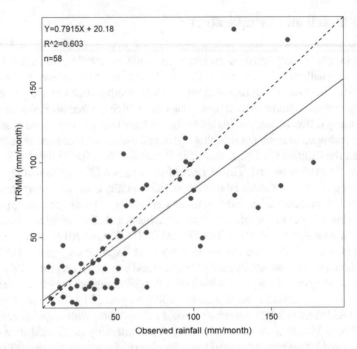

Fig. 8.3 Comparison of observed rainfall from Erbil metrological station and the TRMM data for the pixel located in the same location for the period September 2002-March 2010

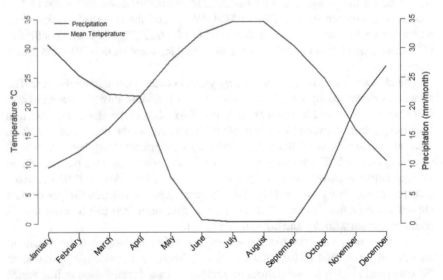

Fig. 8.4 Monthly precipitation and temperature of entire Iraq based on climate change knowledge portal data, from 2003 to 2015

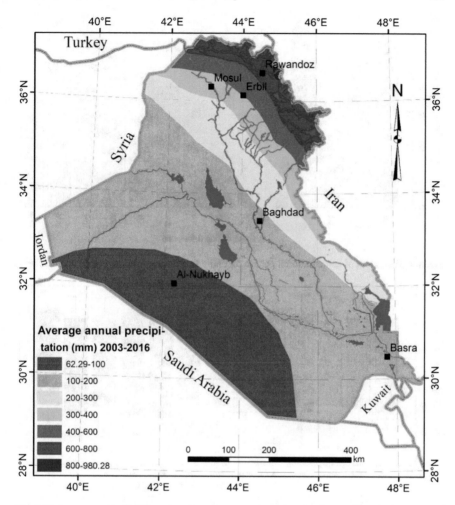

Fig. 8.5   Average annual precipitation map of Iraq for the period (2003–2016) using TRMM data

area receives 400–800 mm of precipitation per year. The mountainous area (mostly in Kurdistan Region) gains more than 800 mm/year of precipitation (Fig. 8.5).

Figure 8.6 shows the variation in the amount of annual precipitation in Iraq collected from TRMM data during the last thirteen years. The amount of annual precipitation in 2006 (306.5 mm/year) is higher than other years, while the amount of annual precipitation in 2008 (152.7 mm/year) is lower than other years. The decline of the linear fit (red color) between the annual precipitation (y-axis) and the year (x-axis) represents an inverse relationship, which means "in general" that the annual precipitation is decreasing yearly.

Figure 8.7 shows that all boxplots of the annual mean temperature during the period from 2003 to 2016 in Iraq have upper quartile values of temperature <32.25 °C,

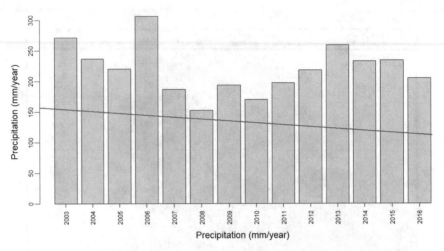

Fig. 8.6 Histogram of annual rainfall pattern for entire Iraq territory for the period 2003–2016

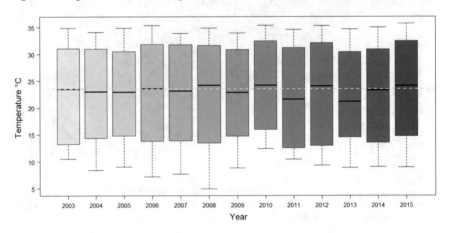

Fig. 8.7 Boxplot of annual mean temperature variation based on the climate change knowledge portal for Iraq during the period 2003–2015

while the lower quartile values are >12.9 °C. It is also shown that the median of the annual mean temperatures for the same period is 23 ± 1.5 °C and the mean is 22 ± 2 °C. The coldest annual mean temperature in the last fourteen years was 5 °C in 2008, while the warmest annual mean temperature was 35.72 °C in 2016. Moreover, there is not any continuous increase or decrease pattern in the annual mean temperature for the last fourteen years in Iraq.

8.4 eMODIS NDVI V6

We used MODIS NDVI V6 with a spatial resolution of 250 m. It is collected by AQUA MODIS sensor. The MODIS data was acquired by the National Aeronautics and Space Administration's (NASA) Earth Observing System (EOS). The data were downloaded from U.S. Geological Survey (https://earthexplorer.usgs.gov/). The MODIS NDVI data has a scaling factor (10^{-4}), and the radiometric resolution is a 16-bit image that is composed of values ranging from 0 to 65,535. Therefore, the derived values of this data range from $-10,000$ to 10,000 providing high sensitivity for detecting small differences in the density or the health of the vegetation. Most of the pixels of vegetation cover have values greater than 2000 (Jenkerson et al. 2010). The NDVI results from the following equation:

$$NDVI = (NIR - R)/(NIR + R) \tag{1}$$

where R and NIR are the spectral reflectances in the red and near-infrared wavebands of the MODIS, respectively.

The eMODIS suite of products used includes 10-day data composited datasets (Jenkerson et al. 2010). For the period between 2003 and 2017, one scene for each year between 6 and 15 April was used. We applied kernel moving window average for each pixel to remove the noise, low quality and the high error such as dropouts in the eMODIS-NDVI data acquired by 2007 and 2017. The scene covers 18,348 samples and 11,556 lines, which equal 13,251,843 km^2. We used the Iraq border shapefile as a mask to extract the NDVI that comes within Iraq.

8.5 Land Cover

MODIS land cover (MCD12Q1) is produced using a decision tree supervised classification algorithm over 8-day MODIS observations (NASA 2013), Nadir BRDF-Adjusted surface Reflectance (NBAR) (Wan et al. 2002) and land-surface temperature (Lina et al. 2008). It is yearly global maps of land cover, with ~463 m spatial resolution since 2001–2013. This data comes with HDF file format and Sinusoidal grid (NASA 2013). This data was downloaded from NASA's LP DAAC (https://lpdaac.usgs.gov/data access).

The land cover types are highly dynamic. There are changes from year to year as a result of the effect of several natural and human factors. Therefore, we used the MOD12Q1 land cover for the period from 2003 to 2013, to identify the predominant land cover types over Iraq. Seventeen classes were found in the study area. Five classes are non-vegetation, which are barren or sparsely vegetated, permanent wetlands, snow and ice, urban and built-up, and water. The rest are vegetation classes, which are closed shrublands, open shrublands, cropland/natural vegetation mosaic, croplands, deciduous broadleaf forest, deciduous needleleaf forest, ever-

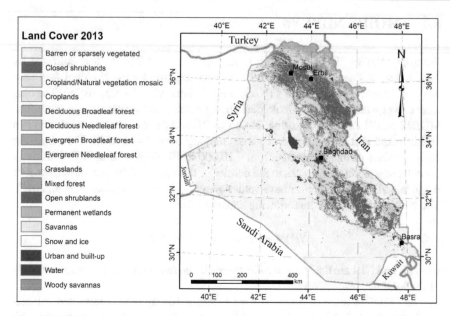

Fig. 8.8 MODIS land cover (MOD12Q1) for Iraq in 2013

green broadleaf forest, evergreen needleleaf forest, mixed forest, grasslands, savannas, woody savannas. Figure 8.8 shows the distribution of classes for the MODIS land cover in 2013. This figure shows that the most dominant vegetation classes in Iraq are shrublands, grasslands and croplands, which are located in the Mesopotamia and the northern part. The western and southwestern parts of Iraq are mostly desert.

Iraq covers 2,037,176 pixels of MODIS land cover (MCD12Q1) map. 71.77% (1,462,036 pixels) of the land cover in 2013 is non-vegetation lands. The maximum number of pixels (1,431,194 pixels) are classified as Barren or sparsely vegetated, which is equal to 70.25% of the entire country. The rest 28.23% is vegetated lands. We determined the permanent pixels, which have permanent vegetation class for the period from 2003 to 2013. These pixels show that the shrublands, grasslands and croplands classes represent about 99.5% of the vegetation lands (Fig. 8.9). Therefore, we merged the rest vegetation classes (which cover 0.5% of the vegetation in Iraq) i.e. deciduous broadleaf forest, deciduous needleleaf forest, evergreen broadleaf forest, evergreen needleleaf forest, mixed forest, savannas, woody savannas together in one class, which we called it an "Other nature vegetation" (Figs. 8.9 and 8.10b).

The shrublands are the most widespread class. It is distributed mostly in the lowlands in the north and in near Basra governorate (Fig. 8.10a) in the south of Iraq. It covers 58.52% of the vegetation lands. Croplands and grasslands cover 27.29 and 13.69% of the vegetation lands, respectively (Fig. 8.9). Most of the croplands located in the center part of Iraq near and surround Baghdad, but grasslands particularly located in the areas that have high elevation in the north (Fig. 8.10b).

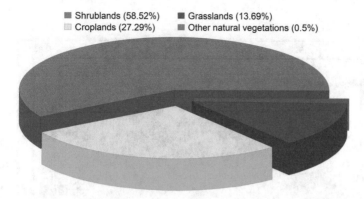

Fig. 8.9 Pie diagram shows the mean percentage of the land cover types in Iraq for the period from 2003 to 2013

8.6 Vegetation Distribution in Iraq

Iraq covers 437,298.86 km^2, 27% of this area (about 11,000 km^2) is considered suitable for cultivation. This equals about 11,000 km^2 of which 4400 km^2 are classified highly suitable, 4,700 km^2 moderately suitable, and 2,000 km^2 less suitable. Half of the lands suitable for cultivation are located in the central and southern parts of Iraq. These lands are irrigated by Tigris and Euphrates Rivers, which supply around 77 billion m^3 to 44 billion m^3, yearly to Iraq. The remaining half mostly located in the north is rain-fed (FAO 2013).

Vegetation land may be defined broadly as land used primarily for the production of food and fiber. It could be classified into two classes agricultural vegetated lands, and natural vegetated lands, which are grown through human activates or naturally, respectively (Anderson et al. 1976).

The natural vegetation in Iraq includes natural forests of different trees and all other natural vegetation as shown in Fig. (8.11). Shrublands mostly comprised of herbaceous plant species that are mainly less than 1 m in height; which occurs either as an individual or small communities on the wide valley floors (Othman and Al-Saady 2010; Al-Ma'amar and Al-Rubaiay 2010).

The agricultural vegetation includes many types shown in Fig. (8.12). In general, agricultural production is characterized by smallholding, especially, in the south and center parts of Iraq (averaging 0.01–0.25 Km^2). The rain-fed farms of north of Iraq seem to be a bit larger than those in the south and middle of Iraq (averaging 0.1–0.3 Km^2). Wheat and barley have been Iraq's most important crops, which accounted for 73% of croplands area in 2002 (Schnepf 2004).

Rain-fed agriculture is located in Ninawah, Erbil, Duhok, and Sulaimaniyah governorates. Vegetable production and fruit orchard occupy most of the high-rainfall zone (700–980.28 mm), wheat predominate in the medium-rainfall zone (400–700 mm), and barley is the main produce in the low-rainfall zone

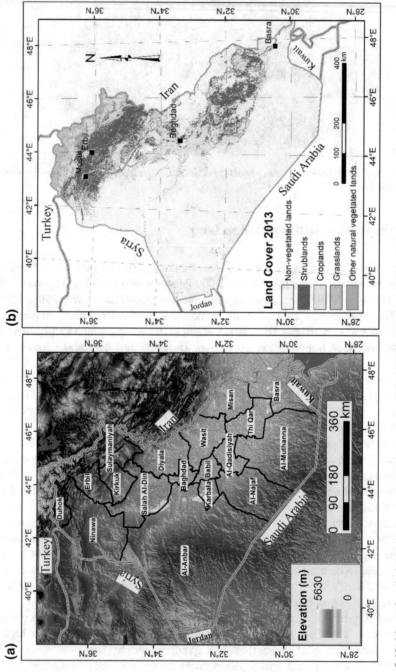

Fig. 8.10 Maps of **a** administrative governorates of Iraq **b** simplify MODIS land cover (MOD12Q1) for Iraq in 2013

Fig. 8.11 Natural vegetated land **a** grasslands near Penjween town in east of Sulaymaniyah governorate, **b** Sedge of Hor Al- Hwaiza between Basra and Misan governorates, **c** dry herbaceous land near Maila anticline near Dokan town northwest Sulaymaniyah governorate, and **d** Shrubs near Al-Ammara town in Misan governorate

Fig. 8.12 Agriculture vegetated land **a** Grape orchard northwest Ranya city in Sulaymaniyah governorate, **b** Wheat field Northeast of Al-Gharraf north of Thi Qar governorate, **c** Palm orchards on the bank of Tigris River near Al-Kut town in Wasit governorate, **d** Cropland near of BaladRuz town in Diyala governorate

(400–150 mm). Wheat and barley are planted in October–November and harvested in April–June.

Iraq's irrigated production zone represents the area where rainfall is less than 150 mm. It runs from the center of Iraq between the Tigris and Euphrates Rivers extending southeastward to the marshlands of the Tigris-Euphrates. It includes wheat and barley, which are produced in winter. Whereas rice, corn, cotton and vegetables are produced in summer. In the major part of Iraq, a single product is planted every year, which has encouraged plant disease and pests. Occasionally, if water is available, in some areas multiple cropping of vegetables and double cropping of barley/rice or wheat/rice are practiced. The irrigated crops are planted in April–May and harvested in September–October. One of the most important irrigation farms is date palm. Date palm trees are located in the south and center of Iraq, especially the area surrounding Basra, and northward to Karbala (Schnepf 2004).

8.7 Vegetation Change Detection in Iraq

The average spread of vegetation in Iraq for the period from 2003 to 2013 is 112,702 km^2. Figure 8.13 shows that the spread of vegetation has lack of regularity. In the years 2005, 2007, 2010, 2011, and 2012, the vegetation lands are less than the average spread of vegetation. It reached its lowest spreading (102,638 km^2) in 2005. In the rest years, the vegetation lands cover more than the average of spreading. The vegetation reached its peak in (123,588 km^2) in 2013. The direct relationship between the area of vegetation and the time is very week. Therefore, it is linear fitting has less than 0.05 correlation.

The vegetation pixels from 2003 to 2013 for the three main classes (shrublands, croplands, and grasslands) of land cover were tested (Fig. 8.14). We draw the time-series graph (Fig. 8.14), to show the change in the vegetation area over time. The

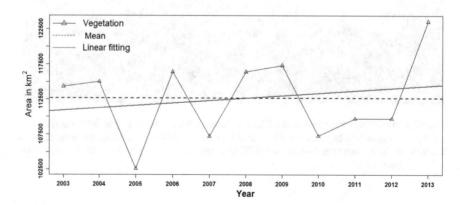

Fig. 8.13 Time-series graph shows the vegetation spreading in Iraq

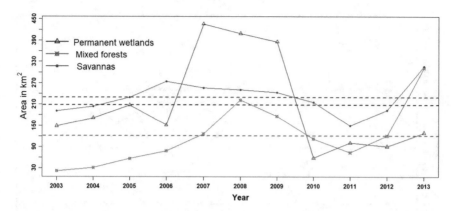

Fig. 8.14 Time-series graph shows the main three vegetation classes i.e. shrublands, croplands, and grasslands spreading in Iraq

average spread of the shrublands for the periods from 2003 to 2013 was 66,122 km². The minimum spread of shrublands was recorded in 2005. It covered of 55,317 km². In the years 2007, 2011, 2012 and 2013, the shrublands covered less than the average of spreading. In 2008, the shrublands were more spread than other years that covered 77,778 km² area. Despite that, the amount of annual precipitation in 2008 was lower than in other years.

The average spread of croplands for the period from 2003 to 2013 was 30,832 km² area. Figure 8.14 shows that the spread of cropland is more stable than the shrublands. However, in 2003 and 2013 the croplands are more than the average of spreading, which reached its peak (38,428 km²) in 2013. The croplands spreading from 2008 to 2010 were less than the average spreading. It has reached its lowest spreading (25,278 km²) in 2010.

The average spread of grasslands for the period from 2003 to 2013 was 15,472 km². Figure 8.14 shows that the spread of cropland is somewhat stable. However, from 2012 the grasslands area was more than the average of spreading, which reached in maximum (23,786 km²) in 2013. The grasslands in 2010 were less than the average spreading that covered 10,294 km².

Change detection of vegetation includes determination of the change in the density of the vegetation within the pixel beside the change in the vegetation area. The pixels that have less NDVI value means it has less vegetation density. For this type of test, we used the permanent pixels, which have the same vegetation class for the period from 2003 to 2013. From 2003 to 2017 one NDVI scene for each year acquired between 6 and 15 April was used. We calculated the mean of the NDVI (NDVImean) for the permanent pixels of each class, separately. Fifteen values of NDVImean were collected (Fig. 8.15) and these values divided by 10,000 to remove NDVI scaling factor.

The average of the NDVImean values for the other natural vegetation is higher than other vegetation types, while the croplands is lower than the other vegetation

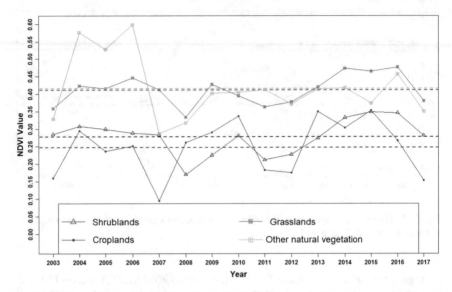

Fig. 8.15 Time-series graph shows the mean of the NDVI values for the scenes acquired from 6 and 15 April in the period between 2003 and 2017 for the main vegetation classes in Iraq

types. The average of the NDVImean values for the other natural vegetation, the grasslands, the shrubland, and the croplands are 0.416, 0.4113, 0.2774, and 0.2473, respectively. In 2007, the croplands and the other natural vegetation gained the minimum NDVImean values, while in the next year, i.e. in 2008 the grasslands and the shrublands collected the minimum NDVImean values. In 2015, the shrublands and croplands obtained the maximum NDVImean values, the other natural vegetation and the grasslands have the maximum value in 2006 and 2016, respectively. The NDVI mean of all vegetation types has declined in 2017 (Fig. 8.15).

8.8 Relationship Between Vegetation Distribution, Precipitation and Elevation

It is self-evident that there is a relationship between the vegetation distribution and water. The relationship between the natural vegetation distribution and precipitation is self-evident, too. However, quantification of this relationship by this study is an important contribution to estimation of the precipitation and the vegetation distribution area. To find the quantity of these relationships, we tested the correlations between the area of the vegetation distribution for the main three classes and the annual precipitation in Iraq (Fig. 8.16). We calculated the area of each class for the eleven land cover maps between 2003 and 2013. After that, we estimated the annual precipitation of Iraq for the same period. However, the accumulative annual precip-

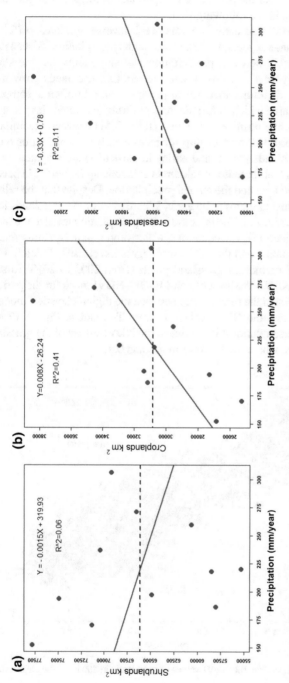

Fig. 8.16 Correlation between the yearly precipitation (mm/year) and the main three vegetation classes i.e. shrublands, croplands, and grasslands spreading in Iraq

itation was calculated for the period from September of the previous year and finish in the end of August in the following year.

Figure 8.16a shows that there is a very weak inverse relationship ($R^2 = 0.06$) between the shrublands vegetated area and the annual precipitation. We can recognize an inverse relationship between precipitation and shrublands, i.e. the shrublands increased in the dry year and this is sensible. As the shrublands grow mainly in the croplands area, therefore, increase in precipitation rate often accompanied by increase croplands area which in turn decreased shrublands areas. It was also found that there is a moderate positive correlation ($R^2 = 0.41$) between precipitation and croplands (Fig. 8.16b). This relationship attributes mainly to the irrigation type where the vast area of croplands (wheat and barley in particular) are irrigated by rainfall. Figure 8.16c shows that there is a weak direct relationship ($R^2 = 0.11$) between the grasslands vegetated area and the annual precipitation. Despite that the relationship is weak, this relationship seems to be logic. Where most of the grasslands located in the areas, which are rain-fed. The increase in the precipitation leads to the increase in the grasslands distributions. The equations, which represent the quantity relationship between the precipitation and the land cover vegetation classes plotted in Fig. 8.16.

We used 17,133 permanent grassland pixels (from 2003 to 2013) to select the values of MODIS-NDVI. We used fifteen MODIS-NDVI maps for the periods from 2003 to 2017. We tested the relationship between the digital elevation model SRTM (resolution 3 Arc) and the NDVI values in Iraq. The plot in Figs. 8.17 and 8.18 illustrate good inverse relationships between the NDVI values of the grasslands and the elevation values, where the R^2 were more than 0.7.

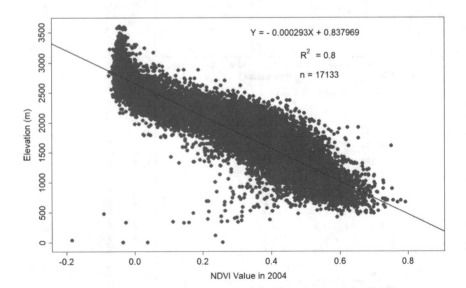

Fig. 8.17 Correlation between the NDVI values of grasslands in 2004 and the elevation in Iraq

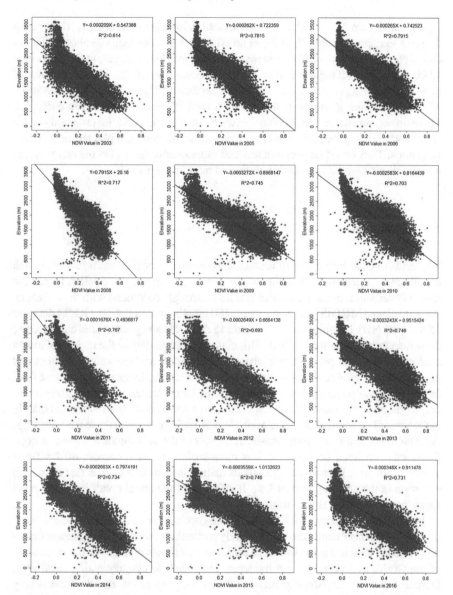

Fig. 8.18 Correlation between the NDVI values of grasslands in different years and the elevation in Iraq

The maximum NDVI value reached around 0.8 while the minimum NDVI value dropped to approximately 0. There is no NDVI value more than 0.2 above 2,600 m elevation and decreased with the increase in the elevation, whereas, at the altitude of 1,000 m and less, all regions had an NDVI value more than 0.35. The NDVI value of more than 0.8 was present only at a few places at the altitude less than 750 m elevation. The slope and the intercept of the equation for each year are plotted in each plot. There are no statistical relationships between the elevation and the croplands or the shrublands. That is because the croplands and shrublands are planted in different elevations. They could be found in low and mountainous lands. On the contrary, the grasslands are spreading in the foothills and mountainous lands.

8.9   Conclusion

The land cover in Iraq territory reflects the controlling of four main factors, which are climate, surface and subsurface water resources, lithology and relief. Hence the spatial distribution of the land cover classes is strongly correlated with these factors. We used eleven MODIS land cover maps, 180 monthly precipitation TRMM data and 15 MODIS-NDVI acquired between 6 and 15 April. Non-vegetated land class occupy more than two third of the territory. This class is mainly extent in the regions of low to moderate relief, low precipitation, absent of permanent surface water resources and high salinity of ground water aquifer or absent of regular investment. The rest third of Iraq is vegetated lands. It is covered by three main classes, which are shrublands, croplands, and grasslands. Shrublands is the most widespread. It is distributed mainly in the areas of agriculture and grasslands. Where its flourish when these lands undergo degradation or abundant from farmers as a result of natural or anthropogenic factors. Croplands restricted between the two main surface water resources of the Tigris and the Euphrates Rivers. The grasslands mainly located in the areas that have high elevation in the northern part of Iraq. It infers that the natural condition is the main factor governing their distribution.

MODIS land cover and MODIS-NDVI are good moderate spatial resolution satellite data for environmental monitoring study particularly, with applying change detection. TRMM data is a good source to evaluate the climate changes in the large scale of the world, as well as for the moderate scale such as country and/or local area. Despite the direct relationship between the area of vegetation and the time is very weak, but the spread areas of grasslands and croplands are somewhat stable during the last decades in Iraq. The most important factors predominate on the distributions grassland are the precipitation and the elevation.

There are good inverse relationships between the NDVI values of the grasslands and the elevation values. There is a moderate direct relationship between the accumulative annual precipitation amount and the croplands area, which reflect that vast area of croplands such as wheat and barley are irrigated by rainfall.

8.10   Recommendation

It is recommended in the future that a monitoring program is set up to track changes in the land cover of Iraq territory using remote sensing data and field stations. A further recommendation for this study is the organization of educational programs to provide information to people in the local area about the significance of land preservation and management and soil conservation methods using latest technology particularly for the farmer and rural communities. As the urbanization and climate change represent the main challenge threatened agricultural land. Therefore, issuing legislations regulating urban land expansion and protection of agricultural land and natural vegetation cover. The land cover issue is necessary to be seriously studied, through the corporation of different scientific fields in order to prevent deterioration of vegetation cover. The main problem suffered by the land cover of Iraq is soil salinity and sabkha. This problem can be avoided by avoidance using of Al-Tharthar Lake as a water reservoir (an artificial reservoir to collect flood waters of the Tigris River). The Lake formed mainly by karstification, due to the dissolving of gypsum rocks of the Fat'ha Formation and its shallow broad depression. This Lake contributes to washing out the salts from the stored water, to the Euphrates River. Hence, it contributes to the increase of water salinity as a result of dissolving salty rocks and also increase the evaporation rate of water. The increase in water salinity leads to increase of the saline in all other land cover classes, which related to the vegetation and agriculture.

Acknowledgements We thank NASA for providing MODIS and TRMM data, and we thank the USGS for providing the MODIS land cover (MCD12Q1). We are grateful to the Geological Survey of Iraq for providing the environmental reports and supporting the fieldwork.

References

Al-Ma'amar AF, Al-Rubaiay AT (2010) Series of land use–land cover maps of Iraq scale 1:250000, Kani Rash quadrangle, sheet NJ–38–10 (LULCM 2). Baghdad
Alexander LV, Allen SK, Bindoff NL, Bréon F-M, Church JA, Cubasch U, Emori S, Forster P, Friedlingstein P, Gillett N, Gregory JM, Hartmann DL, Jansen E, Kirtman B, Knutti R, Kanikicharla KK, Lemke P, Marotzke J, Masson-Delmotte V, Meehl GA, Mokhov II, Piao S, Plattner G-K, Dahe Q, Ramaswamy V, Randall D, Rhein M, Rojas M, Sabine C, Shindell D, Stocker TF, Talley LD, Vaughan DG, Xie SP (2013) Climate change 2013 the physical science basis working group I contribution to the fifth assessment report of the intergovernmental panel on climate change summary for policymakers. Switzerland
Anderson J, Hardy E, Roach J, Witmer R (1976) A land use and land cover classification system for use with remote sensor data. Washington
Bulliet R, Crossley PK, Headrick D, Hirsch S, Johnson L, Northup D (2010). The earth and its peoples
Dawson B, Spannagle M (2008) The complete guide to climate change. Routledge
FAO (2013) Special FAO/WFP crop, food supply and nutrition assessment mission to Iraq
Fearnside PM, Righi CA, Graça PML, de A, Keizer EWH, Cerri CC, Nogueira EM, Barbosa RI (2009) Biomass and greenhouse-gas emissions from land-use change in Brazil's Amazonian "arc

of deforestation": the states of Mato Grosso and Rondônia. For Ecol Manage 258:1968–1978. https://doi.org/10.1016/j.foreco.2009.07.042

Friedl MA, Sulla-Menashe D, Tan B, Schneider A, Ramankutty N, Sibley A, Huang X (2010) MODIS Collection 5 global land cover: algorithm refinements and characterization of new datasets. Remote Sens Environ 114:168–182. https://doi.org/10.1016/j.rse.2009.08.016

GISTEMP (2017) GISS surface temperature analysis (GISTEMP)

GSFC_DAAC (2017) Tropical rainfall measurement mission project (TRMM ; 3B43 V7) [WWW Document]. URL http://disc.gsfc.nasa.gov/datacollection/TRMM_3B42_daily_V6.shtml

Hadi SJ, Shafri HZM, Mahir MD (2014) Factors affecting the eco-environment identification through change detection analysis by using remote sensing and GIS: a case study of tikrit. Iraq Arab J Sci Eng 39:395–405. https://doi.org/10.1007/s13369-013-0859-8

Hamad R, Balzter H, Kolo K (2017) Multi-criteria assessment of land cover dynamic changes in Halgurd Sakran National Park (HSNP), Kurdistan Region of Iraq, using remote sensing and GIS. L. https://doi.org/10.3390/land6010018

Hansen J, Ruedy R, Sato M, Lo K (2010) Global surface temperature change. Rev Geophys 48. https://doi.org/10.1029/2010RG000345

Hashemi SM, Yavari AR, Jafari HR (2015) Spatio-temporal analysis of environment quality of ecotonal zones in iranian central plateau using landscape ecological metrics. J Environ Stud 41:201–218

Jabbar MT, Zhou X (2011) Eco-environmental change detection by using remote sensing and GIS techniques: a case study Basrah province, south part of Iraq. Environ Earth Sci 64:1397–1407. https://doi.org/10.1007/s12665-011-0964-5

Jasim MA, Shafri HZM, Hamedianfar A, Sameen MI (2016) Land transformation assessment using the integration of remote sensing and GIS techniques: a case study of Al-Anbar Province. Iraq Arab J Geosci 9:667. https://doi.org/10.1007/s12517-016-2697-y

Jenkerson C, Maiersperger T, Schmidt G (2010) eMODIS: a user-friendly data source. Virginia

Keeling CD, Piper SC, Bacastow RB, Wahlen M, Whorf TP, Heimann M, Meijer HA (2001) Exchanges of atmospheric CO_2 and $13CO_2$ with the terrestrial biosphere and oceans from 1978 to 2000. San Diego

Khiry M (2007) Spectral mixture analysis for monitoring and mapping desertification processes in semi-arid area in North Kordofan state. Sudan, Dresden

Kottek M, Grieser J, Beck C, Rudolf B, Rubel F (2006) World map of the Köppen-Geiger climate classification updated. Meteorol Z 15:259–263. https://doi.org/10.1127/0941-2948/2006/0130

Kummerow C, Barnes W, Kozu T, Shiue J, Simpson J (1998) The tropical rainfall measuring mission (TRMM) sensor package. J Atmos Ocean Technol 15:809–817. https://doi.org/10.1175/1520-0426(1998)015%3c0809:TTRMMT%3e2.0.CO;2

Lacis A (2010) CO_2: the thermostat that controls earth's temperature [WWW Document]. NASA. URL https://www.nasa.gov/topics/earth/features/co2-temperature.html

Lina H, Wanga J, Jia X, Bo Y, Wang D, Wang Z (2008) Evaluation of modis land cover product of east china. In: International geoscience and remote sensing symposium (IGARSS), p IV762–IV765. https://doi.org/10.1109/IGARSS.2008.4779834

Liu D, Zhua Q, Li Y (2016) Land cover change detection in Chinese Zhejiang Province based on object-oriented approach. In: Proceedings of SPIE—the international society for optical engineering. https://doi.org/10.1117/12.2241175

NASA (2013) User guide for the MODIS land cover type product (MCD12Q1)

Othman AA, Al-Saady YI (2010) Series of land use–land cover maps of Iraq scale 1:250 000, Karbala quadrangle, sheet NI–38–14 (LULCM 26). Baghdad

Othman AA, Al-Saady YI, Al-Khafaji AK, Gloaguen R (2014) Environmental change detection in the central part of Iraq using remote sensing data and GIS. Arab J Geosci 7:1017–1028

Qader SH, Atkinson PM, Dash J (2015) Spatiotemporal variation in the terrestrial vegetation phenology of Iraq and its relation with elevation. Int J Appl Earth Obs Geoinf 41:107–117. https://doi.org/10.1016/j.jag.2015.04.021

Qader SH, Dash J, Atkinson PM (2018) Forecasting wheat and barley crop production in arid and semi-arid regions using remotely sensed primary productivity and crop phenology: a case study in Iraq. Sci Total Environ 613–614:250–262. https://doi.org/10.1016/j.scitotenv.2017.09.057

Schnepf R (2004) Iraq agriculture and food supply: background and issues. Washington D.C

Shao Y, Fan X, Liu H, Xiao J, Ross S, Brisco B, Brown R, Staples G (2001) Rice monitoring and production estimation using multitemporal RADARSAT. Remote Sens Environ 76:310–325. https://doi.org/10.1016/S0034-4257(00)00212-1

The Climate Change Knowledge Portal (2017) Climate data [WWW Document]. URL http://sdwebx.worldbank.org/climateportal/index.cfm?page=downscaled_data_download&menu=historical

Usman M, Liedl R, Shahid MA, Abbas A (2015) Land use/land cover classification and its change detection using multi-temporal MODIS NDVI data. J Geogr Sci 25:1479–1506. https://doi.org/10.1007/s11442-015-1247-y

Wan Z, Zhang Y, Zhang Q, Li Z-L (2002) Validation of the land-surface temperature products retrieved from terra moderate resolution imaging spectroradiometer data. Remote Sens Environ 83:163–180. https://doi.org/10.1016/S0034-4257(02)00093-7

WMO (2012) WMO statement on the status of the global climate in 2012

Chapter 9
Effects of Land Cover Change on Surface Runoff Using GIS and Remote Sensing: A Case Study Duhok Sub-basin

Hasan Mohammed Hameed, Gaylan Rasul Faqe and Azad Rasul

Abstract A catchment area is an area from which runoff is resulting from precipitation flows. Land cover changes are the most significant factors that directly impact the runoff process. Much research on runoff response has focused on projected climate variation, while the endemic catchment area is directly affected by urban growth. Due to urbanization, the continuous growth in urban areas has led to a significant transformation of land cover pattern in built-up areas which has significantly affected the surface runoff behavior in the urban realm. This chapter aims to examine the impact of projected land cover changes on runoff in the Duhok sub-basin during the study period of 1990–2016. The study describes long-term hydrological responses within the quickly developing watershed of Duhok. Rainfall data from five meteorological rainfall stations in and around the study area were used. Historical data of land cover changes were mapped from 1990 to 2016, using a temporal satellite image (Landsat) to identify land cover and land use changes. Land use and land cover were integrated with a hydrological model SCS-CN to compute the runoff volume from the catchment area. Geographic Information Systems (GIS) was used to prepare different layers belonging to rainfall spatial distribution using Inverse Distance Weighting (IDW) tools, and various land covers were determined from remotely sensed data. Yearly average storm event rainfall data was used as a hydrological input to the SCS-CN model to estimate surface runoff over the years due to changes in the land cover of the Duhok watershed. The chapter indicates that the urban growth of the watershed increased from 10% in 1990 to 70% in 2016. Surface runoff volume increased from 12% in 1990 to 36% in 2016, while the vegetation land decreased from 47 to 14% in the same period. The chapter points out that an increment in built-up area or barren land cover and a decrease in vegetation cover have occurred, leading to larger surface runoff volumes in urban catchment areas.

Keywords Land cover change · Runoff · SCS-CN · GIS · Remote sensing

H. M. Hameed (✉)
Civil Engineering Department, Soran University, Erbil, Iraq
e-mail: hasan.hamid@soran.edu.iq

H. M. Hameed · G. R. Faqe · A. Rasul
Geography Department, Soran University, Erbil, Iraq

© Springer Nature Switzerland AG 2020
A. M. F. Al-Quraishi and A. M. Negm (eds.),
Environmental Remote Sensing and GIS in Iraq, Springer Water,
https://doi.org/10.1007/978-3-030-21344-2_9

205

9.1 Introduction

The need to provide shelter, water and food to people all over the world has led to changes in land cover, such as agricultural lands, urbanization, forests, etc. The great interest in land cover changes results from their direct relationship to many fundamental characteristics of the land and processes, such as geomorphological processes, land productivity and hydrological cycles (Endreny 2005; Githui et al. 2009). Evaluating the effect of land cover and land use changes on hydrologic conditions is vital for catchment area management and development (Woldesenbet et al. 2017). Many studies showed that the significant increase in global runoff in the 20th century was because of precipitation and land cover changes (Wang et al. 2017). A suitable understanding of land cover dynamics is fundamental for anticipating the future changes in urban runoff volume (Suribabu and Bhaskar 2015; Meierdiercks et al. 2010). The ramifications of anthropogenic activities on land cover and water resources have become clear. Generally, investigations demonstrated that transformations in catchment hydrology happen principally because of alterations in interception, infiltration, evapotranspiration and groundwater recharge which would join land cover transformations (Meierdiercks et al. 2010; Hameed et al. 2015). Evaluating the effect of land cover changes remains a significant step in catchment area management strategies as well as water resources planning and conservation measures. Conversion of land cover such as deforestation, pasturelands and grasslands for urbanization and agricultural intensification because of rapid population growth has become a critical environmental issue (Gyamfi et al. 2016). Unplanned urban growth is now a major problem in several countries. In developing countries, there is often a lack of consistent and reliable data including spatial data, and that situation is highly evident in Iraq where political conflicts and factional instability caused subsequent internal displacement and migration. However, improved economic opportunities and population growth are often the driving forces behind the high level of land cover and land use changes, particularly in the cities of Iraqi Kurdistan Region that enjoys a complete state of security and economic stability (Hameed 2017).

The change in land cover has a considerable impact on the nature of runoff. Land cover in some areas remains unchanged for long periods, while in other regions land cover sees drastic changes more frequently. Economic and social activities such as urban growth and agricultural development have a direct effect on land cover changes (Letha et al. 2011). Urban growth increases impervious covers (Guan et al. 2015) that lessen infiltration capacity and prompt downstream flooding (Karamage et al. 2016; Pan et al. 2017; Kulkarni et al. 2014). A catchment or basin is an area from which overflow or runoff resulting from rainfall flows past a signal point into lakes, streams, rivers or seas (Hamdi et al. 2011). Land cover describes the material such as rock cover, vegetation or water body which dominate the surface, whereas land use refers to how natural land is converted to a new use such as agriculture, settlements or industry etc. It composes key environmental information for many sciences, natural resources such as water resources, also for a range of human activities (Letha et al. 2011). Urban growth is one of the major environmental problems of a catchment area.

The urban growth process includes an unsustainable use of natural land cover and causes numerous problems both within and outside the city (Zare et al. 2016). The urban hydrological system of a catchment area increases the fluctuation amount of surface water which may become extremely high during periods of rainfall (Kulkarni et al. 2014). Hydrological impacts of the increased impervious surface area typically result in higher flow peak and increasing runoff or streamflow volume (Perrin and Bouvier 2004), which lead to shifts in subsurface flow to surface flow and increases in flood frequency (Li and Wang 2009).

The objectives of this chapter are to assess the impact of land cover and land use changes on the runoff surface in sub-basin of Duhok city and to determine the contribution of changes in individual land cover and land use to changes in the major runoff component.

9.1.1 Land Cover and Land Use

Land use/cover throughout the world is mounting rapidly and puts the high risk on environmental components (such as water, soil, rainfall, and temperature and runoff rate) particularly in metropolitan areas. Therefore, the issues of land use/cover change demands wide interests (Guo et al. 2012). Changes in land cover are mainly man-made as the consequences of population growth, and they have an impact on the environment. The physical characteristics of earth's surface are called land cover, and they include physical features of land created solely by human activities such as settlements, based on the distribution of vegetation, water, soil and other features. However, the way land is used by humans is called land use, and includes land utilized for economic activities (Rawat and Kumar 2015).

Urbanization plays a major role in changes of land use/cover by replacing the natural land cover with inhabited sites, vegetation for economic purposes and infrastructure. Urbanization refers to changing the land cover by civil engineering constructions such as roads, housing etc. Urban expansion can be harmful to a region and ultimately to a large extent if it moves too far and is unbound, and it emphasizes the need for a cautious method to urban expansion planning (Du et al. 2014). The process of urbanization has a significant hydrological effect regarding manipulating the run-off nature (Goudie 1990). Human activities are the main geomorphic cause that impacts the earth surfaces; another effect of urban growth is that farming is the key procedure currently affecting the earth (Sharp 2010; Rasul et al. 2018). Nowadays, over 50% of the population in the world live outside of cities, and it is predicted that by 2025, this will decrease to less than 33% (Zhang et al 2013).

Remote Sensing and GIS techniques have been extensively applied to observe land use/cover changes through gathering, storing, evaluating, and awarding of geospatial data for diverse aims (Belaid 2003; Fadhil 2011). To acquire differences and update the data on the state, feature, pattern and growth of urban situations, advanced techniques are required (Esch et al. 2010). Remote sensing (RS) and Geographic Information systems (GIS) are effective techniques for conducting studies such as detection of land cover change and forecasting future situations (Ahmed and Ahmed 2012; Fadhil 2013). Remote sensing data broadly adapt in different fields such as urbanization monitoring, land use/land cover mapping, analysis of urban heat island, estimation of impervious area and evaluation of urban ecological security (Yang 2011; Al-Quraishi 2004; Fadhil 2006). Image satellite is now very useful for analyzing urban areas that are integrally appropriate to analysis urban land use land cover changes at different spatial and temporal scales (Phinn et al. 2002; Wilson et al. 2003). In current years, a geographic information approach has been well-recognized for urban land use. Geographical planning has been used as a tool for geospatial simulation and land use change modeling, and its forecast has been confirmed in earlier studies of Land use change as efficient. There are several studies pointed to geospatial simulation and land use change modeling (Wang and Zhang 2001; Araya et al. 2010; Suribabu et al. 2012; Wu et al 2006; Bharath et al. 2013; Griffiths et al. 2010; Dewan and Yamaguchi 2009; Xiao and Weng 2007; Buyadi et al 2013; Jiang and Tian 2010; López et al. 2001; Lu et al. 2004; Mitsova et al 2011).

Geometrics technology with urban land use was applied in several studies in Duhok city to evaluate LULC changes and their effects on different fields, such as (Mohammed and Ali 2014) who attempted to assess urban expansion and changes in land use/cover in Duhok city between 2003–2012 utilizing multi-dates. The study documented that residential areas have expanded remarkably while other classes have decreased. Also, (Mustafa et al. 2012) implemented supervised classification with maximum likelihood to detect urban land use cover changes between (1989–2012). The study area was divided into five classes: forest, soil areas, grassland, water, and urban/built-up areas. The result showed that the urban land increased whereas all the other classes decreased.

9.1.2 Soil Conservation Service Curve Number (SCS-CN) Method

Easy methods for estimating runoff from catchment areas are particularly important in hydrologic engineering that is applied in many hydrologic uses, such as water balance estimation and flood design. The SCS-CN method is widely used for estimating runoff of depth and volume for a given rainfall event (Zhan et al 2005). The CN factor value depends on the various soil, land use, land cover and climate conditions. The method was originally developed by the Soil Conservation Services (United State Department of Agriculture) to estimate water surface volumes for a given precipitation event. It is the most popular technique among engineers. It is also simple to obtain and well documented environmental input, and it computes for many of the factors impacting runoff generation, incorporating them in an individual CN parameter (Soulis and Valiantzas 2012; Mockus 2007).

The procedure estimates the runoff depth from the pixels of the watershed as a function of existing soil type, land cover, land use and climate condition. Land cover and land use characteristics in the Soil Conservation Service approach is explained by Curve Number (Li and Wang 2009). It is possible to estimate a composite CN for an entire catchment by area weighted average of CN for sub-basin units. Runoff volume from entire sub-basin can easily be estimated by composite CN for a variety of rainfall inputs. The SCS-CN model is simple, predictable, stable and accurate to reflect the catchment area characteristic for depth of runoff estimation. Also, the SCS-CN approach is more suitable for heavy precipitation of longer duration rather than light rainfall, and it is suitable for small watersheds. On the other hand, the application of the SCS-CN model illustrates the possibility of using this model for continuous time series precipitation-based estimation of runoff depth and to assess the impact of land cover and land use changes on runoff volume in a catchment area (Kowalik and Walega 2015).

9.2 Methodology and Data

9.2.1 Study Area

Duhok city covers the center of Duhok province which is located on latitudes 37° 6′ 15″N; 43° 49′ 51″E in the north of Iraq and is about 585 m above the sea level Fig. 9.1 (Abas 2012). The Study area has a strategic site, being close to the tri-junction of the borders of Iraq, Turkey and Syria in the north-west of Kurdistan region. The topography has caused a linear shape due to being comprised by two mountain chains, the Bekher and Zawa Mountains from the north to the south respectively. The study area is surrounded by Edit Township of Mamseen Mountain from the east and Semel agricultural plain in the west. The city of Duhok also has two rivers passing through it, they are named the Duhok and Heshkarow Rivers, and they meet in the south-west of the study area. These stream waters are utilized for irrigating the agricultural area; also, Duhok dam was constructed in the northern part of the city in the period 1980–1988 for irrigating considerable rain fed agrarian lands, which are situated in the west of the city and lengthen to neighboring Semel area (Brendan et al 2005; Dawn 2010).

9.2.2 Climate Conditions

The climate of Duhok city is comparable to the Mediterranean climate, with the effect of comparatively high altitude of the surrounding mountains. The main characteristics of the Mediterranean climate are dry, extremely hot summers with bright sunlight and low comparative moisture (Koeppe and De 1959). Whereas, stated that higher comparative humidity and similarly lower quantity of sunshine mark the winter season as compared with the summer season. The study area in the winter

Fig. 9.1 Location of the study area

season is considered cold with snowfalls on the highlands. The temperature and evapotranspiration in the winter season are least (1.56 °C and 1.20 °C mm), and the precipitation is maximum (358.41 mm). The hot season starts in June until the beginning of September when the heat and evapotranspiration are maximum (43.30 °C and 13.70 mm), whereas the precipitation is least (0.00 mm. The mean annual precipitation was around 514.00 mm for the period (2000–2015). However, the average yearly temperature for the same period was around 20.44 °C.

9.2.3   Satellite and Rainfall Data

Five stations in and around the study area were employed to estimate surface runoff from rainfall data (Duhok, Malta, Duhok dam, Zawita and Sumel stations). The stations of the study area recorded continuous data from 2001 to 2016. Daily and monthly rainfall data are available for every year. The minimum average rainfall, which was recorded in Sumel, is 430 mm/year, and the maximum average rainfall in Zawita station is 747.8 mm/year. GIS tools were applied to estimate the spatial distribution of rainfall. Grids of rainfall were calculated and mapped for a selected rainfall depth per pixel. Inverse distance weighting (IDW) was used to determine the spatial distribution of rainfall and transfer the data to a raster layer in 30 m resolution.

This study adapted both primary and secondary data. Multi-temporal remote sensing data was required to research changes in land use/land cover over the study area. The primary data was collected from the United States geological survey (USGS). In the study, two Landsat images were comprised: TM-5 (02 August 2000) and (21 August 2015) with 30 m resolution. Secondary data were obtained from the Governorate and Municipality of Duhok city which provided maps such as the municipal boundaries, geographical, wards, and master plan maps.

Data preprocessing: Before conducting the study, Landsat data was examined and processed to generate land use maps and their effect on runoff. Images were geometrically corrected and projected to the Universal Transverse Mercator (UTM) coordinates zone 38 N, WGS 1984. Furthermore, to improve the visual interpretability, image enhancement was used to aggregate the difference among features.

9.2.4 Images Classification

Examining the effects of anthropological actions on land cover/use classification is essential for revealing land use/cover changes in the period of the study. Earlier training samples were developed based on additional information from different sources such as higher resolution Google map and aerial photographs. Several samples were selected from the image for each category; each class had 50 training samples of 50 pixels. Then, the supervised classification was utilized for each satellite to extract the detailed changes. A maximum likelihood classifier was adapted for initial classification as it shows higher accuracy classification (Lillesand et al. 2008; Faqe 2015). Six land use/cover categories were designated in this chapter, namely vegetation area, barren land, forest; ruck, water, and built-up area. Finally, both classified images were evaluated again for accuracy. For measuring remotely sensed data, the procedure of supervised classification is commonly used (Richards 1999). Supervised classification is a method of categorizing pixels through a well-known computer algorithm and the current statistical descriptors of land cover (Lillesand et al. 2008). It depends on the prior information on the study site (Ahmed and Ahmed 2012). For digitizing polygons around each training site, a chosen color composite (RGB = 432) was used for similar land cover. Then each known land cover type was assigned to a unique identifier. To achieve precise classification results, maximum likelihood supervised classification is a suitable approach (Gómez et al. 2008) and it provides the best consistent Landsat band mixture. The maximum likelihood method offers a more accurate result of classification than other methods such as (Parallelepiped) thus the current study selected maximum likelihood classifier.

9.2.5 SCS-CN Model and the Proposed Method

The SCS-CN is an event-based rainfall-runoff model that is widely used in the design of major hydraulic structures. The SCS model uses the CN values as input parameters to estimate runoff depth (DeFries and Eshleman 2004). The CN is estimated per pixel for a watershed using land cover, land use and soil types that are reclassified into Hydrologic Soil Groups (HSG) (see Table 9.1) (Hameed 2013). The SCS approach divides soils in any watershed area into four HSG according to the United States

Table 9.1 Soil groups and corresponding soil texture

Soil group	Runoff description	Soil texture
A	Low runoff potential because of high infiltration rates	Sand, loamy sand and sandy
B	Moderately infiltration rates are leading to a moderate runoff potential	Silty loam and loam
C	High/moderate runoff potential because of slow infiltration rates	Sandy clay loam
D	High runoff potential with very low infiltration rates	Clay loam, silty clay loam, sandy clay, silty clay and clay

Geology Survey land use and land cover classification system (A, B, C, and D). Soil classification to HSG relies on infiltration rates and the soil texture composition(Day 2010). Table 9.1 illustrates the HSG depends on the United States Geology Survey classification System.

Depending on Table 9.1 that explains soil texture and its soil group to find the HSG of the study area, Fig. 9.2 shows the spatial distribution of hydrologic soil groups in the study area. The hydrologic soil group of the study area is distributed into three main types; A, B and C. Depending on Table 9.1, the study area could be divided into low and high-moderate runoff potential. The soil texture consists of three main zones; Deep, Lime-rich and gravelly silty clay to clay with surface cracks, shallow to moderately deep, well-drained loamy to clayey soils, and deep, well-drained lime-rich soils with a variable texture and gravel content. The soil types were generated from reconnaissance soil maps of the three northern governorates, Iraq 2001. As shown in table 1, runoff potential of the study area is distributed between low, moderate and high or moderate runoff potential.

9.2.6 Estimate Runoff Depth

Evaluation of runoff depth is most important for assessing the potential water supply during rainfall events. The Soil Conservation Service and Curve Number model assumes that the ratio of actual soil retention after water surface begins to reach potential maximum retention is equal to the ratio of water surface to available rainfall,

$$\frac{F}{S} = \frac{Q}{P - I_a} \qquad (9.1)$$

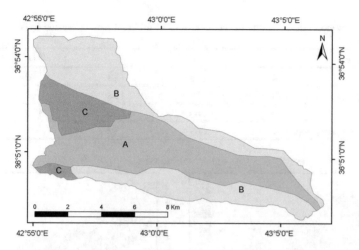

Fig. 9.2 Classified soil map into hydrologic soil groups (A, B and C)

where F is cumulative infiltration, S is the potential maximum retention (mm), Q is a runoff (mm), P is storm rainfall (mm), and I_a is an initial abstraction. The value of I_a is set equal to 0.2S in Eq. (9.1). Therefore, the SCS-CN model can be expressed as follows:

$$Q = \frac{(P - 0.2S)^2}{(P + 0.8S)} \quad P \geq 0.2S \tag{9.2}$$

For the application of Eq. (9.2), S is expressed in the form of a dimensionless runoff curve number that represents the runoff potential of the land cover soil characteristics governed by soil antecedent moisture condition, soil texture and land cover and land use.

$$S = \frac{25,400}{CN - 254} \tag{9.3}$$

where S is expressed in mm. The values of CN can be taken from tables, and they are dependent on the hydrologic soil group, land cover and land use, hydrologic condition and land treatment. The CN high values point to higher surface runoff, while the CN lower values refer to lower runoff (Zare et al. 2016). The hydrologic condition indicates the impact of land cover and represents the surface conditions in the catchment area in relation to infiltration and runoff (Weng 2001). The land cover and land use that are extracted from satellite images in remote sensing can be used together with a map of hydrologic soil group in GIS to match the hydrologic soil group with land cover and land use. Table 9.2 generated the curve number dependent on the USGS classification system (A, B, C and D).

Table 9.2 Runoff curve number for combinations of different land cover and hydrological soil groups

Land cover	A	B	C	D
Bare soil	74	83	88	90
Built up	98	98	98	98
Forest	36	60	73	79
Grass and open wood	49	69	79	84

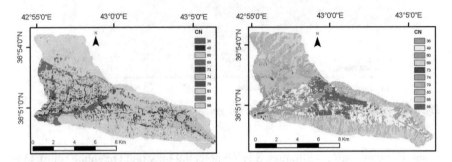

Fig. 9.3 Curve number map for the study area

Estimating the spatial and temporal variations runoff depth each pixel of the study area uses the SCS-CN model. The rainfall distribution layer was used as the main parameter to estimate runoff depth. Inverse Distance Weighting was applied to interpolate determined pixel values of rainfall in a linearly weighted combination of a set of station points. Runoff for each pixel and each studied year was estimated by applying the SCS-CN model using as inputs rainfall depth and the corresponding CN values raster layers. The resultant runoff rater layers were then intersected with land cover and land use to estimate the runoff volumes corresponding to each land cover category for each year as shown in Fig. 9.3. Distribution of CN in the study area shows that there are different values of CN. The minimum value of CN is 36, while the maximum value of CN is 98. This difference completely depends on the soil types and land cover to display the capability of the study area in runoff and infiltration as shown in Fig. 9.3 and also Fig. 9.4 illustrates the methodology of the study to detect the affect of land use / land cover changes on the runoff volume.

9.3 Results and Discussion

9.3.1 Changes in Land Cover

The watershed of Duhok city experienced several significant land cover changes between 1990 and 2016. Table 9.4 and Fig. 9.5 illustrate the spatial distribution

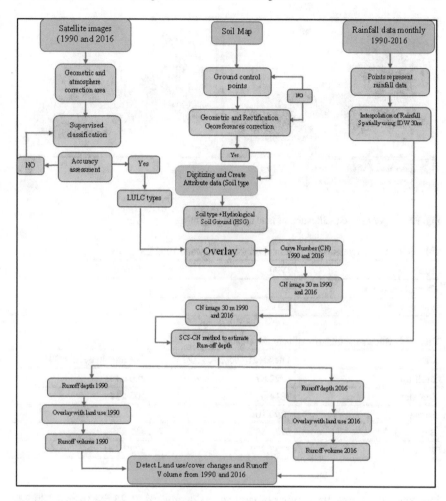

Fig. 9.4 The flowchart illustrates the methodology

of land cover changes in the study area. The supervised classification was used to produce the land use land cover map in 1990 and 2016 with high accuracy as seen in Table 9.3. The overall area of interest is about 8737.8 ha; as seen in Table 9.3. Figure 9.5 displays the land cover changes that occurred in Duhok watershed through the study period. As illustrated, the spatial spread of the land cover types and the urban land growth amounts varied considerably in the period of the study. The detailed hectares of land cover change in Duhok watershed is documented in Table and chart 1. The study selected four classes (Built-up, grassland, forest and bare land).

Table 9.4 proved that the dominant area in 1990 was grassland that made up around 3,814.6 ha in 1990, and it dropped to 1,048.34 ha in 2016. Thus, forest area decreased from 290.16 ha to 211.209 ha in 2016. Noticeable alteration occurred in

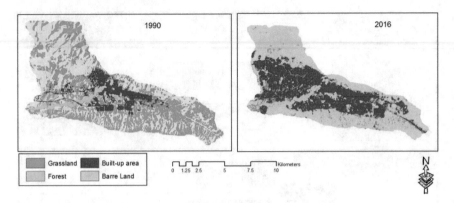

Fig. 9.5 Supervised classification of land use/cover map

Table 9.3 Accuracy assessment of land use/cover between 1990 and 2016

Year	1990	2016
Overall accuracy	93	90
Kappa index	0.91	0.89

Table 9.4 Shows the quantity of land use change

Class	The area in hectare 1990	The area in hectare 2016
Built-up	822.87	2804.44
Grassland	3814.6	1048.34
Forest	290.16	211.209
Bare land	3775.68	4673.79
Total	8737.779	8737.779

total vegetation area, these types of land covers decreased by 23.6% from 4,139.229 to 1,259.549 ha between the years 1990 and 2016. In contrast, the finding indicated that bare land significantly increased from 3,737.68 ha to 4,673.79 ha from 1990 to 2016. The study revealed that exceptional changes took place in the quantity of the vegetation class between 1990 and 2016. The modifications might be the outcome of many factors such as urban growth and climate change. The study area experienced drought seasons in 1999 and 2002. Due to most of the vegetation types in the study site being reliant on precipitation, the increase and decline of precipitation results in alterations in vegetation area. Also, urban classes (inhabited, profitable, road, education and other administrative building) remarkably increased by (822.87 ha) (2,804.44 ha) and between 1990 and 2016. This excessive change in urban areas is an outcome of the political and economic modifications that have occurred in the region. The finding of the study agrees with the results of (Al Rawashdeh and Saleh 2006; Shupeng et al. 2000; Weng 2014) who posited that economic and political factors contribute to urban land use change.

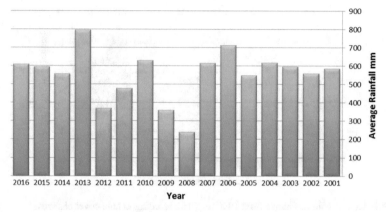

Fig. 9.6   Comparison of rainfall between 2001–2016 and 2016

9.3.2   Rainfall Variable in Time

The total rainfall received in the study area from 2001 to 2016 was highly variable from one year to another. The variable depends on the type of climate and the length of the considered period. The climate of the study area changes between the warm climate region and Mediterranean climate of the steps. The characteristic of this type of clime is dry and wet periods (Faqe et al. 2016). Comparisons of the average monthly rainfall by 5 stations in and around the study area were made on a yearly basic. Figure 9.6 illustrates the variation of the average rainfall between different years. The average annual rainfall was around 554.5 mm/year. The depth of rainfall fluctuated from 2001 to 2016. The minimum mean annual rainfall in the same period was 239 mm in 2008, while the maximum mean annual rainfall was 797 and 713 mm in 2013 and 2006 respectively. These differences in the amount of rainfall caused changes in runoff potential. Therefore the volume of runoff potential could be changed from year to year. Spatial distribution values of rainfall in the study area increased from south-west to north-east. This difference in the spatial distribution values of rainfall is linked with the climate condition and topography (Rijabo and Bleej 2010).

9.3.3   Changes in Potential Runoff

The results showed in Fig. 9.5, and Table 9.4 demonstrate the feasibility of converting land cover parameters so that all parameter divisions could be simulated. A set of rainfall data was used for a different land cover and land use condition to analyze the effect of land cover changes on runoff generation. Many studies have shown that vegetation land cover can play a positive role in decreasing runoff potential.

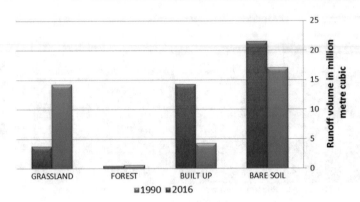

Fig. 9.7 Runoff volume change from 1990 to 2016 according to land cover changes

The spatial distribution of impervious areas and bare soil in the sub-basin watershed of Duhok had a significant impact on the generation of surface water. Figure 9.7 illustrates the impact of land cover changes on the volume of water surface from 1990 to 2016. The main change in the volume of runoff in the same period was clearly shown in built-up, bare soil and grassland cover. Runoff volume increased from 4.2 to 14.3 million cubic meter from 1990 to 2016 respectively, and runoff volume of bare soil also increased from 17 to 21 million cubic meter in the same period. The impervious area provided most of the runoff. The increased urbanization in the study area led to a decrease in grassland and forest area; therefore, the decline of vegetation area resulted in the increased water surface. The grassland land and forest area decreased from 3,814.6 to 1,016.5 ha and 288.1 to 211.2 ha respectively as shown in Fig. 9.5.

The transformation in land cover from vegetation area to built-up and bare land from 1990 to 2016, that is shown in Table 9.4, indicates the changing trends of urban growth in Duhok watershed and their role in surface water increase. As can be seen, the increment in runoff volume is largely correlated with the shift in land cover. Specifically, the built-up and bare soil land increased from 4,598.6 to 7,478.2 ha. This variation is primarily attributed to the alteration in the type and area of land cover. The runoff potential increased following the reduction in natural vegetated land cover, in particular, grassland coverage. The economic development of the last decades had resulted in widespread residential and commercial expansion, especial after 2003 when the Iraqi government changed.

The effect of land cover change runoff was tested by comparing the predicted runoff depth in 1990 with that in 2016. The runoff depth image of 2016 was subtracted from that of 1990. The resulting image of change indicated that the average runoff depth had increased by 48 mm during the 26-year period due to land cover changes. This number points to a uniform runoff depth for the whole study area, and it has a standard deviation of 65.5. In order to understand the impact of land cover change on surface water production, the spatial pattern of the runoff depth changes image was reclassified into three categories as shown in Fig. 9.8. The first category refers

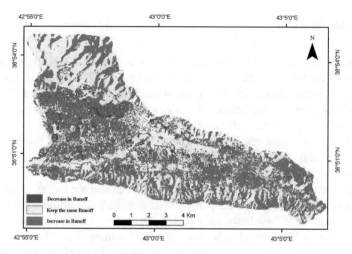

Fig. 9.8   The runoff volume changes from 1990 to 2016

to values of runoff depth less than zero which indicate decreasing surface water due to the expansion of vegetation land, while the second zone indicates no change in runoff depth from 1990 to 2016. The third area indicates increasing runoff depth because of land cover change to barren soil and built-up areas. The study concluded that 50.3% of the Duhok sub-basin belongs to the third category. These changes will directly affect the amount of runoff volume.

The Duhok watershed has been exposed to significant land cover change, in particularly urban growth and bare land in the period from 1990 to 2016. The land cover had changed over the study period. Grassland and forest land decreased from 44.1% to 12% and from 3.3 to 2.3 respectively, in the catchment area from 1990 to 2016, while urban growth and bare land increased from 9.4 to 32.1% and 43.2 to 53.5% respectively, in the same period. Both urban areas and bare land caused an increase in runoff of around 89%. There is a positive correlation between the increase in impervious area and runoff simulation results. Urban growth increased around 241%, more than that of population growth (82%). The average increase in land growth being faster than the population growth is due to the investment in land for building, which increased significantly from 1990 to 2016. The land cover factor directly affected the surface water (runoff), where the volume of runoff increased 30% from 1990 to 2016. Also, land cover changes caused changes in the stream networks of the watershed and changed the flow direction of surface runoff. This process also leads to an increase in runoff volume in some areas and a decrease in other areas.

9.4   Conclusions

This chapter focused on some possible method to use the SCS-CN approach to estimate runoff for monthly rainfall events. A GIS-based SCS-CN model was applied
to estimate surface runoff in Duhok sub-watershed. According to the simulation
results of two events of rainfall-runoff analysis, the land cover condition is an important factor that affects the volume of runoff. The Duhok sub-basin was subjected
to significant land cover changes in the period from 1990 to 2016. Satellite images
were regrouped into four classes (built up, grassland and open wood, forest and barren soil). The increase in urbanized areas and bare land and the decrease in other
land covers are significant reasons for the increase in surface runoff. Built-up land
and bare land increased from 53% in 1990 to 86% in 2016, while vegetation land
decreased from 47 to 14% in the same period. The research has also shown that both
built-up and bare land caused an increase in runoff volume from 59 to 89% in 2016,
while the vegetation area decreased in runoff volume from 41 to 11%. The result
of this chapter shows that the sub-basin of Duhok has experienced major changes
which have affected the environment of the watershed and changed the land cover
pattern. The result of this chapter supports the idea that decision-makers would be
well advised to improve plans for increased vegetation cover particularly upstream
of the watershed to decrease surface runoff.

9.5   Recommendations

The chapter recommends that the future studies should use high-resolution satellite
images and different techniques to clarify impact of climate change and urban growth
on the Run-off. The new studies should be adopted object-based analysis to classify
imagery. It is called a segment that is a cluster of the pixel with comparable spectral,
spatial and texture attributes for urban detection. Additionally, the studies focus on
the relationship of raise the rate of the urban runoff and chemical contaminants and
make diseases.

References

Abas KA (2012) Analysis of climate and drought conditions in the Fedral Region of Kurdistan.
    College of Basic Education, Salahaddin University, Erbil, Iraq
Ahmed B, Ahmed R (2012) Modeling urban land cover growth dynamics using multi-temporal
    satellite images: a case study of Dhaka, Bangladesh. ISPRS Int J Geo Inf 1(1):3–31. APA
Al Rawashdeh S, Saleh B (2006) Satellite monitoring of urban spatial growth in the Amman area,
    Jordan. J Urban Plann Dev 132(4):211–216
Al-Quraishi AMF (2004) Design a dynamic monitoring system of land degradation using Geoinfor-
    mation technology for the Northern Part of Shaanxi Province, China. J Appl Sci 4(4):669–674.
    https://scialert.net/abstract/?doi=jas.2004.669.674

Araya Yikalo H, Cabral Pedro (2010) Analysis and modeling of urban land cover change in Setúbal and Sesimbra, Portugal. Remote Sens 2(6):1549–1563

Belaid M (2003) Urban-rural land use change detection and analysis using GIS and RS technologies. In: Proceedings of the 2nd FIG regional conference, December

Bharath HA, Vinay S, Ramachandra TV (2013) Prediction of land use dynamics in the rapidly urbanising landscape using land change modeller. In: Proceedings of the international conference on advances in computer science, vol 1314. AETACS, NCR Delhi, India

Brendan O, McGarry J, Salih K (2005) The Future of Kurdistan in Iraq, University of Pennsylvania Press, Philadelphia

Buyadi SNA, Mohd WMNW, Misni A (2013) Impact of land use changes on the surface temperature distribution of area surrounding the National Botanic Garden, Shah Alam. Proc Soc Behav Sci 101: 516–525

Chen S, Zeng S, Xie C (2000) Remote sensing and GIS for urban growth analysis in china. PE RS Photogram Eng Remote Sens 66:593–598

Dawn Chatty (2010) Displacement and dispossession in the modern Middle East. Cambridge University Press, New York, pp 231–278

Day CA (2010) Using remote sensing imagery to determine the impact of land cover changes on potential runoff for the mid-cibolo creek watershed, texas. Geocarto Int 25(7):543–554

DeFries R, Eshleman KN (2004) Land-use change and hydrologic processes: a major focus for the future. Hydrol Process 18(11):2183–2186

Dewan AM, Yamaguchi Y (2009) Land use and land cover change in greater Dhaka, Bangladesh: using remote sensing to promote sustainable urbanization. Appl Geogr 29:390–401

Du P, Liu P, Xia J, Feng L, Liu S, Tan K, Cheng L (2014) Remote sensing image interpretation for urban environment analysis: methods, system and examples, pp 9458–9474

Endreny T (2005) Land use and land cover effects on runoff processes: urban and suburban development. Encycl Hydrol Sci (2002):29

Esch T, Esch T, Taubenböck H, Heldens W, Thiel M, Wurm M, Geiss C, Dech S, (2010). Urban remote sensing—how can earth observation support the sustainable development of urban environments? In: Proceedings of the 15th international conference on urban planning, regional development and information Society, Vienna, Austria, 18–20, pp 837–847

Fadhil AM (2006) Environmental change monitoring by Geoinformation technology for Baghdad and its neighboring areas. In: Proceeding of international scientific conference of Map Asia 2006: The 5th Asian conference in GIS, GPS, aerial photography and remote sensing. Bangkok, Thailand, 28 Aug–1 Sept 2006. http://www.gisdevelopment.net/application/environment/conservation/ma06-103.htm

Fadhil AM (2011) Drought mapping using geoinformation technology for some sites in the Iraqi Kurdistan region. Int J Digit Earth 4(3):239–257

Fadhil AM (2013) Sand Dunes monitoring using remote sensing and GIS techniques for some sites in Iraq. In: Proceeding SPIE 8762, PIAGENG 2013: intelligent information, control, and communication technology for agricultural engineering, 876206 (March 19, 2013); doi:10.1117/12.2019735; http://dx.doi.org/10.1117/12.2019735

Faqe GR, Ibrahim Saied, Pakiza A, Hameed HM (2016) Urban Growth prediction using cellular automata Markov: a case study using Sulaimaniya city in the Kurdistan Region of North Iraq. IISTE Humanit Soc Sci 6:108–118

Faqe GR, Ibrahim (2015) Urban expansion monitoring in Erbil city. Utilizing remote sensing tools in the Kurdistan Region, GRIN Verlag

Githui FF, Mutua, Bauwens W (2009) Estimating the impacts of land-cover change on runoff using the soil and water assessment tool (SWAT): case study of Nzoia catchment, Kenya/Estimation des impacts du changement d'occupation du sol sur l'écoulement à l'aide de SWAT: étude du cas du bassi. Hydrol Sci J 54(November 2017):899–908

Goudie A (1990) The human impact on the natural environment, 3rd edn. The MIT Press, Cambridge, Massachusetts

Griffiths P, Hostert P, Gruebner O, Linden SVD (2010) Mapping mega city growth with multi-sensor data. Remote Sens Environ 114:426–439

Guan M, Sillanpää N, Koivusalo H (2015) Modelling and assessment of hydrological changes in a developing urban catchment. Hydrol Process 29(13):2880–2894

Guo Z, Wanga SD, Chengc MM, Shub Y (2012) Assess the effect of different degrees of urbanization on land surface temperature using remote sensing images. Proc Environ Sci 13:935–942

Gyamfi C, Ndambuki JM, Salim RW (2016) Hydrological responses to land use/cover changes in the Olifants Basin, South Africa. Water (Switzerland) 8(12)

Gómez D, Montero J, Biging G (2008) Improvements to remote sensing using fuzzy classification, graphs and accuracy statistics. Pure Appl Geophys 165(8):1555–1575

Hamdi R, Termonia P, Baguis P (2011) Effects of urbanization and climate change on surface runoff of the Brussels capital region: a case study using an urban soil-vegetation-atmosphere-transfer model. Int J Climatol 31(13):1959–1974

Hameed HM (2017) Estimating the effect of urban growth on annual runoff volume using GIS in the Erbil sub-basin of the Kurdistan Region of Iraq. MDPI Hydrol 4:12

Hameed HM, Faqe G Rasul, Qurtas S Sharif, Hashem H (2015) Impact of Urban growth on ground-water levels using remote sensing—case study: Erbil City, Kurdistan Region of Iraq. IISTE Nat Sci Res 5(18):72–84

Hameed HM (2013) Water harvesting in Erbil governorate, Kurdistan region, Iraq detection of suitable sites using geographic information System and remote sensing, no 271, p 68

Jiang J, Tian G (2010) Analysis of the impact of land use/land cover change on land surface temperature with remote sensing. Proc Environ Sci 2:571–575

Karamage F, Karamage C, Zhang X, Fang T, Liu F, Ndayisaba L, Nahayo A Kayiranga, Nsen-giyumva JB (2016) Modeling rainfall-runoff response to land use and land cover change in Rwanda. Water (Switzerland) 9(2):2017

Koeppe CE, De Long GC (1959) Weather and climate. Q J R Astron Soc 85(365):318–319

Kowalik T, Walega A (2015) Estimation of CN parameter for small agricultural watersheds using asymptotic functions. Water (Switzerland) 7:939–955

Kulkarni A, Kulkarni T, Bodke SS, Rao EP, Eldho TI (2014) Hydrologic impact on change in land use/land cover in an urbanizing catchment of Mumbai: a case study. ISH J Hydraul Eng 20(3):314–323

Letha J, Thulasidharan Nair B, Amruth Chand B (2011) Effect of land use/land cover changes on runoff in a river basin: a case study. WIT Trans Ecol Environ 145:139–149

Li Y, Wang C (2009) Impacts of urbanization on surface runoff of the Dardenne Creek Watershed, St. Charles County, Missouri. Phys Geogr 30(6):556–573

Lillesand TM, Kiefer RW, Chipman JW (2008) Remote sensing and image interpretation. Hoboken, NJ, Wiley

Lu Dengsheng, Mausel P, Brondizio E, Moran E (2004) Change detection techniques. Int J Remote Sens 25(12):2365–2401

López E, Bocco G, Mendoza M, Duhau E (2001) Predicting land-cover and land-use change in the urban fringe: a case in Morelia city, Mexico. Landscape Urban Plann 55(4):271–285

Meierdiercks KL, Smith JA, Baeck ML, Miller AJ (2010) Heterogeneity of hydrologic response in urban watersheds. J Am Water Resour Assoc 46(6):1221–1237

Mitsova D, Shuster W, Wang X (2011) Cellular automata models of land cover change to integrate urban growth with open space conservation. Landscape Urban Plann 99(2):141–153

Mockus V (2007) National engineering handbook. Chapter 7 Hydrologic soil groups. Natural Resources Conservation Service

Mohammed HD, Ali MA (2014) Monitoring and prediction of urban growth using GIS techniques: a case study of Duhok city Kurdistan region of Iraq. Int J Sci Eng Res 5:1480–1488

Mustafa YT, Ali RT, Saleh RM (2012) Monitoring and evaluating land cover change in the Duhok city, Kurdistan region-Iraq, by using remote sensing and GIS. Int J Eng Invent 1(11):28–33

Pan S, Liu D, Wang Z, Zhao Q, Zou H, Hou Y, Liu P, Xiong L (2017) Runoff responses to climate and land use/cover changes under future scenarios. Water 9(7):475

Perrin J, Perrin, Bouvier C (2004) Rainfall–runoff modelling in the urban catchment of El Batan, Quito, Ecuador. Urban Water J 1(4):299–308

Phinn S, Stanford M, Scarth P, Murray AT, Shyy PT (2002) Monitoring the composition of urban environments based on the vegetation-impervious surface-soil (VIS) model by subpixel analysis techniques. Int J Remote Sens 23:4131–4153

Rasul A, Balzter H, Faqe GR, Hameed HM (2018) Applying built-up and bare-soil indices from landsat 8 to cities in dry climates. MDPI Land 7:1–13

Rawat JS, Kumar M (2015) Monitoring land use/cover change using remote sensing and GIS techniques: a case study of Hawalbagh block, district Almora, Uttarakhand, India. Egypt J Remote Sens Space Sci 18(1):77–84

Richards J ed (1999) Remote sensing digital image analysis. 3rd edn. Verlag Berlin, Heidelberg. New York, Springer

Rijabo WIA, Bleej DA (2010) Variation of rainfall with space and time in Duhok government, no 23

Sharp JM (2010) The impacts of urbanization on groundwater systems and recharge. Aqua Mundi 1:51–56

Soulis KX, Valiantzas JD (2012) SCS-CN parameter determination using rainfall-runoff data in heterogeneous watersheds-the two-CN system approach. Hydrol Earth Syst Sci 16:1001–1015

Suribabu CR, Bhaskar J (2015) Evaluation of urban growth effects on surface runoff using SCS-CN method and Green-Ampt infiltration model. Earth Sci Inform 8(3):609–626

Suribabu CR, Bhaskar J, Neelakantan TR (2012) Land use/cover change detection of Tiruchirapalli city, India, using integrated remote sensing and GIS tools. J Indian Soc of Remote Sens 40(4):699–708

Wang F, Ge Q, Yu Q, Wang H, Xu X (2017) Impacts of land-use and land-cover changes on river runoff in yellow river basin for period of 1956–2012. Chin Geogr Sci 27(1):13–24

Wang Y, Zhang X (2001) A dynamic modeling approach to simulating socioeconomic effects on landscape changes. Ecol Model 140:141–162

Weng Q (2001) Modeling urban growth effects on surface runoff with the integration of remote sensing and GIS. Environ Manage 28(6):737–748

Weng QA (2014) Remote sensing and GIS evaluation of urban expansion and its impact on surface temperature in the Zhujiang Delta, China. Int J Remote Sens 22. 2001

Wilson JS, Clay M, Martin E, Stuckey D, Vedder-Risch K (2003) Evaluating environmental influences of zoning in urban ecosystems with remote sensing. Remote Sens Environ 86:303–321

Woldesenbet TA, Elagib NA, Ribbe L, Heinrich J (2017) Hydrological responses to land use/cover changes in the source region of the Upper Blue Nile Basin, Ethiopia. Sci Total Environ 575:724–741

Wu Qiong, Li H, Wang R, Paulussen J, He Y, Wang M, Wang B, Wang Z (2006) Monitoring and predicting land use change in Beijing using remote sensing and GIS. Landscape Urban Plann 78(4):322–333

Xiao H, Weng Q (2007) The impact of land use and land cover changes on land surface temperature in a karst area of China. J Environ Manage 85(1):245–257

Yang X (2011) Remote sensing and geospatial technologies for coastal ecosystem assessment and management. Springer, Berlin, German

Zare M, Zare AAN, Samani, Mohammady M (2016) The impact of land use change on runoff generation in an urbanizing watershed in the north of Iran. Environ Earth Sci 75(18)

Zhan Y, Wang C, Niu Z, Cong P (2005) Remote sensing and GIS in runoff coefficient estimation in Binjiang Basin. In: Proceedings 2005 IEEE international geoscience and remote sensing symposium 2005. IGARSS 05, vol 6, pp 4403–4406

Zhang Hao, Qi Z, Ye X, Cai Y, Ma W, Chen M (2013) Analysis of land use/land cover change, population shift, and their effects on spatiotemporal patterns of urban heat islands in metropolitan Shanghai, China. Appl Geogr 44:121–133

Part V
Land Degradation, Drought, and Dust Storms

Chapter 10
Monitoring and Mapping of Land Threats in Iraq Using Remote Sensing

Ahamd Salih Muhaimeed

Abstract Iraq with a total land of 438,317 km^2 is located between longitudes 38° 45′ and 48° 45′ E, and between 29° 5′ and 37° 15′ N. It was indicated that the crop land is losing its inherent productivity due to poor agricultural practices and over exploitation. The direct loss of agricultural land is most acute around urban centres, where established agricultural land is being lost to alternative uses, including urbanization. Four processes of land degradation are usually recognized in Iraq including salinization, erosion, sand dunes, and urbanization. Many studies have shown that salinity is one of the most serious degradation processes in the central and southern Iraq lands. More than 70% of the irrigated agriculture lands in the central and southern Iraq have been abandoned in the recent years and causing yield declined between 30 and 60% as a result, mainly, of salt accumulation by salinization process. About more than 25% of the land area of Iraq has a serious erosion problem. More than 20% of the total area, mainly in southern Iraq was seriously affected by water lodging. So, most of Iraqi agricultural lands are highly affected by one or more of the desertification processes due to poor management practices, dry climatic conditions and to the effects of socio-economic.

Keywords Land degradation · Salinity · Soil erosion · Monitoring · Sand dunes

10.1 Introduction

In the last few decades, the soil science community has made great efforts to develop regional and global natural resources and threats affecting their quality including soil and vegetation databases. Recently, FAO-UNESCO (2008) developed several geo-referenced soil databases available at map scales smaller than 1:250,000; namely the Harmonized World Soil Database at a map scale of 1:5 M (million). Lack of comprehensive information about national land resources increases the risk of releas-

A. S. Muhaimeed (✉)
College of Agriculture, Baghdad University, Baghdad, Iraq
e-mail: profahmad1958@yahoo.com

Soil Classification and Management, ACSAD, Damascus, Syria

© Springer Nature Switzerland AG 2020
A. M. F. Al-Quraishi and A. M. Negm (eds.),
Environmental Remote Sensing and GIS in Iraq, Springer Water,
https://doi.org/10.1007/978-3-030-21344-2_10

ing uninformed policy decisions, avoidable continued degradation of land, water resources and land cover. The viability and cost of vital infrastructure is affected by this information shortage just as much as the food and water security and response to environmental change (Van Engelen 2008). Remote sensing offers a unique opportunity in monitoring, assessing and empirically quantifying spatial and temporal changes in soil and vegetation taking place during a long period. The benefits of satellite remote sensing surveillance are not only in terms of its cost-effectiveness, obtaining a global perspectives and timeliness, but are usefulness in evaluating ongoing changes. Under the existing security situation in Iraq, implementation of large scale field studies to monitor changes in vegetation is extremely difficult and therefore use of satellite images can be seen as an effective technique to monitor land cover changes, especially in areas where reliable field data is unavailable (Evans et al. 2002).

Remote sensing may offer possibilities for extending existing soil survey data sets. The data it provides can be used in various ways. Firstly, it may help in segmenting the landscape into internally more or less homogeneous soil–landscape units for which soil composition could be assessed by sampling using classical or more advanced methods. Secondly, remotely sensed data could be analysed using physically-based or empirical methods to derive soil properties. Moreover, remotely sensed imagery can be used as a data source supporting digital soil mapping (Ben-Dor 2008; Slaymaker 2001). Finally, remote sensing methods facilitate mapping inaccessible areas by reducing the need for extensive time-consuming and costly field surveys.

Land degradation is one of the most serious ecological problems in the world. It entails two interrelated, complex systems: the natural ecosystem and the human social system. Causes of land degradation are not only biophysical, but also socio-economic (marketing, income, human health, institutional support, poverty), undermining food production and political stability (UNEP 1991; UNCCD 2013). Land degradation is a process in which the value of the biophysical environment is affected by a combination of human-induced processes acting upon the land. Land degradation leads to a significant reduction of the productive capacity of land. Human activities contributing to land degradation include unsustainable agricultural land use, poor soil and water management practices, deforestation, removal of natural vegetation, frequent use of heavy machinery, overgrazing, improper crop rotation and poor irrigation practices. Land degradation is the result of complex interactions between the physical, chemical, biological, socio-economical, and political issue of local, national or global nature (McDonagh et al. 2006; Bationo et al. 2006). The direct and indirect causes of land degradation are linked by a chain of causes and effect called vicious cycle/causal nexus. Limited land resources and increase in rural population are the two external or deriving forces for poverty. Land shortage and poverty, taken together, lead to non sustainable land management practices which are the direct causes of land degradation. Poverty is very likely to contribute to land degradation for many reasons. When people lack access to alternative source of livelihood, there is a tendency to exert more pressure on a few resources that are available to them.

About more than 25% of the land area of Iraq has a serious erosion problem. More than 20% of the total area, mainly in southern Iraq was seriously affected by water lodging and more than 70% was affected by salinization (Muhaimeed et al. 2013; Wu et al. 2014). So, most of Iraqi agricultural lands are highly affected by one or more of the desertification processes due to poor management practices, dry climatic conditions and to the effects of socio-economic degradation of land and ecosystems may be the result of numerous factors or a combination there of, including anthropogenic (human-related) activities such as unsustainable land management practices and climatic variations. The dominant types of land degradation in arid regions including Iraq are: salinization, erosion, sand dunes and sand storms, water lodging, pollution and overgrazing. The effect of some degradation processes was studied in detail, and other studied on small scales. The following sections will demonstrate some information about the extent and changes of some land degradation types using remote sensing. Data about Iraqi natural resources have been collected are mostly at small-to-medium scales, using different standards and methods which leading to lack digital and accessible soil information, while Larger scale digital data are limited. Remote sensing and GIS have been utilized for studies in land degradation and factors that causes land degradation, such as drought in some sites of Iraq (Fadhil 2009, Almamalachy et al. 2019; Al-Quraishi et al. 2019). In this chapter, an attempt was done to demonstrate the output of some works which have been done on monitoring and mapping spatial and temporal changes of Iraqi resources using remote sensing and GIS techniques.

10.2 Soil Salinity

10.2.1 Monitoring and Mapping Soil Salinity

The saline soils from the agricultural point of view are those, which contain sufficient neutral soluble salts in the root zone to adversely affect the growth of most crops. For definition, saline soils have an electrical conductivity of saturation extracts of more than 4 dS m^{-1} at 25 °C (Richards 1954). As salinity levels increase, plants extract water less easily from the soil, aggravating water stress conditions. High soil salinity can also cause nutrient imbalances, which then result in the accumulation of elements toxic to plants, and reduce water infiltration if the level of one salt element (like sodium) is high. In many areas, soil salinity is the factor limiting plant growth.

The extent of salt-affected areas on a wide world vary and it estimated, in general close to 1 billion ha, which represents about 7% of the earth's continental extent (Ghassemi et al. 1995). In addition to these naturally salt-affected areas, about 77 million ha have been salinized as a consequence of human activities, with 58% of these concentrated in irrigated areas. On average, salts affect 20% of the world's irrigated lands, but this figure increases to more than 30% in countries such as Egypt, Iran and Argentina (Ghassemi et al. 1995). According to estimates by FAO and UNESCO, as much as half of the world's existing irrigation schemes are more or less under the influence of secondary salinization and waterlogging. About 10 million hectares of irrigated land are abandoned each year because of the adverse effects of irrigation, mainly secondary salinization and alkalinization (Szabolcs 1987).

In Iraq, salinity is the major problem in the agriculture area, mainly in Mesopotamia plain. It is reported that approximately 60% of the cultivated land has been seriously salinized, of which 20–30% has been abandoned (Buringh 1960; FAO 2011) due to irrational land management (e.g., over irrigation and poor drainage) and other natural factors (e.g., flooding, drought, and impermeability of the under- lying formation). More than 70% of the irrigated agriculture lands in the central and southern Iraq have been abandoned in the recent years and causing yield declined between 30 and 60% as a result, mainly, of salt accumulation by salinization process (Muhaimeed 2013). Salinization is a common problem for agriculture in dryland environments and it has greatly affected land productivity and even caused cropland abandonment in Central and Southern Iraq (Wu et al. 2013, 2014). It is one of the most serious degradation processes in the central and southern Iraq lands due to irra- tional land management (e.g., over irrigation and poor drainage) and other natural factors (e.g., flooding, drought, and impermeability of the underlying formation).

Remote sensing has been widely applied in salinity mapping and assessment in the recent decades and a number of achievements have been obtained (Mougenot et al. 1993; Mustafa et al. 2019; Metternicht and Zinck 2003; Wu et al. 2018, 2019), e.g. the relationship between salinity and vegetation indices (Steven et al. 1992; Zhang et al. 2011) and the best band combination for salinity assessment (Eldeiry and Garcia 2010; Dwivedi and Rao 1992).

10.2.2   Dynamic Salinity Changes

In Iraq, a few studies had been done on measuring and mapping salt affected soils using remote sensing on small scales (Muhaimeed et al. 2013; Wu et al. 2014). Part from Iraqi soil salinity management project funded by ACIAR (Australian Center for International Agriculture Research) in cooperation with ICARDA (Wu et al. 2014) studied the status of soil salinity level and its temporal changes in the most agricultural land in Iraq represented by the Mesopotamia plain based on field survey, a rational and operational remote sensing salinity mapping (Fig. 10.1).

10.2.3   Methodology for Monitoring and Mapping Soil Salinity

In order to study temporal changes for salinity level in Mesopotamia plain, in middle and southern Iraq, five pilot areas were selected represent all variations for land use and land cover with in Mesopotamia plain. Field survey data (including EM38 read- ings, soil laboratory analytical results) with a multi-resolution satellite dataset was

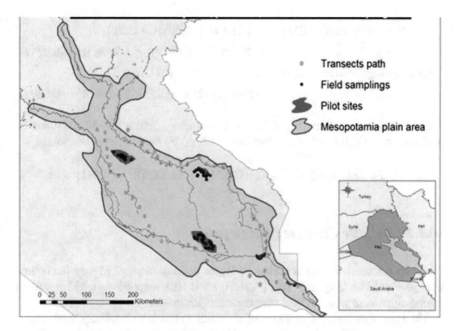

Fig. 10.1 Location of the study area, Mesopotamia, Iraq

prepared including 33 spring (Feb–Apr) and summer (Aug) Landsat ETM+ images in the period 2000–2012, four SPOT and three Rapid Eye imagery respectively dated Mar 2010–Apr 2012, time-series of MODIS vegetation indices data (MOD13Q1), and land surface temperature (LST, MOD11A1 and A2) from 2000 to 2012 were obtained.

Atmospheric correction using FLAASH for all Landsat ETM+, SPOT and Rapid Eye imagery. A set of mostly applied VIs such as NDVI (Normalized Difference Vegetation Index) (Rouse et al. 1973), SAVI (Soil Adjusted Vegetation Index) (Huete 1988), SARVI (Soil Adjusted and Atmospherically Resistant Vegetation Index) (Kaufman and Tanré 1992), EVI (Enhanced Vegetation Index) (Huete et al. 1997) were produced from the atmospherically-corrected and reflectance-based satellite imagery; a new vegetation index was used, the Generalized Difference Vegetation Index (GDVI) developed by Wu (2014) and in the form of:

$$\text{GDVI} = \left(\rho_{NIR}{}^{n} - \rho_{R}{}^{n}\right)/\left(\rho_{NIR}{}^{n} + \rho_{R}{}^{n}\right) \tag{1}$$

where ρ_{NIR} is the reflectance of the near-infrared band and ρ_{R} is that of the red band, and n is the power.

After separation of the vegetated and non-vegetated area, a multiple linear least-square regression analysis was undertaken to couple the EM38 measurements with VIs for vegetated area and with NVIs for bare soils. The following remote sensing salinity models were obtained:

$$\text{For vegetated area: EMV} = -824.134 + 918.536 * \text{GDVI}$$
$$- 754.204 * \ln(\text{GDVI}) \ (\text{Multiple } R^2 = 0.925)$$
$$\text{For non-vegetated area: EMV} = 2{,}570{,}683.24 + 1821.24 * \text{ST}$$
$$- 546{,}476.07 * \ln(\text{ST}) \ (\text{Multiple } R^2 = 0.829)$$

where EMV—vertical reading of EM38, ST—spring surface T in K. EMV can be converted into EC (electrical conductivity in dS/m) by the following relationships:

$$\text{EC} = 0.0005\text{EMV}^2 - 0.1007\text{EMV} + 15.632 \ \left(R^2 = 0.841\right)$$

10.2.4 Salinity Changes in Mesopotamia

Based on the created models (Table 10.1), the salinity maps of Mesopotamia area were generated for 2000 and 2010 (Figs. 10.2 and 10.3, respectively), which were in a good agreement with the field measurement ($R^2 = 0.811$).

The results indicate the presence of six salinity classes with salinity level ranged from less than 4 dS m^{-1} to more than 50 dS m^{-1} in Mesopotamia plain soils. In general, salinity level shows dynamic changes with time from one region to other due mainly to poor management agricultural practices. Table 10.2 shows the net dynamic changes for salinity classes between 2000 and 2010 in Mesopotamia plain. The areas and percentages of low and high salinity level classes increase, while the areas and percentages of moderately level classes 3 and 4 decrease with time.

The results indicate that during the last two decades, low and moderate salinity level classes have been increased by 4 and 8%, respectively. While, high salinity level classes decrease by 14 and 3%, respectively. In general, the main salt sources in Mesopotamia plain are saline shallow ground water, poor quality of irrigation water and poor drainage system. Irrigation water always contains some salt and incorrect methods may lead to accumulation of this salt. When waterlogging is present at some depth the water evaporates again and the salt transported with the water from else-

Table 10.1 Salinity models for Mesopotamia area

Scale	Type	Salinity models	Error scope	Multiple R^2
Regional-scale	Vegetated area	$\text{EM}_V = 66.338 - 258.114 * \ln(\text{GDVI})$	±88.882	0.717
	Non-vegetated area	$\text{EM}_V = 3{,}055{,}497.34 + 2161.09 * \text{LST} - 649{,}347.93 * \ln(\text{LST})$	±92.524	0.695

Note EM$_V$ and EM$_H$ can be converted into EC (dS/m) from the regional transect sampling, i.e. $\text{EC} = 0.0005\text{EM}_V^2 - 0.0779\text{EM}_V + 12.655$ ($R^2 = 0.8505$); $\text{EC} = 0.0002\text{EM}_H^2 + 0.0956\text{EM}_H + 0.0688$ ($R^2 = 0.7911$)

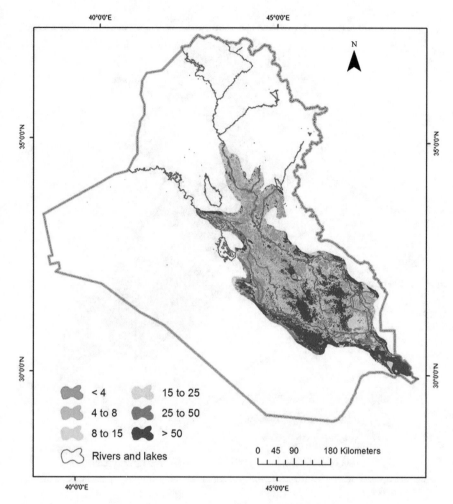

Fig. 10.2 Spatial distribution of salinity level in Mesopotamia plain, 2000 (After Wu et al. 2014)

where is left be-hind. It is seen that the salinity has experienced significant changes Mesopotamia plain in the past decades which are closely related to land use and management (e.g. land abandonment) by farmers that are associated with both macro and micro socioeconomic environment in Iraq apart from the natural factors such as high salt concentration of groundwater, and dry climate. For a better understanding of the causes of salinization, spatially explicit modeling incorporating natural, socioeconomic, and climate data to reveal both the spatial and human determinants and the impacts of climate change on salinization is required in the future.

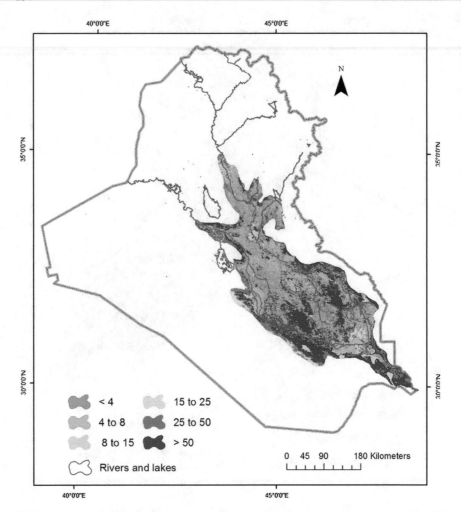

Fig. 10.3 Spatial distribution of salinity level in Mesopotamia plain, 2010 (After Wu et al. 2014)

Table 10.2 Salinity level net dynamic changes 2000–2010 in Mesopotamia plain

Salinity range (dS m^{-1})	2000		2010		Change 2000–2010	
	Area (ha)	%	Area (ha)	%	Area (ha)	%
<4	1,457,716	16	1,817,371	20	359,655	4
4–8	1,857,269	20	2,578,802	28	721,533	8
8–15	1,551,738	17	229,928	3	−1,321,809	−14
15–25	806,020	9	514,689	6	−291,331	−3
25–50	1,114,529	12	1,526,283	17	411,754	4
>50	2,397,990	26	2,518,080	27	120,090	1
Total area (ha)	9,185,261	100	9,185,153	100		

10.2.5   Effect of Soil Salinity on Vegetation Changes in Mesopotamia Plain

Vegetation conditions was created using MODIS images for the period 2000–2011 for the Mesopotamian Plain is presented in Fig. 10.4. The map demarcates locations with consistent bare soil and vegetation (low, moderate and high) and trends in the changes in vegetation conditions (degraded, improved or unstable). The results show that more than 44% of the Mesopotamian Plain falls under bare soil class. In general, vegetation is sparse and improvements in vegetation over the studied period are marginal. Only 20% of the area has low stable vegetation and 19.5% is degraded. Inter-annual changes of the area for land cover classes (Fig. 10.5) indicate that the most favorable conditions for vegetation growth were observed in 2004, whereas in 2000 and during the period 2008–2011 vegetation condition was the poorest. There are variations of vegetation condition in different years for analyzed governorates and test sites. One year (2004) was very favorable for vegetation growth and two years (2000 and 2008) were very unfavorable. The factors, influenced on vegetation condition in arid areas are the climate (air temperature and rainfall), the surface water availability and the soil salinity.

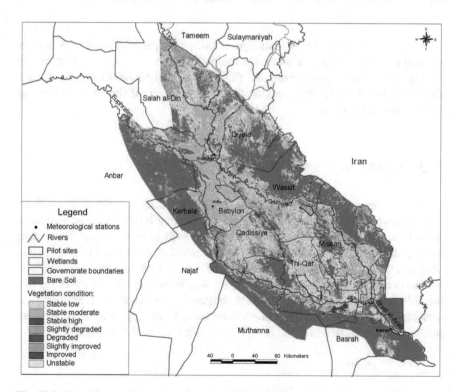

Fig. 10.4 Vegetation conditions over the period 2000–2011 in the Mesopotamia

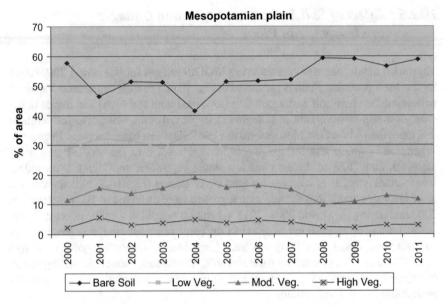

Fig. 10.5 The inter-annual changes of the area for land cover classes in Mesopotamian Plain

In the Mesopotamian Plain, limited availability of water, combined with shallow and saline groundwater conditions are the major causes of the expansion of soil salinity. The coarse spatial resolution (250 m) of MODIS NDVI images did not allow us to link directly the vegetation condition with soil salinity for whole area of the Mesopotamian Plain.

10.3   Land Cover/Land Use

Land use refers to what people do on the land surface, such as agriculture, commercial, residential development, and transportation (Jensen 2005), while the land cover is the type of material present on the landscape such as natural vegetation, water bodies, rock/soil, manmade features and others resulting due to land transformation (Roy and Giriraj 2008). Since the inception of the Earth observation satellites, land use/land cover map produced by remote sensing technique has been serving as a valuable resource to support the decisions of the planners, economist, ecologist and decision-maker involved in the process of sustainable development for the territory (John and Chen 2004; Mustapha et al. 2010; Malinverni et al. 2010).

10.3.1 Land Cover/Land Use Dynamic Changes

Image classification for the land use and land cover mapping is one of the important parts in many remote sensing applications. However, it is not easy to generate a satisfactory result for land use/land cover classification from remotely sensed data because of the limitation of data, image processing techniques and complexity of the land use/land cover types. Many factors are to be considered in order to get a good classification accuracy, which are the characteristics of the study area, availability of suitable remotely sensed data and ground reference data, proper use of the variables and the classification algorithms, the producer's experience, and the time constraint (Lu and Weng 2005). Landsat images may be the most common data source for land use-land cover classification because of its long history of space-based data collection on a global scale. However, its coarse spatial resolution often cannot meet the specific requirements of the LULC classification, especially the complex urban-rural interface (Jensen and Cowen 1999). Besides Landsat, high resolution satellite sensors such as IKONOS can provide high resolution imagery with multispectral and panchromatic data for the LULC classification. Spectral, texture, and structural information can be extracted from high resolution images to investigate the characteristic of complex land surfaces and greatly reduce the mixed-pixel problem (Lu and Weng 2009). Apart from the spectral data, spatial image processing (texture processing) has greater importance for LULC mapping (Shivashankar and Hiremath 2011).

Land cover undergoes continuous changes around the world, especially in highly populated areas. This phenomenon can be attributed to human activities including population growth and the need for more housing. Development and population growth have triggered rapid changes to Earth's land cover over the last two centuries, these rapid changes are superposed on long-term dynamics associated with climate variability. Land cover change can affect the ability of the land to sustain human activities through the provision of multiple ecosystem services and pave a path to monitor the long-term trends as well as inter annual variability and at a level of spatial detail to allow the study of human-induced changes.

Using remote sensing techniques for monitoring and mapping LULC in Iraq was applied on the small and moderate scale by many researchers (Al-Daghastani 2008a, b; Kareem et al. 2014). They pointed out that the dominant land covers in Iraq including forests, shrubs, grasses, bare land, cultivated land, and water bodies. The area of all these classes is changing with time. These changes in the nature of LULC, in time and space, due to climatic condition and land management activities. Few LULC studies, where few of them done on a large scale. One of the most work about the distribution of land cover in Iraq was done by USDA (2000) created by using Landsat data (Fig. 10.6). The results indicated that the bare land Class is the most dominant in the whole country, following by agricultural land. While, water bodies show the lowest percentage from the total area of Iraq. These results are in agreement with the results of Jaradat work done in 2003 (Fig. 10.7). The most recent work about the types and the distribution of land use land cover in Iraq which was

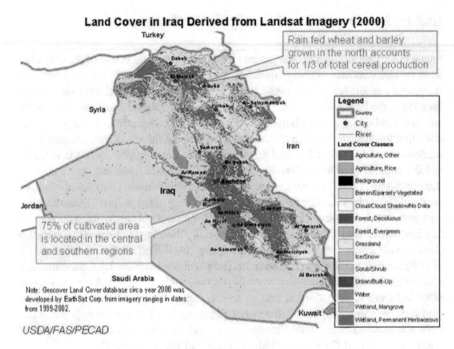

Fig. 10.6 Land cover types in Iraq (USDA 2000).

done by Iraq Land degradation neutrality project using remote sensing indicated the spatial and temporal distribution pattern for LULC as in the following section.

10.3.2   Methodology for Monitoring and Mapping LULC Changes in Iraq

Series of digital Land Use and Land Cover (LULC) maps for whole Iraq were created by Iraqi LDN project. Normalized difference vegetation index (NDVI) and different indices can be depicted inventories of change detection (Kontoes 2008; Mas 2010). These maps can be implemented depending on different types of satellite imagery, as well as field observations. Series of LULC map is very important a preliminary tool for land management, predictive environmental risk and management. The dataset for Iraqi land cover maps—V2.0.7 was obtained from the ESA CCI-LC (http://maps.elie. ucl.ac.be/CCI/viewer/download.php) and cropped depending on nationality boundary of Iraq for each epoch 2000/2015 as showing in Figs. 10.8 and 10.9. These maps were created by Iraqi Land Degradation Neutrality (LDN) project in 2017. According to European Space Agency Climate Change Initiative Land Cover (ESA CCI-LC) dataset (Land Cover Maps—v2.0.7), which is consists of consistent global land cover maps at 300 m spatial resolution, on an annual basis from 1992 to 2015.

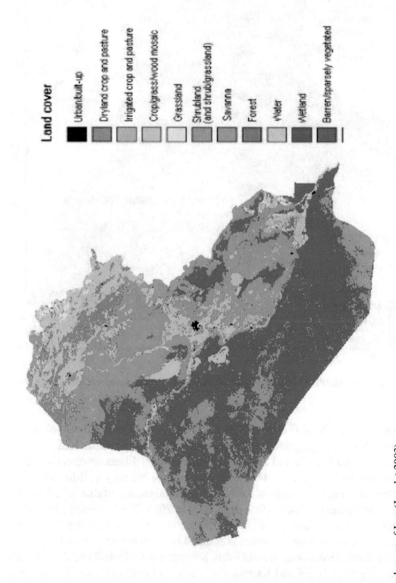

Fig. 10.7 Land cover of Iraq (Jaradat 2003)

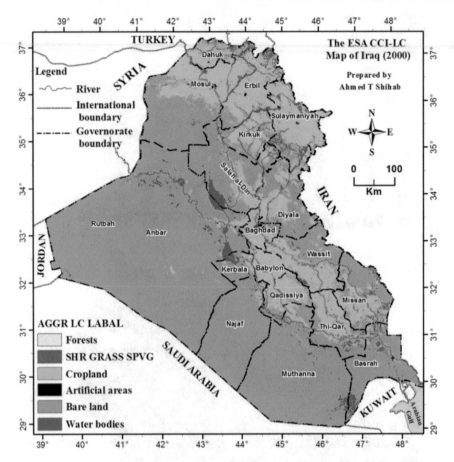

Fig. 10.8 Spatial distribution of the aggregate LC classes of Iraq for the epoch 2000 (MoA 2017)

Moreover, the ESA CCI-LC classes 2000 and 2015 (Land Cover Maps—v2.0.7) of Iraq's territory were aggregated into 6 LC groups according to Table 10.3.

Based on ESA LC 2000 and 2015 the most widespread land cover class is Bare Land covering 74.1573 and 74.8467% of the national territory in 2000 and 2015 respectively showing an increase of 0.6893%. Artificial areas also showed a 0.3416% increase of the country's land surface from 2000 to 2015. The increased coverage of the artificial areas has reached 1,493.9 km², which represents more than twice as it was in 2000. However, most of which is situated in the northern part of Iraq within Kurdistan region. Particularly, around Erbil, Sulaymaniyah, Duhok and Zakho cities as well as Baghdad, Najaf and Karbala cities toward the central and southern part of Iraq, due to they are expansion at the expense of Cropland land, whereas the class value 3 change to 5 (red color) as showing in Fig. 10.10. Moreover, the shrubs, grasslands and sparsely vegetated areas reduced significantly by 4449.8 km², which is change to Bare Land in 2015, due to the passage of Iraq periods of drought during the

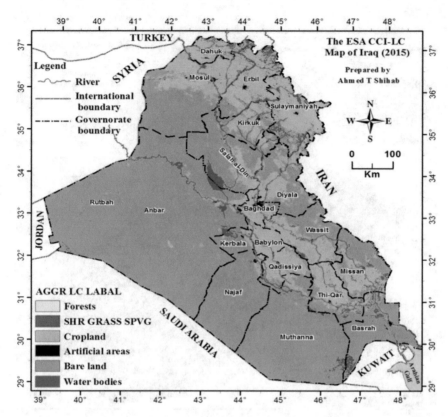

Fig. 10.9 Spatial distribution of the aggregate LC classes of Iraq for the epoch 2015 (MoA 2017)

Table 10.3 Area and percent of land cover change in Iraq between 2000 and 2015

VALUE	Description	Area (Km²) in 2000	Area (Km²) in 2015	Change (Km²) (2000-2015)	Percentage in 2000	Percentage in 2015	Change in percentage (2000-2015)
1	Forests	272.12	287.47	15.34	0.0622	0.0657	0.0035
2	Shrubs, grasslands and sparsely vegetated areas	19085.64	14635.83	-4449.81	4.3641	3.3466	-1.0175
3	Cropland	86831.01	87404.99	573.97	19.8546	19.9858	0.1312
4	Wetlands	—	—	—	—	—	—
5	Artificial areas	888.65	2382.55	1493.90	0.2032	0.5448	0.3416
6	Bare lands	324315.81	327330.53	3014.72	74.1573	74.8467	0.6893
7	Water bodies	5941.46	5293.34	-648.13	1.3586	1.2104	-0.1482

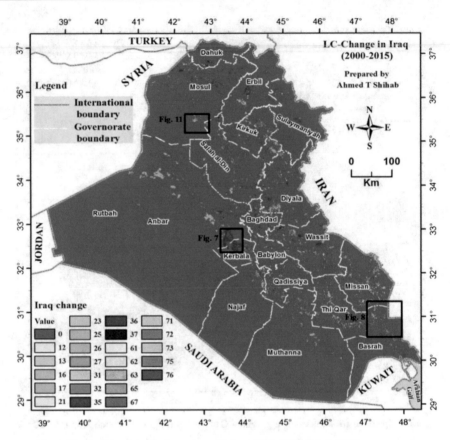

Fig. 10.10 Distribution of the land use change in Iraq 2000–2015 (MoA 2017)

past decades. It is worth to mention that the Water bodies also reduced by 648.13 km^2 most of which is situated around Al-Razzaza Lake, due to the water policy adopted by neighboring countries and the significant decrease in the water resources of the Tigris and Euphrates Rivers.

10.4 Sand Dunes

The phenomenon of sand dunes is considered to be one of the most dangerous consequences of desertification, due to its negative impact on every vital aspect of life. Degradation of Land and ecosystems may be the result of numerous factors or a combination there of, including anthropogenic (human-related) activities such as unsustainable land management practices and climatic variations. The dominant types of land degradation in arid regions including Iraq are: salinization, erosion, sand dunes and sand storms, water lodging, pollution and overgrazing. In general, there are three main regions for sand dunes in Iraq: sand dunes region in the west side of the Euphrates river, sand dunes region in the east side of the Tigris river and sand dunes region between the Tigris and the Euphrates rivers. Most of Iraqi lands are affected by wind erosion and formation and movement of sand dunes, particularly in the middle and south of Iraq. The total area covered by sand dunes is approximately 2 millions hectares. And there are other millions of hectares threatened by sand dunes (Kaufman and Tanré 1992). In Iraq, there are two types of sand dunes. The first type is pseudo sand dunes that contain high percentages of silt and clay, such as the sand dunes in the middle and south of Iraq. The second type is the true sand dunes containing a high percentage of sand, such as the sand dunes located in Baiji. The sand dunes in the middle and south of Iraq have been invistigated using remote sensing and GIS techniques to monitor and map the existence of sand dunes, whereas Landsat TM and ETM+ images were utilized to depict the sand dunes encroachments (Fadhil 2009, 2013).

The problems have become worse since the imposition of economic sanctions in 1990. Poor soil/water management and severe climatologically factors have changed extensive agricultural lands in Iraq's alluvial plain into the present bare, water logged soils covered with aeolian sand sheets and pseudo sand dunes. Iraq faces a severe desertification problem that jeopardizes its food security through the effects of soil salinity, water logging, loss of vegetative cover, shifting sand dunes and severe sand/dust storms. All of these problems need to be addressed to halt the threat. To combat these problems Iraq has launched programs to rectify soil salinity, to develop natural vegetative cover and to halt the encroachment of sand dunes, as well as reduce the frequency and severity of sand and dust storms.

10.4.1 Dynamic Changes for Sand Dunes Areas During 2006–2016

Images for Landsat 7 and 8 curried through 2006 to 2016 were used GIS staff member DFD to monitoring the dynamic changes in the area for sand dunes and sand sheet in Iraq. Figures 10.11 and 10.12 show the locations and spatial distribution of sand dunes in Iraq for 2006 and 2016, respectively.

The dynamic changes in the area of sand dunes are shown in Fig. 10.13 and Table 10.4. The results indicated the presence of four classes for the areas affected by sand including sevy sand, sand dunes, fixed sand dunes and sand sheet. Sand dunes class occupied the largest areas in both years (67,822 and 732,005 ha in 2006 and 2016, respectively), and increase rate of 8,963.2 ha/yr. While, all other sand affected classes include sevy sand, and sand sheet increase at rate of 87,089.96 and 39,354.4 ha/yr. These results lustrate the serious degradation phenomena facing the agriculture land in Iraq due, mainly to dry climatic condition and poor vegetation which allow the wind to carry sand particle from one area to other. On other side,

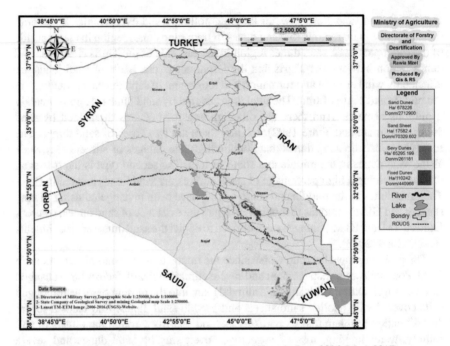

Fig. 10.11 Spatial distribution of sand dunes and sand sheet in Iraq, 2006 (MoA 2017)

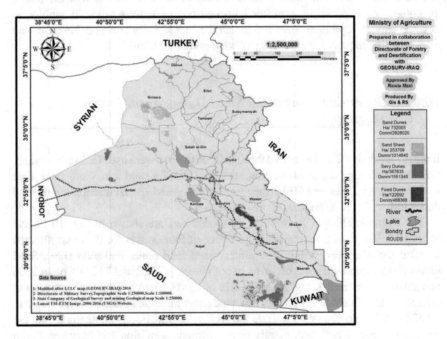

Fig. 10.12 Spatial distribution of sand dunes and sand sheet in Iraq, 2016 (MoA 2017)

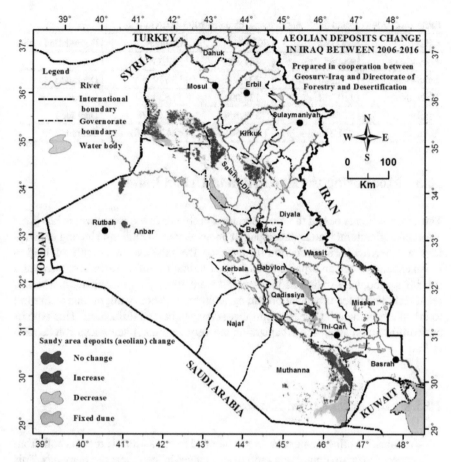

Fig. 10.13 Areas of sand dune and sand sheet changes between 2006 and 2016

there is some efforts done by some conservation fixing sand dunes project carried the ministry of agriculture aimed to minimize the movement of sand dune and sand sheet. The results shown in Fig. 10.13 and Table 10.4 show that the area of fixed sand dune increase from 110,242 ha to 112,092 ha in 2006 and 2016, respectively with increasing rate of 1,975 ha/yr.

In general, degradation of land and ecosystems may be the result of numerous factors or a combination there of, including anthropogenic (human-related) activities such as unsustainable land management practices and climatic variations.

Table 10.4 Area of sand dunes net dynamic changes 2006–2016 in Iraq

Class	2006 ha	2016 ha	Changes	Degradation status
Sevy sand	65,295.199	387,835	322,539.801	Degraded
Sand dunes	678,226	732,005	53,779	Degraded
Fixed dunes	110,242	122,092	11,850	Improve
Sand sheet	17,582.4	253,709	236,126.6	Degraded
Total	871,356.599	1,495,641	624,295.401	Degraded

10.5 Monitoring the Expansion of Urban Land

The urbanization is one of the most common problems in the world which has great effect on agricultural land in developed or undeveloped counties and having negative impacts on agriculture production for instance, from the loss of agricultural land to urban expansion and an urban bias in public funding for infrastructure, services and subsidies. About 1.9 billion ha in the world were affected by urbanization. There is relationship between urbanization and agriculture, with heavy migrations from rural to urban areas, there have been significant changes in land utilization. This type of basic information about the urbanization are very common phenomena can be seen in Iraq.

10.5.1 Impacts of Urbanization

1. Urbanization brings major changes in demand for agricultural products both from increases in urban populations and from changes in their diets and demands. This has brought and continues to bring major changes in how demands are met and in the farmers, companies, corporations, and local and national economies who benefit (and who lose out). It can also bring major challenges for urban and rural food security.
2. Decrease in the total area of agriculture land causing low food production.
3. Pollution of agriculture land due to non-planning urban land.
4. Effects on ecological systems due to the reduction in the green area.
5. Increase in air concentration from CO_2 with increase air temperature due to the decrease of vegetation cover.

10.5.2 Causes of Urbanization

1. Increase of the total population with the increase in the housing demand.
2. Increase of enplaning urban and settlements.

Table 10.5 Temporal change in the urban area in Baghdad Governorate

Year	1976	1985	1990	2003	2014	% change
Area (ha)	76,879	93,302.3	108,452	121,134	1,250,667	10.6

3. Rural to urban people migration is happening on a massive scale due to population pressure and lack of resources in rural areas. people hope for well paid jobs, the greater opportunities to find casual or 'informal' work, better health care and education.
4. Social development: people living in rural areas believe that the standard of living in urban areas will be much better than in rural areas.
5. Toward the industrial production than the agricultural production.
6. Looking for governmental jobs rather than the agricultural work.
7. Growth of industries has great contribution to the growth of the cities.
8. Economic effects: people moving in response to better economic opportunities in urban areas, or to the lack of prospects in their home farms or villages.

   Iraq is one of the undeveloped countries facing high rate of population with mean annual increasing of 3.4%. Some works indicated that Iraq have lost about 538,100 ha from its agriculture lands during 1957–1976, while the total population in the urban is higher than rural during 1985–2010. According to the case study for using remote sensing done by Muhaimeed and Ali (2015) regarding the status of urbanization in Baghdad governorate and its effects on the degradation of agriculture land during 1976–2014, there was a continuity increase in the area of urban land and decrease in agriculture area (Table 10.5). They used four different Images for MSS, TM, ETM+ and OLI Landsat sensors queried during 1976, 2003 and 2014, respectively. Temporal change for Land cover/Land use and Urban lands were detected using NDVI, SAVI and NDBI indices.

   The total percentage of urban area increased from 18% in 1976 to 33.7% from the total area in 2014, while, there was great decrease in the percentage of agricultural land with time. The area of the built-up land has increased by 3.55% (9.48 km^2) due to mainly the increase in the total population and the expansion in the area of Baghdad city which caused the conversion of agricultural land to urban land during the last four decades. Results indicated that the main land degradation processes are urbanization and bad management practices reflected by the formation of salt affected soils. According to data of the Ministry of planning (2015) the total population for Baghdad governorate increase from 3,189,700 in 1977 to 7,665,292 in 2014 (Fig. 10.14). High increase in the population allow to increase in the area of urbanization and decrease in agricultural land. Figure 10.15 shows the spatial and temporal changes in the urban land in Baghdad governorate taken between 1976 and 2014.

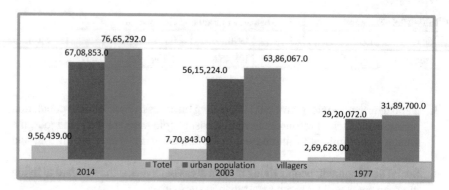

Fig. 10.14 Temporal changes in the population of urban and rural land in Baghdad between 1977 and 2014 (Muhaimeed and Ali 2015)

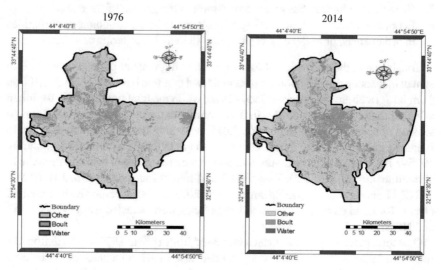

Fig. 10.15 Temporal change in the urban area for Baghdad Governorate (Muhaimeed and Ali 2015)

10.6   Conclusions

Remote sensing data have been used extensively to identify and mapping land threats on a large scale in the world. Multispectral satellite sensors are the preferred method for mapping and monitoring land threats, due to the low coast of such imagery and the ability to map extreme surface expressions of land threats. Several spectral indices were developed by researchers to identify the type of land threats as a single index may not be perform best for all types of land threats. Each site needs to be assessed regarding the strength and the weakness of the proposed indices before appropriate remote sensing based indices are used for land threat mapping and assessing. Thus, the present studies illustrate that remote sensing is an important technologies for temporal analysis and quantification of spatial phenomena.

10.7 Recommendations

There is a great need to monitor the natural resources and their properties in practical place and time. The application of remote sensing data have proven useful technology to monitor and mapping the natural resources and their properties and improving the accuracy and consumed time, effort and coast.

References

Al-Daghastani H (2008a) Land use and land cover map of Nineveh governorate using remote sensing data. Iraqi J Earth Sci 8(2):17–29

Al-Daghastani H (2008b) Land use and cover change assessment using remote sensing and GIS: Dohuk City, Kurdistan, Iraq (1998–2011). Int J Geomatics Geosci 3(3):552–569

Almamalachy YS, Al-Quraishi AMF, Moradkhani H (2019) Agricultural drought monitoring over Iraq utilizing MODIS products. In: Al-Quraishi AMF and Negm AM (eds) Environmental remote sensing and GIS in Iraq. Springer Water

Al-Quraishi AMF, Qader SH, Wu W (2019) Drought monitoring using spectral and meteorological based indices combination: a case study in Sulaimaniyah, Kurdistan region of Iraq. In: Al-Quraishi AMF, Negm AM (eds) Environmental remote sensing and GIS in Iraq. Springer Water

Bationo A, Hartermink A, Lungu O, Naimi M, Okoth P, Smaling E, Thiombiano L (2006) African soils: their productivity and profitability of fertilizer use. In: Background paper prepared for the African fertilizer summit, Abuja, Nigeria

Ben-Dor E (2008) Imaging spectrometry for soil applications. Adv Agron 97:321–392

Buringh P (1960) Soils and soil conditions in Iraq. Ministry of Agriculture of Iraq, Iraq, 337 p

Dwivedi RS, Rao BRM (1992) The selection of the best possible Landsat TM band combination for delineating salt-affected soils. Int J Remote Sens 13:2051–2058

Eldeiry AA, Garcia LA (2010) Comparison of ordinary kriging, regression kriging, and cokriging techniques to estimate soil salinity using Landsat images. J Irrig Drainage Eng 136:355–364

Evans C, Jones R, Svalbe I, Berman M (2002) Segmenting multispectral Landsat TM images into field units. IEEE Trans Geosci Remote Sens 40(5):1054–1064

Fadhil AM (2009) Land degradation detection using geo-information technology for some sites in Iraq. J Al-Nahrain Univ Sci 12(3):94–108

Fadhil AM (2013) Sand dunes monitoring using remote sensing and GIS techniques for some sites in Iraq. In: Proceedings SPIE 8762, PIAGENG 2013: intelligent information, control, and communication technology for agricultural engineering, p 876206. http://doi.org/10.1117/12.2019735

FAO (2011) Country pasture/forage resource profiles: Iraq. FAO, Rome, Italy, p 34

FAO, IIASA, ISRIC, ISS-CAS, JRC (2008) Harmonized world soil database (version 1.0). FAO, Rome, Italy and IIASA, Laxenburg, Austria

Ghassemi F, Jakeman AJ, Nix HA (1995) Salinisation of land and water resources: human causes, extent, management and case studies. The Australian National University, Canberra, Australia and CAB International, Wallingford, Oxon, UK

Huete AR (1988) A soil adjusted vegetation index (SAVI). Remote Sens Environ 25:295–309

Huete AR, Liu HQ, Batchily K, van Leeuwen W (1997) A comparison of vegetation indices global set of TM images for EOS-MODIS. Remote Sens Environ 59:440–451

Jaradat A (2003) Agriculture in Iraq: resources, potentials, constraints, research needs and priorities. Food Agric Environ 1(2):160–166

Jensen JR (2005) Introductory digital image processing: a remote sensing perspective, 3rd edn. Prentice Hall, Upper Saddle River, New Jersey

Jensen JR, Cowen DC (1999) Remote sensing of urban/suburban infrastructure and socioeconomic attributes. Photogram Eng Remote Sens 65:611–622

John R, Chen DM (2004) Remote sensing technology for mapping and monitoring land-cover and land-use change. Prog Plann 61:301–325

Kareem NA, Venkatesh MJ, Abed Ali AH (2014) Land use and land cover changes of Al-Kut City in Iraq using remote sensing and GI techniques. Int J Sci Eng Res 3(17):3555–3560

Kaufman YJ, Tanré D (1992) Atmospherically resistant vegetation index (ARVI) for EOS-MODIS. IEEE Trans Geosci Remote Sens 30:261–270

Kontoes CC (2008) Operational land cover change detection using change vector analysis. Int J Remote Sens 29:139–152

Lu D, Weng Q (2005) Urban classification using full spectral information of Landsat ETM+ imagery in Marion County, Indiana. Photogram Eng Remote Sens 71:1275–1284

Lu D, Weng Q (2009) Extraction of urban impervious surface from an IKONOS image. Int J Remote Sens 30(5):1297–1311

Malinverni ES, Tassetti AN, Bernardini A (2010) Automatic land use/land cover classification system with rules based both on objects attributes and landscape indicators. Ghent, Belgium

Mas JF (2010) Monitoring land-cover changes: a comparison of change detection techniques. Int J Remote Sens 20:139–152

McDonagh J, Stocking M, Lu Y (2006) Global impacts of land degradation. Overseas Development Group, University of East Anglia, Norwich, United Kingdom

Metternicht GI, Zinck JA (2003) Remote sensing of soil salinity: potentials and constraints. Remote Sens Environ 85:1–20

Ministry of Agriculture (MoA) (2017) Iraqi LDN national report, Office of Forest and Desertification, Iraq (unpublished report)

Ministry of Planning (2015) Division of statistic. Baghdad, Iraq

Mougenot B, Pouget M, Epema G (1993) Remote sensing of salt-affected soils. Remote Sens Rev 7:241–259

Muhaimeed AS (2013) Soil resources of Iraq. In: Yigini Y, Panaggos P, Montanarella L (eds) Soil resources of the Mediterranean and Caucasus Countries. Extension of the European soil database

Muhaimeed AS, Ali Z (2015) Monitoring degradation of agricultural land in Baghdad Province using remote sensing and GIS. Int J Geosci Geomatics 3(1):34–40

Muhaimeed AS, Wu W, AL-Shafie W, Ziadat F, Al-Musawi H, Kasim AS (2013) Use remote sensing to map soil salinity in the Musaib Area in Central Iraq. Int J Geosci Geomatics 1(2):34–41

Mustafa BM, Al-Quraishi AMF, Gholizadeh A, Saberioon M (2019) Proximal soil sensing for soil monitoring. In: Al-Quraishi AMF, Negm AM (eds) Environmental remote sensing and GIS in Iraq. Springer Water

Mustapha MR, Lim HS, Mat Jafri MZ (2010) Comparison of neural network and maximum likelihood approaches in image classification. J Appl Sci 10:2847–2854

Richards LA (1954) Diagnosis and improvements of saline and alkali soils. U.S. Salinity Laboratory DA, US Department of Agriculture Hbk 60, 160 p

Rouse JW, Haas RH, Schell JA, Deering DW (1973) Monitoring vegetation systems in the Great plains with ERTS. In: Proceedings of the third ERTS-1 symposium, NASA SP-351, vol 1, pp 309–317

Roy PS, Giriraj A (2008) Land use and land cover analysis in Indian context. J Appl Sci 8:1346–1353

Shivashankar S, Hiremath PS (2011) PCA plus LDA on wavelet co-occurrence histogram features for texture classification. Int J Remote Sens 3(4):302–306

Slaymaker O (2001) The role of remote sensing in geomorphology and terrain analysis in the Canadian Cordilleran. J Appl Earth Obs Geoinf 3(1):7

Steven MD, Malthus TJ, Jaggard FM, Andrieu B (1992) Monitoring responses of vegetation to stress. In: Cracknell AP, Vaughan RA (eds) Remote sensing from research to operation. Proceedings of the 18th annual conference of the remote sensing society, United Kingdom

Szabolcs I (1987) The global problems of salt-affected soils. Acta Agron Hung 36:159–172

United Nations Convention to Combat Desertification (UNCCD) (2013) The economics of desertification, land degradation and drought: methodologies an analysis for decision making. Background document. In: UNCCD 2nd scientific conference. http://2sc.unccd.int/fileadmin/unccd/upload/documents/Background_documents/Background_Document_web3.pdf

UNEP (1991) Desertification: a global threat. Desertification control bulletin no. 20. UNEP, Nairobi

USDA/FAS/PECAD (2000) Geocover land cover developed by Earthsat crop from 1990–2002

Van Engelen V (2008) e-SOTER, Annex 1—description of work. EU 211578, ISRIC, Wageningen, The Netherlands

Wu W (2014) The generalized difference vegetation index (GDVI) for dryland characterization. Remote Sens 6:1211–1233

Wu W, Al-Shafie WM, Muhaimeed AS, Dardar B, Ziadat F, Payne W (2013) Multiscale salinity mapping in Central and Southern Iraq by remote sensing. In Second international conference on agro-geoinformatics. Fafax, VA, USA, pp 470–476

Wu W, Al-Shafie WM, Muhaimeed AS, Ziadat F, Nangi V, Paye W (2014) Soil salinity mapping by multiscale remote sensing in Mesopotamia. IEEE J Sel Top Appl Earth Observations Remote Sens 7(11):4432–4443

Wu W, Muhaimeed AS, Al-Shafie AM, Al-Quraishi AMF (2019) Using radar and optical data for soil salinity modeling and mapping in Central Iraq. In: Al-Quraishi AMF, Negm AM (eds) Environmental remote sensing and GIS in Iraq. Springer Water

Wu W, Zucca C, Muhaimeed AS, Al-Shafie WM, Al-Quraishi AMF, Nangia V, Zhu M, Liu G (2018) Soil salinity prediction and mapping by machine learning regression in Central Mesopotamia, Iraq. Land Deg Dev 29(11):4005–4014

Zhang T, Zeng S, Gao Y, Ouyang Z, Li B, Fang C, Zhao B (2011) Using hyperspectral vegetation indices as a proxy to monitor soil salinity. Ecol Indic 11:1552–1562

Chapter 11
Agricultural Drought Monitoring Over Iraq Utilizing MODIS Products

Yousif S. Almamalachy, Ayad M. Fadhil Al-Quraishi
and Hamid Moradkhani

Abstract Iraq is a country that was well known for its agricultural production and fertile soil. However, agricultural drought overshadows the vegetative cover in general and cropland specifically in Iraq as it represents a creeping disaster. The arable lands in Iraq experienced increasing drought events that led to land degradation, desertification, economic losses, food insecurity, and deteriorating environment, particularly in recent years. Remote sensing dataset and techniques were employed in this chapter for drought monitoring over Iraq. Four different spectral drought indices; Vegetation Health Index (VHI), Vegetation Drought Index (VDI), Visible and Shortwave infrared Drought Index (VSDI), Temperature–Vegetation Dryness Index (TVDI) were utilized, each of them is derived from MODIS dataset of Terra satellite. Agricultural drought maps were produced from 2003 to 2015 after masking the vegetation cover. The results of the current study in this chapter revealed that 2008 was the most severe drought year in the period from 2003 to 2015, whereas drought covered 37% of the vegetated lands, while years 2009, 2011, and 2012 were the less-severe drought years dominated by mild or moderate drought with an areal coverage of 44, 50, and 48.5%, respectively. The VDI was found to be the more suitable drought index for Iraq because of temperature integration in its structure, in addition to meeting the assumptions it was based on. The results also indicated the capability of remote sensing in fulfilling the need for an early warning system for agricultural drought over such a data-scarce region.

Y. S. Almamalachy
National Center for Water Resources Management, Ministry of Water Resources, Baghdad, Iraq
e-mail: yousif_samir@yahoo.com

A. M. F. Al-Quraishi (✉)
Department of Environmental Engineering, College of Engineering, Knowledge University, Erbil 44001, Kurdistan Region, Iraq
e-mail: ayad.alquraishi@gmail.com; ayad.alquraishi@knowledge.edu.krd

H. Moradkhani
Department of Civil, Construction, and Environmental Engineering, University of Alabama, Tuscaloosa, USA
e-mail: hmoradkhani@ua.edu

© Springer Nature Switzerland AG 2020
A. M. F. Al-Quraishi and A. M. Negm (eds.),
Environmental Remote Sensing and GIS in Iraq, Springer Water,
https://doi.org/10.1007/978-3-030-21344-2_11

Keywords Agricultural drought · Remote sensing · MODIS · Drought indices · Iraq

11.1   Introduction

Although the primary cause of drought is attributed to meteorological anomalies represented by a shortage of precipitation below normal levels, anthropogenic activities can also be a factor in the occurrence of drought (Van Loon et al. 2016). Humans contribution in drought manifestation can either be direct such as ground water abstraction or indirect such as the greenhouse gas (GHG)-induced global warming (Gustard et al. 1997; Zhao and Dai 2015).

UNESCO (2014) highlighted that Iraq has suffered from several drought events in the period of 2003–2012, where different factors contributed in the occurrence of such events including low rainfall rates, higher temperatures, lower water income from upstream countries, and low efficiency in water utilization. These factors caused a multidimensional effect on the region such as lower discharge of Tigris and Euphrates, less available and more saline groundwater, population migration, and agricultural degradation (UNESCO 2014). Despite all the aforementioned effects of drought and land degradation over Iraq, and the fact that drought is known to be the costliest extreme natural phenomenon (Dai and Zhao 2016; Van Loon 2015; Yan et al. 2017), only few studies have addressed this phenomenon in a comprehensive and updated manner in this region, such as Fadhil (2011, 2013), Eklund and Seaquist (2015), Hameed et al. (2018), Al-Quraishi et al. (2019).

Agriculture in Iraq is known to be the second contributor to revenues after oil; and it is still responsible for recruiting about 20% of the workforce even after the decline of its share in the Gross Domestic Product (GDP) from 9% in 2002 to 3.3% in 2008 (FAO and World Bank 2011). Agricultural drought can be defined as the inadequacy of soil moisture needed to fulfil vegetation needs resulting from a prolonged shortage of precipitation, known as the meteorological drought and usually leads to crop failure. A study by Al-Timimi and Al-Jiboori (2013) assessed the meteorological drought in Iraq between 1980 and 2010 and concluded that drought has shifted from normal to extreme conditions in the last decade, while 2008 was the driest year. Albarakat et al. (2018) found in his study that in the last 36 years the AVHRR and Landsat images illustrated a decrease in water and vegetation coverage, which in turn has led to an increase in barren lands. Hameed et al. (2018) used the Standardized Precipitation Evapotranspiration Index (SPEI) as a multi-scalar drought index to account for the effects of temperature variability on drought over Iraq. They analyzed different characteristics including intensity, frequency, duration and spatio-temporal extent of drought. Their results showed a significant drought intensification over Iraq during the period of 1998–2009. Fadhil, 2011 mentioned in his study that there was a significant decrease in the vegetation cover in Erbil, Kurdistan region of Iraq in 2008 by 56.7%, in addition to a 29.9% decline in the soil/vegetation wetness compared to the status in 2007.

The fact that Iraq experienced serious meteorological drought events combined with decreasing annual flow from Tigris and Euphrates leads by nature to diminished vegetation cover for the period of 2003–2012 (UNESCO 2014). Agricultural drought can have huge economic impacts, for example, California State loss in 2016 was found to be 603 million USD plus 4,700 job losses (Medellen-Azuara et al. 2016). The two main causes for the 2012 drought in central US were the precipitation deficits and high temperatures (Hoerling et al. 2014). Combination of substantial precipitation deficits and high temperatures caused the soil moisture decrease rapidly and the drought propagated from meteorological drought to agricultural drought (Yan et al. 2017). Also, Canadian prairies suffered from agricultural drought during growing seasons in 2001 and 2002 resulting in 3.6 billion USD with 41,000 job losses (Wheaton et al. 2005).

The impact of the 2007–2009 drought in vegetation was evaluated by Trigo et al. (2010) with the Normalized Difference Vegetation Index (NDVI), which obtained from the VEGETATION instrument. It is shown that large sectors of south-eastern Turkey, eastern Syria, northern Iraq and western Iran present up to six months of persistently stressed vegetation (negative NDVI anomalies) between January and June 2008. While, Notaro et al. (2015) mentioned that the dried soils and diminished vegetation cover in the Fertile Crescent, as evident through remotely sensed enhanced vegetation indices, supported greater dust generation and transport to the Arabian Peninsula in 2007–2013. As identified both in increased dust days observed at weather stations and enhanced remotely sensed aerosol optical depth. UNESCO (2014) report highlighted that 40% of cropland in Iraq was lost in 2008–2009 drought years. The results of that study showed that a decrement in vegetation cover referred to as desertification is increasing with time. The agricultural drought was monitored over three adjacent governorates (Mosul, Kirkuk, and Salah al-Din) from the year 2000 to the year 2010 using NDVI as a monitoring index (Muhaimeed and Al-Hedny 2013). The study found that the drought conditions in 2007–2008 growing season were the most severe with the lowest value of NDVI. In his study to investigate the land degradation in the upper central part of Iraq, Fadhil 2009 indicated that drought is one of the main causes of the land degradation in Iraq. Eklund and Seaquist (2015) studied meteorological, agricultural, and socioeconomic drought over Duhok governorate, Kurdistan region in northern Iraq using the Enhanced Vegetation Index (EVI) to monitor agricultural drought between 2000 and 2011. They concluded that the study area experienced agricultural drought between 2007 and 2009, emphasizing in the year 2008 as the peak of the drought event.

The objective of this chapter is to utilize remote sensing capabilities to conduct a nationwide agricultural drought monitoring for the period between August 2002 and December 2015 over a data-inaccessible region such as Iraq.

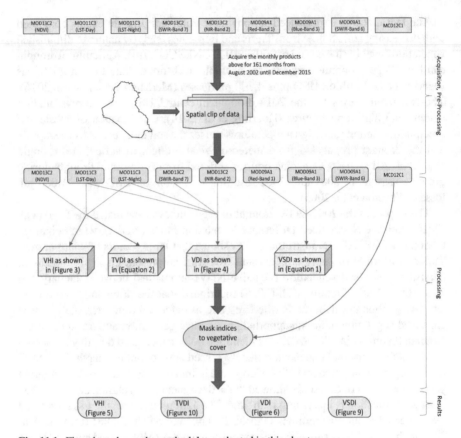

Fig. 11.1 Flowchart shows the methodology adopted in this chapter

11.2 Materials and Methods

Figure 11.1 shows the methodology, data, and indices which were adopted in this chapter.

11.2.1 Selection of the Study Area

The Republic of Iraq (Fig. 11.2) is located in southwestern Asia, forming the eastern boundary of the Arab homeland. Surrounded by Turkey from the north, Iran from the east, Syria and Jordan from the west, Saudi Arabia and Kuwait from the south. According to the Food and Agriculture Organization (FAO) of the United Nations (Omer 2011), the topography of the country can be divided into four regions: Mountainous region (21%), Sedimentary plain (30%), Desert plateau (39%), and

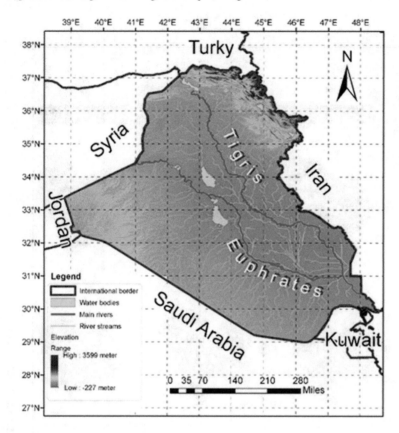

Fig. 11.2 Topography map of Iraq

Undulating terrain (10%). Iraq's climate mainly ranges between arid and semi-arid condition except the northeastern part.

This is verified in Fig. 11.3 by using De Martonne's aridity index, an aridity index that incorporates temperature and precipitation data as climatic variables (De Martonne 1926).

Climatically, the spatial pattern of rainfall distribution in Iraq can be summarized into the western desert receiving less than 100 mm per year, the Mesopotamian flood plain and Jezira area stretching from northwest to the southeast receiving 100–300 mm per year. On the other side, the foothills in the mid-northern part receiving about 300–700 mm per year, and the mountainous region in the far north receives more than 700 mm per year (Jassim and Goff 2006).

Rainfall spatial patterns are closely related to the topography of the country where each of the four patterns above has unique elevation range; the elevation distribution can be seen in Fig. 1. About 90% of the precipitation is received between November and April, which makes this period in addition to October and May; a growing season of agriculture in Iraq (UNESCO 2014; World Bank 2006). In addition to this

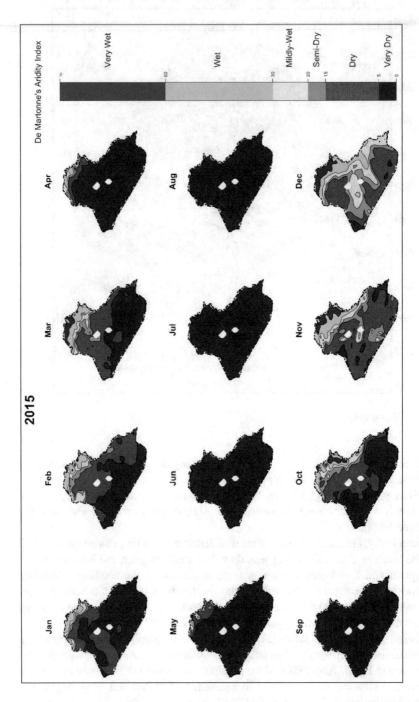

Fig. 11.3 Spatial distribution of aridity in Iraq in 2015 using De Martonne's aridity index

variability in precipitation, temperature also varies seasonally from more than 48 °C in July to less than 0 °C in January (World Bank 2006). Climatic factors such as, high temperature, low rainfall, and high wind speed escalate evaporation rate in the region to high levels. For instance, Abu Dibbis lake in the central part of the country has an evaporation rate of about 2,170 mm per year (Omer 2011). Tigris and Euphrates are the two main rivers in the study area, and they stretch along the region from northwest to southeast as shown above in Fig. 11.2.

11.2.2 Data Collection and Processing

11.2.2.1 Masking Vegetation

Since we are targeting agricultural drought in this study, each of the above indices is masked to vegetative areas, because these indices are designed to monitor vegetation in the first place, and misleading results might occur if non-vegetated areas are included. The produced indices were also examined (not shown) and they showed that the desert area always demonstrates drought condition. Agricultural drought indices that mainly monitor vegetation and moisture indicate a constant drought condition in the desert area encompassing about 39% of the total area in Iraq (Omer 2011).

In order to isolate vegetative areas, the MCD12C1 product is utilized. The product is provided annually at 0.05 degree spatial resolution. The land cover dataset of 2002 is used, and the University of Maryland (UMD) type 2 classification scheme is chosen following Ahmadalipour et al. (2017b). For masking purposes, all vegetation classes are combined and isolated from other non-vegetated classes (e.g., urban areas).

11.2.2.2 Precipitation

The response of VDI and VHI with rainfall is examined in this study. Precipitation data are obtained from the Global Land Data Assimilation System (GLDAS): Noah land surface model version 1 monthly data with 0.25 × 0.25 degree spatial resolution (Rodell et al. 2007). Because of the difference in spatial resolution between rainfall and both VHI and VDI, the latter ones are resampled to match the rainfall data, and Spearman correlation coefficient is calculated spatially over each pixel.

11.2.2.3 Remotely Sensed Dataset

The Moderate Resolution Imaging Spectroradiometer (MODIS) onboard TERRA satellite products are utilized in this study, and some spectral drought indices (were presented in the following sections) for agricultural drought monitoring over Iraq were also used for this study.

11.2.2.4 Spectral Indices

Vegetation Health Index (VHI)

The Normalized Difference Vegetation Index (NDVI) is one of the earliest remote sensing indices that has been used to monitor drought starting in the 1980s (Anyamba and Tucker 2012). However, it has certain limitations in drought monitoring, such as saturation of vegetation canopy, cloud cover, and soil background effect, so it is advisable to combine NDVI with other parameters (Sruthi and Aslam 2015).

In order to enhance drought monitoring, Kogan (1995) proposed Vegetation Health Index (VHI), a vegetation index that relies on NDVI and temperature as its main components. It is important to mention that VHI is based on the assumption that the relation between vegetation and temperature is negative, that is, vegetation experience stress when temperatures go high. The VHI has been widely used over different regions of the world with different environmental conditions (Rojas et al. 2011), and is considered a potent indicator for monitoring agricultural drought from a vegetation stress perspective (Ahmadalipour et al. 2017b).

In order to calculate the VHI, the NDVI and temperature must be developed into Vegetation Condition Index (VCI) and Temperature Condition Index (TCI), respectively. The objective of VCI is to separate the fluctuations of NDVI resulting from short-term weather-related conditions and the fluctuations resulting from long-term ecosystem effects (Kogan 1995). While TCI's goal is to monitor vegetation stress from a temperature perspective rather than greenness in order to prevent the mentioned false drought signal (Owrangi 2011). Also, the temperature component in TCI is less sensitive to water vapor than visible light bands, which in turn lessens the effect of cloud cover on TCI and then VHI as a result (Rojas et al. 2011).

Since NDVI and temperature are the main components of VHI, NDVI data are obtained from MODIS product MOD13C2 (Didan 2015). The temperature data are obtained from MODIS product MOD11C3 (Wan 2015). Both of these datasets with a spatial resolution of 0.05 degree were collected on a monthly basis from August 2002 to December 2015 (161 months). MOD11C3 provides temperature as Land Surface Temperature (LST) for day time and nighttime separately. The formulation of VHI following Kogan (1995) is shown in Fig. 11.4. For each grid cell, $NDVI_i$ represents the value of the ith month, where i ranges from 1 to 161 months. $NDVI_{max}$ and $NDVI_{min}$ are the maximum and minimum NDVI values at each grid cell over all months. The VCI_i is then calculated for each grid cell and each month. The TCI computation is similar to VCI, and VHI is the combination of both where $\alpha1$ and $\alpha2$ are equal to 0.5 because moisture and temperature contribution during a vegetation cycle is unknown.

Therefore, we assume that the shares of VCI and TCI in VHI are equal (Kogan 2001). Following (Tran et al. 2017) VHI is classified into 5 different classes where each represents a level of drought severity as shown in Table 11.1.

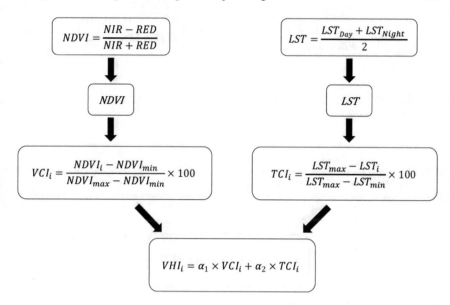

Fig. 11.4 VHI calculation method

Table 11.1 Classification of agricultural drought based on VHI

Drought severity class	VHI
Extreme drought	≤ 10
Severe drought	≤ 20
Moderate drought	≤ 30
Mild drought	≤ 40
No drought	>40

Vegetation Drought Index (VDI)

The VDI was proposed by Sun et al. (2013) as an agricultural drought-monitoring index, and it was successfully applied over China. In that study, VDI and VHI were compared on different levels such as their fulfilment to the assumptions they were built on, their performance with respect to 3-months SPI and 6-months SPI from 753 weather stations, and their performance with respect to in situ crop yield data.

The VDI came out as an upgrade to VHI because the assumption of VHI that NDVI and LST are negatively correlated is not always true (Sun et al. 2013). This implies that there is an inherent uncertainty in VHI's results where positive NDVI-Temperature relation exists. Although VDI has similar assumption to VHI where its components are assumed to be negatively correlated, Sun et al. (2013) showed that VDI components have a more stable negative correlation than VHI, which makes VDI less uncertain in drought monitoring.

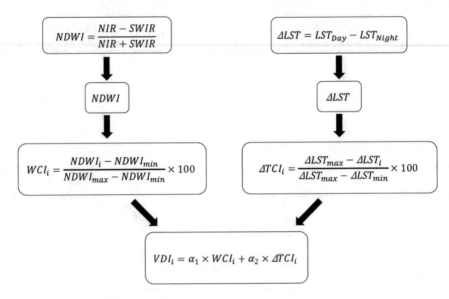

Fig. 11.5 VDI calculation method

According to Sun et al. (2013), VDI is dependent on Temperature and Normalized Difference Water Index (NDWI) that was first proposed by Gao (1996). As shown in Fig. 11.5, in calculating of VDI, instead of average temperature, the difference between day and night temperatures is used because temperatures difference has a high correlation with soil moisture (Sun et al. 2013). Also, Gu et al. (2007) showed that the response of NDWI to drought conditions is more sensitive (rapid) than NDVI.

Similar to the VHI, the NDWI is developed into Water Condition Index (WCI) and the temperature difference is developed into day-night Temperature Difference Condition Index (ΔTCI). Both WCI and ΔTCI correlates positively with drought, where higher values reflect good conditions and lower values indicate severe conditions. In general, WCI represents the water deficit in vegetation while ΔTCI represents the water deficit in soil (Sun et al. 2013). The VDI calculation is pixel by pixel based, where $NDWI_i$ represents the value of a specific pixel in the ith month, and i ranges from 1 to 161 months. $NDWI_{min}$ and $NDWI_{max}$ are the minimum and maximum values of NDWI for the same pixel over all months. ΔTCI is calculated on the same basis.

In addition to temperature data used for VHI analysis, the NDWI requires Near Infrared (NIR) band centered at 857 nm and Short Wave Infrared (SWIR) band centered at 1230 nm. This information are acquired on a monthly basis from MODIS product MOD13C2 (Didan 2015). $\alpha 1$ and $\alpha 2$ are set to 0.5 because WCI and ΔTCI are considered to equally contribute in measuring drought severity. Sun et al. (2013) classified drought severity into five classes based on VDI values as shown in Table 11.2.

Table 11.2 Classification of drought based on VDI

Drought severity class	VDI
Extreme drought	≤13
Severe drought	≤22
Moderate drought	≤32
Mild drought	≤41
No drought	>41

Visible and Shortwave Infrared Drought Index (VSDI)

The VSDI index was proposed as an agricultural drought monitoring index where its validity was tested over Oklahoma, USA by comparing VSDI performance with other remote sensing indices including Land Surface Water Index (LSWI), Surface Water Capacity Index (SWCI), and Normalized Multi-band Drought Index (NMDI) (Zhang et al. 2013a). All these indices including VSDI were compared with a reference index that relies on in situ soil moisture noted as Fractional Water Index (FWI). The VSDI was also compared with the results from the United States Drought Monitor (USDM).

The results of that study showed that VSDI outperformed LSWI, SWCI, and LSWI given the highest correlation with the in situ reference index (FWI). The results also showed that VSDI was capable of monitoring moisture in both soil and vegetation, and unlike other indices, it is applicable over different land cover types (Zhang et al. 2013a). Besides, a good agreement was found between VSDI and USDM. However, VSDI like any other index has limitations. The limitation of VSDI is represented by the fact that it is mainly dependent on spectral bands without incorporating temperature, which might affect the index performance in drought areas constrained by temperature rather than precipitation (Zhang et al. 2013a). Later in 2013, VSDI was further validated by applying it over China, and it was shown that it outperformed six other optical drought indices (Zhang et al. 2013b). In addition, the study shed light on important merit of VSDI as a robust index insensitive to cloud cover effect. VSDI employs the SWIR, Red, and Blue bands in order to monitor drought. It was found that SWIR was the most sensitive band to water content variation in vegetation, followed by the Red band, while the Blue band was found to be the least sensitive to vegetation moisture content (Zhang et al. 2013a). Equation 11.1 expresses VSDI calculation where the difference between sensitive and insensitive bands to vegetation moisture is utilized. The blue band is considered as the reference band, and the deviation of SWIR and Red bands from the Blue band represents the variation in vegetation moisture content.

$$VSDI = 1 - ((SWIR - Blue) + (Red - Blue)) \qquad (11.1)$$

This equation is applied to each pixel at each time step. The bands and data were obtained from MODIS product MOD09A1 (Vermote 2015). It is important to note that SWIR here is band 6 not 7 similar to the one used previously in VDI calculation.

Table 11.3 Classification of drought based on VSDI

Drought severity class	VSDI
Extreme/exceptional drought	<0.64
Severe drought	≥0.64
Moderate drought	≥0.68
Abnormally dry	≥0.71
No drought (normal)	≥0.75
Water/snow	>1

Unlike previously used MODIS products, MOD09A1 has an 8-day temporal resolution and 500 m spatial resolution, which made the calculation of VSDI for the whole period of study challenging given the massive data and long processing time required to achieve this goal. In order to overcome this problem, we chose a month that is more sensitive than other months to vegetation variation.

Growing season in Iraq starts from October–September until April–May of each year (World Bank 2006). Since April is at the end of the growing season and that vegetation is expected to be at its peak, we chose this month to be the representative month of the year for vegetation monitoring purposes. This also has been verified by plotting different months using VHI and VDI (not shown here). Finally, April features a clear sky condition, which reduces uncertainty in remote sensing monitoring for agricultural drought monitoring over the region.

Therefore, VSDI data is downloaded for the 8-day composite of April 15–23 of each year for the period of the study. MOD09A1 is also different from MOD13C2 and MOD11C3 in terms of scene size. The latter covers Iraq with one scene, while MOD09A1 covers Iraq by 4 scenes. Numerically, MOD09A1 covers Iraq by 2,075,486 gird cells, which are about 120 times the amount of pixels that covers Iraq in MOD13C2 and MOD11C3. Zhang et al. (2013a) proposed the classification scheme of drought based on VSDI. In addition to drought severity classes, VSDI can also distinguish water/land surfaces as shown in Table 11.3. Six drought severity classes were originally proposed; however, this study merges the first two classes (Extreme and Exceptional droughts) so that one can have five classes comparable with other indices in this study.

Temperature-Vegetation Dryness Index (TVDI)

The TVDI was proposed by Sandholt et al. (2002) as a soil moisture index that utilizes vegetation and temperature information. Similar to VHI and VDI, this index is subject to the vegetation-temperature space concept, which assumes that a negative correlation exists between vegetation and temperature. TVDI has wide applications in different regions of the world. For example, Wang et al. (2004) evaluated TVDI over China and concluded that by combining temperature and vegetation, TVDI is more accurate in monitoring soil moisture than Crop Water Stress Index (CWSI),

Table 11.4 Classification of agricultural drought based on TVDI

Drought severity class	TVDI
Extremely dry	≤ 1
Dry	≤ 0.8
Normal	≤ 0.6
Wet	≤ 0.4
Extremely wet	≤ 0.2

which is solely dependent on temperature. Holzman et al. (2014) utilized TVDI to predict crop yield 1–3 months before harvest in Argentina. This index has also played a role in drought monitoring in the Mekong River (Son et al. 2012), western India (Dhorde and Patel 2016), and northwest China (Zhang and Bai 2016).

To calculate TVDI, the pixel values of NDVI at each time step (1 month in this study) is plotted against the values of temperature. Given the nature of the relationship between vegetation and temperature, the scatter plot usually forms an approximate shape of a triangle (see Sandholt et al. 2002). The NDVI-Temperature space is then divided into slices based on NDVI. In this study, 100 slices are formed at 0.01 intervals. For each slice, the maximum temperature, its corresponding NDVI value, and minimum temperature are recorded. Three lists of these variables are constructed depending on the number of slices. From the maximum temperature and its corresponding NDVI values, a linear regression model is developed, while the mean of the minimum temperatures is calculated. The linear regression model between NDVI and maximum temperature (LST_{max}) forms the hypotenuse of the triangle, and the mean of minimum temperatures (LST_{min}) forms the base of the triangle. Equation 11.2 then can be used to calculate TVDI.

$$TVDI_{i,j} = \frac{LST_{i,j} - (LST_{min})_j}{(LST_{max})_{i,j} - (LST_{min})_j} \tag{11.2}$$

where

i is the spatial index for pixels from 1 to 17,212.
j is the temporal index for time steps in months from 1 to 161.
LST_{max} is the dry edge and is equal to ($a_j - NDVI_{i,j} \times b_j$), and LST_{min} is the wet edge.
a, and b represent the linear regression coefficients for the jth time step.

Each time step has only one value of LST_{min}, a, and b, while $LST_{i,j}$ and $NDVI_{i,j}$ vary temporally and spatially. Then, TVDI is calculated for each pixel at each time step, and it ranges from 0 to 1 where 1 is the dry edge at which the soil is in the driest condition, and no evaporation is occurring; and 0 is the wet edge where moisture supply to the soil is unlimited, and evaporation is occurring at maximum rate (Sandholt et al. 2002). The data required for calculating TVDI is the same as VHI. Following Cao et al. (2014), TVDI can be classified as shown in Table 11.4.

11.3   Results and Discussion

Agricultural drought is quantified in this study by comparing the areal coverage of drought severity classes over the years for the vegetated area in Iraq. In this study, we defined a drought year as any year where "No drought" condition is not spatially dominant. We choose April as the representative month of the year because it is the time of the year corresponding to the end of the growing season where vegetation is at peak. Also, April has a prevailing clear sky condition, which reduces uncertainty in the results (see Sect. 11.2.2.4).

The VHI and VDI demonstrate similar results to some extent by characterizing 2008 as a drought year as shown in Figs. 11.6 and 11.7. However, the drought severity level is different between the two. The VHI shows that 2008 was dominated by moderate drought while VDI shows severe drought as the prevailing drought class.

In 2012, VHI showed that about 55% of the vegetated area is experiencing either mild or moderate drought, while VDI, on the other hand, shows that mild or moderate drought is dominant in 2009, 2011, and 2012 by 44, 50, and 48.5%, respectively. For both VHI and VDI, all other years either are dominated by no drought condition or are equally distributed between no drought and mild/moderate drought classes. Other than 2008, severe or extreme drought is not dominant in any year; in fact, these two classes are at minimal areal coverage in general.

The VHI and VDI are compared with precipitation data in order to examine their accordance with rainfall. Both indices show a good agreement with precipitation, with VHI in the lead as shown in Fig. 11.8. The resulted agreement with precipitation is achieved with no lag time between vegetation and precipitation. Although the results might indicate the immediate response of vegetation to precipitation, it is not always true. Iraq is known to have rainfed farming in the northern parts and irrigated farming system in the middle and southern parts of the country (Hussein et al. 2016; Omer 2011; Schnepf 2003). This is clearly shown on the correlation distribution in Fig. 11.8, where southern parts are less correlated with vegetation than northern parts.

However, each index is tested for meeting its assumptions by calculating the Spearman correlation coefficient between its components. As mentioned previously, VHI assumes a negative relation between NDVI and LST. Similarly, VDI assumes a negative relationship between NDWI and ΔLST. If a positive relation is found in the study area, then we should expect uncertainty introduced in the drought results. For this purpose, the correlation analysis is conducted over three periods: on January, a month in the middle of the growing season, for the growing season, and for the whole period of study (Fig. 11.9). It is clearly shown that VDI has outperformed VHI in terms of meeting the assumptions, which makes VDI more reliable than VHI in monitoring drought over Iraq.

The VSDI, on the other hand, shows that 2008, 2009, 2012, and 2013 are dominated by extreme/exceptional drought, while abnormally dry severity class is dominant in 2005 (see Fig. 11.10). All other years had the normal condition as the prevailing class. Although these years are dominant by normal condition, abnormally dry and extreme/exceptional drought classes also show significant portions of areal

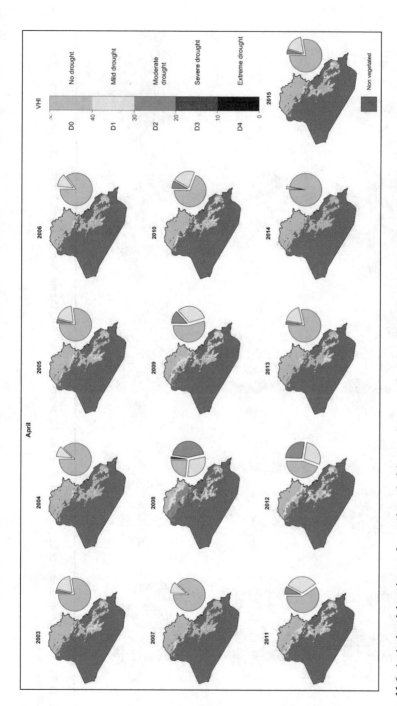

Fig. 11.6 Agricultural drought over Iraq at the end of the growing season based on VHI. The pie chart next to each subfigure shows the spatial distribution of drought classes

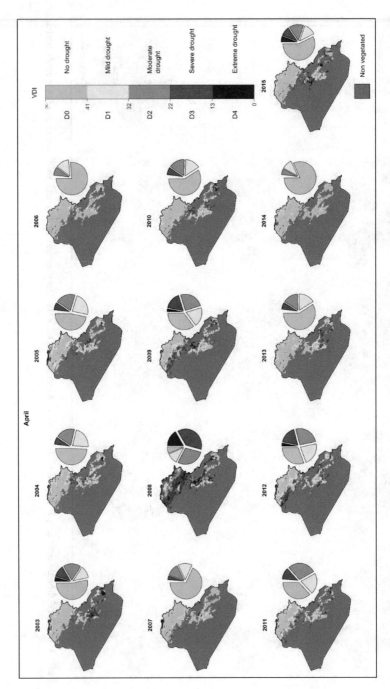

Fig. 11.7 Agricultural drought over Iraq at the end of the growing season based on VDI. The pie chart next to each subfigure shows the spatial distribution of drought classes

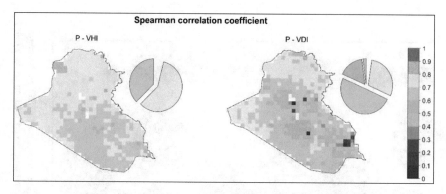

Fig. 11.8 Spearman correlation coefficient between VHI and precipitation (left side) and VDI and precipitation (right side) for the study period (161 months). The pie chart next to each subfigure shows the spatial distribution of correlation classes

coverage where each of these classes covers around 20% of the vegetated area. In general, VSDI presents more intense severity classes than VHI and VDI, and this result can be attributed to two factors: the nature of the data incorporated in VSDI and its resolution. As aforementioned, VSDI is dependent on spectral bands in vegetation monitoring without incorporating temperature in its calculation. This might be the reason that VSDI showed high values in this study, knowing that temperature has played as a quality control factor in drought monitoring using VHI, and VDI (see Sect. 11.2.2.4.1). Unlike other indices in this study, VSDI uses 8-day composites in its calculation instead of monthly data, which means that data variability is higher than other indices. This conclusion is reached by the fact that data with smaller temporal scale shows higher variability than larger temporal scale (Guijarro 2014); and averaging the data is smoothing it depending on the averaging window (Brown and Mac Berthouex 2002; Langbein 2006). It is expected that the spatial resolution of VSDI also affects the results since it is finer than other indices. These facts lead to the perception that extreme values are relatively suppressed in other indices by larger spatial and temporal resolutions.

Another factor, which is less likely to happen, is the cloud contamination. It has been clarified previously that VSDI is less effected by the cloud cover than other indices in this study. This merit might be the reason that VSDI showed more severe drought conditions than VHI and VDI. However, this factor is uncertain because Iraq has a clear sky condition prevailing in April (Ahmad et al. 1983). Although TVDI has wide applications in soil moisture monitoring, it has limitations that may affect its performance, and several studies have addressed these limitations. For example, Kimura (2007) highlighted that defining dry and wet edges in TVDI has an empirical nature because NDVI-Temperature space is dependent on the size of study area, and in order to define dry and wet edges correctly, a large study area need to be investigated so that it covers moisture extremes for a wide variety of vegetation surfaces. The TVDI assumes that air temperature is constant over the investigated area

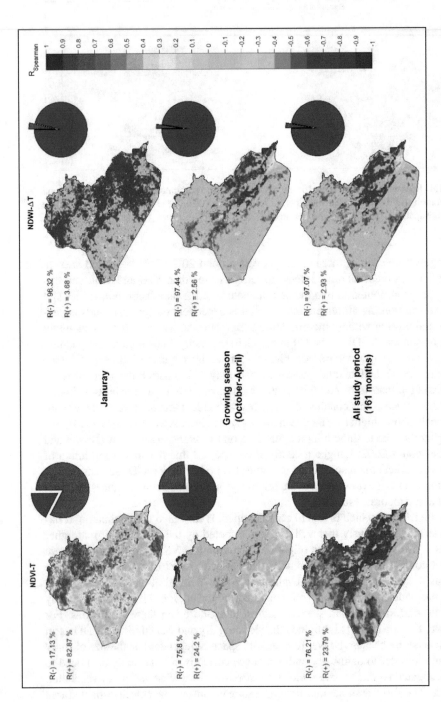

Fig. 11.9 Spearman correlation coefficient between VHI components (NDVI and temperature on the left side) and VDI components (NDWI and temperature on the right side). The pie chart next to each subfigure shows the distribution of negative correlation (blue) and positive correlation (red), while the text next to each subfigure represents the percentage of positive and negative correlations in that subfigure

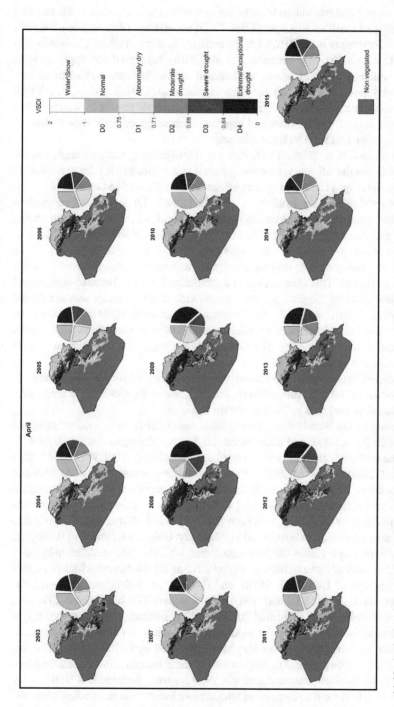

Fig. 11.10 Agricultural drought over Iraq at the end of the growing season based on VSDI. The pie chart next to each subfigure shows the spatial distribution of drought classes

which is not true and introduces uncertainty in the results, in addition to the fact that heterogeneity of earth's surface amplifies TVDI uncertainty (Rahimzadeh-Bajgiran et al. 2012; Srivastava et al. 2016). On the contrary to Kimura (2007), Rahimzadeh-Bajgiran et al. (2012), and Srivastava et al. (2016) suggested the application of TVDI on small-scale areas where variation in air temperature and earth surface heterogeneity (e.g., topography) are less. Garcia et al. (2014) investigated TVDI accuracy and concluded that TVDI showed more accurate results in water-limited regions than energy-limited ones and suggested the usage of daily data instead of 8-day composite data for TVDI calculation.

As a soil moisture index, TVDI can not distinguish specific drought events. The TVDI shows that all years in the study period experienced either dry or extremely dry conditions (Fig. 11.11). This poor performance might be due to multiple limitations mentioned previously. In addition to the fact that TVDI depends on vegetation-temperature space concept represented by NDVI and LST, which has been shown to be not accurate over Iraq (see Fig. 11.6). In order to test the validity of each of the indices above, we compare the results with the findings of other studies, as we are trying to employ remote sensing to monitor agricultural drought over the data-inaccessible region. This comparison is a challenge by itself because agricultural drought has not been targeted yet on a nationwide scale. A study was conducted by Trigo et al. (2010) over the Fertile Crescent area, utilized NDVI and showed that northern part of Iraq experienced a persistent pattern of vegetation stress for 6 months in 2008 and 5 months in 2009. The same study highlighted a degradation in total cereal production in those years.

Total cereal production degradation was also observed by the FAO statistics database downloaded from (http://faostat.fao.org), where the recession in cereal production was also observed in 2012 as shown in Fig. 11.12.

According to the Standardized Precipiation Index (SPI), three evident droughts took place in 2006, 2008, and 2009, where the last two droughts caused about 40% loss for the cropland, especially in the northern part of Iraq (UNESCO 2014). That report pointed that a major declination in vegetation cover occurred between 2009 and 2012 especially in Diyala, Salah al Din, and Basrah governorates where the vegetation cover loss was estimated at 65, 47, and 41% respectively. Comparing those results with the findings of this study, it is seen that VSDI and VDI are more applicable over Iraq than other indices introduced in this study. Unlike VHI and TVDI, they are capable of detecting the 2008–2009 drought event, which has been confirmed by other studies. Also, both of these indices identified 2012 as a drought year which coincides with the findings of UNESCO (2014), and FAO statistics database. However, the only difference between VDI and VSDI results is that VDI highlighted 2011 as a drought year while VSDI identified 2013 and 2005 instead. This difference might be attributed to the difference in the input data and calculation methodology of each index. Since no other studies over Iraq have addressed agricultural drought in this manner, it may not be practical to judge which of these indices is more accurate given the limited observational data and analysis. We, however, speculate that VDI is more suitable than VSDI for Iraq because of temperature integration in its calculation, the main factor that plays a vital role in vegetation condition and climatic variables. In

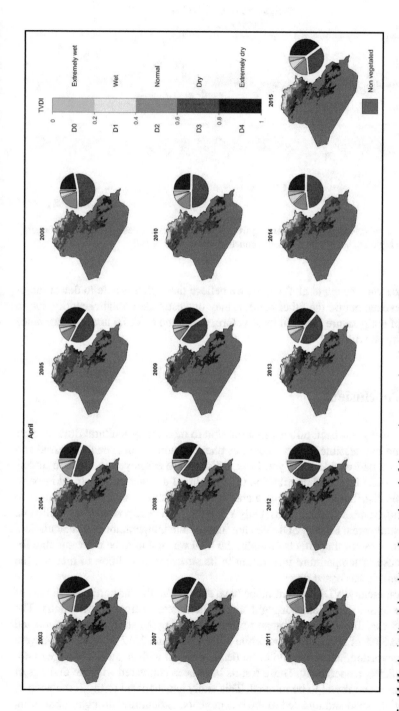

Fig. 11.11 Agricultural drought over Iraq at the end of the growing season based on TVDI. The pie chart next to each subfigure shows the spatial distribution of drought classes

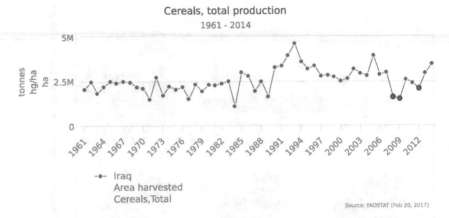

Fig. 11.12 Total cereals production in Iraq from FAO statistical database (http://faostat.fao.org). Red circles highlight years with reduced production in the study period

accordance with Zhang et al. (2013a), we believe that VSDI is able to detect major drought events, but on the other hand, it may overestimate drought severity due to the lack of temperature factor in its structure, which led to VSDI underperformance in comparison with VDI.

11.4 Conclusions

Remote sensing has been utilized in a mission to monitor agricultural drought over Iraq, a data-inaccessible region where this phenomenon has not been addressed and studied on a nationwide scale yet. Four indices were employed in order to understand the spatio-temporal distribution of agricultural drought in Iraq. In addition to geographic distribution of drought, a classification scheme was provided in order to understand the most affected areas. Only VSDI and VDI were proved to be applicable over the study area. Since VSDI does not incorporate temperature in its calculation, an important factor that affects drought, the VDI was found to be more suitable for Iraq because of temperature integration in its structure in addition to meeting the assumptions it was based on.

Geographically, VDI drought maps have shown that the far northeastern part of Iraq is the least to experience drought among the vegetated areas of the country. The year 2008 was found to be the most severe year during the study period dominated by around 37% of severe drought, while 2009, 2011, and 2012 were the less-severe drought years dominated by mild or moderate drought with an areal coverage of 44, 50, and 48.5%, respectively. These results have been compared with the findings of other studies and found to be rational. This study is conducted to fulfill the need of a nationwide scale and updated to the recent years, agricultural drought monitoring system that can be utilized as an early warning system. Given the latest advances made in hydrologic/land surface modeling, availability of other satellite data including those with soil moisture which can characterize the antecedent soil moisture condition, and also state of the art data assimilation, Iraq can benefit from new approaches for drought monitoring and forecasting and recovery (e.g., Madadgar and Moradkhani 2013; Yan et al. 2017, 2018; Abbaszadeh et al. 2019; Ahmadi et al. 2018).

Acknowledgements The first author would like to acknowledge the financial support and the scholarship those provided by the Higher Committee for Education Development (HCED) in Baghdad, Iraq for his M.Sc. study at the Portland State University, United states, under the supervision of Prof. Dr. Hamid Moradkhani.

References

Abbaszadeh P, Moradkhani H, Zhan X (2019) Downscaling SMAP radiometer soil moisture over the CONUS using an ensemble learning method. Water Resour Res (in press)

Ahmad I, Al-Hamadani N, Ibrahim K (1983) Solar radiation maps for Iraq. Solar Energy 31(1):29–44. Retrieved from https://doi.org/10.1016/0038-092X(83)90031-2

Ahmadalipour A, Moradkhani H, Yan H, Zarekarizi M (2017) Remote sensing of drought: vegetation, soil moisture, and data assimilation. In: Lakshmi V (ed) Remote sensing of hydrological extremes (pp. 121–149)

Ahmadi B, Ahmadalipour A, Moradkhani H (2018) Hydrological drought persistence and recovery in the CONUS: a multi-stage framework considering water quantity and quality. Water Res 150:97–110. https://doi.org/10.1016/j.watres.2018.11.052

Albarakat R, Lakshmi V, Tucker CJ (2018) Using satellite remote sensing to study the impact of climate and anthropogenic changes in the Mesopotamian Marshlands, Iraq. Remote Sens 10(10):1524. https://doi.org/10.3390/rs10101524

Al-Quraishi AMF, Qader SH, Wu W (2019) Drought monitoring using spectral and meteorological based indices combination: a case study in the Iraqi Kurdistan region. In: Al-Quraishi AMF, Negm AM (eds) Environmental Remote Sensing and GIS in Iraq. Springer Water.

Al-Timimi Y, Al-Jiboori M (2013) Assessment of spatial and temporal drought in Iraq during the period 1980–2010. Int J Energy Environ 4(2):291–302. Retrieved from http://www.ijee.ieefoundation.org/vol4/issue2/IJEE_12_v4n2.pdf

Anyamba A, Tucker C (2012) Historical perspectives on AVHRR NDVI and vegetation drought monitoring. In: Remote sensing of drought. CRC Press, Boca Raton, pp 23–50. Retrieved from http://www.crcnetbase.com/doi/abs/10.1201/b11863-4, https://doi.org/10.1201/b11863-4

Brown L, Mac Berthouex P (2002) Statistics for environmental engineers, 2nd edn. CRC Press, Boca Raton. Retrieved from https://doi.org/10.1201/9781420056631

Cao X, Feng Y, Wang J, Gao Z, Ning J, Gao W (2014) In: Gao W, Chang N-B, Wang J (eds) The study of the spatiotemporal changes of drought in the Mongolian Plateau in 40 years based on TVDI. p 92210X. Retrieved from https://doi.org/10.1117/12.2059062

Dai A, Zhao T (2016) Uncertainties in historical changes and future projections of drought. Part I: estimates of historical drought changes. Climatic Change. Retrieved from https://doi.org/10.1007/s10584-016-1705-2

De Martonne E (1926) Géographie physique. Aréisme et indice d'aridité. Gauthier-Villars

Dhorde AG, Patel NR (2016) Spatio-temporal variation in terminal drought over western India using dryness index derived from long-term MODIS data. Ecol Inform 32:28–38. Retrieved from https://doi.org/10.1016/j.ecoinf.2015.12.007

Didan K (2015) MOD13C2 MODIS/terra vegetation indices monthly L3 global 0.05deg CMG V006. Retrieved from https://doi.org/10.5067/modis/mod13c2.006

Eklund L, Seaquist J (2015) Meteorological, agricultural and socioeconomic drought in the Duhok Governorate, Iraqi Kurdistan. Nat Hazards 76(1):421–441. Retrieved from https://doi.org/10. 1007/s11069-014-1504-x

Fadhil AM (2009) Land degradation detection using geo-information technology for some sites in Iraq. J Al-Nahrain Univer Sci 12(3):94–108

Fadhil AM (2011) Drought mapping using geoinformation technology for some sites in the Iraqi Kurdistan region. Int J Digital Earth 4(3):239–257

Fadhil AM (2013) Sand dunes monitoring using remote sensing and GIS techniques for some sites in Iraq. In: Proceedings SPIE 8762, PIAGENG 2013: intelligent information, control, and communication technology for agricultural engineering, 876206 (March 19, 2013). http://dx.doi. org/10.1117/12.2019735

FAO and World Bank (2011) Iraq agriculture sector note, Technical report. Retrieved from http:// www.fao.org/docrep/017/i2877e/i2877e.pdf

Gao BC (1996) NDWI—a normalized difference water index for remote sensing of vegetation liquid water from space. Remote Sens Environ 58(3):257–266. Retrieved from https://doi.org/10.1016/ s0034-4257(96)00067-3

Garcia M, Fernández N, Villagarcía L, Domingo F, Puigdefábregas J, Sandholt I (2014) Accuracy of the temperature vegetation dryness index using MODIS under water-limited vs. energy-limited evapotranspiration conditions. Remote Sens Environ 149:100–117. Retrieved from https://doi. org/10.1016/j.rse.2014.04.002

Gu Y, Brown JF, Verdin JP, Wardlow B (2007) A five-year analysis of MODIS NDVI and NDWI for grassland drought assessment over the central Great Plains of the United States. Geophys Res Lett 34(6):L06407. Retrieved from https://doi.org/10.1029/2006gl029127

Guijarro J (2014) Quality control and homogenization of climatological series. In: Handbook of engineering hydrology. CRC Press, Boca Raton, pp 501–513. Retrieved from https://doi.org/10. 1201/b15625-25

Gustard A, Blazkova S, Brilly M, Demuth S, Dixon J, Van Lanen H, Llasat C, Mkhandi S, Servat E (1997) FRIEND'97: regional hydrology: concepts and models for sustainable water resource management (No. 246). International Association of Hydrological Sciences, UK

Hameed M, Ahmadalipour A, Moradkhani H (2018) Apprehensive drought characteristics over Iraq: results of a multidecadal spatiotemporal assessment. Geosci 8(2):58

Hoerling MP, Eischeid J, Kumar A, Leung, Mariotti A, Mo K, Schubert S, Seager R (2014) Causes and predictability of the 2012 Great Plains drought. Bull Amer Meteor Soc 95:269–282

Holzman M, Rivas R, Piccolo M (2014) Estimating soil moisture and the relationship with crop yield using surface temperature and vegetation index. Int J Appl Earth Obs Geoinf 28:181–192. Retrieved from https://doi.org/10.1016/j.jag.2013.12.006

Hussein MH, Amien IM, Kariem TH (2016) Designing terraces for the rainfed farming region in Iraq using the RUSLE and hydraulic principles. Int Soil Water Conserv Res 4(1):39–44. Retrieved from https://doi.org/10.1016/j.iswcr.2015.12.002

Jassim SZ, Goff JC (2006) Geology of Iraq. DOLIN, sro, distributed by Geological Society of London

Kimura R (2007) Estimation of moisture availability over the Liudaogou river basin of the Loess Plateau using new indices with surface temperature. J Arid Environ 70(2):237–252. Retrieved from https://doi.org/10.1016/j.jaridenv.2006.12.021

Kogan F (1995) Application of vegetation index and brightness temperature for drought detection. Adv Space Res 15(11):91–100. Retrieved from https://doi.org/10.1016/0273-1177(95)00079-t

Kogan F (2001) Operational space technology for global vegetation assessment. Bull Am Meteorol Soc 82(9):1949–1964. Retrieved from https://doi.org/10.1175/1520-0477(2001)082%3c1949: ostfgv%3e2.3.co;2

Langbein L (2006) Public program evaluation. Routledge, London. Retrieved from https://doi.org/ 10.4324/9781315497891

Madadgar S, Moradkhani H (2013) A Bayesian framework for probabilistic drought forecasting. J Hydrometeorology, special issue of Adv Drought Monit 14:1685–1705. https://doi.org/10.1175/jhm-d-13-010.1

Medellen-Azuara J, MacEwan D, Howitt RE, Sumner DA, Lund JR (2016) Economic analysis of the 2016 California drought on agriculture, Technical report. California Department of Food and Agriculture, Davis, California. Retrieved from https://watershed.ucdavis.edu/files/DroughtReport20160812.pdf

Muhaimeed A, Al-Hedny S (2013) Evaluation of long-term vegetation trends for northeastern of Iraq: Mosul, Kirkuk and Salah al-Din. IOSR J Agric Vet Sci 5(2):67–76. Retrieved from http://iosrjournals.org/iosr-javs/papers/vol5-issue2/K0526776.pdf?id=8064

Notaro M, Yu Y, Kalashnikova O (2015) Regime shift in Arabian dust activity, triggered by persistent Fertile Crescent drought. J Geophys Res Atmos. https://doi.org/10.1002/2015jd023855

Omer T (2011) Country pasture/forage resource profile, Iraq. Retrieved 2016-01-01, from http://www.fao.org/ag/agp/agpc/doc/counprof/iraq/iraq.html

Owrangi MA (2011) Drought monitoring methodology based on AVHRR images and SPOT vegetation maps. J Water Resour Prot 03(05):325–334. Retrieved from https://doi.org/10.4236/jwarp.2011.35041

Rahimzadeh-Bajgiran P, Omasa K, Shimizu Y (2012) Comparative evaluation of the Vegetation Dryness Index (VDI), the Temperature Vegetation Dryness Index (TVDI) and the improved TVDI (iTVDI) for water stress detection in semi-arid regions of Iran. ISPRS J Photogrammetry Remote Sens 68:1–12. Retrieved from https://doi.org/10.1016/j.isprsjprs.2011.10.009

Rodell M, Beaudoing HK, NASA/GSFC/HSL (2007) GLDAS Noah Land Surface Model L4 Monthly 0.25 x 0.25 degree, version 001. Greenbelt, Maryland, USA. Retrieved from https://doi.org/10.5067/7NP2052IA62C

Rojas O, Vrieling A, Rembold F (2011) Assessing drought probability for agricultural areas in Africa with coarse resolution remote sensing imagery. Remote Sens Environ 115(2):343–352. Retrieved from https://doi.org/10.1016/j.rse.2010.09.006

Sandholt I, Rasmussen K, Andersen J (2002) A simple interpretation of the surface temperature/vegetation index space for assessment of surface moisture status. Remote Sens Environ 79(2–3):213–224. Retrieved from https://doi.org/10.1016/s0034-4257(01)00274-7

Schnepf R (2003) Iraq's agriculture: background and status, Technical report. Congress. Retrieved from http://nationalaglawcenter.org/wp-content/uploads/assets/crs/RS21516.pdf

Son N, Chen C, Chen C, Chang L, Minh V (2012) Monitoring agricultural drought in the Lower Mekong Basin using MODIS NDVI and land surface temperature data. Int J Appl Earth Obs Geoinf 18:417–427. Retrieved from https://doi.org/10.1016/j.jag.2012.03.014

Srivastava P, Petropoulos G, Kerr Y (2016) Satellite Soil Moisture Retrieval, 1st, Elsevier

Sruthi S, Aslam MM (2015) Agricultural drought analysis using the NDVI and land surface temperature data; a case study of Raichur District. Aquat Procedia 4:1258–1264. Retrieved from https://doi.org/10.1016/j.aqpro.2015.02.164

Sun H, Zhao X, Chen Y, Gong A, Yang J (2013) A new agricultural drought monitoring index combining MODIS NDWI and daynight land surface temperatures: a case study in China. Int J Remote Sens 34(24):8986–9001. Retrieved from https://doi.org/10.1080/01431161.2013.860659

Tran HT, Campbell JB, Tran TD, Tran HT (2017) Monitoring drought vulnerability using multispectral indices observed from sequential remote sensing (case study: Tuy Phong, Binh Thuan, Vietnam). GIScience Remote Sens 54(2):167–184. Retrieved from https://doi.org/10.1080/15481603.2017.1287838

Trigo RM, Gouveia CM, Barriopedro D (2010) The intense 2007–2009 drought in the Fertile Crescent: impacts and associated atmospheric circulation. Agric Forest Meteorol 150(9):1245–1257. Retrieved from https://doi.org/10.1016/j.agrformet.2010.05.006

UNESCO (2014) Integrated drought risk management, DRM: national framework for Iraq, an analysis report, Technical report. UNESCO Office Iraq (Jordan), Amman. Retrieved from http://www.unesco.org/new/fileadmin/MULTIMEDIA/FIELD/Iraq/pdf/Publications/DRM.pdf

Van Loon AF (2015) Hydrological drought explained. Wiley Interdisc Rev Water 2(4):359–392. Retrieved from https://doi.org/10.1002/wat2.1085

Van Loon AF, Gleeson T, Clark J, Van Dijk AIJM, Stahl K, Hannaford J, Di Baldassarre G, Teuling AJ, Tallaksen LM, Uijlenhoet R, Hannah DM, Sheffield J, Svoboda M, Verbeiren B, Wagener T, Rangecroft S, Wanders N, Van Lanen HAJ (2016) Drought in the anthropocene. Nat Geosci 9(2):89–91. Retrieved from https://doi.org/10.1038/ngeo2646

Vermote E (2015) MOD09A1 MODIS/terra surface reflectance 8-day L3 global 500 m SIN grid V006. Retrieved from https://doi.org/10.5067/modis/mod09a1.006

Wan Z (2015) MOD11C3 MODIS/terra land surface temperature/emissivity monthly L3 global 0.05deg CMG V006. Retrieved from https://doi.org/10.5067/modis/mod11c3.006

Wang C, Qi S, Niu Z, Wang J (2004) Evaluating soil moisture status in China using the temperature vegetation dryness index (TVDI). Can J Remote Sens 30(5):671–679. Retrieved from https://doi.org/10.5589/m04-029

Wheaton E, Wittrock V, Kulshretha S, Koshida G, Grant C, Chipanshi A, Bonsal B (2005) Lessons learned from the Canadian drought years of 2001 and 2002: synthesis report, Technical report. Saskatchewan Research Council, Saskatchewan. Retrieved from http://www.agr.gc.ca/eng/programs-and-services/list-of-programs-and-services/drought-watch/managing-agroclimate-risk/lessons-learned-from-the-canadian-drought-years-2001-and-2002/?id=1463593613430

World Bank (2006) Iraq: country water resource assistance strategy: addressing major threats to people's livelihoods, Technical report. World Bank, Washington. Retrieved from http://siteresources.worldbank.org/INTWAT/Resources/Iraq.pdf

Yan H, Moradkhani H, Zarekarizi M (2017) A probabilistic drought forecasting framework: a combined dynamical and statistical approach. J Hydrol 548:291–304. Retrieved from https://doi.org/10.1016/j.jhydrol.2017.03.004

Yan H, Zarekarizi M, Moradkhani H (2018) Toward improving drought monitoring using the remotely sensed soil moisture assimilation: a parallel particle filtering framework. Remote Sens Environ 216:456–471. https://doi.org/10.1016/j.rse.2018.07.017

Zhang J, Bai J (2016) The spatial-temporal dynamic monitor of spring drought based on TVDI model in Guanzhong area. In: 2016 fifth international conference on agro-geoinformatics (agro-geoinformatics), IEEE, pp 1–6. Retrieved from https://doi.org/10.1109/agro-geoinformatics.2016.7577685

Zhang N, Hong Y, Qin Q, Liu L (2013a) VSDI: a visible and shortwave infrared drought index for monitoring soil and vegetation moisture based on optical remote sensing. Int J Remote Sens 34(13):4585–4609. Retrieved from https://doi.org/10.1080/01431161.2013.779046

Zhang N, Hong Y, Qin Q, Zhu L (2013b) Evaluation of the visible and shortwave infrared drought index in China. Int J Disaster Risk Sci 4(2):68–76. Retrieved from https://doi.org/10.1007/s13753-013-0008-8

Zhao T, Dai A (2015) The magnitude and causes of global drought changes in the twenty-first century under a low moderate emissions scenario. J Clim 28(11):4490–4512. Retrieved from https://doi.org/10.1175/jcli-d-14-00363.1

Chapter 12
The Aeolian Sand Dunes in Iraq: A New Insight

Arsalan Ahmed Othman, Younus I. Al-Saady, Ahmed T. Shihab
and Ahmed F. Al-Maamar

Abstract Desertification can be considered as the major challenge in the arid and semi-arid regions, particularly in the last decades. Aeolian sediments arise as one of the main factors of desertification in term of extent and movement in the Iraqi territory. Aeolian sand dunes are one of the most amazing natural features on Earth. Understanding how aeolian sediments (i.e. sand dunes) form and move has long been a research topic in Earth surface processes. This chapter describes a remote sensing approach utilized to monitor temporal and spatial changes of aeolian sand dunes in Hor Al-Dalmaj area, which is classified according to climatology as an arid area. The aeolian sand dunes in Hor Al-Dalmaj area characterized by NW-SE direction, make them parallel to the fold axes extend. The study area is located in the central part of Iraq, where the growth of desertification has been observed. Two primary types of sensors: passive and active have been utilized in this chapter. The aeolian sand dunes were extracted from Landsat TM and OLI acquired in 2000 and 2016, respectively. This result shows that desertification has been increased in the study area. Moreover, we used two C-Band SAR Sentinel-1A (2 ascending) data to monitor the aeolian sand dunes in Hor Al-Dalmaj area between March 2015 and August 2015. The SARPROZ, SNAP and SNAPhu software were used to recognize and monitor aeolian sand dunes movements using Differential Interferometry Synthetic-Aperture Radar (DInSAR) technique.

Keywords Aeolian sand dunes · DInSAR · Iraq · Change detection

A. A. Othman (✉)
Iraq Geological Survey, Sulaymaniyah Office, Sulaymaniyah, Iraq
e-mail: arsalan.aljaf@gmail.com

Y. I. Al-Saady · A. T. Shihab · A. F. Al-Maamar
Iraq Geological Survey, P.O. Box 986, Baghdad, Iraq

© Springer Nature Switzerland AG 2020
A. M. F. Al-Quraishi and A. M. Negm (eds.),
Environmental Remote Sensing and GIS in Iraq, Springer Water,
https://doi.org/10.1007/978-3-030-21344-2_12

12.1 Introduction

The Mesopotamia is an ancient region located in the east of the Mediterranean, corresponding to most of Iraq between the Tigris–Euphrates rivers system (Bulliet et al. 2010). It was called the black land due to the spread of agriculture since 5000 BC. Therefore, the historical Mesopotamia was free of aeolian sand dunes (Sarnthein 1978). In the present day, the aeolian sand dunes are distributed in Mesopotamia. The dry climate in the Mesopotamia and the reduction in surface water upstream countries (Iran, Turkey and Syria) lead to an increase the desertification processes in Iraq (Othman et al. 2014). Desertification process is the persistent degradation of land in arid and semi-arid environments due to the drought (particularly the global warming) and overexploitation of soil through human intervention (e.g., overgrazing, over-cultivation) (Lam et al. 2011). One of the famous indicator of desertification is aeolian sand dunes. It is resulted from the erosion of agricultural soils, which allows to increase of dust storms, and the awakening of aeolian sand dunes (Huggett 2007).

In long-term, wind erosion may impact on humans and human activities. It may cause to lose agricultural and recreational lands. The infrastructures such as roads, railways, and irrigation canals are affected by movable aeolian sand dunes (Huggett 2007). The migration of aeolian sand dunes is one of the most speedily existing cases of geomorphological change. The motion of some types of aeolian sand dunes; such as barchans is several meters per year, makes their monitoring easy to be detected. The availability of high temporal and spatial resolution satellite imagery helps the researchers to measure aeolian sand dunes movement (Lorenz et al. 2013). In Iraq, few works about aeolian sand dunes have been published (Hadeel et al. 2011, 2010a, b; Jabbar and Zhou 2011, 2012; Othman et al. 2014). They used passive data of remote sensing represented by Landsat imagery. These works, in general, focused on environmental changes monitoring for land covers. They measured the aeolian sand dunes change in addition to other land cover types.

Since the 1970s, several remote sensing techniques were used to monitor the aeolian sand dunes' changes. Early studies used mainly the passive optical systems (Brera and Shahrokhi 1978; Carlisle and Marrs 1982; Kolm 1982). In the last century, Differential Interferometry Synthetic-Aperture Radar (DInSAR) is the most widely used microwave sensor for geological studies and periodic monitoring surveys (Del Ventisette et al. 2014). DInSAR is one of the most important techniques for geodesy. Few researchers have successfully used DInSAR technique to monitor the aeolian sand dunes movement (Barbat et al. 2008; Havivi et al. 2018). It is allowing mm-accuracy monitoring of individual structures through multi-temporal analysis of SAR images (Ferretti et al. 2001; Sowter et al. 2013; Vajedian et al. 2015).

In this chapter, we utilized the two-primary types of sensors: passive and active. The main objective of this chapter is to detect the aeolian sand dunes changes between 2000 and 2016 using Landsat imagery for entire Iraq. Moreover, we selected a specific sample area in the central part of the Mesopotamia to determine the activity of the aeolian sand dunes' movement by applying DInSAR technique for the period between March 2015 and August 2016 using Sentinel 1A.

12.2   Type of Aeolian Sand Dunes and Their Distribution in the World

Aeolian sand dunes are hills of sand accumulated by the wind. The maximum length of the individual dune could be 1 m, while in some areas reaches to several tens of kilometers. Its height ranges from few centimeters to 150 m. Sand dune chains or dune networks are formed by linking the dunes together (Pye and Tsoar 2008). Aeolian sand dunes are most dynamics phenomena in nature due to the lack of cohesion between the sand grains, which ranges between 0.062 and 2.0 mm in diameter. Therefore, the movement of the sand grains is easy to be carried by the wind. The wind directions impact to form the shape of aeolian sand dunes (Tsoar 2001). The shape, degree of form mobility, and number and orientation of slip-faces relative to the prevailing wind were used to classify the aeolian sand dunes (Hunter et al. 1983; Mainguet 1983; Thomas 1989; Wasson and Hyde 1983).

According to Pye and Tsoar (2008) the simple aeolian sand dunes can be subdivided into three essential groups:

1. Sand accumulation related to topographic obstacles, which includes three subclass types:

   a. Windward accumulation, which is subdivided into climbing, and echo dunes
   b. Leeward accumulation, which is subdivided into lee, and falling dunes
   c. Cliff-lop accumulation.

2. Sand accumulation related to vegetation, which can be subdivided into three types, includes parabolic, vegetated linear, and hummock dunes.
3. Sand accumulation related to bed roughness changes or aerodynamic fluctuations. It can be subdivided into:

   a. Dunes composed of fine sand, which includes barchans, transverse (barchanoid) ridges, linear dunes, dome dunes, and star dunes.
   b. Forms composed of poorly sorted and bimodal sand includes two types of aeolian sand dunes, which are sand sheet and Zibars types.

According to Sarnthein (1978) (Fig. 12.1) the sand dunes' types cover about 19.54 million km^2. The land surface area of Earth was around 148.94 million km^2 (Martin 2011). Therefore, the sand dunes represent 13.12% of the world's land. More than 77% of the dunes in the world (which represent ~15.06 million km^2) exist between 30°S and 49°N latitude, and between 15°W and 115°E longitude, which extend from North Africa through the Arabic countries in Asia, Iran and Pakistan (Fig. 12.1). About 86.6% (16.93 million km^2) of these dunes are located in the northern hemisphere.

The Sahara Desert in Africa, Rub Al'Khali in Asia and Takla Makan in Xinjiang Province, western China are the most famous accumulation of aeolian sand dunes in the northern hemisphere . The great Sandy Desert and Gibson Desert in Central

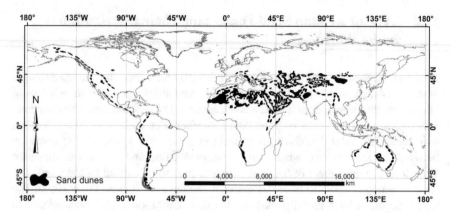

Fig. 12.1 Active sand dune fields in the present (Sarnthein 1978)

Australia, and the Namib, and the Kalahari sand sea in the southwest Africa represent the major aeolian sand dunes accumulation in the southern hemisphere (Pye and Tsoar 2008).

12.3 Climate and Aeolian Sand Dunes in Iraq

The evolution of aeolian sand dunes results from three main factors, which are wind strength, vegetative cover, and sand supply. The vegetation cover is related to the precipitation as well (Hack 1941). Therefore, it is important to understand the climate of Iraq before discussing the nature of the sand dunes. The climate of Iraq is characterized by greater changes throughout the year. The climate is described by long hot summer with occasional dust storms and short winter with limited and seasonal rainfall. According to the Köppen–Geiger climate classification system (Kottek et al. 2006), the climate of Iraq can be classified as warm temperate with dry and hot summer (Csa).

Aeolian processes are active in arid areas where vegetation cover is rare, the wind is strong enough to carry the sand grains, and there are many sand sources to form aeolian sand dunes (Marzolf 1988; Pye and Tsoar 2008). In general, the primary determinant of dune forms in any given area is the nature of the wind regime of that area. Most dune classification schemes are based primarily on the direction and intensity of the winds carrying the sand. Several studies identify wind regime and wind power as the most important physical factors in determining the mobility of dunes (Fryberger and Ahlbrandt 1979). The wind factors, which have to be considered are wind velocity (it should be more than the required speed to keep sand in saltation) and directionality (Levin et al. 2009). "Sand mobility is a function of wind power related to the cube of the wind speed above the threshold speed" (Bagnold 1941). We used the speed and direction of the wind for eighteen metrological stations to

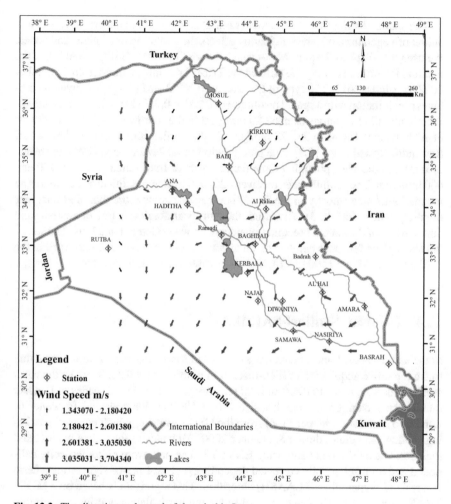

Fig. 12.2 The direction and speed of the wind in Iraq

build a map of speed and direction of the wind in Iraq (Fig. 12.2). These data cover the period from 2000 to 2013. The annual mean wind direction and wind speed were calculated. The Inverse Distance Weighted (IDW) method was implemented to estimate the speed and direction for the areas, which do not have metrological stations. Figure 12.2 shows that the maximum speed of the wind is about 3.7 m/s in the south eastern part of Iraq. While the minimum wind speed is about 1.34 m/s in the north of Iraq. Most of the speed direction is to the south, south-west and sometimes to the west. In Iraq, the wind speed in the Mesopotamian Plain is the highest; almost. Moreover, Iraq suffered from many dust storms in the last decade (Halos et al. 2017; Shubbar et al. 2017). These dust storms reach to 4.93 days/year (Shubbar et al. 2017). The dust storms, definitely increase the aeolian sand dunes formation.

The spatial distribution of the annual rainfall is important to understand the distribution of vegetation, which significantly effects the evolution of aeolian sand dunes. We used the Tropical Rainfall Measuring Mission (TRMM) (GSFC DAAC 2017) to measure the rainfall in Iraq. We utilized the TRMM rainfall acquired from September 2002 to August 2017. The type of the data used is TRMM (3B43-V7), which combines precipitation with a spatial resolution of $0.25° \times 0.25°$ (Kummerow et al. 1998). Precipitation in Iraq varies from 62.29 mm/year in the southwest (~70 km southwest of Al-Nukhayb Town) to 980.28 mm/year in the northeast of Iraq (~30 km east of Rawandoz Town) (Fig. 12.3). The precipitation in Al-Jazeera area (west-northwest of Iraq) and the Mesopotamian Plain (central part of Iraq) ranges between 100 and 400 mm/year. The foothills area receives 400–800 mm of precipitation per year. The mountainous area (mostly in the Kurdistan Region) gains more than 800 mm/year of precipitation (Fig. 12.3). Therefore, the Kurdistan Region is free of aeolian sand dunes. Most of the annual precipitation occurs between October and May. The snowfall takes place few days per year on an average between November and April in the north of Iraq. Above 1500 (a.s.l.), heavy snowfall occurs in the winter.

12.4 Landsat, Sentinel-SAR Data

In the last decades, Landsat data are widely used for environmental monitoring due to the repetitive acquisition of high-resolution multispectral data. It covers the earth surface since the early 1970s. The Iraqi territory is covered by 32 scenes of Landsat multispectral data. Each scene has size 170×180 km. We collected 16 scenes of Landsat 5 Thematic Mapper (TM) from the USGS website, which covers the existing aeolian sand dunes during September 2000. All bands of these data were staked together; except thermal band using ENVI 5.1. The data were radiometrically calibrated and atmospherically corrected depending on Fast Line-of-sight Atmospheric Analysis of Hypercubes (FLAASH) algorithm. This algorithm is based on a MODTRAN4 approach to eliminate the molecular and particulate scattering, absorption, and adjacency effects to retrieve at surface reflectance values (ρ) (Felde et al. 2003; Othman and Gloaguen 2017). However, the applications of workflow classification and supervised classification were conducted depending on the Spectral Angle Mapper (SAM) algorithm (Kruse et al. 1993). The thematic classification maps were converted to vectored type, using ArcGIS software. Aeolian sand dunes' classes were exported as an individual layer from each scene. The resulted layers were merged to achieve the coverage area of aeolian sand dunes in Iraq. Thus, we obtained the first coverage area of aeolian sand dunes during September 2000 in Iraq, while the second dataset was collected over Iraq country in 2016 (Remote Sensing Division of GEOSURV-IRAQ 2016).

There are many objectives of Sentinel-1 mission; such as land monitoring of forests, water, soil and agriculture. Emergency mapping supports in the event of natural disasters; marine monitoring of the maritime environment; sea ice observations and iceberg monitoring; production of high-resolution ice charts; forecasting ice con-

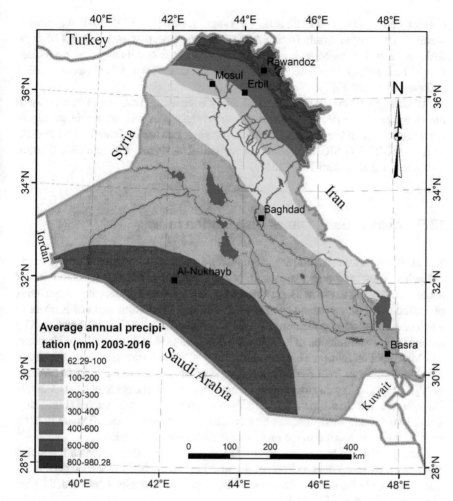

Fig. 12.3 Average annual precipitation map of Iraq for the period (2003–2016) using TRMM data

ditions at sea; mapping oil spills; sea vessel detection; climate change monitoring. Sentinel-1A was launched on April 03, 2014 its payload type is C-SAR (C-band Synthetic Aperture Radar) (ESA 2014).

We processed four Single Look Complex (SLC) scenes in ascending geometry of Sentinel-1A sensor. These images are Interferometric Wide (IW) mode acquired by the European Space Agency (ESA). The spatial resolution of these images are (5 × 20) m, and the swath width is 250 km. Sentinel-1 data has either vertical-vertical (VV) and vertical-horizontal (VH) or VV polarization (ESA 2014).

We used SARPROZ, to process all SAR data. This software is developed by Perissin (2017). It is written in MATLAB for processing SAR data (Perissin et al. 2012). Since SARPROZ is based on graphical interfaces and does not need a knowledge of coding (Perissin et al. 2012), a simple technique of DINSAR was implemented. The output result from SARPROZ can appear as raster or KMZ files. SARPROZ is a research tool, and for this reason, it is continually under development. The source code of SARPROZ is not public, despite that, it is not a commercial software (Perissin 2017).

All GIS operations (change detection application using ArcGIS software between the two datasets to determine aeolian sand creeping, as well as newly aggregate deposits, base map preparation, shapefile creating, and area calculation) were done using ArcGIS10 (ESRI 2011) and all statistical analyzes were conducted using R-based scripts (R Core Team 2017).

12.5 Aeolian Sand Dunes Distribution in Iraq

The aeolian sand dunes in Iraq are accumulated as related to topographic or vegetation obstacles. They exist near plateaus, hills and mountains. The first accumulation of aeolian sand dunes exists as a long strip along Abu Jeer Fault Zone, where huge fields of aeolian sand dunes extend to near the Euphrates River from north of Karbala to south of Al-Muthanna governorates. This accumulation has a NW-SE direction. The second main accumulation of aeolian sand dunes located in the foothills of Hemreen and Mak'hool anticlines. This accumulation has NW-SE direction, too. The dunes are extending from Iraqi-Syrian borders west of Mosul city to the borders between Salah Al-Din and Diyala governorates, where these dunes disappear in most of Diyala Governorate and then appear again in Wassit and Missan governorates (Fig. 12.4). These two aeolian sand dunes accumulations are old (more than 20 years in age). Both of them are located in the margin of the Mesopotamia Plain.

Geologically, the first accumulation is located in the margin between the Stable Shelf and the Mesopotamian Zone of the Arabian Plate, while the second accumulation is located in the margins between the Mesopotamian and the Low Folded Zones of the Arabian Plate (Fig. 12.5).

There is a new accumulation began to form in the center of the Mesopotamian Zone. This accumulation has NW-SE direction, too. It includes three big fields located from the NW to SE in the east of Baghdad, south of Al-Qadissiya, and south of Thi-Qar governorates (Figs. 12.4 and 12.5). Moreover, there are some small fields located in Al-Anbar and Al-Muthanna governorates without any strip and direction due to the lack of any vegetation on topographic obstacles.

The three huge accumulations of aeolian sand dunes in Iraq include almost all of the aeolian sand dunes' types. Among the three types of sand accumulation related to topographic obstacles (i.e. windward, Leeward and cliff-top accumulation), the Windward dunes and Leeward existing in Iraq near the plateaus, hills and mountains. Figure 12.6 shows a good example of windward dunes (echo dunes). The echo dunes are commonly existing in front of cliffs (Pye and Tsoar 2008).

The sand accumulation related to vegetation; such as vegetated linear, and hummock dunes exist in Iraq. Figure 12.7a shows an example of the vegetated linear

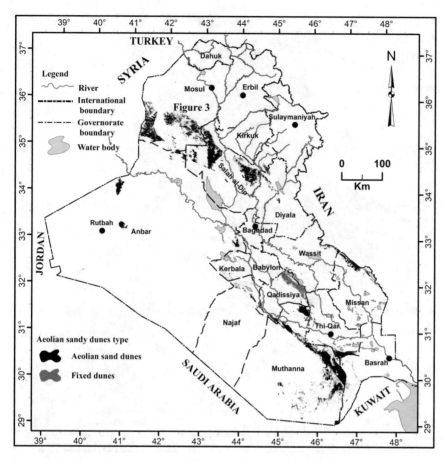

Fig. 12.4 Aeolian sand dunes over Iraq in 2016 (Modified after Remote Sensing Division 2016)

dunes. This type is sometimes described wrongly as a seifs type (for instance Walker 1986). Vegetated linear dunes are very close to seifs, but they have a rounded cross-sectional profile. They vary in height and length (Pye and Tsoar 2008). The most common type of the sand accumulations related to vegetation in Iraq is hummock dunes. It is a sand dune type, which is formed around vegetation (Fig. 12.7b). It is characterized by irregular shape, forms a mound of sand, slightly or densely vegetated (El-Sheikh et al. 2010). The old and Arabic name of hummock is "nabkha", which is used by some authors "nabkha" too (El-Sheikh et al. 2010; Langford 2000).

Almost all of the sand accumulation which is related to bed roughness changes or aerodynamic fluctuations are distributed in Iraq. The types of this accumulation are very common in Iraq. Barchan dune: a free transverse dune with a crescent plan-shape in which the crescent opens downwind (Bishop et al. 2002). The simple form of Barchans is an individual crescent feature most of them are formed on the pre-Quaternary and Quaternary sediments and extend along the international Iraqi-

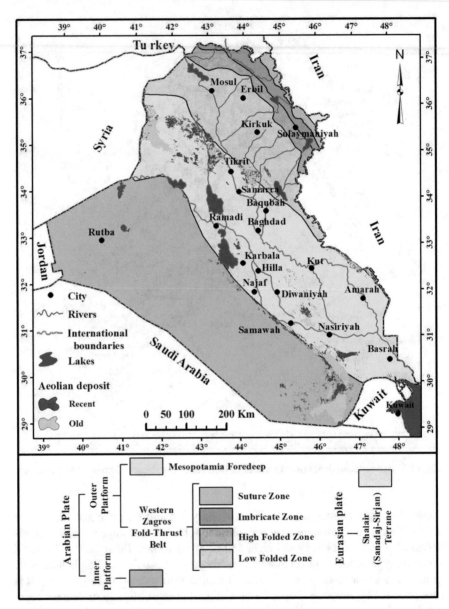

Fig. 12.5 Tectonic zones of Iraq showing the aeolian sand dunes distribution (Modified after Fouad 2012)

Fig. 12.6 Small echo dunes developed in front of the cliff of the valley in Midan village near Iraqi-Iranian borders, south-east Wasit governorate

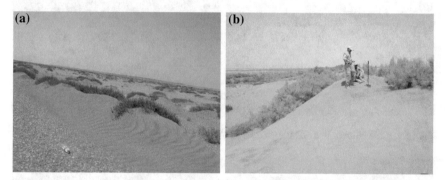

Fig. 12.7 Sand accumulation related to vegetation **a** small vegetated linear dunes SE Jassan town near Iraqi-Iranian borders, **b** hummock (Nabkha) dunes near the borders between Al-Qadissiya, Wasit and Babylon governorates

Iranian border as a strip. They display a wide range of sizes with some having great widths and some only a few tens of meters (Fig. 12.8). Generally, dunes trend NW–SE, which reflect the dominated prevailing wind direction influence.

Merk (1960) described these complex transverse dunes as consisting of the large primary; also called (transverse) dunes that have slip faces on the east, formed by winds blowing from the west. Smaller sharply defined secondary dunes that have to slip faces on the west that were shaped by winds blowing from east to west. The transverse (barchanoid) ridges are observed in the field checking with variation in distribution and density from place to another. The linear or seif dunes are common in Iraq (Fig. 12.9). They are characterized by sinuous, sharp-crested, and linear-shape (Bishop et al. 2002). The dome dunes, and star dunes are existing in Iraq, but not common like barchan and transverse dunes. Sand sheet and Zibars, which are poorly sorted and bimodal sand types of aeolian sand dunes. They are characterized by coarse-grained bedforms of low relief with no slip faces.

Fig. 12.8 Barchan Dunes near Kharkhar River NE Wasit governorate

(a)   **(b)**

Fig. 12.9 Linear dunes in **a** NE Babylon and **b** Ali Al-Gharbi area

Fig. 12.10 Aeolian sand dunes are destroying farms and croplands areas in NE Babylon

Most of the sand dunes are still active in the movement and have a long-term neg-ative impact on the agriculture land, oil extraction companies; roads, human health and etc. (Figs. 12.9b and 12.10). The aeolian sand dunes contributed to degradation of the soil. Several farms and cropland areas have been destroyed due to aeolian sand dunes creeping.

12.6 Aeolian Sand Dunes Monitoring in Iraq

Figure 12.11 figures out the distribution of the aeolian sand dunes' changes over Iraq between 2000 and 2016. The increase and decrease in aeolian sand dunes in 2016 appear in red and green colors, respectively. The fixed dunes and permanent aeolian sand dunes (no changes) since 2000 appear in gray and Turquoise colors, respectively. The fixed dunes (stabilized aeolian sand dunes) are noticed to be scattered in several parts in Iraq. Those dunes are covered by either a clayey layer cover or some types of plants, which have the ability to adapt hard environmental conditions to fix them. There are many attempts to fix the sand dunes, sand sheets and nabkha by using clayey bed cover and plantation. These attempts; however, are succeeded in some places and threaten to fall down in other places. The clay layer is crashed or destroyed in some places as a result of its thin thickness and absence of maintenance work in addition to severe climate conditions.

Table 12.1 shows the areas of the aeolian sand dunes in 2000 and 2016. These dunes are classified into increase and decrease, fixed and no change. The coverage of the aeolian sand dunes' areas was about 9,891.29 km^2 in 2000; increased to 14,419 km^2 in 2016. The total increased area of the aeolian sand dunes in Iraq between 2000 and 2016 is 4,527.69 km^2. The existing area of the aeolian sand dunes in 2016 and did not exists in 2000 is 8,998.21 km^2. The existing area of the aeolian sand dunes in 2000 and not exists in 2016 is 4,470.52 km^2.

Figure 12.12 is a good example of aeolian sand dunes' changes. It shows the color composite of the R: ρ (2.08–2.35 μm), G: ρ (0.76–0.90 μm) and B: ρ (0.52–0.60 μm) for Landsat 5 and R: ρ (2.107–2.294 μm), G: ρ (0.851–0.879 μm) and B: ρ (0.533–0.590 μm) for Landsat OLI. It is displayed the change in aeolian sand dunes in the north western part of Iraq at two periods (2000 and 2016). The aeolian sand dunes appear in red in 2016 scene, while the scene of 2000 is free of the sand dunes.

12.7 Sand Dunes Movement in the Center of Mesopotamia

12.7.1 Preparing SAR Data for DInSAR

In the last decades, remote sensing data (e.g. SAR) and techniques (e.g. differential interferometry) have become an essential method to measure infrastructure displacements. In this study, a simple DInSAR technique is applied to process Sentinel A1 images for monitoring the aeolian sand dunes near Hor Al-Dalmaj area.

An interferometric processing technique was used to compute two ascending and four descending Single Look Complex (SLC) scenes for monitoring the aeolian sand dunes in the central part of Iraq. Unfortunately, the descending scenes do not give good results; therefore, we ignored them. Only the ascending two scenes were considered in this work. At the first step, we read the SLC scenes, provided by

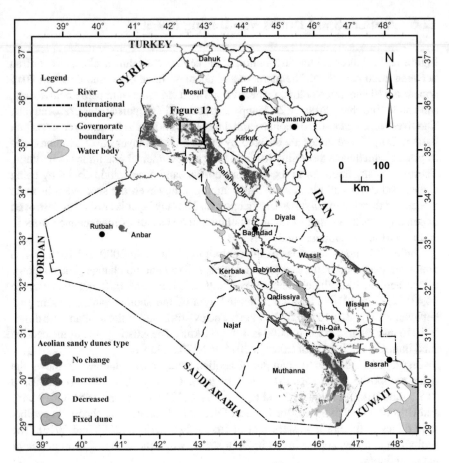

Fig. 12.11 Aeolian sand dunes' changes over Iraq between 2000 and 2016 (Modified after Remote Sensing Division 2016)

Table 12.1 Distribution of aeolian sand dunes over Iraqi territory

Aeolian sand dunes and change type	Coverage area (km²)
Aeolian sand dunes 2000	9891.29
Aeolian sand dunes 2016	14,419
Fixed dunes	1245.22
No change	5420.77
Decreased	4470.52
Increased	8998.21

Fig. 12.12 Landsat images are showing sand sheet creeping forming new desertification region in the northwestern part of Iraq (For location see Fig. 12.11)

Sentinel 1A. We used only the VV polarization to monitor the aeolian sand dunes. The VV polarization is quite enough for this type of processing. The master (acquired on March 27, 2015) and slave (acquired on August 30, 2015) images were extracted using an area of interest. The area of interest was selected from WI2 of these two scenes. The normal baselines of these two scenes Selection two scenes are 24 m, which avoid additional noise due to spatial decorrelation corrupts the signal (Zebker and Villasenor 1992).

The signal of the interferometric phase ($\Delta\Phi_{int}$) resulted from multiple contributions shows in Eq. (12.1; Hooper et al. 2004):

$$\Delta\Phi_{int} = \Delta\Phi_{topo} + \Delta\Phi_{displ} + \Delta\Phi_{atmo} + \Delta\Phi_{noise} \qquad (12.1)$$

where $\Delta\Phi_{topo}$ is the residual topographic height (H) without the Digital Elevation Model (DEM), $\Delta\Phi_{displ}$ is the searched displacement (D) information, $\Delta\Phi_{atmo}$ is the disturbance caused by the atmospheric phase screen, and $\Delta\Phi_{noise}$ is the non-removable phase disturbance. The D and H values can be estimated by choose a set of PSCs using an ASI threshold to select the more stable pixels in terms of the amplitude signal over the analyzed period.

An area (~1,862 km^2) from the second swath was selected (Fig. 12.13). This area consists of 11,000 samples and 3240 lines. It is located in the central part of Mesopotamian Plain. It covers part of Wassit, Qadissiya and Babylon governorates. To process SLC products precisely, we ignored restituted orbit, and information orbit, and used precise orbit information. Precise co-registration of SAR images is a strict requirement and an essential step for accurate determination of interferometric deformation analysis (Milillo et al. 2016). We co-registered the slave image to the master image. After that, we estimated the flat-earth phase using the orbital and metadata information and then subtracted its results from the complex interferogram.

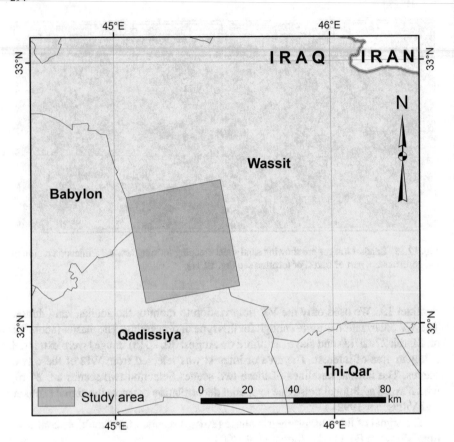

Fig. 12.13 Location map of the selected area

We removed the zero Doppler time within the subswath using deburst operator. After that, we applied Goldstein adaptive filter (Goldstein and Werner 1998) to decrease the phase noise (thermal noise, temporal change, and baseline geometry), and improved both measurement accuracy and phase unwrapping for SAR interferometry. To compute the actual altitude variation between the two scenes and solves the ambiguity (interferometric phase within 2π), we unwrapped the interferometric phase. The final result is the displacement in mm (Fig. 12.14a). This displacement is calculated parallel to the line of sight (LOS) of Sentinel; therefore, it is necessary to convert it to vertical displacement (Fig. 12.14b) using Eq. 12.2:

$$\mathrm{VD} = LOSD * \cos_{in} \tag{12.2}$$

where VD is the vertical displacement, and LOSD is the line of sight displacement, and $\cos_{in}$ is the cosine of the incidence angle.

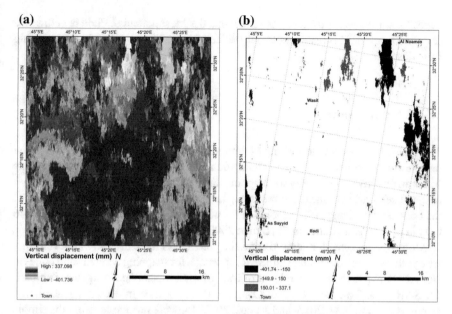

Fig. 12.14 **a** Vertical displacement **b** classified vertical displacement

12.7.2   Results and Discussion of DInSAR

DInSAR technique was applied to determine the movement of the aeolian sand dunes in a part of the Mesopotamian Plain. Figure 12.14 shows the interferogram of Hor Al-Dalmaj marshland and the surrounding area within a 157-day. The variation in the vertical displacement was measured in millimeter. The range of the variation in the vertical displacement was between 337.098 and −401.736 mm (Fig. 12.14a). The vertical displacement was classified into three classes, which are: Class (A) more than 150 mm, could be the area of accumulation of new aeolian sand dunes. Class (B) is ranging between 150 and −150 mm, which is the neutral area. Class (C) represents the pixels that have displacement less than −150 mm, which could be those areas that totally or partly eroded from aeolian sand dunes (Fig. 12.14b).

In Fig. 12.14b, the old and new dunes are characterized by the longitudinal shape, NW-SE direction, which supports the possibility that these areas are aeolian sand dunes. The area of the new possible aeolian sand dunes is about 19.5 km^2, while the area of the possible eroded from aeolian sand dunes is about 102.9 km^2. The volume of the new possible aeolian sand dunes is about 0.0036 km^3, while the volume of the possible eroded aeolian sand dunes is about 0.02124 km^3.

This work can demonstrate that the Mesopotamian Plain is affected by aeolian sand dunes and they are active. In spite of that, the areas of the possible eroded from aeolian sand dunes are more than the possibility of new aeolian sand dunes deposition. However, this does not reflect that the Mesopotamian Plain is getting rid of desertification. This result could only reflect that the Mesopotamia is still suffering from desertification.

12.8 Conclusion

This chapter deals with monitoring of the aeolian sand dunes in Iraq using active and passive remote sensing data. The application of multitemporal remote sensing data offers an effective opportunity for mapping desertification processes in Iraq and other arid and semi-arid countries, at relatively low cost. The supervised classification of the Landsat data using ASM algorithm for whole Iraq shows an increase about 1.46 time in aeolian sand dunes in 2016 than 2000.

Moreover, this chapter demonstrates the potential of the Differential Interferometry Synthetic-Aperture Radar (DInSAR) approach for mapping aeolian sand dunes at Hor Al-Dalmaj area, the central part of Iraq using European Space Agency Sentinel-1A SAR data. Two Single Look Complex (SLC) scenes in ascending geometry, acquired on March 27, 2015, and August 30, 2015, were processed using the simple DInSAR technique to determine the displacement estimation. The vertical displacement ranges between 337.098 and −401.736 mm. The volume of the new possible aeolian sand dunes is less ~5.9 times than the volume of the possible eroded aeolian sand dunes. This result does not reflect that the Mesopotamian Plain is getting rid of desertification. It could only reflect that the Mesopotamia is still suffering from desertification.

12.9 Recommendations

As the climate change, building dams in the rivers upstream countries, and poor and old agriculture practices are the mean cause of increased desertification in Iraq territory. Therefore, all these factors should be taken into consideration to control the aeolian deposits movement and expansion. There are several recommendations can be proposed to mitigate desertification in Iraq territory such us (1) Design policies to mitigate the impacts of wind erosion, particularly, in the abundant agricultural areas. Where the increase in the area of abundant agricultural lands is one of the most important factors contribute to the expanded desertification in Iraq, particularly, in the Mesopotamian Plain. Moreover, such areas require immediate attention for remedial or reclamation these lands. (2) Applying new irrigation method in order to adopted with the current conditions in term of the decrease in water supplies from upstream river countries (i.e. Turkey, Iran, Syria), as well as increased salt content in agriculture land because of old irrigation habits. (3) Avoidance use Al-Tharthar Lake as a water reservoir (an artificial reservoir to collect flood waters of the Tigris River). The Lake formed mainly by karstification, due to the dissolving of gypsum rocks of the Fat'ha

Formation and its shallow broad depression. Hence, it contributes to the increase of water salinity as a result of dissolving salty rocks and also increase the evaporation rate of water. Moreover, it contributes to washing out the salts from the stored water, to the Euphrates River. (4) Stabilize the mobile sand dune through the development of natural herbaceous cover under the protection of a plantation using special perennial species can tolerance hard climate condition and salty soil. (5) Avoidance of topsoil damage by overgrazing of animals and other destructive activities, which is likely to be more important than planting new species. Overall the impact of anthropogenic activities on the increase of aeolian desertification and degradation of lands is much more active than that from natural factors. (6) Planting trees fence around main cities and towns to mitigate the effect of the dust storm, as well as around the local sources of Aeolian deposits. Overall, establishing an international center to develop contribution between the relating countries can actually contribute to mitigating desert storms. (7) Preventing destroyed palm orchards and other agriculture lands by urban expansion particularly around main cities and increase current vegetation areas.

Acknowledgements We thank NASA for providing TRMM data, and we thank the USGS for providing the Landsat data. Many thanks to the European Space Agency for providing the sentinel SAR data. We are grateful to the Geological Survey of Iraq for providing the environmental reports and supporting the fieldwork.

References

Bagnold RA (1941) The physics of blown sand and desert dunes. London

Barbat A, Gloaguen V, Moine C, Sainte-Catherine O, Kraemer M, Rogniaux H, Ropartz D, Krausz P (2008) Structural characterization and cytotoxic properties of a 4-O-methylglucuronoxylan from *Castanea sativa*. 2. Evidence of a structure-activity relationship. J Nat Prod 71:1404–1409

Bishop SR, Momiji H, Carretero-González R, Warren A (2002) Modelling desert dune fields based on discrete dynamics. Discret Dyn Nat Soc 7:7–17. https://doi.org/10.1080/10260220290013462

Brera AM, Shahrokhi F (1978) Use of landsat imagery to monitor desert encroachment in arid areas, pp 58–68

Bulliet R, Crossley PK, Headrick D, Hirsch S, Johnson L, Northup D (2010) The earth and its peoples

Carlisle WJ, Marrs RW (1982) Eolian features of the southern high plains and their relationship to windflow patterns. Spec Pap Geol Soc Am

Del Ventisette C, Righini G, Moretti S, Casagli N (2014) Multitemporal landslides inventory map updating using spaceborne SAR analysis. Int J Appl Earth Obs Geoinf 30:238–246. http://dx.doi.org/10.1016/j.jag.2014.02.008

El-Sheikh MA, Abbadi GA, Bianco PM (2010) Vegetation ecology of phytogenic hillocks (nabkhas) in coastal habitats of Jal Az-Zor National Park, Kuwait: role of patches and edaphic factors. Flora Morphol Distrib Funct Ecol Plants 205:832–840. https://doi.org/10.1016/j.flora.2010.01.002

ESA (2014) Sentinel-1 [WWW Document]. European Sp. Agency. URL https://earth.esa.int/web/guest/missions/esa-operational-eo-missions/sentinel-1

ESRI (2011) ArcGIS

Felde GW, Anderson GP, Cooley TW, Matthew MW, Adler-Golden SM, Berk A, Lee J (2003) Analysis of Hyperion data with the FLAASH atmospheric correction algorithm. In: 2003 IEEE international geoscience and remote sensing symposium, IGARSS '03. Proceedings, vol. 1, pp. 90–92. https://doi.org/10.1109/IGARSS.2003.1293688

Ferretti A, Prati C, Rocca F (2001) Permanent scatterers in SAR interferometry. IEEE Trans Geosci Remote Sens 39:8–20. https://doi.org/10.1109/36.898661

Fouad SF (2012) Tectonic map of Iraq, scale 1:1,000,000, vol 3. Iraq Geological Survey Publications, Baghdad, Iraq

Fryberger SG, Ahlbrandt TS (1979) Mechanisms for the formation of eolian sand seas. Zeitschrift fur Geomorphol 23:440–460

Goldstein RM, Werner CL (1998) Radar interferogram filtering for geophysical applications. Geophys Res Lett 25:4035–4038. https://doi.org/10.1029/1998GL900033

GSFC_DAAC (2017) Tropical rainfall measurement mission project (TRMM ; 3B43 V7) [WWW Document]. URL http://disc.gsfc.nasa.gov/datacollection/TRMM_3B42_daily_V6.shtml

Hack JT (1941) Dunes of the Western Navajo Country. Geogr Rev 31:240–263. https://doi.org/10.2307/210206

Hadeel AS, Jabbar MT, Chen X (2011) Remote sensing and GIS application in the detection of environmental degradation indicators. Geo-spatial Inf Sci 14:39–47. https://doi.org/10.1007/s11806-011-0441-z

Hadeel AS, Jabbar MT, Chen X (2010a) Environmental change monitoring in the arid and semi-arid regions: a case study Al-Basrah Province, Iraq. Environ Monit Assess 167:371–385. https://doi.org/10.1007/s10661-009-1056-9

Hadeel AS, Jabbar MT, Chen X (2010b) Application of remote sensing and GIS in the study of environmental sensitivity to desertification: a case study in Basrah Province, southern part of Iraq. Appl Geomatics 2:101–112. https://doi.org/10.1007/s12518-010-0024-y

Halos SH, Al-Taai OT, Al-Jiboori MH (2017) Impact of dust events on aerosol optical properties over Iraq. Arab J Geosci 10. https://doi.org/10.1007/s12517-017-3020-2

Havivi S, Amir D, Schvartzman I, August Y, Maman S, Rotman SR, Blumberg DG (2018). Mapping dune dynamics by InSAR coherence. Earth Surf Process Landforms. https://doi.org/10.1002/esp.4309

Hooper A, Zebker H, Segall P, Kampes B (2004) A new method for measuring deformation on volcanoes and other natural terrains using InSAR persistent scatterers. Geophys Res Lett 31:1–5. https://doi.org/10.1029/2004GL021737

Huggett R (2007) Fundamentals of geomorphology, Routledge. Fundamentals of physical geography. Taylor & Francis

Hunter RE, Richmond BM, Alpha TR (1983) Storm-controlled oblique dunes of the Oregon coast. Geol Soc Am Bull 94:1450–1465. https://doi.org/10.1130/0016-7606(1983)94%3c1450:SODOTO%3e2.0.CO;2

Jabbar MT, Zhou J (2012) Assessment of soil salinity risk on the agricultural area in Basrah Province, Iraq: using remote sensing and GIS techniques. J Earth Sci 23:881–891. https://doi.org/10.1007/s12583-012-0299-5

Jabbar MT, Zhou X (2011) Eco-environmental change detection by using remote sensing and GIS techniques: a case study Basrah province, south part of Iraq. Environ Earth Sci 64:1397–1407. https://doi.org/10.1007/s12665-011-0964-5

Kolm KE (1982) Predicting the surface wind characteristics of southern Wyoming from remote sensing and eolian geomorphology. Spec Pap Geol Soc Am. https://doi.org/10.1130/SPE192-p25

Kottek M, Grieser J, Beck C, Rudolf B, Rubel F (2006) World map of the Köppen-Geiger climate classification updated. Meteorol Zeitschrift 15:259–263. https://doi.org/10.1127/0941-2948/2006/0130

Kruse FA, Lefkoff AB, Boardman JW, Heidebrecht KB, Shapiro AT, Barloon PJ, Goetz AFH (1993) The spectral image processing system (SIPS)—interactive visualization and analysis of imaging spectrometer data. Remote Sens Environ 44:145–163. http://dx.doi.org/10.1016/0034-4257(93)90013-N

Kummerow C, Barnes W, Kozu T, Shiue J, Simpson J (1998) The tropical rainfall measuring mission (TRMM) sensor package. J Atmos Ocean Technol 15:809–817. https://doi.org/10.1175/1520-0426(1998)015%3c0809:TTRMMT%3e2.0.CO;2

Lam DK, Remmel TK, Drezner TD (2011) Tracking desertification in California using remote sensing: a sand dune encroachment approach. Remote Sens 3:1–13. https://doi.org/10.3390/rs3010001

Langford RP (2000) Nabkha (coppice dune) fields of south-central New Mexico, U.S.A. J Arid Environ 46:25–41. https://doi.org/10.1006/jare.2000.0650

Levin N, Tsoar H, Herrmann HJ, Maia LP, Claudino-Sales V (2009) Modelling the formation of residual dune ridges behind Barchan Dunes in north-east Brazil. Sedimentology 56:1623–1641. https://doi.org/10.1111/j.1365-3091.2009.01048.x

Lorenz RD, Gasmi N, Radebaugh J, Barnes JW, Ori GG (2013) Dunes on planet Tatooine: observation of barchan migration at the Star Wars film set in Tunisia. Geomorphology 201:264–271. https://doi.org/10.1016/j.geomorph.2013.06.026

Mainguet M (1983) Mobile dunes, fixed dunes and covered dunes: a classification according to sand supply balance, wind regime and dynamics of sand structures. Zeitschrift fur Geomorphol Suppl 45:265–285

Martin R (2011) Earth_s_evolving_systems. Jones & Bartlett Learning

Marzolf JE (1988) Controls on late Paleozoic and early Mesozoic eolian deposition of the western United States. Sediment Geol. 56:167–191. https://doi.org/10.1016/0037-0738(88)90053-X

Merk GP (1960) Great sand dunes of Colorado. In: Weimer RJ, Haun JD (eds) Guide to the geology of Colorado. Geological Society of America, Rocky Mountain Association of Geologists, and Colorado Scientific Society

Milillo P, Perissin D, Salzer JT, Lundgren P, Lacava G, Milillo G, Serio C (2016) Monitoring dam structural health from space: insights from novel InSAR techniques and multi-parametric modeling applied to the Pertusillo dam Basilicata, Italy. Int J Appl Earth Obs Geoinf 52:221–229. https://doi.org/10.1016/j.jag.2016.06.013

Othman AA, Al-Saady YI, Al-Khafaji AK, Gloaguen R (2014) Environmental change detection in the central part of Iraq using remote sensing data and GIS. Arab J Geosci 7:1017–1028. https://doi.org/10.1007/s12517-013-0870-0

Othman AA, Gloaguen R (2017) Integration of spectral, spatial and morphometric data into lithological mapping: a comparison of different machine learning algorithms in the Kurdistan Region, NE Iraq. J Asian Earth Sci. https://doi.org/10.1016/j.jseaes.2017.05.005

Perissin D (2017) SARPROZ

Perissin D, Wang Z, Lin H (2012) Shanghai subway tunnels and highways monitoring through Cosmo-SkyMed Persistent Scatterers. ISPRS J Photogramm Remote Sens 73:58–67. https://doi.org/10.1016/j.isprsjprs.2012.07.002

Pye K, Tsoar H (2008) Aeolian sand and sand dunes

R Core Team (2017) R: the foundation for statistical computing

Remote Sensing Division (2016) Series of land use–land cover maps of Iraq scale 1:250,000. Baghdad

Remote Sensing Division of GEOSURV-IRAQ (2016) Series of land use—land cover maps of Iraq scale of 1:250,000. Baghdad

Sarnthein M (1978) Sand deserts during glacial maximum and climatic optimum. Nature 272:43

Shubbar RM, Salman HH, Lee D-I (2017) Characteristics of climate variation indices in Iraq using a statistical factor analysis. Int J Climatol 37:918–927. https://doi.org/10.1002/joc.4749

Sowter A, Bateson L, Strange P, Ambrose K, Fifiksyafiudin M (2013) Dinsar estimation of land motion using intermittent coherence with application to the south derbyshire and leicestershire coalfields. Remote Sens Lett 4:979–987. https://doi.org/10.1080/2150704X.2013.823673

Thomas DSG (1989) Aeolian sand deposits. Arid Zo. Geomorphol 232–261

Tsoar H (2001) Types of Aeolian sand dunes and their formation BT-geomorphological fluid mechanics. In: Balmforth NJ, Provenzale A (eds) Springer Berlin Heidelberg, pp 403–429. https://doi.org/10.1007/3-540-45670-8_17

Vajedian S, Motagh M, Nilfouroushan F (2015) StaMPS improvement for deformation analysis in mountainous regions: implications for the Damavand volcano and Mosha fault in Alborz. Remote Sens 7:8323–8347. https://doi.org/10.3390/rs70708323

Walker AS (1986) Eolian landforms. In: Short NM, Blair RW (eds) Geomorphology from space. NASA, Washington

Wasson RJ, Hyde R (1983) Factors determining desert dune type. Nature 304:337

Zebker HA, Villasenor J (1992) Decorrelation in interferometric radar echoes. IEEE Trans Geosci Remote Sens 30:950–959. https://doi.org/10.1109/36.175330

Chapter 13
Drought Monitoring for Northern Part of Iraq Using Temporal NDVI and Rainfall Indices

Suhad M. Al-Hedny and Ahmad S. Muhaimeed

Abstract Climate change is the major global challenge facing water resources managers. Drought is a natural hazard temporarily affecting almost every region in the world. In this study, the climate change in term of rainfall fluctuation in the northern part of Iraq (Mosul, Kirkuk and Salah Al-Din) has been investigated using a set of data containing monthly precipitation for the period from 1980 to 2010, and the MODIS time series images for the period from 2000 to 2010. All data series have been used to calculate standardized precipitation index (SPI) and Normalized Difference Vegetation Index (NDVI). Monthly rainfall data from 12 stations were used to derive the SPI at several time scales (3, 6 and 12-months), the analysis was carried out for the period from 1980 to 2010. Results of the SPI analyses showed that the year 2007–2008 was an extremely drought year for the whole study governorates (Mosul, Kirkuk and Salah Al-Din) with the lowest SPI-12 values -2.67, -2.07 and -2.0 for the three above mentioned governorates, respectively. The results also pointed to the importance of using short time scales in detecting and monitoring the agricultural drought during the crop growing season. The multiple time scales analyzed in this study reflected a clearer picture of the severity and frequencies of drought events, which happened in the study area. The NDVI results were analyzed to get the agricultural drought risk map. The highest NDVI values were 0.33 in 2001, 0.39 in 2003 and 0.20 in 2001 for Mosul, Kirkuk and Salah Al-Din, respectively. While the lowest NDVI values were 0.10 in, 0.19 and 0.13 in 2008 for the three above mentioned governorates respectively. This study emphasized the use of Remote Sensing and GIS in the field of drought risk evaluation. The results showed that the NDVI is an efficient way to monitor changes in vegetation conditions (weekly or daily) during the growing season, and can be used as simple and cost-efficient drought index to monitor agricultural drought at a small or large scale. The NDVI and rainfall were found to be highly correlated 0.83, 0.70 and 0.72 for Mosul, Kirkuk and Salah Al-

S. M. Al-Hedny (✉)
Department of Environment, Faculty of Environmental Science, Al-Qasim Green University, Babil, Iraq
e-mail: suhad.khudair@environ.uoqasim.edu.iq

A. S. Muhaimeed
Department of Soil and Water Science, College of Agriculture, Baghdad University, Baghdad, Iraq

© Springer Nature Switzerland AG 2020
A. M. F. Al-Quraishi and A. M. Negm (eds.),
Environmental Remote Sensing and GIS in Iraq, Springer Water,
https://doi.org/10.1007/978-3-030-21344-2_13

Din, respectively. Therefore, the temporal variations of NDVI are closely linked with precipitation. Results of statistical correlation analysis between NDVI and SPI (3, 6 and 12-months) time scales showed that the highest correlation coefficients were between NDVI and SPI-6, which verified that the short time scales could be related closely to soil moisture. It was observed that the studied indices (NDVI & SPI) could be effectively used for monitoring and assessing agricultural productions and in that way, proper agricultural policies can be adopted to mitigate drought impacts.

Keywords Drought monitoring · Remote sensing · SPI · NDVI

13.1 Introduction

Drought is classified as the environmental hazard and natural disaster that depreciates the sustainable development of society (Shaheen and Baig 2011). Its long-lasting impacts badly have increased its extent on agricultural production, livestock, physical environment and the overall economy. Over the last three decades, many world regions have suffered from water crises and drought caused serious impacts on local economies (Ghulam et al. 2008). Iraq, Syria, Turkey and Iran, have been dealing with decreased rainfall affected the agricultural sector, livelihood system, employment and water allowable quantity and quality negatively (UNDP 2010).

Drought may be treated as a meteorological, hydrological, or agricultural phenomenon. In each one, the variable representing water availability is different. Meteorological drought is a situation of rainfall shortage from normal precipitation over an area. Agricultural drought occurs when soil moisture and rainfall are inadequate during the growing season. Hydrological drought represents the long-term meteorological drought that causes a decline in reservoirs, lakes, streams, rivers, and groundwater level (Rathore 2004). Over the last decade, many regions of the world have focused their attention on drought mitigation strategy. The first step in formulating a preparedness planning and mitigation process is drought monitoring. By monitoring drought over a long period of time (more than ten years) early drought warning systems can be developed to ensure global food security. Depending on the level of protection, the level of consequences may be reduced by decreasing the magnitude of the water shortage and by improving the public awareness (social factor). Drought is characterized as a multi-dimensional phenomenon (severity, duration, magnitude). Attention has been paid so far to simplify these dimensions to reach a practical way to assess the severity of drought. The Standardized Precipitation Index (SPI) is one among many others of the most popular indices used for assessing the severity of drought (Tsakiris 2010).

Based on the above consideration, drought severity could be represented by the SPI and complementarily by other drought monitoring indices for resulting in meaningful management decisions. Remote sensing is one of the most efficient monitoring methods. Compared with traditional measurements, remote sensing methods provide more reliable drought information over large geographic areas. In the agricultural drought monitoring, the most effective indicators that are responsive to vegetation health and soil moisture status. The overall effect of rainfall and soil moisture on crops could be effectively reflected by indices such as Normalized Difference Vegetation Index (NDVI), which plays an important role in early warning of drought events. The Moderate Resolution Imaging Spectroradiometer (MODIS) satellite series images provide spatial information on the amount of vegetation present on the ground. Based on the slow onset of the disaster, drought allows a warning time between the first indication and the point where the population will be affected. Thus, this information can be used as the first indicator of drought occurrence, and could be used as a simple remote-sensing tool to map drought conditions for crops (Shahabfar and Eitzinger 2011).

In semi-arid areas covering a large part of northern Iraq, crop failures due to agricultural drought have been common. According to the USDA report, 2003, three years of drought from 1999 to 2001 reduced production to be down 50% from the 1990 to 1991 level. Wheat and barley production declined by 51% compared with the previous year production (USDA 2008). From this view, it is critical for drought vulnerability regions (arid and semi-arid conditions) to understand their drought climatology (probability of drought at different levels of intensity and duration) and develop national drought policies and preparedness plan. A good agricultural drought monitoring system should establish comprehensive drought information that integrates climate, soil, crop and social factors to reduce the consequences of drought.

Drought stress on agricultural sector is one of the most important and lowest studied issues at the country level. This chapter aims to using the NDVI and SPI indices to detect the appearance and severity of the drought events for the northern part of Iraq.

13.2 Background

13.2.1 Drought as a Concept and Definition

Drought is a concept with such a wide scope of influence and meaning, generally, it is a temporary aberration which occurs in all climate regimes. This phenomenon is usually related to climate change and its characteristics may be different from one region to another. As a consequence, there is no wide accepted idea of what drought is, whether it is a physical phenomenon or natural event, how it can be measured, is the end of drought determined by a return to normal precipitation or by mitigating the negative impacts on society and the environment? For a better understanding of how the definition is a critical factor to describe and quantify drought, it is necessary to start with drought definition and concept from different perspectives.

Different definitions exist regarding drought. According to the American Meteorological Society Glossary of Meteorology defined drought as: "a period of abnormally dry weather sufficiently prolonged for the lack of water to cause a serious hydrologic imbalance in the affected area." (Huschke 1959).

Explanation of drought as conceptual or operational definitions was an idea put forward by Wilhite and Glanz (1985). Conceptually, drought is defined as a slow departure from normal precipitation over a period that causes a lack of water supply and crop damage (NDMC 2006). This crawling disaster is a normal part of the climate which can occur in all regions at any time of the year (Tallaksen and Van Lanen 2004). Due to its crawling nature, its effects may take weeks or months to appear in a reduction of surface/ground water to support crop growth and human activities. The severity of drought is often associated not only with the deficiency of precipitation, but also with other climatic factors such as high temperature, high wind and low humidity. However, there is an agreement that the definition of drought should be region and application-specific (WMO 2006).

From an operational standpoint, the definition explains drought from a meteorological perspective to socio-economic perspective. According to NDMC (2006) the operational definitions "specify the degree of departure from the average of precipitation or some other climatic variable over some period". The beginning of drought is determined by comparing the current situation of precipitation to the historical average (over long time 20–30 years). Meteorological, agricultural, hydrological, and socioeconomic drought are also used in literature as defining drought according to "disciplinary" perspective. The definitions described by Dracup et al. (1980), Wilhite and Glanz (1985), Tate and Gustard (2000) and considered droughts are related to precipitation (meteorological), stream flow (hydrological), soil moisture (agricultural), and water supply (socio-economic drought). The four types of drought are classified based on physical, biological, and socio-economic variables regionally set. These parameters reflect differences in zone climatic characteristics.

So, meteorological drought is the origin of all drought types, it relates to the degree of precipitation's departure from normal (intensity) over a period (duration). These parameters are highly dependent on climatic regimes, which vary from region to region (NDMC 2006).

Hydrological drought is defined by Yevjevich et al. (1977) as "A period of below average water content in streams, reservoirs, ground-water aquifers, lakes and soils". The amount and status of water in surface and sub-surface water reflects single or multi-year drought events. For this reason hydrological drought is considered as a function of both intensity and duration of drought.

Agricultural drought reflects extremely meteorological and hydrological drought impacts. According to Rosenberg (1979) drought is defined as "A climatic excursion involving a shortage of precipitation sufficient to adversely affect crop production or range production". Plant water requirements depend on current weather conditions, plant's characteristics, its age, and soil properties. Thereupon, the precipitation deficiency may cause a fast depletion of soil moisture leading to reduce crop production and increase the probability of forest fires. Generally, agricultural drought is defined by the availability of soil water to meet plant water needs in growing season: for example deficient of the moisture at planting stage, leads to reduce final yield by hindering germination, while deficient subsoil moisture at this stage may not affect final yield. As a result, the determination of agricultural drought is based on multi-variable include a combination of various parameters: meteorological (precipitation and evaporation), plant factors (biological characteristics and growth stage), and soil factors (physical, chemical, and biological characteristics) (Heidorn 2007).

Crop water requirements highly depend on soil moisture status; therefore, soil texture, structure, and organic matter are also important factors in controlling how much water can be held by soil. Thus, a definition that suits the agricultural sector has to take into consideration both plant and soil indicators in addition to the meteorological indicators (Legesse 2010).

A meteorological drought can further develop into socio-economic drought defined as the relationship between the supply and demand for some commodities, such as water, which depends on precipitation (Tallaksen and Van Lanen 2004). Socio-economic drought is usually used to refer to the impacts of drought on population and economy. It differs from the types above of drought (meteorological, hydrological, and agricultural drought), because it highly depends on both time and scale processes of supply and demand. The shortfall of water supply occurs due to: natural variability of climate, increased population and per capita consumption, and increased demand more rapidly than supply (Okorie 2003).

Drought characteristics: intensity, duration, and spatial coverage. Intensity refers to the deviation of precipitation from normal amount. It is usually measured by the Standard Precipitation Index (SPI) (World Meteorological Organization (WMO) 2006). Drought duration of an event is defined as the time between the negative and positive values. The impacts of drought are strongly related to the period of the precipitation shortfall, subsequently, droughts can develop rapidly in some climate regimes. The third feature of drought is spatial characteristic, where this regional phenomenon differs in its intensity from region to region.

From the combination of all the above mentioned concepts, on a very general level, drought is a reduction in precipitation which has a negative effect on the environment and human activities. Mitigation of drought impacts is based on a combination of physical nature of drought, which is measured by: intensity, duration, and spatial extent, and social characteristics, which measured by their ability to: anticipate, deal with, and recover from the drought. The impacts of drought can either be reduced or aggravated based on how these systems are managed (WMO 2006).

13.2.2 Impact of Drought on Soil Properties

In arid and semi-arid regions, intense drought events could increasingly lead to reducing the productivity of soil, which increases land degradation and desertification of a wide area (Prince 2002; Fadhil 2013; Wessels et al. 2004). Soil properties are the key to the biological activity and productivity. Heat waves, high soil temperatures and low water content could cause long-lasting changes in soil quality. Several studies have shown that soil degradation affects agronomic drought through the reduction of soil ability to keep and release water for the plant, as Reich and Eswaran (2004), Bot and Benites (2005) and Lal (2009) reported that the soil capacity for water retention is related to soil's biological, chemical and physical properties. Soil Salinization is generally caused by natural and agricultural factors. Climate change is one of the natural factors and consequently, could be an agricultural factor which causes soil Salinization. Rainfall shortages lead to use improper irrigation water and methods which magnify soil salinization risk in the future.

Organic matter (especially its humus fractions) is one of the most important factors to evaluate soil quality in semi-arid regions. The soil humic acid represents the major and more stable fraction of the soil organic matter, since it improves the absorption of nutrients by plants and soil microorganisms, has a positive effect on dynamics of N and P in soil, and favors the formation of soil aggregates (Hayes and Swift 2011). The results of Hueso et al. (2012) study on the effect of severe drought stress on the characteristics of semi-arid soil humic acids showed that drought exerts some measurable effect on the composition and properties of humic acid. Same study proved that the addition of the organic amendment improves of the soil resistant to drought conditions.

In drought conditions, lower levels of soil moisture affect the behavior of microbes that decompose organic matter in the soil, which is made of waxes and oils from fungi, and decomposed plant materials. That could increase soil water repellency to a maximum value; this soil will not be wettable again even when critical soil water content point reached (Goebel et al. 2011).

A drought is primarily driven by unbalanced precipitation and evaporation, depending on its status the soil will swell and shrink with soil moisture changes that can cause serious damage in soil subsidence. According to Wuest et al. (2011) and Corti et al. (2009) soil subsidence increases infrequency and severity with climate change, they defined it as a hidden risk of climate change and soil should be a key objective of any climate adaptation strategy.

For all these reasons, we can conclude that, to minimize the impact of drought, soil need to capture much of the water for future plant use that can be achieved through a strategy for drought management. Well-aggregated soil, high content of organic matter, and eliminating tillage are keys to drought-proofing a soil.

13.2.3 Drought Monitoring

Drought monitoring is done to predict the occurrence and severity of drought. This information is very important to identify drought impacts and water supply trends. Intensity, duration, and spatial coverage are the most distinguishing features of drought. Drought intensity refers to the amount of precipitation departure from normal (WMO 2006). Duration is based on the climate regime; it can quickly develop in some climate regimes. The third feature depends on the intensity and duration of a drought event. There are many tools to identify drought characteristics, the choice is subjective to: drought type, hydro-climatology of the region, the purpose of the study and the availability of data (Zhang et al. 2017; Yao et al. 2011; Hisdal et al.

2004). Thus, meteorological, vegetative as well as soil moisture status can be used to detect patterns of seasonal drought (Martiny et al. 2006).

For all these reasons, no single index is sufficient for drought monitoring, instead a combination of monitoring indices are integrated together to quantify and measure the severity of drought events.

13.2.3.1 Meteorological-Based Drought Index

Drought indices are functions of measurable hydro-meteorological variables like rainfall, temperature, and other variables. McKee et al. (1993), developed the Standardized Precipitation Index (SPI) as a tool to define and monitor drought. According to Thavorntam and Mongkolsawat (2006), the SPI is one of the most commonly used indexes to examine drought characteristics in the given region. Hayes et al. (1999) pointed out that compared with PDSI (Palmar Drought Severity Index 1965) SPI is a less complex tool, and the onset of the drought has been detected one month in advance of the PDSI. With minimum data requirements (rainfall data), SPI offers a quick, practical and simple way to quantify the impacts of precipitation in both wet and drought periods (Komuscu 1999). Different studies have indicated the effectiveness of the SPI to identify different drought types (Smakhtin and Hughes 2004; Al-Quraishi et al. 2019; Almamalachy et al. 2019; Vicente-Serrano and Opez-Moreno 2005).

The SPI has the ability to describe the impacts of drought on the availability of the different water resources through the different time scales of precipitation anomalies (Cancelliere et al. 2007). Originally, McKee et al. (1993) calculated the SPI for 3-, 6-, 12-, 24-, 48-month time scales. Precipitation anomalies on short scales can be used to detect agriculture drought (soil moisture conditions), while stream flow, groundwater, and reservoir storage are reflected by longer-term precipitation anomalies. According to Edwards and Mckee (1997) the mean SPI for the location and studied period is zero, this value is obtained by applying a suitable transformation of the long-term precipitation record from probability distribution into a normal distribution.

The SPI was formulated to provide a brief overall picture of drought, McKee et al. (1993) used the classification system shown in Table 13.1 to define dry and wet events.

As McKee et al. (1993) reported that the drought event starts when the SPI value reaches −1.0 or less and ends when it becomes positive again. Thus, the drought event has a duration identified by its beginning and the end, and intensity for each month the drought event continues. The SPI can be an excellent tool to policymakers in the monitoring and analysis of droughts. Although it is quite a modern index, it was used extensively, in addressing drought-related issues worldwide: in Turkey (Komuscu 1999); in China (Wu et al. 2001); in Poland (Labedzki 2007) and in Iran (Tabrizi et al. 2010).

13.2.3.2 Remote Sensing-Based Drought Index

In 1973, Rouse suggested the Normalized Difference Vegetation Index (NDVI) as an indicator of vegetation health and density (Rouse et al. 1974). Mathematically, NDVI is defined as:

$$NDVI = \frac{(NIR - RED)}{(NIR + RED)}$$

where, NIR and RED are the reflectances in the near infrared and red bands.

The main reasons behind choosing these two bands are: they are most affected by the absorption of chlorophyll in leafy green vegetation, and the contrast between soil and vegetation is at a maximum in red and near-infrared bands. Contamination of data from different sources (clouds and other land surfaces reflections) considerably reduces NDVI. Although, the technique of compositing data for 7 days and numerous algorithms correction of noise due to different sources are developed, data can still be contaminated (Gutman 1991; Kogan and Sullivan 1993; Kogan 1995, 1997).

The NDVI value varies between -1 and $+1$, rock and bare soils have a similar reflectance in both red and near-infrared bands and result in NDVI near zero, vegetated land have values which range from 0.1 to 0.7 while values lesser than 0.1 indicating no vegetation. Clouds, water and snow yield negative values due to larger visible reflectance than of near-infrared reflectance.

The NDVI has become the most important tool for monitoring and detecting drought impacts on agriculture (Singh et al. 2003; Yagci and Deng 2014; Fern et al. 2018). Since climate is a key factor affecting vegetation conditions, the NDVI has been widely used at regional and global scales to identify weather impacts on crop growth conditions and yields (Jain et al. 2009). Each region has its own characteristics, which should be accounted when using vegetation indices. That means, different vegetation indices have a different suitability for different use (Xue and Su 2017). Fadhil (2011) pointed to the useful of utilizing the NDVI to detect drought impacts in the Kurdistan region of Iraq. The study showed a significant decrease in the vegetative cover by 56.7% and a decline in the soil/vegetation moisture by 29.9%.

Table 13.1 Classification of drought based on the SPI index

SPI	Classification
2 or more	Extremely wet
1.5 to 1.99	Very wet
1 to 1.49	Moderately wet
0.99 to -0.99	Near normal
-1 to -1.49	Moderately dry
-1.5 to -1.99	Severely dry
-2 and less	Extremely dry

Ji and Peter (2003) focused on the relationship between NDVI and SPI, whereas their results showed that the 3-month SPI time scale is the best way for determining drought severity and has the highest correlation to the NDVI. Based on the positive and significant correlation between the NDVI and rainfall anomalies, several studies concluded that the NDVI was the most common form of vegetation index can be used effectively in drought early warning system (Martiny et al. 2006; Murthy et al. 2009; Quiring and Ganesh 2010).

13.2.4 Remote Sensing and GIS: Their Relation to Drought Issues

The mitigation of drought impact requires rapid and continuous real-time data. Remote sensing technology represents an excellent tool to collect data in digital form rapidly and repetitively at various levels (global and regional levels). The space technology has outstanding possibilities to provide baseline data on natural resources, soil degradation, climate change, and another important area of concern. In recent years, the development in space technology to address drought issues (drought detection, monitoring, and assessment) have been dealt with the current, before, during, and after-situation of a drought event. According to Kogan (1990) drought can be detected 4–6 weeks earlier than before, and its impact can be diagnosed far in advance of the most critical stage of plant growth (harvest stage). As a slow onset disaster, drought can lead to socio-economic instability especially in developing countries by reducing crop production levels and setback in the agricultural sector. For that reason, there is a need for monitoring and reporting of economical and environmental impacts of drought in vulnerable areas.

Vegetation is the first feature can be affected by drought; as a result, remote sensing indices have been developed for the quantification of drought based on brightness values of the land cover types. Many of the vegetation indices are introduced using ratios of visible, near-infrared, and mid-infrared portions of the electromagnetic spectrum (Tucker 1979; Fadhil 2011; Fadhil 2009; Yan et al. 1998). The newly developed indices from remote sensing data powered the historical drought indices (meteorological indices) by providing a comprehensive view of regions with a spatial resolution from a few hundred meters to few kilometers.

13.2.5 Moderate Resolution Imaging Spectroradiometer (MODIS)

The Moderate Resolution Imaging Spectroradiometer (MODIS) is the primary sensor for monitoring the terrestrial ecosystems for the NASA Earth Observing System (EOS) program (Justice and Townshend 2002). MODIS is the key instrument aboard the Terra and Aqua satellites. The instrument was integrated on the Terra; it was successfully launched on December 1999. The second instrument was integrated on the Aqua and launched on May 2002. Both Terra MODIS and Aqua MODIS are viewing the entire earth surface every one to two days, and so weather events are much less of an obstacle. MODIS sensor acquires data in 36 spectral bands, in the variable resolution of 250–1000 m, in a narrow bandwidth and 12 bits. Land products of MODIS are produced at various temporal resolutions based on the instrument's orbital cycle: daily, 8-days, 16-days, Monthly and yearly. MODIS has been extensively used in drought studies, as it monitors earth surface continuously, freely accessible and furthermore it has broadly recognized around the world (Persendt 2009).

In recent years, the space techniques have been used widely to provide a comprehensive view on drought situation. Chopra (2006) pointed that, the RS and GIS can be effectively used for monitoring and assessing the food grain production and he stressed upon the use of RS and GIS in the field of drought risk evaluation. According to Jain et al. (2010) the integrated analysis of ground measured data and satellite data has a great potential in drought monitoring. In general, remote sensing and GIS have played a key role in studying different types of drought and obtained the risk map for the area facing drought, thereby management plans can be formulated by the government authorities to cope with the disastrous effects.

13.2.6 Drought Mitigation: Options and Implementation

Ensuring water security worldwide faced a number of challenges: population growth, urbanization, common and intense drought, and the expected climate change. Drought long-term risk is directly associated with the consequences. The level of consequences depends on the level of protection. In other words, the consequences level is enclosed in an envelope between the non-protected and well- protected system. The components of drought mitigation to reduce long-term drought risk are the following: prediction, monitoring, and impact assessment (Wilhite 2009).

Kutson et al. (1998) designed a guide to identify a step-by-step process to reduce drought-related impacts before occurrence of the drought. The guide consists of sex steps: step 1 begins with making sure that the right people are supplied with adequate data. Both steps 2 and 3 focus on identifying drought-related impacts relevant to the user's location and activity. Step 4 demonstrates the environmental, economic, and social causes of the identified impacts. Finally, steps 5 and 6 bring together all of the previous information to identify feasible, cost-effective, and equitable actions that can be taken to reduce drought-related impacts and risk. A risk assessment approach is assisted by multi-criteria methods used by Tsakiris (2010) for selecting the optimal strategy to face drought and water scarcity in future. The options to face drought are grouped by Tsakiris and Spiliotis (2007) into three categories: demand reduction measures, system improvements, and emergency water supplies. They pointed that, drought is the cause of creating water shortages. Thus, the water shortage in each sector reflected the vulnerability of each system and considered as a key to determine water demand for improving the system under drought risk. They also pointed to the importance of public input and consideration because the burden of implementing these plans can be undertaken by regional organizations.

13.3   Materials and Methods

13.3.1   Study Area

The Mosul, Kirkuk, and Salah Al-Din governorates are located on longitude 41 to 43 and latitude 34 to 36. The region shares its borders with Syria in the west, Kurdistan region in the north, Diyala in the east, and Baghdad and Anbar in the south (Fig. 13.1). Mosul is the Iraq's third largest city located in northern Iraq, its area is 37,323 km^2 (8.6% of Iraq). Agriculture is a key component of Mosul economy, particularly cereal production (IAU Report 2010). Kirkuk is one of the ancient provinces in Iraq with significant geographic location linking between central and northern Iraq, it is situated in the northeast of Iraq. The governorate's area is 9,679 km^2 (PRT/USAID/RTI 2007). Salah ad-Din is located north of Baghdad, its area is 24,075 km^2. Salah ad-Din's population is one of the most rural in Iraq. Agriculture provides 36% of the jobs in governorate (IAU Report 2010).

The climate of the study area is classified as semi-arid, subtropical, Mediterranean climate, which is characterized by a hot dry summer and cold winter. Most of the rain falls in winter and spring (October through May). There is no rainfall during summer

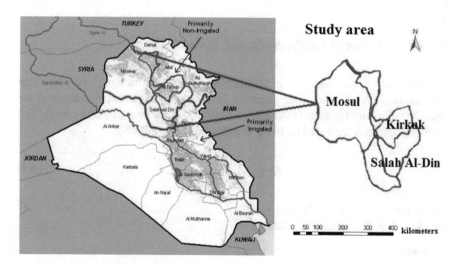

Fig. 13.1 Location and extent of the study area

Table 13.2 Average monthly climatic data for the period from 1980 to 2010 taken from meteorological stations in the study area

Governorate	Jan	Feb	Mar	Apr	May	June	July	Aug	Sep	Oct	Nov	Dec
	Precipitation (mm)											
Mosul	61.5	54.3	58.9	31.1	16.2	1.2	0.1	0.0	0.6	16.9	40.7	57.3
Kirkuk	61.9	56.3	43.8	36.0	11.7	0.4	0.3	0.1	0.7	15.2	43.1	51.9
Salah Al-Din	33.0	29.7	28.1	17.0	9.7	0.2	0.0	0.0	0.7	9.3	25.6	32.7
Min. temperature (°C)												
Mosul	3.4	4.6	8.0	12.9	18.3	23.7	27.0	26.6	22.1	16.6	9.3	4.9
Kirkuk	4.9	6.3	9.8	14.9	21.0	25.6	28.2	27.8	24.1	18.6	11.2	6.4
Salah Al-Din	4.2	5.8	9.7	15.3	20.9	25.1	27.9	27.3	23.1	17.8	10.3	5.8
Max. temperature (°C)												
Mosul	11.5	13.7	18.4	24.5	31.6	37.5	41.3	40.8	36.0	29.2	20.1	14.2
Kirkuk	14.1	16.1	20.9	27.0	34.3	40.3	43.3	42.9	38.2	31.6	22.7	16.5
Salah Al-Din	14.6	17.2	22.5	28.9	35.6	40.7	43.9	43.6	39.5	32.6	23.3	16.6

(the period from June to September). The climatic data for the period from 1980 to 2010 of Mosul governorate are taken from 7 meteorological stations: Mosul, Sinjar, Rabiha, Telafr, Telabta, Al-Baaj and Mkhmoor. Kirkuk and Tuz stations provided climatic data for Kirkuk and Sammara, Baiji and Tikrit stations provided climatic data for Salah Al-Din (Table 13.2).

13.3.2 Standardized Precipitation Index (SPI)

Drought indices are functions of measurable hydro-meteorological variables like rainfall, temperature, and other variables. Generally, SPI is a transformation of the monthly precipitation into standardized normal distribution (Z-distribution). The basic calculations are based on using the equations (Borg 2009):

$$SPI_{i,k} = \frac{(W - \overline{W})}{S_{wk}} \tag{13.1}$$

where i = 1, 2, ... (Hydrological year)
and K = 1, 2, 3, 4 (Reference period)

where

$$w_{ik} = \ln(R_{ik}) \tag{13.2}$$

$i = 1, 2, \ldots$ and $k = 1, 2, 3, 4$

$\overline{Wk}$: is the average of the Wi, k's

S_{wk} : is the standard deviation of the Wi, k's

$P_{i,j}$: is the value of Rainfall in mm for the jth month of the ith year

$i = 1, 2, \ldots$ Hydrological year

$j = 1, 2, \ldots$ (i.e. 3, 6, 9 or 12 months depending on the value of k)

and $k = 1, 2, 3,$ or 4 (reference period number)

$$R_{ik} = \sum_{j=1}^{3k} P_{ij} \tag{13.3}$$

Monthly rainfall data for the period from 1980 to 2010 in seven rainfall stations in Mosul, two rainfall stations in Kirkuk, and three rainfall stations in Salah Al-Din were used as an input in SPI program. The Drinc (Drought indices calculator) software was developed at the laboratory of Reclamation Works and Water Resources Management of National Technical University of Athens (Tsakiris et al. 2007). A free downloadable version of the software can be found at http://www.ewra.net/drinc.

For all stations, input data were created and SPI values had been calculated for each station on the four reference periods: October to December, October to March, October to May, and October to September (3, 6 and 12-month time scales). Considering to 3-months reference period the SPI had been calculated on three periods per year October to December, January to March, and April to June. The database was produced for SPI results from 1980 to 2010 using the equation shown above.

13.3.3 Pre-processing of Satellite Images

The satellite data that were used were derived from the MODIS sensor. Tile (h21, v5), 16-day composites images at 250 m resolution were directly downloadable from the USGS data center (http://glovis.usgs.gov). For each governorate ten 16-day composites (7–23 April) were downloaded for the period from 2000 to 2010. All of the images were re-projected into GIS format (TIFF) using MODIS reprojection tool from USGS. 16-day compositing periods minimize cloud contamination, hence most images downloaded had less than 10% cloud cover, and hence the data is cloud free.

Most of the image pre-processing was already on the downloaded MODIS; hence images were only geo-referencing. Through the using of rectification algorithm for the image, the column and row coordinates of the image could be fitted to the geodetic datum WGS84. Once all ground control points were compiled, error checking was used to gauge the efficiency of points used.

Table 13.3 Drought severity classification based on the NDVI anomalies

Percent of NDVI anomalies	Class
0 to −10%	Slight drought
−10 to −20%	Moderately drought
−20 to −30%	Severe drought
above −30%	Very severe drought

Source Chopra (2006)

13.3.4 NDVI Calculations

All of the downloaded MODIS images were used to calculate:

$$NDVI = \frac{(NIR - RED)}{(NIR + RED)} \tag{13.4}$$

$$Average\ NDVIy = \frac{(NDVI_1 + NDVI_2 + \cdots + NDVI_{10})}{10} \tag{13.5}$$

where NDVIy is NDVI across study period, NDVI1 (7–23 April 2000), NDVI2 (7–23 April 2001), NDVI10 (7–23 April 2010).

$$NDVI\ Anomaly\ i = \frac{(NDVIi - Mean\ NDVI)}{(Mean\ NDVI)} * 100 \tag{13.6}$$

where NDVIi = NDVI in the year and Mean NDVI = long term mean NDVI in the period study.

The resulting NDVI anomaly percentage assigned to the respective grid cell was reclassified into five drought severity classes in Table 13.3 (Chopra 2006).

Finally, correlation analysis was performed between the values, annual rainfall, and SPI (3, 6 and 12-months time scales).

13.3.5 Software Used

The study used software such as ESRI ArcGIS 10.0, and ENVI 4.0, which were the main image processing and analyzing software. Tableau 7.0 and Microsoft Excel were the main software used for arrangement data.

13.4   Result and Discussion

13.4.1   Drought Classification Based on Meteorological Data

The overall meteorological drought vulnerability in the three governorates (Mosul, Kirkuk and Sala Al-Din), have been assessed by modernizing historical occurrences of droughts at varying time steps and drought categories with the SPI approach. The basic idea is that this can be a guide to the decision makers in Iraq to develop strategies of water resources management in the context of drought. The present study computed the SPI values over multiple time scales (3, 6 and 12-months) in order to achieve different goals in the study. Since, a no rainfall event is usually experienced during the summer season at the study area; results of the dry season (June, July, August, and September) are not given. Figure 13.2 shows the drought severity by 12-months SPI values, several drought events occurred in all studied governorates (Mosul, Kirkuk and Salah Al-Din) and the year 2007–08 was the most drought year. The results detected 11 drought years for Mosul, 14 years for Kirkuk, and 15 years for Salah Al-Din. Based on Table 13.2, the years can be classified as one extremely dry (2007–08) for all three governorates, one severely dry (1998–99) for Mosul and Salah Al-Din and two severely dry (1983–84 and 2008–09) for Kirkuk, two moderately dry (1999–00 and 2008–09) for Mosul and Kirkuk, (1986–87 and 1999–00) for Salah Al-Din, while 7, 8, and 11 years were normal, but with negative values for Mosul, Kirkuk, and Salah Al-Din, respectively.

Regarding the 6-months SPI Fig. 13.3, analyses showed that the rainy season 2007–08 was extremely dry from October to March for all studied governorates, and severely dry year from June to September for Kirkuk and Salah Al-Din. Analyses exhibited that the 6-month SPI increases the number of drought events in which every drought category was observed even if it is small. This can be clearly observed from Fig. 13.3, that the years with negative values increased from 11 to 18 for Mosul, from 14 to 18 for Kirkuk, and from 15 to 19 years for Salah Al-Din. Drought category based on SPI-6 represents the true picture of the drought situation during the years (1983–84, 1986–87, 1988–89, 1990–91, and 2002–03). Six months (especially for October–March) drought events may provide useful information for agriculturalist and agricultural risk analyzers. The SPI value through the end of March showed critical drought conditions; 12, 15, and 14 drought events were experienced out of 30 years for Mosul, Kirkuk and Salah Al-Din, respectively. Thus, using several time scales could be useful when applying the drought management plan; because sometimes a drought event could be detected using a specified time step, but the same event could not be detected using another time scale.

Figure 13.4 demonstrates the SPI values based on the 3-months timescale. According to SPI values of October, November and December, year 2007–08 was classified as extreme dry year in Mosul and Kirkuk with SPI values of −2.86 and −2.18 respectively. In Salah Al-Din the same year was classified as severely dry with SPI value of −1.71, while year 2005–06 was extremely dry year with SPI value of − 2.44. The years with negative values but reach up to a value of −1 were considered

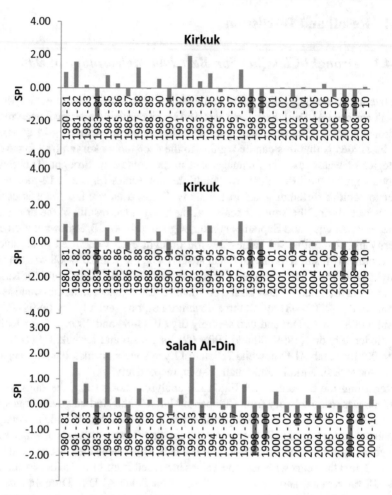

Fig. 13.2 Drought severity showed by 12-months (Oct–Sept) SPI for Mosul, Kirkuk, and Salah Al-Din

as near normal years. Figure 13.4 shows clear decline in the number of the normal year. 3 months SPI for months January, February, and March indicated that the year 2008–09 was considered as an extremely dry year for Mosul and Kirkuk. Regarding Salah Al-Din, the years (1986–87, 1999–00, and 2002–03) were classified as extremely dry years. The SPI value through the end of June (April, May, and Jun) showed only one extremely drought event with SPI −2.90 for the year 2007−08 in Mosul. While the same year was a severely dry year in Kirkuk and Salah Al-Din.

Figure 13.5 illustrates the accumulated magnitude of the negative values of SPI based on 3, 6, and 12 months time scales. The negative values of the SPI have been aggregated, based on the above mentioned timescales. The main objective of this figure is to apply a guide for the selection of the driest years (an actual drought

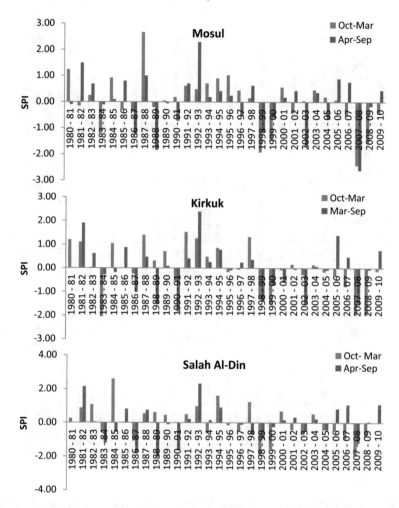

Fig. 13.3 Drought severity showed by 6-months SPI values for Mosul, Kirkuk, and Salah Al-Din

event) and to reflect the cumulative rainfall anomalies based on 3 and 6 months time scales of SPI analyses. According to this figure, several years (such as 1983–84, 1986–87, 1990–91, 1998–99, 2002–03, and 2007–08) exposed to extreme drought in the study area. It is clear from the same figure that, these years are well reflected by SPI 3 and SPI 6, but not by SPI 12. The SPI 12 seems not to reflect well the quick development of agricultural drought especially in the rain-fed based cropping system, where the winter grains are grown during the wet months from November through April. The amount and timing of rainfall are critical and cause wide variations in production. The results are in good agreement with the results obtained by Bussay et al. (1998), Szalai and Szinell (2000) and Boken et al. (2005) who assessed that

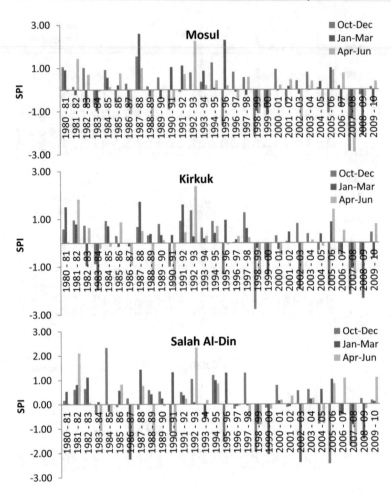

Fig. 13.4 Drought severity showed by 3-months SPI values for Mosul, Kirkuk, and Salah Al-Din

agricultural drought (described by soil moisture content) followed best the SPI on a scale of 2–3 months.

The percentages of the wet and dry years were classified based on SPI 12 for the period (1980–2010). Each percentage was obtained by taking the ratio of event occurrence in any time scale to the total event occurrence in the same timescale and event category. The total percentage of wet years were 10% for Mosul, 16% for Kirkuk, and 13% for Salah Al-Din, while the percentage of the dry years were 16% for Mosul, 17% for Kirkuk, and 13% for Salah Al-Din. These percentages could be used as a guide to evaluate the drought risk in a given region. A drought emergency is declared when there is a reasonable probability. Due to this fact, it is uncommon that drought in any one year causes major hardship. It is the sequence of low rainfalls that creates difficulties. For example, in Mosul, Kirkuk and Salah Al-Din, the drought

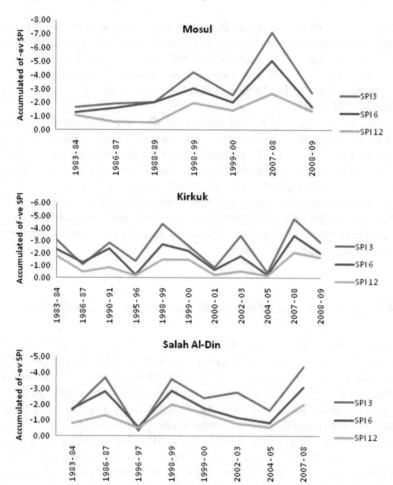

Fig. 13.5 Magnitude of drought based on the accumulated negative values of the SPI at different time scales

of 1999–00 was really caused by the low rainfalls in the preceding year, while the drought of 2007–08 was the result of the low rainfalls from 1998 to 2007. The SPI therefore needs more attention to include drought sequences as recommended by Agnew (2000).

The main advantages of applying the SPI approach is to compare the rainfall of two areas with different rainfall characteristic in terms of how badly they are experiencing drought conditions. It can clearly be noticed from the results (Fig. 13.4) that the year 2005–06 was an extremely dry year in Salah Al-Din with SPI 3 (Oct–Dec) value reaches −2.44, while the same year was moderately dry with value −1.09 in Mosul, and near normal with value −0.32 in Kirkuk. Through SPI, we can identify drought-sensitive regions in order to know the region which suffered extremely dry or other intensity of drought event. With the help of SPI-3 it is also possible to detect the onset of drought event, that is mean which region suffer from early drought and which suffer from late one. Salah Al-Din suffered from early extreme drought in 2005–06 and late severe drought, whereas Mosul suffered from early and late drought in the year 2007–08. Rain-fed wheat and barley constitute 40–50% of the total wheat and barley production (FAO 2003).

In view of this, not only rainfall amount but also rainfall distribution is a critical factor for crop production. Under the condition of climate change increasing of drought certainly affects the agricultural production in arid and semi-arid regions such as Iraq. So, more droughts mean more limitation related to water resources and therefore, this affects agricultural yields as well as more tend to accelerate desertification and increase the frequency of sandstorms in the future. In the situation of increasing of the drought, more attention should be focused on precipitation trends and drought distribution during the agricultural development decision making process and drought hazards mitigation.

13.4.2 Drought Classification Based on Remote Sensing Data

Figures 13.6, 13.7 and 13.8 show NDVI maps in April for the period (2000–2010) for Mosul, Kirkuk, and Salah Al-Din. Figures give you an idea about the amount and distribution of vegetations in studied governorate maps which reflect the vegetation situation and greenness. The highest average NDVI values observed were 0.33 and 0.20 in 2000/01 for Mosul and Salah Al-Din respectively and 0.39 in 2002/03 for Kirkuk. The lowest NDVI values observed were 0.10, 0.19, and 0.13 in 2007/08 for Mosul, Kirkuk, and Salah Al-Din respectively. NDVI has been found to be lowest due to the extremely unfavorable weather. The year 2007/08 was a year of drought with precipitation levels much below the normal. Maximum vegetation is developed in years with the optimal weather; since such weather encourages efficient use of ecosystem resources (like an increase in the rate of soil nutrition uptake). In contrast, the lake of water in drought years reduces the amount of soil nutrition uptake which suppresses vegetation growth through a reduction in ecosystem resources. The pattern of change of NDVI (Figs. 13.6–13.8) are generally representing the seasonal fluctuation between the early rainy season (October, November, and December) and the main rainy season (January, February, March, and April). Season 2007/08 started and ended with very unfavorable conditions making the planting of crops difficult and reducing harvest.

During seasons 2000/01, 2002/03 started with near-normal conditions the three governorates received good rains during the first half of the rainy season (October–December) and consequently led to increasing vegetation density. Season 2008/09 reflects low dense vegetation illustrated by the dry conditions from 2007/08 season continue into season 2008/09. Figures 13.6 and 13.7 reflect the fluctuation of NDVI values in relative to the changes in local weather conditions in Mosul and Kirkuk, while Fig. 13.8 clearly shows a little stable NDVI pattern in Salah

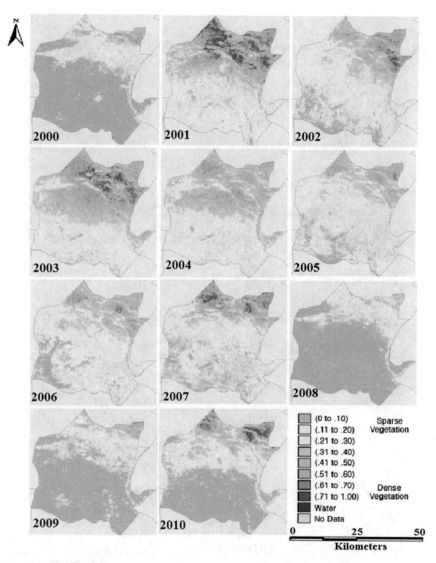

Fig. 13.6 NDVI maps in (07–23 April) for the period from 2000 to 2010 for Mosul

Al-Din. This can be attributed to irrigation farming throughout the year and are not influenced much by variability in rainfall.

The results of the NDVI analysis show the sensitivity of NDVI to detect drought events and seasonal vegetation dynamics across all seasons. These results are in good agreements with many studies of NDVI time series to exam the response of vegetation vigor to climatic variations of variables like rainfall to understand causes

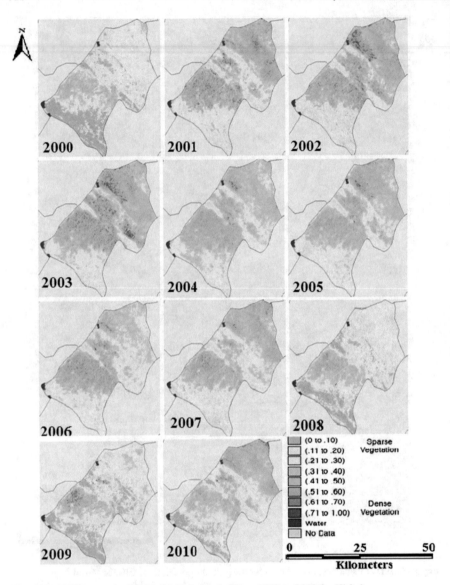

Fig. 13.7 NDVI maps in (07–23 April) for period from 2000 to 2010 for Kirkuk

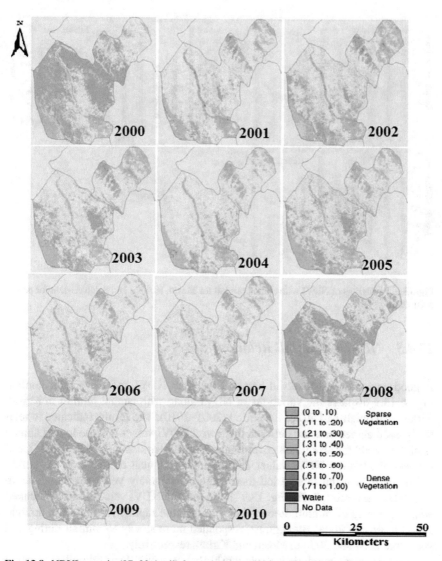

Fig. 13.8 NDVI maps in (07–23 April) for period from 2000 to 2010 for Salah Al-Din

of observed changes in vegetation greenness (Fensholt and Proud 2012; Fensholt and Rasmussen 2011; Eastman et al. 2009).

The results obtained in their study reflect the possibility of using satellite images index (like NDVI) to monitor drought under crop development and measure the degree of stress of crop cover due to water stress conditions.

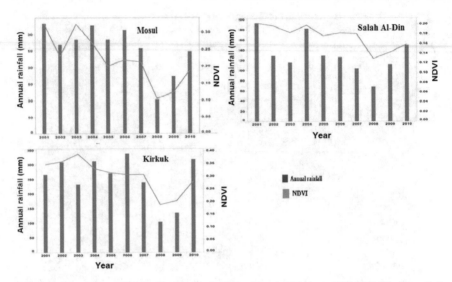

Fig. 13.9 Variations of NDVI with annual rainfall for Mosul, Kirkuk, and Salah Al-Din for period 2001–2010

13.4.3 NDVI—Rainfall Relationship

A good agreement is observed between the peak NDVI and the annual rainfall. Figure 13.9 displays the variations of NDVI along with annual rainfall in Mosul, Kirkuk and Salah Al-Din for the period (2001–2010). The results indicated to relatively good agreements are between the lowest NDVI values and the lowest annual rainfall in 2008 for all studied governorates (Mosul, Kirkuk and Salah Al-Din), and between the highest NDVI values and the highest annual rainfall over the period (2001–2010). During this period there were considerably year-year variations in both NDVI and rainfall. As Fig. 13.9 shows that the NDVI values sharply declined with decreasing of rainfall amount (responded more rapidly to rainfall in drought years) mostly in 2008. By contrast, responded more slowly to rainfall during wet years like 2004 and 2006 in Mosul and Kirkuk respectively.

In order to study the statistical relationship between NDVI and rainfall, correlation analysis was performed between the values of vegetation index (NDVI) and annual rainfall for whole studied governorates during the period (2001–2010). Figure 13.10 exhibits the results of such correlations. The results of the correlation were significant in all governorates with values 0.83, 0.70, and 0.72 for Mosul, Kirkuk, and Salah Al-Din respectively. Results showed that the correlation coefficient for Mosul was higher than both for Kirkuk and Salah Al-Din, these results explain clearly that the agricultural pattern in Mosul mostly depends on rainfall compared with an agricultural pattern in Kirkuk and Salah Al-Din. Overall, this finding confirms the results

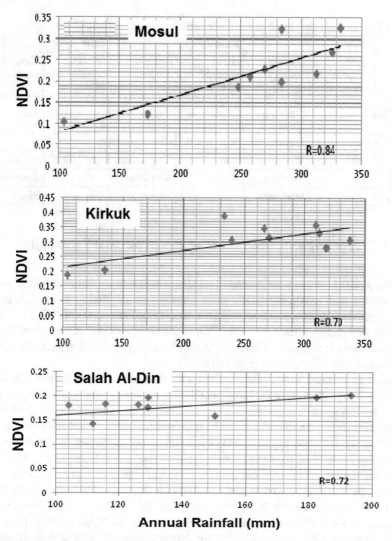

Fig. 13.10 Correlation coefficient (r) between NDVI and annual rainfall for Mosul, Kirkuk, and Salah Al-Din for period 2000–2010

presented by Ghulam et al. (2007) and Qin et al. (2008) for arid and semi-arid conditions that found throughout Iraq, the NDVI can be used as simple and cost-efficient drought indices to monitor crop development stages in relation to rainfall during growing season.

Table 13.4 Drought class based on NDVI anomalies for the period (2001–2010)

Year	Mosul		Kirkuk		Salah Al-Din	
	NDVI anomaly (%)	Drought category	NDVI anomaly (%)	Drought category	NDVI anomaly (%)	Drought category
2005	−9	Slight	4	–	1	–
2006	0	Slight	1	–	4	–
2007	−3	Slight	2	–	3	–
2008	−53	Very severe	−38	Very severe	−26	Severe
2009	−44	Very severe	−33	Very severe	− 19	Moderately
2010	−15	Moderately	−7	Slight	−9	Slight

13.4.4   Agricultural Drought Risk Based on NDVI Anomaly

Monthly NDVI images were generated for all growing season (October through July). The month with maximum NDVI value (April) was selected to assess vegetation anomalies during the specific growing season (year). Time series of NDVI anomaly used to detect agricultural drought (Chopra 2006; Murad and Saiful Islam 2011). The threshold values used in this chapter to classify agricultural drought risk using NDVI anomalies that presented in Table 13.4.

Table 13.4 shows the NDVI anomalies for the study area (Mosul, Kirkuk, and Salah Al-Din). It is evident from the figure that during the low rainfall years NDVI values were also low and two major years 2008 and 2009 are classified as very severe/severe drought for three governorates. It can be observed that during 2008–2009, the whole of the study area had negative NDVI anomalies corresponding negative rainfall anomalies Table 13.4. Vegetation shows a good response and NDVI values with rainfall amount, which confirmed that rainfall has a great impact on vegetation conditions. Further, the same figure shows that Mosul had a slight drought during 2005–2007 and moderately drought in 2010, where the years 2005, 2006, and 2007 were near normal years and slight drought in 2010 for Kirkuk, and Salah Al-Din. Thus this study shows that the NDVI value highly depends on average rainfall condition in a region. These results are in agreement with those reported by Li et al. (2002) in China; Chopra (2006) in India and Shahabfar and Eitzinger (2011) in Iran. They found that the NDVI has a positive relation to rainfall and NDVI is good indicator of vegetation vigor. Further NDVI is an excellent pointer to assess agricultural drought from the point of agricultural production and its linkages, so that the drought risk can be marked out taking into consideration the crop yield and describing an area at risk.

13.5   Conclusions

The main conclusions of this study are the followings:

- The SPI results obtained over multiple time scales (3, 6 and 12-months) provided a seasonal estimation of precipitation and indicated that the short-time scales and its application could be closely related to soil moisture.
- The SPI-3 detected the onset of drought events (which region suffered from early drought and which suffered from late one), and the inter-annual precipitation variations during planting period.
- The advantage of using the SPI index is that it provides a spatial and temporal representation of historic drought.
- The findings of the NDVI analysis confirmed the sensitivity of this index to detect drought events and seasonal vegetation dynamics across all seasons.
- The MODIS—derived NDVI can be used as a simple and cost-efficient tool for drought monitoring in Iraq.
- The combination of the SPI and the NDVI indices helps with defining the drought risk-prone area, where the whole of the study area had negative NDVI anomalies corresponding negative SPI values.
- Using the combination of indices can help decision makers to make more informed decisions.

13.6   Recommendations

Despite the fact that the present study deals with satellite and meteorological variables to detect and monitor drought, there are still some of issues which could not be handled due to unavailability of data. The following recommendations are suggested to take up in further researches:

- A detailed study in terms of soil, water availability and crops data combined with climatic data to give map for area that faced high drought risk.
- Agricultural drought severity varies spatially, it is recommended that future studies can build up on the specific region information to get early warning information that helps to reduce the regionally impact of drought.
- Shifting researches towards correct evaluation of the extent of damage are caused by water stress on soil quality.
- Introduction of new management strategies that can be adjusted according to changing trends in rainfall and the spatial patterns of drought frequency.
- More attention should be focused toward archiving and reporting data and information to simulate Iraq drought trends and monitor future drought events.
- Attention must be paid to support of the local population and increase social awareness to cope with drought disaster.

References

Agnew CT (2000) Using the SPI to identify drought. National Drought Mitigation Center, vol 12, no 1

Almamalachy YS, Al-Quraishi AMF, Moradkhani H (2019) Agricultural drought monitoring over Iraq utilizing MODIS products. In: Al-Quraishi AMF, Negm AM (eds) Environmental Remote Sensing and GIS in Iraq. Springer Water

Al-Quraishi AMF, Qader SH, Wu W (2019) Drought monitoring using spectral and meteorological based indices combination: a case study in Sulaimaniyah, Kurdistan region of Iraq. In: Al-Quraishi AMF, Negm AM (eds) Environmental Remote Sensing and GIS in Iraq. Springer Water

Boken VK, Cracknell AP, Heathcote RL (eds) (2005) Monitoring and predicting agricultural drought. Oxford University Press, Oxford, 472 pp

Borg DS (2009) An application of drought indices in Malta, case study. Eur Water (EWRA) 25:25–38

Bot A, Benites J (2005) The importance of soil organic matter: key to drought-resistant soil and sustained food production. FAO Soils Bull 80:94

Bussay A, Szinell C, Hayes M, Svoboda M (1998) Monitoring drought in Hungary using the standardized precipitation index. Annales Geophysicae, Supplement 11 to vol 16, Abstract Book of 23rd EGS General Assembly, C450, Nice, France Apr 1998

Cancelliere A, Mauro GD, Bonaccorso B, Rossi G (2007) Drought forecasting using the standardized precipitation index. Water Resour Manage 21(5):17–22

Chopra P (2006) Drought risk assessment using remote sensing and GIS, a case study in Gujarat, M.Sc. thesis, Dept. of Geo-information Science and Earth Observation, ITC, Netherlands

Corti T, Muccione V, Köllner-Heck P, Bresch D, Seneviratne SI (2009) Simulating past droughts and associated building damages in France. Hydrol Earth Syst Sci 13(9):1739–1747

Dracup JA, Lee KS, Paulson JEG (1980) On the definition of drought. Water Resour Res 16:297–302

Eastman JR, Sangermano F, Ghimire B, Zhu HL, Chen H, Neeti N et al (2009) Seasonal trend analysis of image time series. Int J Remote Sens 30:2721–2726

Edwards DC, McKee TB (1997) Characteristics of 20th century drought in the United States at multiple timescales. Colorado State University: Fort Collins. Climatology Report No. 97-2

Fadhil AM (2009) Land degradation detection using geo-information technology for some sites in Iraq. J Al-Nahrain Univ Sci 12(3):94–108

Fadhil AM (2011) Drought mapping using geoinformation technology for some sites in the Iraqi Kurdistan region. Int J Digital Earth 4(3):239–257

Fadhil AM (2013) Sand dunes monitoring using remote sensing and GIS techniques for some sites in Iraq. In: Proceedings SPIE 8762, PIAGENG 2013: intelligent information, control, and communication technology for agricultural engineering, p 876206. https://doi.org/10.1117/12.2019735

FAO (2003) Special Report: FAO Iraq crop production, 16 Jan 2003

Fensholt R, Proud SR (2012) Evaluation of earth observation based on long term vegetation trends—comparing GIMMS and MODIS global NDVI time series. Remote Sens Environ 119:131–147

Fensholt R, Rasmussen K (2011) Analysis of trends in the Sahelian 'rain-use efficiency' using GIMMS NDVI, RFE and GPCP rainfall data. Remote Sens Environ 115:438–451

Fern RR, Elliott AF, Andrea B, Michael LM (2018) Suitability of NDVI and OSAVI as estimators of green biomass and coverage in a semi-arid rangeland. Ecol Ind 94:16–21

Ghulam A, Qin Q, Kusky T, Li ZL (2008) A re-examination of perpendicular drought indices. Int J Remote Sens 29:6037–6044

Ghulam A, Qin Q, Teyip T, Li ZL (2007) Modified perpendicular drought index (MPDI): a real-time drought monitoring method. ISPRS J Photogrammetry Remote Sens 62:150–164

Goebel M-O, Bachmann J, Reichstein M et al (2011) Soil water repellency and its implications for organic matter decomposition—is there a link to extreme climatic events? Glob Change Biol 17:2640–2656

Gutman GG (1991) Vegetation indices from AVHRR data: an update and future prospects. Remote Sens Environ 35:121–136

Haheen A, Baig MA (2011) Drought severity assessment in Arid Area of Thal Doab using remote sensing and GIS. Int J Water Resour Arid Environ 1(2):92–101

Hayes MHB, Swift RS (2011) Progress towards understanding aspects of composition and structure of humic substances. HIS, University of Adelaide, Australia

Hayes M, Svoboda M, Wilhite D, Vanyarkho O (1999) Monitoring the 1996 drought using the standardized precipitation index. Bull Am Meteor Soc 80(3):429–438

Heidorn KC (2007) Drought: the silent disaster. The Weather Doctor's Weather Almanac [Online]. URL: http://www.islandnet.com/~see/weather/almanac/arc2007/alm07. Accessed 12 June 2011

Hisdal HB, Clausen A, Gustard E, Peters, Tallaksen LM (2004) Events definitions and indices. In: Tallaksen LM, Van Lanen HAJ (eds) Hydrological drought—processes and estimation methods for streamflow and groundwater, developments in water science, vol 48, Elsevier Science B.V., Amsterdam, pp 139–198

Hueso S, Brunetti G, Senesi N, Farrag K, Hernandez T, Garcia C (2012) Semi-arid soils submitted to severe drought stress: influence on humic acid characteristics in organic-amended soils. J Soil Sediments 12:503–512

Huschke RE (ed) (1959) Glossary of meteorology. American Meteorological Society, Boston, 638 p

IAU Report (2010) Inter-Agency Information and Analysis Unit (October 2010), Water in Iraq fact sheet

Jain SK, Keshri R, Goswami A, Sarkar A, Chaudhry A (2009) Identification of drought-vulnerable areas using NOAA-AVHRR data. Int J Remote Sens 30(10):2653–2668

Jain SK, Keshri R, Goswami A, Sarkar A (2010) Application of meteorological and vegetation indices for evaluation of drought impact: a case study for Rajasthan, India. Nat Hazards 54:643–656

Ji L, Peter AJ (2003) Assessing vegetation response to drought in the northern Great Plains using vegetation and drought indices. Remote Sens Environ 87(1):85–98

Justice CO, Townshend J (2002) Special issue on the moderate resolution imaging spectroradiometer (MODIS): a new generation of land surface monitoring. Remote Sens Environ 83(1):1–2

Kogan FN (1990) Remote sensing of weather change impact on vegetation index in non-homogenous areas. Int J Remote Sens 11:1405–1421

Kogan FN (1995) Application of vegetation index and brightness temperature for drought detection. Adv Space Res 15:91–100

Kogan FN (1997) Global drought watch from space. Bull Am Meteorol Soc 78:621–636

Kogan FN, Sullivan J (1993) Development of global drought-watch system using NOAA/AVHRR data. Adv Space Res 13:219–222

Komuscu AU (1999) Using the SPI to analyze spatial and temporal patterns of drought in Turkey. Drought Netw News 11:7–13

Kutson C, Hayes M, Philips T (1998) How to reduce drought risk. Western Drought Coordination Council

Labedzki L (2007) Estimation of local drought frequency in central Poland using the Standardized Precipitation Index SPIy. Irrig Drain 56:67–77. www.interscience.wiely.com

Lal R (2009) Soil degradation as a reason for inadequate human nutrition. Food Sec 1:45–57

Legesse G (2010) Agricultural drought assessment using remote sensing and GIS techniques, M.Sc. thesis, Addis Ababa University

Li B, Tao V et al (2002) Relations between AVHRR NDVI and ecoclimatic parameters in China. Int J Remote Sens 23(5):989–999

Martiny N, Camberlin P, Richard Y, Philippon N (2006) Compared regimes of NDVI and rainfall in semiarid regions of Africa. Int J Remote Sens 27:5201–5223

McKee TB, Doesken NJ, Kleist J (1993) The relationship of drought frequency and duration to time scales. In: Proceedings of the 8th conference on applied climatology. American Meteorological Society, Boston, pp 179–184

Murad H, Saiful Islam AKM (2011) Drought assessment using remote sensing and GIS in north-west region of Bangladesh. In: 3rd international conference on water and flood management (ICWFM)

Murthy CS, Seshasai MVR, Chandrasekar K, Roy PS (2009) Spatial and temporal responses of different crop-growing environments to agricultural drought: a study in Haryana state, India using NOAA AVHRR data. Int J Remote Sens 30:2897–2914

NDMC (2006) Defining drought: overview. National Drought Mitigation Center, University of Nebraska–Lincoln

Okorie FC (2003) Studies on drought in the sub-Saharan Region of Nigeria using remote sensing and precipitation data, JNCASR-Costed Fellowship Programme, University of Hyderabad, India, January–April

Persendt FC (2009) Drought risk analysis using remote sensing and GIS in the Oshikoto—Region of Namibia, M.Sc. thesis, Dept. of Environment and Development, University of KwaZulu-Natal, Pietermaritzburg

Prince SD (2002) Spatial and temporal scales of measurement of desertification//Stafford-Smith M, Reynolds JF, Global desertification: do humans create deserts? Dahlem University, Berlin

PRT/USAID/RTI (2007) Strategic planning of Kirkuk province. Approved by provincial council of Kirkuk province, the future vision of Kirkuk province, government of Iraq, Sept 2011

Qin Q, Ghulam A, Zhu L, Wang L, LI J, Nan P (2008) Evaluation of MODIS derived perpendicular drought index for estimation of surface dryness over northwestern China. Int J Remote Sens 29:1983–1995

Quiring SM, Ganesh S (2010) Evaluating the utility of the Vegetation Condition Index (VCI) for monitoring meteorological drought in Texas. Agric For Meteorol 150:330–339

Rathore MS (2004) State level analysis of drought policies and impacts in Rajasthan, India, Working paper 93, Drought Series. Paper 6, International Water Management Institute, India

Reich P, Eswaran H (2004) Soil and trouble. Science 304:1614–1615

Rosenberg NJ (1979) Drought in the Great Plains—research on impacts and strategies. In: Proceedings of the workshop on research in Great Plains drought management strategies, University of Nebraska, Lincoln, Water Resources Publications, Littleton, Colorado, 225p, 26–28 Mar 1979

Rouse JW et al (1974) Monitoring the vernal advancement and retrogradation (greenwave effect) of natural vegetation. NASA/GSFCT Type III Final report. Greenbelt, MD, USA

Shahabfar A, Eitzinger J (2011) Agricultural drought monitoring in semi-arid and arid areas using MODIS data. J Agric Sci 149:403–414

Singh RP, Roy S, Kogan F (2003) Vegetation and temperature condition indices from NOAA-AVHRR data for drought monitoring over India. Int J Remote Sens 24(22):4393–4402

Smakhtin VU, Hughes DA (2004) Review, automated estimation and analyses of drought indices in South Asia, IWMI Working Paper N 83—Drought Series Paper N 1. IWMI: Colombo, p 24

Szalai S, Szinell C (2000) Comparison of two drought indices for drought monitoring in Hungary—a case study. In: Vogt JV, Somma F (eds) Drought and drought mitigation in Europe. Kluwer, Dordrecht, pp 161–166

Tabrizi AA, Khalili D, Kamgar-Haghighi AA, Zand-Parsa SH (2010) Utilization of time-based meteorological droughts to investigate occurrence of streamflow droughts. Water Resour Manage 24:4287–4306

Tallaksen LM, Van Lanen HAJ (eds) (2004) Hydrological drought—processes and estimation methods for streamflow and groundwater. In: Developments in water sciences, vol 48, Elsevier B.V., Amsterdam

Tate EL, Gustard A (2000) Drought definition: a hydrological perspective. In: Vogt JV, Somma F (eds) Drought and drought mitigation in Europe. Kluwer Academic Publishers, Dordrecht, pp 23–48

Thavorntam W, Mongkolsawat C (2006) Drought assessment and mitigation through GIS and remote sensing

Tsakiris G (2010) Towards an adaptive preparedness framework for facing drought and water shortage. Economic of drought and drought preparedness in climate change context. Options Méditerranéennes, no 95

Tsakiris G, Tigkas D, Vangelis H, Pangalou D (2007) Regional drought identification and assessment—case study in Crete. In: Rossi G, Vega T, Bonaccorso B (eds) Methods and tools for drought analysis and management. Springer, The Netherlands, pp 169–191

Tsakiris G, Spiliotis M (2011) Planning against long term water scarcity: a fuzzy multicriteria approach. Water Resour Manage 25(4):1103–1129

Tucker CJ (1979) Red and photographic infrared linear combinations for monitoring vegetation. Remote Sens Environ 8:127–150

United Nations Development Programme UNDP (2010) Drought impact assessment, recovery and mitigation framework and regional project design in Kurdistan region (KR), Dec 2010 URL: http://www.gisdevelopment.net/application/natural_hazards/drought/ma

USDA foreign agricultural service (2008) IRAQ: drought reduces 2008/09 winter grain production. Commodity Intelligence Report, 9 May 2008

Vicente-Serrano SM, López-Moreno J I (2005) Hydrological response to different time scales of climatological drought: an evaluation of the standardized precipitation index in a Mountainous Mediterranean Basin. Hydrol Earth Syst Sci Discussions 2:1221–1246

Wessels KJ, Prince SD, Frost PE, Van Zyl D (2004) Assessing the effects of human-induced land degradation in the former homelands of northern South Africa with a 1 km AVHRR NDVI time-series. Remote Sens Environ 5(9):47–67

Wilhite DA (2009) Defining drought: the challenges for early warning systems. In: Inter-regional workshop on indices and early warning systems for drought, Nebraska-USA

Wilhite DA, Glanz MH (1985) Understanding the drought phenomenon: the role of definitions. Water Int 10(3):111–120

World Meteorological Organization WMO (2006) Drought monitoring and early warning: concepts, progress and future challenges. WMO-No. 1006

Wu H, Hayes MJ, Welss AHUQ (2001) An evaluation the standardized precipitation index, the china-z index and the statistical z-score. Int J Climatol 21:745–758

Wuest M, Bresch D, Corti T (2011) The hidden risks of climate change: an increase in property damage from soil subsidence in Europe. Swiss Reinsurance Company Ltd. Zurich, Switzerland. http://www.fao.org/ag/agl/aglw/aquastat/main/index.stm

Xue J, Su B (2017) Significant remote sensing vegetation indices: a review of developments and applications. Hindawi J Sens 2017 (Article ID 135691)

Yagci AL, Deng M (2014) The influence of land cover-related changes on the NDVI-based satellite agricultural drought indices. In: 2014 IEEE international geoscience and remote sensing symposium (IGRASS). IEEE, pp 2054–2057

Yan W, Yang L, Merchant JM (1998) An assessment of AVHRR/NDVI ecoclimatological relations in Nebraska, USA. Int J Remote Sens 18(10):2161–2180

Yunjun Y, Qin Q, Fadhil AM, Li Y, Zhao S, Liu S, Sui X, Dong H (2011) Evaluation of EDI derived from the exponential evapotranspiration model for monitoring China's surface drought. Environ Earth Sci 63(2):425–436

Yevjevich V, Hall WA, Salas JD (1977) Drought research needs. In: Proceedings of the conference on drought research needs, Colorado State University, Fort Collins, Colorado, 276 p, 12–15 Dec 1977

Zhang Z, Kang H, Yao Y, Fadhil AM, Zhang Y, Jia K (2017) Spatial and decadal variations in satellite-based terrestrial evapotranspiration and drought over inner Mongolia autonomous region of China during 1982–2009. J Earth Syst Sci 126(8)

Chapter 14
Remote Sensing and GIS for Dust Storm Studies in Iraq

Ali Darvishi Boloorani, Najmeh Neysani Samany, Saham Mirzaei,
Hossein Ali Bahrami and Seyed Kazem Alavipanah

Abstract Dust storms occur when strong winds combine with erodible soils that are mainly located in semi-arid and arid regions of the world. Dust storms are multi-driver phenomena with harmful impacts on health, agriculture, infrastructure, transportation, environment, and economy. These natural-anthropogenic phenomena have been especially influential in south west regions of Asia during two last decades. The phenomenon has major adverse ramifications for countries within the region including Iran, Iraq, Syria, Kuwait, and other countries of the Persian Gulf. During the past two decades, escalations in the impacts of the phenomenon have deemed it necessary to find a suitable solution to tackle these disturbing events. These phenomena are very complex and must be studied in an interdisciplinary framework. Remote Sensing and Geographical Information Systems (RS&GIS), as the inter/multi-disciplinary field of science and technologies, which in combination with other fields of science and technology provide the required knowledge and information for reliable decision making, can be employed to combat against dust storms. In this chapter the problem-solving cases of RS&GIS applications in dust storm studies are mentioned as exemplars to demonstrate the spatial-temporal distributions of sources, causes, and atmospheric and wind patterns of dust storms as well as their environmental circumstances (air-soil-vegetation-water) in Iraq.

A. Darvishi Boloorani (✉) · N. N. Samany · S. Mirzaei · S. K. Alavipanah
Department of Remote Sensing and GIS, Faculty of Geography, University of Tehran, Tehran, Iran
e-mail: ali.darvishi@ut.ac.ir

N. N. Samany
e-mail: nneysani@ut.ac.ir

S. Mirzaei
e-mail: smirzaei67@ut.ac.ir

S. K. Alavipanah
e-mail: salavipa@ut.ac.ir

H. A. Bahrami
Department of Soil Sciences, Faculty of Agriculture, Tarbiat Modares University, Tehran, Iran
e-mail: bahramih@modares.ac.ir

© Springer Nature Switzerland AG 2020
A. M. F. Al-Quraishi and A. M. Negm (eds.),
Environmental Remote Sensing and GIS in Iraq, Springer Water,
https://doi.org/10.1007/978-3-030-21344-2_14

333

Keywords Remote sensing · Geographical information systems · Dust storms ·
Iraq

14.1 Introduction

Dust storms have been the focal point of a vast body of projects and studies during the last decades. Almost all instances of such studies explored the identification of the sources, causes, emission mechanisms, health and socioeconomic impacts, and warning/monitoring as their main objectives. Owing to the fact that dust storms emerge as the final result of numerous composite factors, various methods, data and disciplines in Earth-sciences have been proposed and implemented in this regard. In addition to environmental factors such as land surface vegetation cover, soil, and water resources, dust storms and high speed winds may also originate from special atmospheric conditions. One of the main objectives of this chapter, in aiming to study the dominant factors causing dust storms is to scrutinize RS&GIS, at which point, due to the multi-disciplinary, multi-factor, multi-cause and transnational nature of dust storms, it is necessary to consider several scientific disciplines, simultaneously. Therefore, other fields of study, such as soil sciences and meteorology are also accounted for. Furthermore, this chapter will discuss the potentials of RS&GIS in providing applicable solutions for planning to combat against dust storms. In this regard, the RS&GIS-based problem-solving approaches are discussed. RS&GIS, as interdisciplinary fields of Geo/Earth sciences and technologies deal with spatial and non-spatial data/information as a means for solving Earth-related problems. Geoinformatics acquires data from surveys, field works, remote sensing, GPS, photogrammetry and cartographic maps to synthesize/analyze spatial and non-spatial data/information for a variety of applications including but not limited to sand and dust storms, land-environment degradation, climate change, land use land cover mapping, land use planning, public health, environmental modeling, meteorology and climate change, oceanography, agriculture and urban planning. Combating dust storms is particularly interconnected with planning strategies for sustainable land management. This kind of planning requires access to environmental/natural data such as information on soils, water, climate, evapotranspiration, vegetation cover, land use land cover, socio-economic, and etc., which are captured, stored, manipulated, analyzed, managed, and visualized in the framework of Geoinformatics.

14.2 Remote Sensing for Dust Storm Studies

Dust storm studies are commonly divided into three phases: sources identification and characterization, modeling the transport mechanism, forecasting and deposition measurements. In principle, the applications of remote sensing imagery, with various kinds of images, i.e. optical, thermal, RADAR, LIDAR, etc., are considered adequate

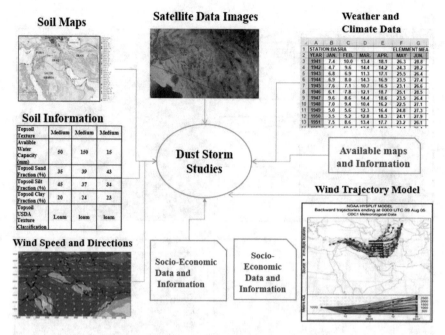

Fig. 14.1 Dust storm study as interdisciplinary procedure

for studying different phases of dust storms. These data and their corresponding image analysis techniques must be applied in the framework of spatial-temporal modeling. In lines with the objective of this chapter a methodology framework for dust storm studies (Fig. 14.1) has been developed. The proposed methodology uses different data/information sources: satellite imagery, soil, weather, vegetation, and socio-economic.

14.2.1 Dust Sources Identification

Darvishi Boloorani et al. (2012) developed a dust sources identification and clustering procedure using Geoinformatics and ancillary data (Fig. 14.2a). They used a spatial clustering approach to discriminate dust sources with similar spatial-temporal patterns of dust emission, wherein wind patterns, soil condition, vegetation temporal patterns, and water situation were used as the main metrics for separating different clusters. Their work has carried out on a regional scale and discriminated three main clusters as well as one secondary cluster of dust sources in west Asia (Fig. 14.2b). Clusters one and two and the secondary are mainly located in Iraq, with some parts in Syria and Jordan, while cluster three is located in Saudi Arabia. Each of the clusters includes several dust emission sources. Cluster one mainly consistes of dust sources

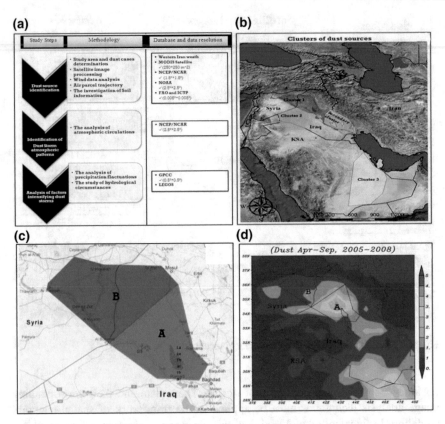

Fig. 14.2 **a** Dust storm study procedure involving different disciplines of Earth science and technologies, **b** four main dust source clusters in west Asia, **c** cluster one as an exemplary case (later presented in this chapter (due to the focus of this book just cluster number one is presented. Interested readers for other three clusters of dust sources are referred to the project report at: https://www.researchgate.net/profile/Ali_Darvishi_Boloorani2/publications)), and **d** monthly average of optical thickness between 2005 and 2008 (*source* Darvishi Boloorani et al. 2012)

between Tigris and Euphrates rivers, with certain sources located in western parts of Euphrates River. They also introduced a secondary dust cluster which did not include very active dust sources and mainly injectes dusts into dust storms originated from clusters one and two. Considering the different patterns of land surface characteristics, cluster one was divided into two northern and southern subclusters of A and B (Fig. 14.2c). Results from dust detections procedures indicated subcluster A as enveloping the most active dust sources in Iraq from 2003 to 2010. The north and north-western areas of Thrathar Lake plays a key role in the dust emission of almost all of the identified regional dust events in cluster one. Acquired maps from GES-DISC (Goddard Earth Sciences Data and Information Services Center)[1]

---

[1] https://earthdata.nasa.gov/about/daacs/daac-ges-disc.

and NASA GES-DISC Interactive Online Visualization and Analysis Infrastructure (GIOVANNI) were used to show the temporal patterns of dust storms. As shown in (Fig. 14.2d) the monthly average of optical thickness between 2005 and 2008 are indicative of the constant formation of dust in subcluster A and in the north west of Tharthar Lake.

Ginoux et al. (2012) made a global-scale map of dust sources based on MODIS Deep Blue optical depth. They used ancillary data such as land-use maps and hydrological information in order to discriminate dust sources in terms of natural and anthropogenic origins. They calculated Dust Optical Depth (DOD) from the Aerosol Optical Depth (AOD) using criteria based on size distribution and optical properties of dust particles; then discriminated the background-DOD using a threshold-DOD, i.e. background aerosols; then used the frequency of occurrences higher than the threshold-DOD to discriminate dust sources; in the next step they compared their outcomes with similar analysis results derived from Total Ozone Mapping Spectrometer (TOMS) and Aura Ozone Monitoring Instrument (OMI) satellite data; they then attributed the sources as anthropogenic or hydrological with respect to the degree of land-use and the presence of ephemeral water bodies; and finally, calculated the contributions of anthropogenic and natural sources to dust emissions using wind speed from a high resolution model. The corresponding global map discriminates 20 main clusters of dust sources in which cluster number 11, shown in Fig. 14.3, is one of the biggest clusters of dust encompassing all kinds of dust sources. The obvious point in the case of Iraq is that the corresponding main dust sources, in this map are attributed as natural non-hydrological dust sources which could be a result of drought and rising temperature in this country during last two decades. On the other hand, the hydrological dust sources are shown in the regions near to Euphrates and Tigris that can be due to the hydropolicy of the countries of this basins in regards to water resources control.

From the viewpoint of dust storm studies in west Asian regions, especially Iraq, Euphrates and Tigris (EuT) basin (Fig. 14.4) plays a significant role. Furthermore, water management throughout the entire basin must also be considered in any kind of dust storm studies in this region. As can be seen in Figs. 14.2 and 14.3 these two rivers pass one of the most active dust sources of west Asia. For instance, Kazemi et al. (2018) used a RS and GIS based model for discriminating dusts sources in EuT basin (Fig. 14.4). They used Normalized Difference Vegetation Index (NDVI), Land Surface Temperature (LST), humidity, wind speed, precipitation, soil texture, soil moisture, and soil erodibility to obtain a map of dust sources in EuT basin from 2000 to 2018. Figure 14.5 illustrates five dust sources, amongst which four are located in Iraq and are somehow associated with water usage and agricultural activities in this country.

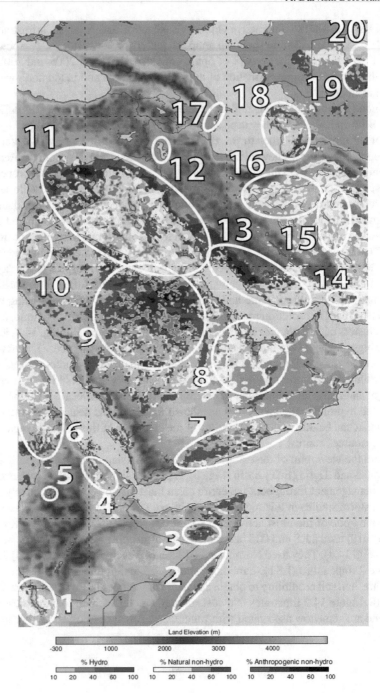

◄Fig. 14.3 Main clusters of dust in North East Africa, Middle East, and Central Asia. **1** Chalbi Desert of Kenya, **2** coastal desert of Somalia, **3** Nogal Valley of Somalia, **4** Danakil Desert of Ethiopia, **5** Lake Tana of Ethiopia, **6** Northeast Sudan, **7** Hadramawt region, **8** Empty Quarter, **9** Highlands of Saudi Arabia, **10** Jordan River Basin of Jordan, **11** Mesopotamia, **12** Urumia Lake of Iran, **13** coastal desert of Iran, **14** Hamun-i-Mashkel, **15** Dasht-e Lut Desert of Iran, **16** Dasht-e Kavir Desert of Iran, **17** Qobustan in Azerbaijan, **18** Atrek delta of Turkmenistan, **19** Turan plain of Uzbekistan **19**, and **20** Aral Sea (*source* Ginoux et al. 2012)

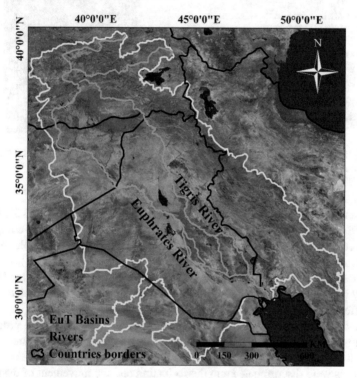

Fig. 14.4 Euphrates and Tigris rivers and basins

14.2.2 Dust Detection Using Satellite Imagery

Dust detection techniques are used to identify the unknown or unclear sources of dust. Most approaches to dust detection and time series satellite image analysis assume that when dust clouds are disconnected from their sources, their density will decrease over time, whereas in certain instances this is not the case since secondary sources intervene through the path, making dust storms wider and denser. Knowing and identifying suspended dust clouds are also convenient for investigating the affected regions on pathways and deposition areas. In order to detect dust storms, Ackerman (1997) and Yue et al. (2017) used bands 20 (3.66–11.28 μm), 31 (10.78–11.28 μm), and 32 (11.77–12.27 μm) of MODIS to extract Brightness Temperature (BT) of dust storms. The difference of brightness temperatures was used to detect dust clouds,

(a)   **(b)**

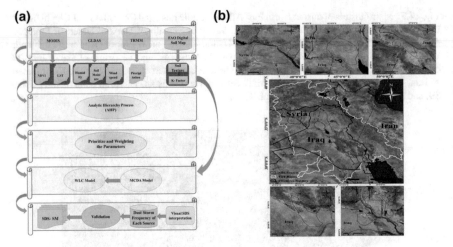

Fig. 14.5 **a** Dust storm sources identification procedure using RS&GIS and **b** the main active dust sources in EuT basins (*source* Kazemi et al. 2018)

for instance: $BTD_{32-31} = BT32 - BT31$ and $BTD_{20-31} = BT20 - BT31$, where, BTD is Brightness Temperature Difference of MODIS bands. Due to the complexity of land covers, several criteria and thresholds must be adapted to be able to detect dust storms under different conditions. For instance, Samadi et al. (2014) developed a threshold-based and somehow complex dust detection methodology for MODIS imagery named Global Dust Detection Index (GDDI) which is able to detect dusts over different land covers (Fig. 14.6).

Dust detection for dust source identification is also a widely used procedure. For example, a dust event occurred in cluster one on August 7 and entered Iran on August 8 2005. As shown in Fig. 14.7a, using GDDI MODIS-based methodology, dust storms were discriminated and the formation and the movement of dust storm from sources in cluster one were analyzed. This event was recorded on August 8 at of Dezful, Ahvaz, and Abadan (southwest of Iran) weather stations, with visibility less than 1000 m. The dense and dark appearance of the dust clouds over this region during the second day (Fig. 14.7b) demonstrates the key role of subcluster A in emission of dusts in cluster number one.

AOD is also a main practical source of information for dust detection and temporal analysis. For instance, the TOMS/Nimbus-7 UV aerosol index can be utilized for discriminating absorbing from non-absorbing aerosols for dust storm studies. As shown in Fig. 14.8 the average TOMS aerosol index values for September 2005 is clearly indicative of map dust concentrations in main global dust sources including the Sahara and West Asia.

Draxler et al. (2001) developed a model for emission of PM_{10} using threshold friction velocity, which is dependent on surface roughness. They applied the method-

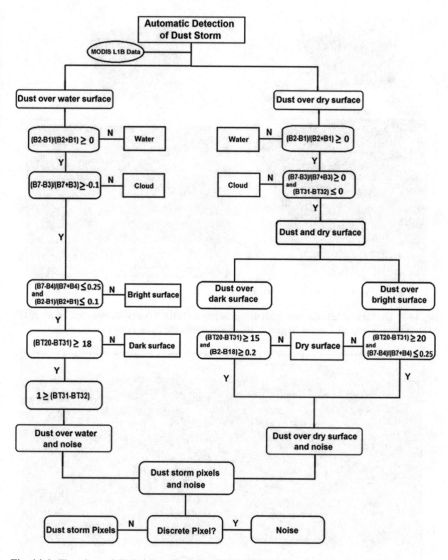

Fig. 14.6 Flowchart of Global Dust Detection Index (GDDI) (*source* Samadi et al. 2014)

ology in Kuwait, Iraq, part of Syria, Saudi Arabia, the United Arab Emirates and Oman. In that work, PM_{10} air concentrations were computed from August 1990 through August 1991. They compared the model predicted PM_{10} (Fig. 14.9a) with the TOMS/Nimbus-7 UV aerosol index (Fig. 14.9b).

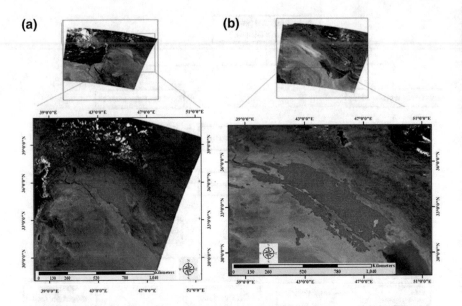

Fig. 14.7 Dust storm detection using MODIS imagery and GDDI (Darvishi Boloorani et al. 2012). **a** Dust storm in August 7 and **b** August 8. Red color shows the dusts storms

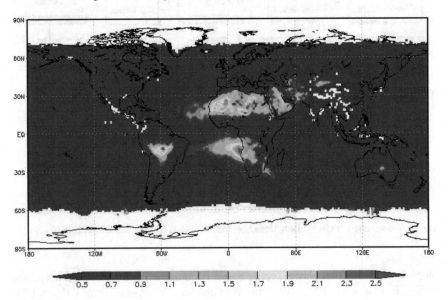

Fig. 14.8 The TOMS/Nimbus-7 UV aerosol index (average for September 2005) (Ahmed et al. 2006)

14.2.3   *Atmospheric Patterns of Dust Storms*

Ever since the launch of the Television InfraRed Observation Satellite (TIROS-1) in 1960, remote sensing instruments have appeared as very advantageous tools for procuring data on Earth-atmosphere systems. Meteorological remote sensing involves a variety of satellites carrying sensors with different functionalities. The obtained data from these satellites are used to monitor atmospheric parameters applied in meteorological and climatological studies, and the information retrieved from satellite-based sensors has greatly enhanced our understanding of the processes and dynamics within the Earth-atmosphere system (Thies and Bendix 2011). Atmospheric patterns, which are activating factors for each dust source, are powerful parameters used in dust storm studies. Wind speed and direction, pressure and temperature, and precipitation are the main information/data that are used in studies of dust formation from corresponding sources. Unquestionably, these data are quite useful in transport and deposition mechanisms of dust clouds as well. For instance, precipitation data can be collected from Global Precipitation Climatology Centre (GPCC) with 0.5° resolution, water surface level measurements can be extracted from Laboratoire d'Etudes en Géophysique et Océanographie Spatiales (LEGOS) database, and land surface temperature can be derived from satellite imagery like MODIS.

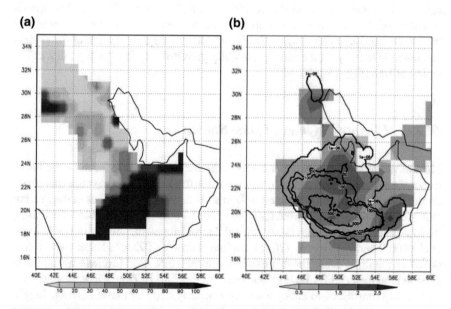

Fig. 14.9 a Percent coverage of active sand sheets and disturbed soils over the potential dust emission sources. **b** TOMS/Nimbus-7 UV aerosol index for February 4, 1991 superimposed on model calculated contours of PM_{10} concentration (mg/m³). AI positive values generally represent absorbing aerosols (dust and smoke) while negative values represent non-absorbing aerosols (*source* Draxler et al. 2001)

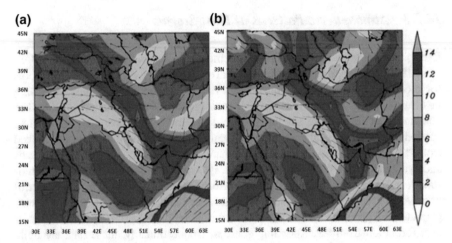

Fig. 14.10 Surface wind speed (m/s) and direction on day of the formation as well as movement of dust storms. **a** and **b** depicting observations on 7 and 8 of August 2005, respectively (Darvishi Boloorani et al. 2012)

For the case of dust storm in Iraq on August 7, 2005, high speed wind streams extended all over Iraq (Fig. 14.10a). These currents could generate dust storms in susceptible source regions in the east of Syria and Northwest of Iraq. The spatial matching of dust storms with the northwest-southeast wind direction is evident in the MODIS imagery (Fig. 14.7a). During the second day, the current of dust storm continued in its way on the same path while tremendously influencing the western region of Iran (Fig. 14.10b).

14.2.4 Climate Regimes of Dust Storms

Most of the times during warm seasons, a thermal low pressure forms in the lower troposphere over Iraq, resulting in wind streams from north west to south east (from north of Iraq to west and southwest of Iran and neighboring countries in the Persian Gulf) referred to as Shamal wind (Aurelius 2008). Furthermore, as altitude increases, the low-pressure system moves east toward Zagros Mountains. According to Zaitchik et al. (2007) and Zarrin et al. (2011), surface warming of Zagros heights was introduced as the main cause for this expansion at 850 hPa level (average height 1500 geopotential meter) (Fig. 14.11a). As shown in Fig. 14.11b Zagros low pressure system mostly transforms to anticyclone circulations at 500 hPa level (average height 5000 geopotential meter) which can be an indication of dominant high pressure over Iran plateau during warm periods of the year (Zarrin and et al. 2011).

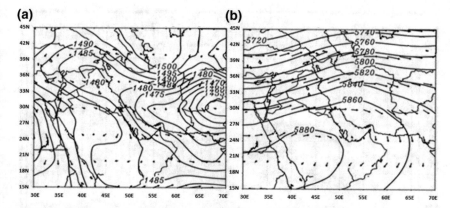

Fig. 14.11 **a** and **b** are the long-term wind current and geopotential height during warm period of a year (May to September) at 850 and 500 hPa, respectively (Darvishi Boloorani et al. 2012)

14.2.5   Dust Storm Tracking Model

NOAA HYSPLIT MODEL[2] is a simulating air parcel trajectory which can extensively be used in atmospheric transport and dispersion models. HYSPLIT (Hybrid Single Particle Lagrangian Integrated Trajectory) has evolved over more than 30 years, from estimating simplified single trajectories based on radiosonde observations to a system accounting for multiple interacting pollutants transported, dispersed, and deposited over local to global scales (https://www.arl.noaa.gov/). The HYSPLIT applications in dust storm studies is an ever-growing area, which contains dust sources identification, dust emission, and dust transport simulation. For example, Fig. 14.12 shows the backward trajectory tracking of dust storm from west of Iran initiated at the end of August 8th and reversed to 00:00 GMT in August 7th 2005. In 8 August 2005, the western parts in Iran took impacts from the dust storm. Along with the aforementioned path, a wind stream from Turkmenistan (northeast) entered the region, which did not have any effect on dust path, a fact verified by MODIS satellite imagery. This dust event was initially formed from cluster number one from northwest of Iraq and east of Syria.

14.2.6   Dust Emission (Soil Erosion) Assessment using Remote Sensing

Soil, with a structure and composition of mineral-chemical-biological materials is directly or indirectly measurable/quantifiable using remote sensing imagery. Accordingly, soil type, spatial distribution, and chemical-physical-biological characteristics

---

[2]https://www.arl.noaa.gov/.

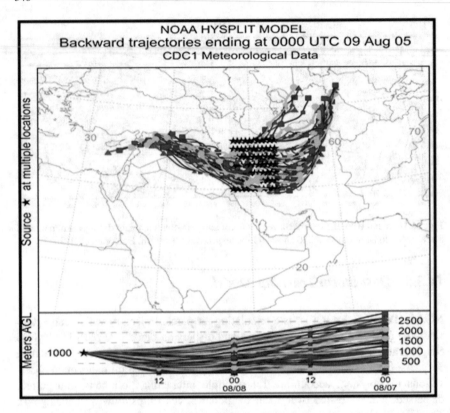

Fig. 14.12 Tracking of wind streams on 7 and 8 of August 2005 (color lines show wind path and star symbols are weather stations. Time distance between every interception lines is 12 h which totally is the interception of dust storm to end of the day (00:00 9 August 2005) from the beginning time (00:00, 7 August 2005) for 48 h). Backward trajectory of wind patterns at 9th August in Iran originating in the north east and the north west on 7 and 8th of August and headed towards Iran

can be surveyed accurately and scientifically in a synoptic view via satellite imagery. Soil data/information on dust source including: soil degradation rates, soil loss, soil drying, and soil-quality have been demonstrated in numerous studies utilizing passive and active remote sensors. Type, texture, and surface coverage of soil are extremely important factors in dust studies (Shoshany et al. 2013). In this regard, soil/surface conditions and characteristics are considered in term of soil erosion which explore direct and indirect indicators from optical, RADAR, LIDAR remote sensing data in combination with ancillary data like DEM and slope. Figure 14.13 shows remote sensing based soil loss modeling using direct and indirect indicators (Shoshany et al. 2013). Coordination of information on the Environment (CORINE) methodology is an exemplary model for combining different data sources in the framework of GIS for different models of soil erosion and risk analysis (Fig. 14.14). Synoptic and repeatable coverage of remote sensing imagery in combination with spatiotemporal modeling of GIS have provided useful tools for dust source identification and

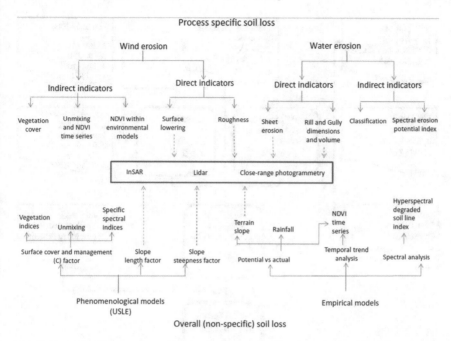

Fig. 14.13 Schematic typology of indicators and models used for monitoring soil loss (*source* Shoshany et al. 2013 (it should be noted that the "wind erosion" in the left part of the diagram is related to dust storms))

early warning of future trends in soil degradation. Figure 14.15 is an example of an integrated modeling which employs spectral behavior of surface, temporal patterns of changes, and phonological models of land vegetation. This model is strongly related to the accuracy of soil information and the feasibility of remote sensing data (Shoshany et al. 2013).

- **Soil map** in cluster one indicates a mixture coverage of gypsum and, in some parts, lime (calcium carbonate) in both subclusters of A and B. These two minerals are mainly in lower parts of the rivers of dry regions which produce gypsisol and calcareous soils with sparse vegetation covers (calcisols) (Fig. 14.16). Most of these minerals are transported to plain regions from seasonal currents of rivers, and emerge on the surface again and form a hard exterior cover of soil due to extreme dryness and evaporation that hindering the growth of vegetation (IUSS & ISRIC & FAO 2007). Soil salinity map of Iraq is another important issue which substantiates the use of satellite imagery. As shown in Fig. 14.17 a very enormous area of Iraq is covered by salty soils ranging from low to high in intensity. These circumstances are indicators of high potentiality for soil erosion, especially wind erosion. As mentioned by Muhaimeed (2017) there are no dialed map of soils in Iraq. There are two projects for acquiring soil maps using remote sensing and GIS modeling (Fig. 14.18). These soil maps are very important for modeling soil

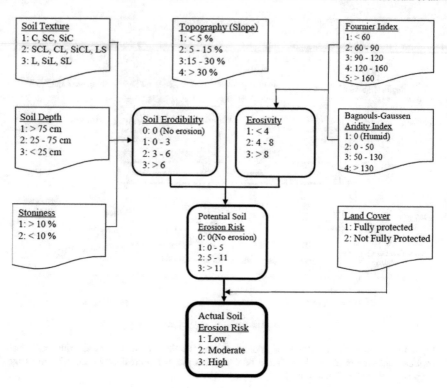

Fig. 14.14 Flow diagram of CORINE methodology (*source* Corine 1992, cited by Ekpenyong 2013)

erosion and dust storm studies. They are also required for finding suitable solutions for dust sources contorting policies.

- **Soil texture** in cluster one has a virtually homogeneous surface, where soil texture consists of 25–23% clay (Fig. 14.20a), 37–54% silt (Fig. 14.19b), and 35–40% sand (Fig. 14.19c). Since about 80% of soil texture in cluster one is clay and silt, which creates the require circumstances for severe erosion of surface soils, therefore dust storms are highly expected in this cluster of sources.

- **Soil available water capacity (field capacity)** is another principal factor that must be considered in dust storm studies. Field capacity map is attainable from remote sensing data/imagery. Field capacity values in subcluster A are between 50 and 75 mm. Following mountainous areas in north of Iraq, which have a naturally low soil available water capacity, area A has the lowest level of water capacity. Whereas in subcluster B, these values range from 75 and to nearly 125 mm (Fig. 14.20). In fact, difference in maintaining humidity can affect the resistance against drought events. Therefore, it seems that area A is more vulnerable to drought and more susceptible to dust storm formation.

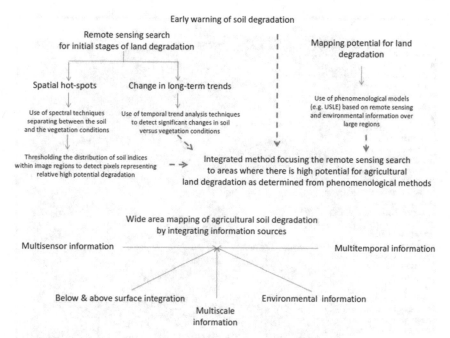

Fig. 14.15 Approaches for early warning and wide-area mapping of soil degradation (*source* Shoshany et al. 2013)

14.2.7 Land Use/Land Cover Mapping Using Remote Sensing

Land use/land cover map and vegetation status are another key parameter in dust storm source studies. As shown in Fig. 14.21, a large area of cluster one is categorized into arid and semiarid classes. Undoubtedly, such characteristics of soil surface with low vegetation cover cannot provide sufficient protection for the region against soil erosion. This information can be obtained in different scales with a variety of details using several kinds of satellite imagery. For instance, Faraj (2013) used Landsat and Rapid Eye imagery for land use/land cover maps and change detection for the Iraqi province of Sulaimaniyah. Jahantab et al. (2018) developed a new method for calibration of land-vegetation degradation modeling in Neinava region, Iraq. They used Landsat satellite images from 1985, 2001, and 2014. They also applied spatial modeling in GIS for defining criteria such as distances from rivers, from lakes, from agricultural areas, from roads, and from residential areas, height, slope, distance from Qanats, distance from wells, erosion, type of climate, and NDVI index for land-vegetation degradation modeling.

Richardson and Hussain (2006) made an investigation into the marshlands of south Iraq. The results were expressive of poor water quality, the presence of toxic materials, and high saline soil conditions in the study area. They made a composite view of the region using Landsat images from 1973, 2000 and 2005. As shown in Fig. 14.22,

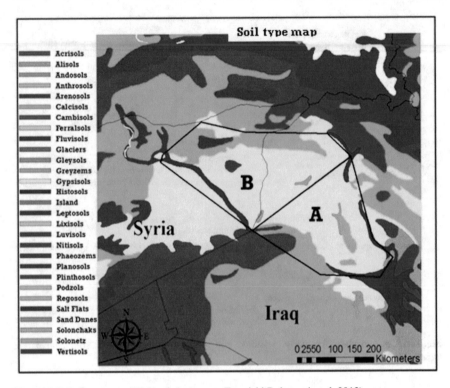

Fig. 14.16 Soil type map (FAO) of cluster one (Darvishi Boloorani et al. 2012)

dense marsh vegetation has dramatically decreased during three decades from 1973 to 2005. These patterns in vegetation cover changes are the result of decreases in available water resources in the marshes. Therefore, agricultural activities will also be distorted, turning this area into one of the main sources of dust storms in west Asia.

14.2.8 Morphological Unite Maps

Morphological units are also central for modeling dust emission from sources. Wind erosion features (like nebkha and dunes) cover other parts of the dust sources of cluster one. In such regions, severe drought can intensify the erosion process which in turn manifests itself in the decrease vegetation cover, reduction of surface humidity of soil and no persistence against wind erosion. The wind erosion in Aljazeera is not precedent and has only emerged during the last geological period of Holocene. As a matter of fact, in previous time spans, main activities were only carried out by water currents, while in the mentioned epoch (2500–3000 years ago) surface

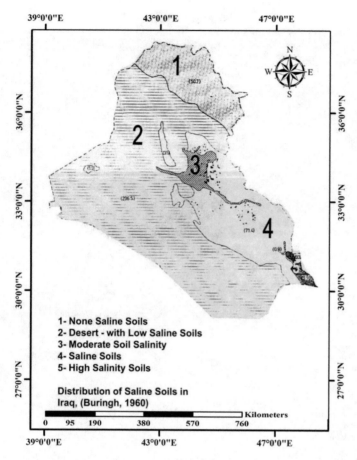

Fig. 14.17 Distribution of saline soils in Iraq (*source* Buringh 1960 cited by Muhaimeed 2017)

deposits were eroded by strong wind currents and transported to other regions in the form of sand and dust storms. Consequently, soil, morphological, and geological data certify the potentiality of cluster one, especially A, in the formation of massive dust storms. Analysis of surface coverage in the region shows that most areas in cluster one (Fig. 14.23), especially in southern regions are parts of the Aljazeera desert. These areas are simultaneously faced with wind and water erosion. The existence of temporary rivers (dried rivers) carrying particles play a major role in deposition of salt and sandy alluviums that make the area highly prone to erosion and dust storms (Ma'ala 2009).

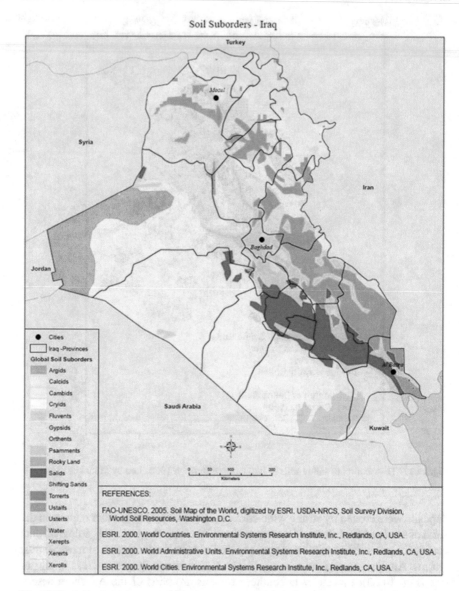

Fig. 14.18 Soil suborders maps of Iraq (*source* FAO-UNESCO, 2005, cited by Muhaimeed 2017)

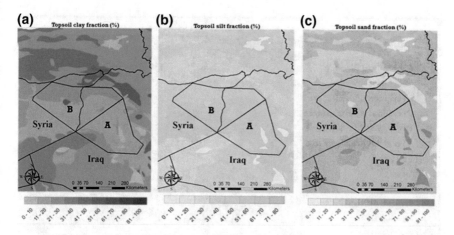

Fig. 14.19 The texture of soil in cluster one. Percentage of clay (**a**), silty (**b**), and sandy soils (**c**) (for more discrimination, the values are exaggerated by adding 10) (Darvishi Boloorani et al. 2012)

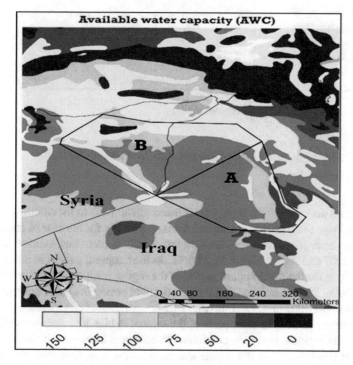

Fig. 14.20 Soil water capacity in subclusters **A** and **B** (Darvishi Boloorani et al. 2012)

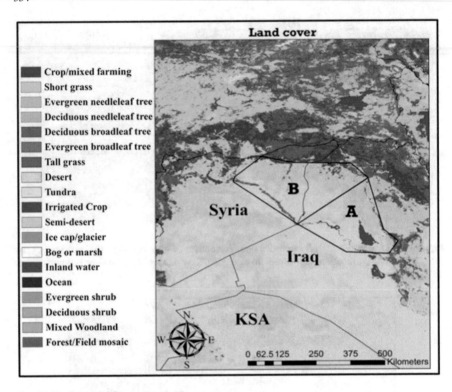

Fig. 14.21 Type of soil coverage in cluster one

14.2.9 Drought and Dust Storms Studies using Remote Sensing

Drought and dust storms are tightly interlinked phenomena in the Middle East. The results of Trigo et al. (2010) in Iraq and Syria, especially for the case of cluster one, showed an emergence of severe drought and its consequent environmental effects between 2007 and 2008.[3] Figure 14.24 presents the temporal pattern of precipitation in cluster one for the precipitation period of the region from end of November 2001 to April 2009. Results indicate that the first drought period in cluster one took place from November 2001 to April 2002 (Fig. 14.24a). Amount of precipitation decreased by a maximum of 20% in most parts of the region. While 40% decreases observed in the certain regions. However, conditions in 2003 and 2004 were different and the amount of precipitation in the region was close to the long time average and even higher (Fig. 14.24b and c). Therefore, in 2002 and 2004, no widespread dust storm was recorded in west of Iran. High correlation between precipitation changes and dust storms in this cluster in 2006 hint at more clues in Fig. 14.24e. During this year

---

[3]This drought has been continued in 2009 and 2010.

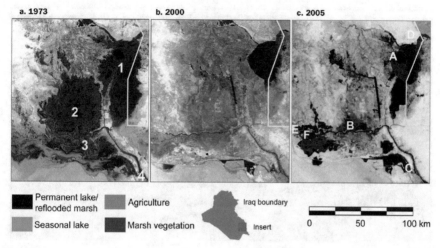

Fig. 14.22 Land use/cover map changes in marshlands of Iraq from 1973 to 2005 (*source* Richardson and Hussain (2006). **a** A composite view of the Mesopotamian marshlands from a mosaic of four Landsat 1 images and two false-color, near-infrared images, 1973–1976. Dense marsh vegetation (mainly *Phragmites australis*) appears in dark red, seasonal lakes in blue, agriculture in pink, and permanent lakes in black. The red elongated patches along riverbanks are date palms. The three main marsh areas are Al-Hawizeh, Central, and Al-Hammar, labeled 1, 2, and 3, respectively. The city of Basrah is located at number 4. Modified from Richardson et al. (2005). **b** A Landsat 7 Enhanced Thematic Mapper mosaic taken in 2000. Most of the drained marshes appear as grayish-brown patches, indicating dead marsh vegetation or low desert shrubs and dry ground. The white and gray patches indicate bare lands with no vegetation and, in some areas, salt shells covering the bottoms of former lakes. By 2000, 85% of the 8,926 square kilometers (km^2) of permanent marsh in 1973 marshlands had been destroyed. Only 3% of the Central marsh and 14.5% of the Al-Hammar remained. A canal known as the Glory River (shown as a straight line across the top and down the east side of the Central marsh), constructed in 1993, completely dried up the Central marsh by stopping water inflow from the Tigris river. The largest expanse (approximately 1,025 km^2) of remaining natural marsh, the Al-Hawizeh, near the Iranian border, is shown in dark red. Modified from Richardson et al. (2005). **c** False-color image of the remaining Mesopotamian marshlands, taken 2 September 2005, shows in black the areas newly reflooded since the war. Reflooded areas adjacent to Al-Hawizeh, the western area of Al-Hammar, and waterways in the northern and southern parts of the Central marsh are also visible in black. Al-Hawizeh (called Hawr Al-Azim in Iran) is the best remaining natural marsh in the region. It straddles the Iraq–Iran border (yellow line). During a field survey in February 2004, we discovered an Iranian dike under construction that is now nearly completed and will traverse directly through the Al-Hawizeh marsh, along the Iraq–Iran border, and, as a result, will significantly reduce the water input from the Karkheh and Karun rivers to the marsh. The ecological effects of this massive water diversion are unknown, but it will significantly affect the last remaining natural marsh system in the region. Sampling sites: A, Al-Hawizeh; B, Central; C, Al-Hammar; D, Al-Sanaf; E, Abu Zarag; F, Suq Al-Shuyukh. MODIS satellite image courtesy of the United Nations Environment Programme, Iraq Marshlands Observation System

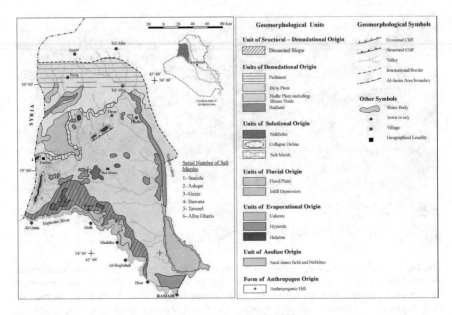

Fig. 14.23 Morphological units of Aljazeera area (Ma'ala 2009)

as most of the regions experienced high amounts of precipitation more than average, only one widespread dust storm was recorded in cluster one.

In 2009, a severe drought occurred which according to Trigo et al. (2010) was an unprecedented precipitation decrease in the region since 1940 (from the first access to precipitation data in the region). During precipitation season in 2007 (Fig. 14.24f), the first signs of this drought appeared as 20–40% decreases in normal precipitation, which resulted in two widespread dust storms in west of Iran and adjacent regions. Finally, the most severe precipitation decrease took place from November 2007 to April 2008 (Fig. 14.24g), during which the amount of precipitation in some areas reached less than 10% of average (i.e. 90% decrease) and decreased to 50% for most parts of cluster one. This drought continued with less intensity in 2009 (Trigo et al. 2010) and is demonstrated in Fig. 14.24h.

Concluding the precipitation analysis, it can be mentioned that the emergence of drought in the regions like Iraq and Syria with numerous dust-prone sources has resulted in extensive dust storms. In fact, the lack of water resources not only deteriorates natural vegetation cover but can also convert agricultural lands into erodible bare lands susceptible to becoming dust storms sources. By applying satellite data from Gravity Recovery and Climate Experiment (GRACE), Voss et al. (2013) analyzed water loss between 2003 and 2009 in a part of the Middle East. Results showed that the influence of drought on water storage of cluster one was significant. In this research the volume of fresh water loss in the region (from 2003 to the end of 2009) was estimated as 143.6 cubic km (km^3). Trend analysis of this period show a dramatic decrease in the volume of fresh water (Fig. 14.25). They used GRACE

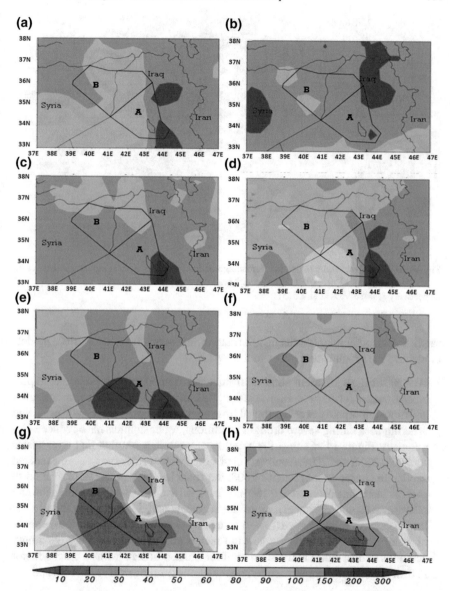

Fig. 14.24 Precipitation anomaly between November and April of **a** 2001–2002, **b** 2002–2003, **c** 2003–2004, **d** 2004–2005, **e** 2005–2006, **f** 2006–2007, **g** 2007–2008, and **h** 2008–2009 in cluster one in percentage (numbers more than 100 indicate the increase of precipitation and lower values show a decrease to less than long time average) (Darvishi Booloorani et al. 2012)

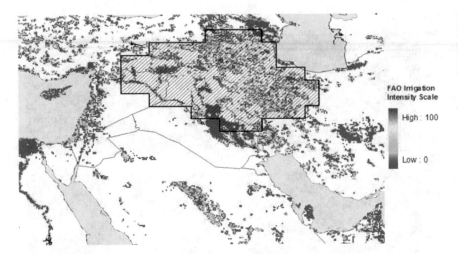

Fig. 14.25 Small grid squares display percentage of land under irrigation (*source* Siebert et al. 2007, cited by Voss et al. 2013). Blue to red gradient represents intensity on a 0–100% scale, respectively (*source* Voss et al. 2013)

satellite data, other remote-sensing information, and output from land surface models to identify the groundwater losses in the region.

- **Water storage equivalent** obtained from remote sensed imagery is another important parameter for dust emission analysis. As a result of intense surface evaporation and more importantly excess removal from aquifers for agricultural, industrial, and drinking uses during severe drought conditions, water supplies have reduced significantly. Figure 14.26 shows the monthly water storage (mm) of EuT river basins for 2005, 2007, and 2008. Comparison of stored water values shows an unexpected decline in cluster one. In August 2008, the equal height of water storage in north part of cluster one decreased to 150 mm less than average. By approaching the end of warm season (September), all of the segments within cluster one showed the lowest value of water storage (lowest value is in October). In addition, the decrease of water storage of the region was not necessarily coincidental with maximum activities of dust storm in the sources of the cluster. In fact, it was a crucial factor in dust formation when atmospheric circulations forming dust storms existed.

14.2.10 Remote Sensing Change Detection

Remote sensing change detection is the process of categorizing differences in the state of phenomena by observing them at different times of image acquisition. Construction of hydraulic structures over Tigris and Euphrates rivers as an important factor influencing dust storms in Iraq and West Asia is evident through satellite imagery. The

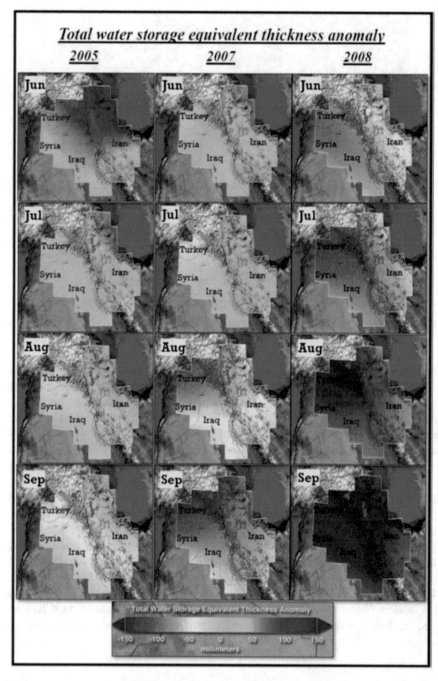

Fig. 14.26 Monthly anomaly of equal height of water storage (mm) in south of Turkey, northwest of Iran, northern part of Iraq and east of Syria [Darvishi Boloorani et al. 2014]

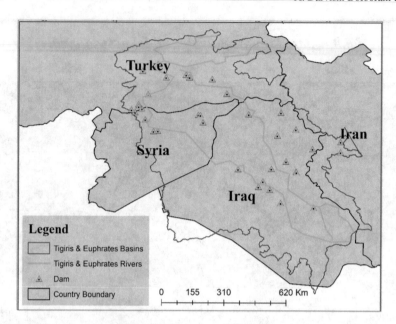

Fig. 14.27 Main dams on Tigris and Euphrates rivers

first dam for these watersheds, named Hindiya was constructed in 1914 on Euphrates river by the government of Iraq. Following Iraq's lead, Syria, Turkey, and Iran also constructed certain dams to fulfill their demands in industry, agriculture, electricity, and fresh water supply (Fig. 14.27). Such hydraulic constructions will undoubtedly have harmful effects such as the reduction of vegetation cover, desertification of agricultural lands and finally dust storms will happen.

In Fig. 14.28 water level data of Al-Qadisiyah lake formed by the Haditha dam in Iraq are shown. Water level has been declining during the last decade, with a steeper rate after mid-2007. However, it is worth mentioning that, after drought period in the region, Al-Qadisiyah lake has showed no increase in water level. Regarding the repeated events in 2009 and 2010, it can be concluded that the constructed dams in the upper stream of Euphrates, which are used to store water especially during drought periods, will prevent the downstream lakes from being revived and, as a result, the dam beds will become prone to dust storm. For instance, satellite imagery from Lake Al-Qadisiyah in 7 September 2006 (Fig. 14.28b) have been compared to the satellite image from 15 September 2009 (Fig. 14.28b). As can be seen the lake bed was heavily dried out and was prone to become a very active source of dust.

(a)   (b)

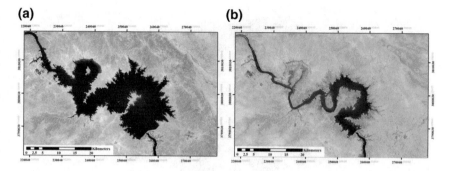

Fig. 14.28 Satellite images of Qadisiyah lake. **a** 7 September 2006 and **b** 15 September 2009

14.3 GIS for Dust Storm Studies

Geographical Information System (GIS) is the study of structures and computational methods to capture, represent, process, and analyze geographic information. GIS has applications in many disciplines in relation to the Earth phenomena. For instance, it can provide geo-solutions for various problems from macro-scale global climate change problems to micro-scale patterns of disease and crimes. GIS is able to simulate the real world by mapping and combining phenomena in the form of raster and vector layers (Fig. 14.29).

Dust storms result in the removal of soil particles such as clay, organic matter, and the nutrient rich finest particles of soil, thereby reducing agricultural productivity. Other effects that may have impacts on the economy are: reduced visibility affecting aircraft and road transportation; reduced sunlight reaching the surface; effects on human health (breathing dust) as well as water pollution that leads to food insecurity and other diseases. In this regard, GIS can be employed as a beneficial tool for handling various tasks as follows (Fig. 14.30).

14.3.1 Dust Sources Modeling with GIS

There are two main GIS approaches to discriminate and model the behavior of the sources of dust storms including: spatial data mining and knowledge-based methods.

14.3.1.1 Spatial Data Mining

Spatial data mining extracts abstract knowledge from raw data. The data are explicit though, usually, the knowledge is implicit. The abstraction might be a repetitive procedure between the user and the dataset. The concrete data are raw explanations of spatial entities. Since the data might be dirty, they should be cleaned, sampled,

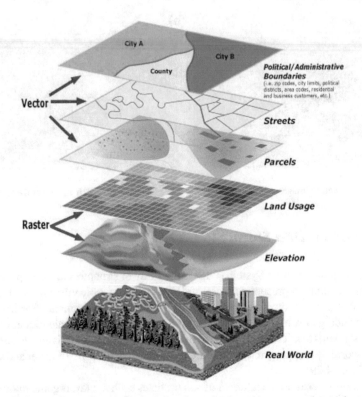

Fig. 14.29 Schematic framework for modeling the reality by using vector and raster layers in GIS (*source* http://shawneepsi.com/gis-designs/)

and transformed in accordance with the spatial data mining dimensions and the user-designated thresholds (Fig. 14.31). The extracted knowledge includes general configurations performing as a set of rules and expressions. The configurations are further interpreted by specialists after they are developed in order to be used for data referenced decision-making with numerous necessities (Fig. 14.32). Considering the availability of huge amounts of remotely-sensed data and information for erodible lands, e.g. time-series of vegetation, evapotranspiration, and soil characteristics; dust storm sources could be effectively detected by spatial data mining analysis in GIS environment (Table 14.1) (Li and Wang 2015).

14.3.2 Knowledge-Based Approaches

Knowledge-based methods rely on the knowledge of experts in decision making process. There are three basic methods in this field: multi-criteria decision analysis, fuzzy inference systems, and expert systems.

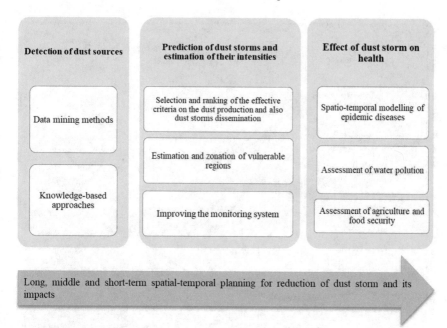

Fig. 14.30 Application of GIS in handling different dust-storm related tasks

Fig. 14.31 Spatial data
mining techniques (adapted
from Li and Wang 2015)

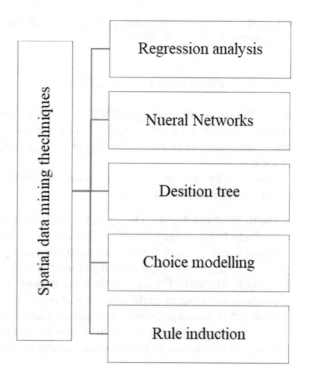

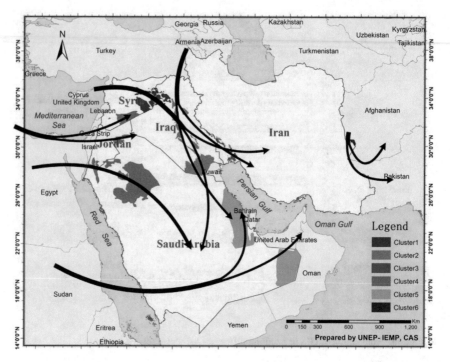

Fig. 14.32 Sand and dust storm main paths and source clusters in West Asia using spatial data mining approaches (Cao et al. 2015)

14.3.2.1 Multi-Criteria Decision Analysis

Multi-Criteria Decision Analysis (MCDA) has been widely used in numerous applications, especially in relation to GIS. Its role in various areas has improved meaningfully (Velasquez and Hester 2013). Figure 14.33 presents the schematic procedure of MCDA in GIS.

Due to the application of these techniques in different aspects of dust storms they are summarized as follows:

- **Multi-Attribute Utility Theory (MAUT)** is basically an extension of Multi-Attribute Value Theory (MAVT) (Keeney and Fishburn 1974) and is "a more rigorous methodology for how to incorporate risk preferences and uncertainty into multi criteria decision support methods" (Loken 2007, p. 1587).
- **Analytic Hierarchy Process (AHP)** is "a theory of measurement through pairwise comparisons and relies on the judgments of experts to derive priority scales" (Saaty 2008, p. 83). It uses pairwise comparisons which allow decision makers to weight coefficients and compare alternatives with relative case. It is scalable, and can easily regulate in size to accommodate decision making complications due to its

Table 14.1 GIS-based knowledge for dust storm studies (adapted from Li and Wang 2015)

Knowledge	Interpretations	Examples
Association rule	A logic association amongst altered sets of objects	Dust storm (location, intensity of dust storm) $\Rightarrow$ sources (location, event), support is 76%, confidence rate of 98%, and interest of 51%
Characteristics rule	A generic characteristic of an entity, or a set of entities	Characterize similar ground matters in a huge set of remotely sensed images
Discriminate rule	A different characteristic that separates one entity from another	Compare land degradation in an arid area with land degradation in urban areas
Clustering rule	A segmentation rules that groups a set of objects by their similarity without having prior knowledge of what caused the cluster and how many clusters there are	Grouping potential locations to find distribution patterns
Classification rule	A rule that describes whether an entity relates to a specific class or a set of classes	Classification of vulnerable locations
Serial rule	A temporal constrained instruction that relates objects or the functional dependency among the factors in a time sequence	Occurrence of dust storms during autumn
Predictive rule	An internal trend that forecasts future values of some variables when the temporal or spatial center is moved to another coordinate	Forecasting the movement trend of dust storm based on available monitoring data
Exception	An outlier that is isolated from the common rules or is the result of other substantially observed data	A monitoring point for detecting exceptional movement for predicting dust storm

hierarchical arrangement. The AHP has shown a variety of applications in issues related to dust storms.

- **Case-Based Reasoning (CBR)** is a kind of MCDA technique that recovers cases like a problem from an existing database of cases, and proposes a solution to a decision-making problem based on the most similar cases (Daengdej et al. 1999). It requires little effort in terms of acquiring additional data. It also needs little upkeep as the database will have previously been created and requires little maintenance. One major advantage that it has over most MCDM methods is that it can improve over time, especially as more cases are added to the database. It can also adapt to changes in environment with its database of cases. Its major drawback is the sensitivity to inconsistency in data (Daengdej et al. 1999). Previous cases could be invalid or special cases may result in invalid answers. Sometimes similar cases may not always be the most precise in terms of solving the problem at hand. CBR is used in industries where a substantial number of prior cases already

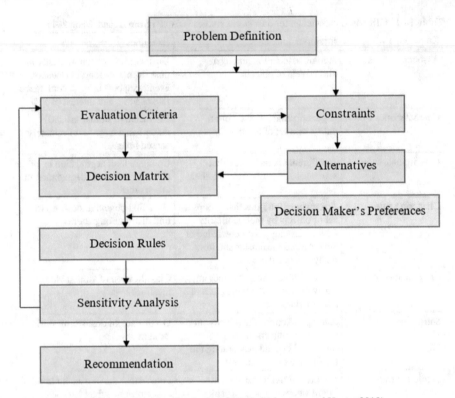

Fig. 14.33 Multi-Criteria Decision Analysis (MCDA) (Velasquez and Hester 2013)

exist such as in financial situations; human preference-oriented prediction and data-oriented prediction. Li and Sun (2008) provide a new method for predicting financial distress in companies one year prior to actual distress using CBR.

- **Data Envelopment Analysis (DEA)** applies a linear programming method to assess the relative efficiencies of alternatives (Thanassoulis et al. 2012). It ranks the effectiveness of alternatives against each other, with the most efficient alternative having a score of 1.0, while the other alternatives being a portion of 1.0. This method is able to manage multiple inputs and outputs and the efficiency can be investigated and computed. It is able to uncover the hidden relationships between the factors. A main drawback of this method is that it does not deal with imprecise data and assumes that all input and output data are exact. However, in real world conditions, usually this assumption is not true (Wang et al. 2005). The outcomes are sensitive dependent on the inputs and the outputs.

- **Goal Programming** is a practical programming that could handle large-scale problems. Its capability to produce infinite alternatives provides a specific benefit and the only disadvantage is its failure to weight coefficients. Many applications find it necessary to use other methods, such as AHP, to correctly rank the coefficients. Goal programming has seen applications in production planning, schedul-

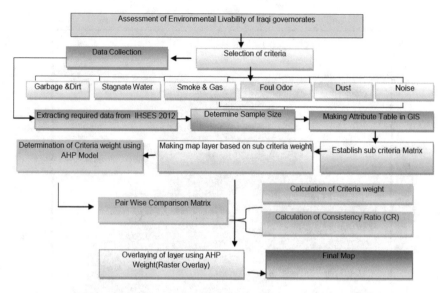

Fig. 14.34 GIS-AHP model developed to assess and map the environmental livability of Iraq (*source* Hassan and Foroughi 2018)

ing, health care, transportation, energy planning, water reservoir management, disaster management, and also assessment of the effects of sand and dust storms. Although some applications could be acted more appropriate in combination with other means to accommodate good weighting. In this regard, it removes one of its faintness while still being able to select from infinite.

- **ELECTRE** (Elimination and Choice Expressing Reality), is a MCDA with many iteration and an outranking method based on concordance analysis. As the major advantage, it considers uncertainty and vagueness. However, its process and outcomes can be described hardly in human terms (Konidari and Mavrakis 2007). ELECTRE has been used in energy, economics, environmental, water management, and transportation problems. Like other methods, it also takes uncertainty and vagueness into account, which many of the mentioned applications appear to need.

All of the above mentioned techniques are able to model different issues of dust storms such as vulnerability mapping, soil erosion, land and environment degradation etc. For instance, Hassan and Foroughi (2018) developed a GIS model for assessing the environmental livability of Iraq (Fig. 14.34). They used AHP model to create an environmental livability map using the percentage of households of Iraqis that are affected by smoke and gases, dust storms, bad odor, noise, garbage and dirt, and stagnant water. The final map classifies the governorates into five classes in terms of livability (Fig. 14.35) (relatively unsuitable, unsuitable, strongly unsuitable, very strongly unsuitable, and extremely unsuitable).

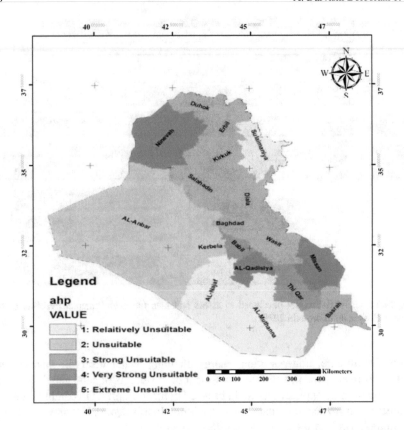

Fig. 14.35 The map of environmental livability of Iraq (*source* Hassan and Foroughi 2018)

14.3.2.2   Fuzzy Inference Systems

A Fuzzy Inference System (FIS) (Fig. 14.36) is a system that applies fuzzy set theory to relate inputs to outputs. The strength of FIS relies on its ability to model linguistic concepts. Also, it is able to handle nonlinear mappings between inputs and outputs. These FISs are comprised of fuzzy rules constructed from expert knowledge and they are named fuzzy expert systems or fuzzy controllers, dependent on their ultimate usage (Guillaume and Charnomordic 2012). Due to the uncertainty of the data and maps including soil, vegetation, land cover, and evapotranspiration maps, the FIS is highly recommended in implementing any kind of spatial analysis such as soil erosion modeling. For instance, Sarmad et al. (2017) used fuzzy logic to assess reference evapotranspiration in Basrah City of Iraq from 1990–2012. They used air temperature, relative humidity, wind speed and solar radiation as inputs of the fuzzy logic system. They concluded that Fuzzy logic models have high efficiency in estimating the reference evapotranspiration values.

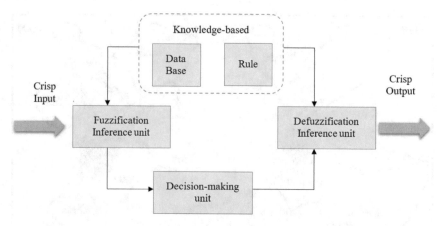

Fig. 14.36 Fuzzy inference system components (adapted from Guillaume and Charnomordic 2012)

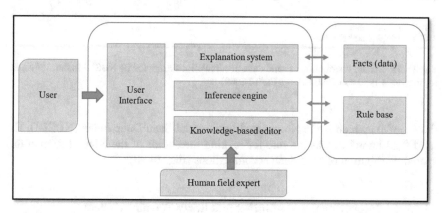

Fig. 14.37 Expert systems components (adapted from Cornelius 2002)

14.3.2.3 Expert Systems

An expert system is a computer system that models the decision-making capability of a human expert (Cornelius 2002). Expert systems are planned to resolve compound problems by reasoning over groups as if–then rules rather than over conventional procedural code. An expert system is distributed into two subsystems including the inference engine and the knowledge base (Fig. 14.37). The knowledge base characterizes facts and rules. The inference engine uses the rules to realize new facts. Furthermore, inference engines contain explanation and debugging capabilities.

Fadhil (2009) sought to assess and map land degradation in the upper Mesopotamian plain of Iraq from 1990 to 2000 and adapted simple rules to show the severity levels of the land degradation risks in the relative study area. He also used satellite based indices such as NDVI, The Normalized Differential Water Index (NDWI), Tasseled Cap Transformation Wetness (TCW), and developed a new sand

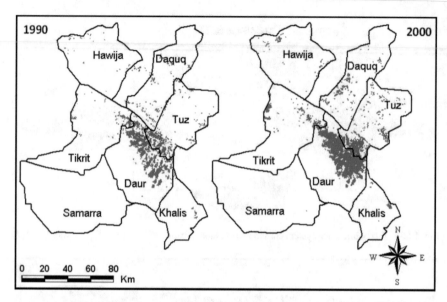

Fig. 14.38 Spatial distribution of sand dune accumulations extracted by NDSDI in the study area for the years 1990 and 2000 (*source* Fadhil 2009)

dunes index, which is the Normalized Differential Sand Dune Index (NDSDI). He used field knowledge as the rule for making thresholds of the NDSDI to map the spatial distribution of sand dune accumulations (Fig. 14.38).

14.3.2.4 Improving the Monitoring and Early Warning System

Sand and dust storms are scattered with great speed and extensive spatial changes, with complex interdependences between running processes. This creates a lot of challenges for timeliness, accuracy and high quality of decisions and actions. Basic for the implementation of these requirements must be an early warning system with incorporated information technologies (Fig. 14.39). On account of this, an early warning system will be able to obtain, analyze and systematize information and knowledge, which will later be submitted to all authorities for decision making in the beginning of disasters (Morabbi 2011).

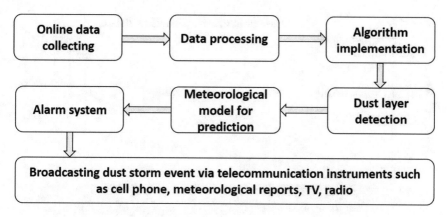

Fig. 14.39   Early warning system structure for dust storm prediction (Morabbi 2011)

14.3.3   GIS for Modeling the Effect of Dust Storms on Health

One of the main human and environmental aspects of dust storms is its effects on the health. In this regard, three disciplines are considered as the basic GIS fields. Epidemic models usually define the spread of a transmissible disease in a population. Often, a compartmental sight of the population is engaged, placing people into one of the three status (S)usceptible, (I)nfectious, or (R)emoved. GIS models could efficiently simulate and assess the relationships between dust storms and epidemic diseases (Meyer et al. 2017). Figure 14.40 shows the relations between the number of hospital admissions for chronic obstructive pulmonary disease and respiratory mortality according to the PMs using spatio-temporal and statistical methods.

14.4   Summary

In this chapter, we have tried to summarize the main characteristics of dust storms in Iraq. Our method is based on the synthesis of RS&GIS knowledge and information to assess the situation of dust storms in Iraq. This chapter is structured as follows: (1) introduction, (2) remote sensing for dust storm studies including: dust sources mapping, detection of dust events using satellite imagery, wind speed and direction, trajectory of suspended dust particles, soil data and land surface coverage map/information, atmospheric patterns of dust storms. (3) GIS for dust storm studies including: dust sources modeling with GIS, knowledge-based approaches, fuzzy inference systems, expert systems, GIS for improving the monitoring and early warning system, and spatio-temporal modelling of epidemic diseases. With regard to the scope of the book we have mentioned some examples of case studies and projects in Iraq.

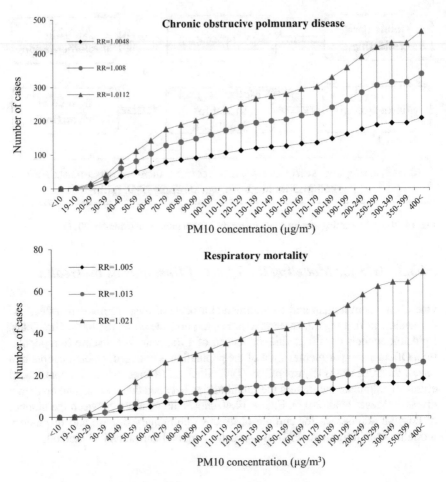

Fig. 14.40 Number of hospital admissions for chronic obstructive pulmonary disease and respiratory mortality according to the PMs (*source* Omidi et al. 2017)

Acknowledgements Certain examples were borrowed from the project: "Dust Storms in West Asia" funded by headquarter of water resources development, drought, erosion and environment technologies (vice-president of science and technology affairs of Islamic republic of Iran) and UNEP-ROWA. We are especially grateful to the vice-president of science and technology affairs of Islamic Republic of Iran and UNEP-ROWA. The authors are thankful to Dr. Seyed Omid Nabavi, Professor. Hossein Mohammadi, and Dr. Reza Khandan who contributed to the "Dust Storms in West Asia Region" project. We are also thankful to the scholars and scientists whose research results and publications were used in this chapter.

References

Abbas SA, Hassan AA, Al-Rekabi WS (2017) Estimation of mean reference evapotranspiration in Basrah City, South of Iraq using fuzzy logic. J Babylon Univ/Eng Sci 25(1):98–108

Ackerman SA (1997) Remote sensing aerosols using satellite infrared observation. J Geophys Res 102:17069–17079

Ahmad SP, Torres O, Bhartia PK, Leptoukh G, Kempler S (2006) Aerosol index from TOMS and OMI measurements. NASA Goddard Space Flight Center, Greenbelt, MD Greenbelt, MD 20771, USA. https://ams.confex.com/ams/Annual2006/techprogram/paper_104496.htm

Aurelius L (2008) The impact of Shamal winds on tall building design in theGulf. Dubai Building, Government of Dubai

Buringh P (1960) Soils and soil conditions in Iraq. The Ministry of Agriculture, Baghdad, Iraq, p 337

Cao H, Amiraslani F, Liub J, Zhoue N (2015) Identification of dust storm source areas in West Asia using multiple environmental datasets. Sci Total Environ 502:224–235

Corine (1992) Soil erosion risk and important land resources in the southeastern regions of the European Community. EUR 13233, Luxembourg, Belgium

Cornelius L (2002). Expert systems: the technology of knowledge management and decision making for the 21st century, pp 1–22. ISBN 978-0-12-443880-4

Daengdej J, Lukose D, Murison R (1999) Using statistical models and case-based reasoning in claims prediction: experience from a real-world problem. Knowl-Based Syst 12(5–6):39–245

Darvishi Booloorani A, Nabavi SO, Bahrami HA, Alavi Panah SK, Mohammadi H, Nezammahalleh MA (2012) Dust Storms in the West Asia Region. Tech Rep, p 215.

Darvishi Booloorani A, Nabavi SO, Bahrami HA, Alavi Panah SK, Mohammadi H, Khandan R (2014) Primary investigation of dust storm sources in West Asia Region (With an Emphasis on Storms Came to Iran). Technical Report, p 89. https://doi.org/10.13140/2.1.2434.7845

Draxler RR, Gillette DA, Kirkpatrick JS, Heller J (2001) Estimating PM_{10} air concentrations from dust storms in Iraq, Kuwait, and Saudi Arabia. Atmos Environ 35:4315–4330

Ekpenyong Robert Etim (2013) An assessment of land cover change and erosion risk in Akwa Ibom State of Nigeria using the Coordination of information on the Environment (CORINE) methodology. Greener J Phys Sci 3(3):076–089

Fadhil AM (2009) Land degradation detection using Geo-information technology for some sites in Iraq. J Al-Nahrain Univ 12(3):94–108. https://doi.org/10.22401/JNUS.12.3.13

Faraj AA (2013) Land use/ land cover and change detection for the Iraqi Province of Sulaimaniyah in using remote sensing. M.Sc. Thesis of George Mason University. http://mars.gmu.edu/bitstream/handle/1920/8704/Faraj_thesis_2013.pdf?sequence=1&isAllowed=y

Ginoux P, Prospero JM, Gill TE, Hsu NC, Zhao M (2012) Global-scale attribution of anthropogenic and natural dust sources and their emission rates based on MODIS Deep Blue aerosol products. Rev Geophys 50:1–36. RG3005. https://doi.org/10.1029/2012rg000388

Guillaume S, Charnomordic B (2012) Fuzzy inference systems: an integrated modeling environment for collaboration between expert knowledge and data using FisPro. Expert Syst Appl 39:8744–8755

Hassan MM, Foroughi S (2018) Environmental livability's assessment of Iraqi governorates using AHP model and Geography Information System (GIS). J Duhok Univ 21(1) (Humanities and Social Sciences):401–410

IUSS Working Group WRB (2007) World reference base for soil resources 2006, first update 2007. World Soil Resources Reports No. 103. FAO, Rome

Jahantab Z, Ale Sheikh AA, Darvishi Booloorani A, Bagheri K (2018) A new method for calibration of land-vegetation degradation modeling. J Geospatial Inf Technol 6(2):87–104

Kazemi Y (2018) Satellite modeling of spatiotemporal behavior of dust sources. Master of Science thesis in remote sensing and GIS, University of Tehran

Keeney R, Fishburn P (1974) Seven independence concepts and continuous multiattribute utility functions. J Math Psychol 11(3):294–327

Konidari P, Mavrakis D (2007) A multi-criteria evaluation method for climate change mitigation policy instruments. Energy Policy 35(12):6235–6257

Li D, Wang S (2015) Spatial data mining, theory and application. Springer-Verlag, Berlin Heidelberg, XXVIII, 308 p

Li H, Sun J (2008) Ranking-order case-based reasoning for financial distress prediction. Knowledge-Based Syst 21(8):868–878

Loken E (2007) Use of multi-criteria decision analysis methods for energy planning problems. Renew Sustain Energy Rev 11(7):1584–1595

Ma'ala KA (2009) Geomorphology of Al-Jazira area. Iraqi Bull Geol Min. Special Issue 3:5–31

Meyer S, Held L, Höhle M (2017) Spatio-temporal analysis of epidemic phenomena using the R package surveillance. J Stat Softw 77(11):1–55

Morabbi M (2011) Risk warning and crisis management for dust storm effects on western border of Iran rapid Response mapping for Dust Storm Crisis United Nations. In: International conference on space-based technologies for disaster risk management "best practices for risk reduction and rapid response mapping", Beijing, China, 22–25 November

Muhaimeed AS (2017) Status of soil information in Iraq. http://www.fao.org/fileadmin/user_upload/GSP/docs/Presentation_NEMA_Inception/Ahmad_Irak.pdf

Omidi KY, Daryanoosh SM, Amrane A, Polosa R, Hopke PhK, Goudarzi Gh, Mohammad MJ, Sicard P, Armin H (2017) Impact of Middle Eastern Dust storms on human health. Atmos Pollut Res 8:606–613

Richardson CJ, Hussain NA (2006) Restoring the garden of Eden: an ecological assessment of the marshes of Iraq. Bioscience 56(6):477–489

Richardson CJ, Reiss P, Hussain NA, Alwash AJ, Pool DJ (2005) The restoration potential of the Mesopotamian marshes of Iraq. Science 307:1307–1311

Saaty T (2008) Decision making with the analytic hierarchy process. Int J Serv Sci 1(1):83–98

Samadi M, Darvishi Boloorani A, Alavipanah SK, Mohamadi H, Najafi MS (2014) Global dust Detection Index (GDDI); a new remotely sensed methodology for dust storms detection. J Environ Health Sci Eng 12(1):1–14

Shoshany M, Goldshleger N, Chudnovsky A (2013) Monitoring of agricultural soil degradation by remote-sensing methods: a review. Int J Remote Sens 34(17):6152–6181

Siebert S, Döll P, Feick S, Hoogeveen J, Frenken K (2007) Global map of irrigation areas version 4.0.1. Johann Wolfgang Goethe University, Frankfurt am Main, Germany/Food and Agriculture Organization of the United Nations, Rome, Italy

Thanassoulis E, Kortelainen M, Allen R (2012) Improving envelopment in data envelopment analysis under variable returns to scale. Eur J Oper Res 218(1):175–185

Thies B, Bendix J (2011) Review satellite based remote sensing of weather and climate: recent achievements and future perspectives. Meteorol Appl 18:262–295

Trigo RM, Gouveia C, Barriopedro D (2010) The intense 2007–2009 drought in the Fertile Crescent: impacts and associated atmospheric circulation. Agric For Meteorol 150:1245–1257

Velasquez M, Hester PT (2013) An analysis of multi-criteria decision making methods. Int J Oper Res 10(2):56–66

Voss KA, Famiglietti JS, Lo M, de Linage C, Rodell M, Swenson SC (2013) Groundwater depletion in the Middle East from GRACE with implications for transboundary water management in the Tigris-Euphrates-Western Iran region. Water Resour Res 49:904–914

Wang Y, Greatbanks R, Yang B (2005) Interval efficiency assessment using data envelopment analysis. Fuzzy Sets Syst 153(3):347–370

Yue Huanbi, He Chunyang, Zhao Yuanyuan, Ma Qun, Zhang Qiaofeng (2017) The brightness temperature adjusted dust index: an improved approach to detect dust storms using MODIS imagery. Int J Appl Earth Obs 57:166–176

Zaitchik BF, Evans JP, Smith RB (2007) Regional impact of an elevated heat source: the Zagros
    Plateau of Iran. J Clim 20:4133–4146
Zarrin Azar, Ghaemi H, Azadi M, Mofidi A, Mirzaei E (2011) The effect of the Zagros Mountains
    on the formation and maintenance of the Iran Anticyclone using RegCM4. Meteorol Atmos Phys
    112(3–4):91–100

Chapter 15
Drought Monitoring Using Spectral and Meteorological Based Indices Combination: A Case Study in Sulaimaniyah, Kurdistan Region of Iraq

Ayad M. Fadhil Al-Quraishi, Sarchil H. Qader and Weicheng Wu

Abstract Drought has dramatically affected Iraq throughout the last decades, which were characterized by a large drop in rainfall, and its main rivers discharge in general. Three spectral indices were derived from the Landsat images of 1990, 2007, and 2008 as indices of soil, vegetation, and moisture to monitor the drought and its impacts. The derived indices were the Normalized Difference Vegetation Index (NDVI), Land Surface Temperature (LST), and the Normalized Differential Water Index (NDWI). The fourth drought index was the Standardized Precipitation Index (SPI), which has been used as a meteorological drought index. The aim of this chapter is to investigate the role of integration of vegetation indices (NDVI in this study) and SPI as a combined index (NDVI-SPI) for drought monitoring in Sulaimaniyah, Kurdistan region of Iraq in 1990, 2007, and 2008. The results showed a significant decrease in the vegetative cover by 28.6% in 2008 in compared with that of 2007. However, results of the combined NDVI-SPI indices maps emphasized the harsh impact of drought on the vegetative cover, which occurred in 2008. In particular, the results revealed a significant increase in areas of the extreme, severe, moderate drought classes in 2008 by percentage of 81.2% more than in 2007. On the other hand, Dukan Lake's surface area in the study site suffered a significantly shrunk by 16.5 and 32.5% in 2007 and 2008, respectively, compared with its total size in 1990. The study concluded that the use of a combination of NDVI-SPI indices provides more reliable results for drought monitoring than any single index in the study area.

A. M. F. Al-Quraishi (✉)
Department of Environmental Engineering, College of Engineering, Knowledge University, Erbil 44001, Kurdistan Region, Iraq
e-mail: ayad.alquraishi@gmail.com; ayad.alquraishi@knowledge.edu.krd

S. H. Qader
Geography and Environment Department, Faculty of Social and Human Sciences, University of Southampton, Southampton, UK

W. Wu
Key Laboratory of Digital Land and Resources, East China University of Technology, Nanchang 330013, Jiangxi, China

© Springer Nature Switzerland AG 2020
A. M. F. Al-Quraishi and A. M. Negm (eds.),
Environmental Remote Sensing and GIS in Iraq, Springer Water,
https://doi.org/10.1007/978-3-030-21344-2_15

Keywords Drought monitoring · Remote sensing · Drought indices · Combined NDVI-SPI · GIS · Sulaimaniyah · Iraq

15.1 Introduction

Iraq has suffered in the last two decades an environmentally difficult period, where the serious decline in water discharges of its two main Tigris and Euphrates and their tributaries, as well as the significant drop in the precipitation averages throughout the country. That reduction has disastrously affected the agricultural areas, the quantities of available water for drinking, industrial and agricultural uses. The situation of drought has been reported recently in Iraq since the annual rainfall has dramatically decreased in the past few years. Drought is a deceitful natural hazard that results from "a lack of precipitation from expected or "normal" such that when it is extended over a season or longer period of time" (Wilhite and Glantz 1985; Fadhil et al 2004; Tucker and Choudhury 1987; Wilhite2005; Ye et al. 2018; Al-Mamalachy et al. 2019). Subsequently, the amount of precipitation is insufficient to meet the demands of human activities and the environment. Several indices have been used for drought monitoring and assessment of its impacts (Fadhil 2011; Hayes 2012; Khalili et al. 2011; Mohammad et al. 2018; Thenkabail and Rhee 2017). A drought index value is typically a single number, far more useful than raw data for decision-making (Hayes 1999; Beg and Al-Sulttani 2019; Fadhil 2009; Dutta et al. 2015). Drought index can be derived from hydro-meteorological data and remote sensing dataset. Meteorological drought indices are normally continuous functions of rainfall and/or temperature, river discharge or other measurable variable. Rainfall data were broadly used to calculate drought indices. One of the meteorological drought indices is the Standardized Precipitation Index (SPI), which can be utilized to assign a single numeric value as an indicator of drought (Inota et al. 2016; Moradi et al. 2015; Portila et al. 2017; Yagoub et al. 2017).

Regarding the relationship between soil moisture and reflectivity, Bowers and Hanks (1965) found that the increase in soil moisture in bare soil would lead to the decrease in soil reflectivity. The remote-sensing-based NDVI and the rainfall-based SPI have been respectively used for drought monitoring in many regions or countries (Al-Quraishi 2004; Fadhil 2006; Vicente-Serrano et al. 2006: Hammouri and El-Naqa 2007; Murad and Saiful Islam 2011; Fadhil 2013). However, the different attributes of vegetation in various areas, make it hard to detect the influences of drought on vegetation cover, and especially, on agriculture.

Severe drought has affected the Kurdistan region as well as the other parts of Iraq throughout the last years, which was characterized by a significant drop in the rainfall amounts (Fadhil 2011; Eklund and Seaquist 2014; Eklund and Thompson 2017; Rasul and Ibrahim 2017; Hameed et al. 2018). Particularly, Kurdistan region was considered the Iraqi's historical breadbasket, where rain-fed wheat is grown (Fadhil 2011; Notaro et al 2015). The winter wheat crop in Iraq depends on the rain that falls between October and April months. Since the crop has little access to other sources of water in the region, the vegetation situations show a direct response to the lack of rainfall at the end of the critical rainfall period in April. In the cropping year 2007–2008, there was a continuous high air temperature and approximately 41% decrease in precipitation compared with the rainfall average in this region and the other parts of Iraq, which led to severe drought (Fadhil 2011).

This chapter aims to investigate the role of the integration of NDVI and SPI indices for drought monitoring in Sulaimaniyah, Kurdistan region of Iraq during the years of 1990, 2007, and 2008.

15.2 The Study Area

The study area (Fig. 15.1) is situated in Sulaimaniyah governorate, which is one of the three governorates in the Kurdistan region of Iraq, and it encompasses a total area of 6,864.5 km², extending from lat. 35°04′ to 36°30′ N and long. 44°50′ to 46°16′ E, accounting 1.6% of the entire area of Iraq. Sulaimaniyah is located in the northeastern part of Iraq, about 331 km northeast to Baghdad and home of 963,390 inhabitants. Its hilly landscape becomes increasingly mountainous towards the eastern border with Iran.

The study area in this chapter includes most of the districts of Sulaimaniyah governorate including the fertile lands, such as the Sharazor and Peshdar plains, which are respectively distributed in southern and northeastern parts of the governorate, and

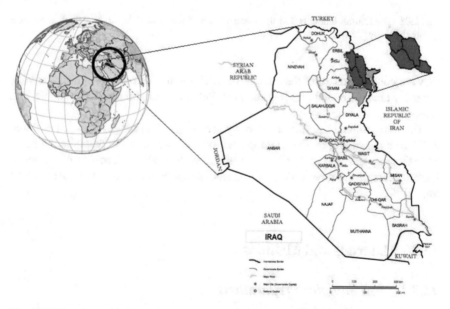

Fig. 15.1 Location map of the study area in Sulaimaniyah Governorate

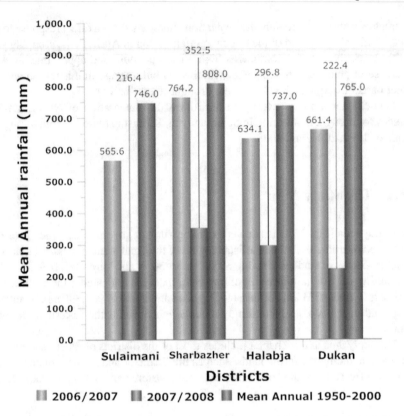

Fig. 15.2 The rainfall amounts and the mean annual of the rainfall in the four studied districts during the study periods

they are considered two of the most fertile plains in northern Iraq. Administratively, the study area is divided into four districts: Sulaimaniyah, Sharbazher, Halabja, and Dukan.

The mean annual rainfall in the study area (Fig. 15.2) ranges between 600 and 1,200 mm depending on the geographical location, whereas the precipitation starts from October and ends in April. It increases from the southwestern toward the northeastern parts of the governorate. The mean annual temperature varies between 12 and 20 °C.

15.3 Materials and Methods

15.3.1 Soil Samples Preparations

In order to characterize the chemical and physical properties of soil, as well as to correlate them with its spectral characteristics in the studied area, soil samples were taken from 22 sites at a depth of 0–15 cm in the Sharazor and Peshdar plains. Their locations coordinates were collected using a GPS receiver (Garmin Rino 120). At the laboratory, the soil samples were air dried before being sieved with a 2 mm sieve, and then prepared for the chemical and physical analyses. The analyses included soil particles size distribution, soil moisture content, bulk density, electrical conductivity (Ec), soil reaction (pH), total calcium carbonate ($CaCO_3$), organic matter (OM), cation exchange capacity (CEC), and total nitrogen (N).

15.3.2   Remotely Sensed Datasets

Three Landsat images (path 168/row 35) were downloaded from the USGS data server (glovis.usgs.gov) and used in this study. The first and second images were Landsat 5 Thematic Mapper (TM) acquired on 13 Sep. 1990 and 12 Sep. 2007, respectively, while the third imagery was a Landsat 7 Enhanced Thematic Mapper (ETM+) acquired on 22 Sep. 2008. From viewpoint of the study objectives, Landsat images were acquired during the dry season in Iraq to best show land features, particularly, vegetation and soil moisture those concerning the occurrence of drought and to avoid overshadowing by too much vegetation.

15.3.3   Preprocessing of the Landsat Images

The Landsat 7 ETM+ image was a SLC-off type dataset. Therefore a gap-filling technique with a SLC-on image of the same study area was adopted using ERDAS Imagine model maker. A model particularly created to perform this task was used. Image to image registration of multi-temporal remotely sensed datasets is crucial for change detection (Stow 1999). Change detection analysis was executed on a pixel-by-pixel basis; thus, any misregistration larger than one pixel will result in an inaccurate result of that pixel. To defeat this problem, the root mean square (RMS) error between any two dates must not exceed 0.5 pixels (Lunetta and Elvidge 1998), so the RMS error between different images was less than 0.25 pixel. Moreover, the three Landsat images were reprojected into the datum WGS84 and projection UTM zone N38 using the first order (linear) of the polynomial function and nearest neighbor rectification re-sampling, which was selected in order to preserve the radiometry and spectral information in the imagery (Richards and Jia 1999).

15.3.4   Drought Indices

Drought indices can be extracted from meteorological and remotely sensed dataset. It is vital to employ some of the drought indices related to the vegetation and soil moisture status for this purpose to track and monitor the drought situation in this study, The NDVI as a spectral-based drought index, and the SPI as a meteorological-based drought index were utilized in this chapter.

15.3.5 SPI

Rainfall data alone may not express the nature of drought-related situations, but they can provide an expedient solution in data-poor regions. The SPI was formulated by McKee et al. (1993) to specify a single numeric value to the precipitation that can be compared across territories with obviously different climates. It was developed for drought monitoring and requires only one input variable. The SPI computation for any site depends on the long-term precipitation record that is fitted to a probability distribution then transformed into a normal distribution to ensure that the mean SPI for the site and the desired period is zero (Edwards and McKee 2006). A drought event occurs at any time when the SPI is continuously negative and reaches an intensity of -1.0 or less, while the drought event ends when the SPI becomes positive.

Conceptually, the SPI is equivalent to the z-score often used in statistics:

$$Z - score = (X - Average)/Standard\ Deviation \qquad (15.1)$$

where the Z-score expresses the X score's distance from the average in standard deviation units (Giddings et al. 2005). For determination of the SPI values in this study, the available monthly rainfall data from 19 and 17 meteorological stations in the hydrological years 2006–2007 and 2007–2008, respectively were assembled and used to calculate the SPI values for the studied area. An MS-Excel based application was developed to perform the SPI computations. In order to get the spatial pattern of drought in the study area, the GIS-Kriging interpolation for SPI in a GIS environment was conducted using the rainfall data of the two hydrological years 2006/2007 and 2007/2008.

15.3.6 Remote Sensing Based Drought Indices

Drought indices contain information on rainfall, stored soil moisture or water supply but without much local spatial detail. To monitor and track drought situation in the study area some remotely sensed-based drought indices related to the vegetation, soil, and water status in this study have been utilized. The Normalized Difference Vegetation Index (NDVI), Land Surface Temperature (LST), and the Normalized Differential Water Index (NDWI) have been used in the study.

15.3.7 NDVI

The NDVI was first proposed by Rouse et al. (1973) though the concept was already discussed by Kriegler et al. (1969), and extended by Tucker (1979). It is an index of plant "greenness" or photosynthetic activity and is one of the most commonly used vegetation indices. It was found that the NDVI is sensitive to the rainfall and there is a positive relationship between them (Kogan 2008). The NDVI thematic images were produced using the following formula:

$$NDVI = \frac{(NIR - RED)}{(NIR + RED)} \qquad (15.2)$$

where NIR is the near-infrared band (0.78–0.90 μm), and RED is the visible-red (0.63–0.69 μm) band.

15.3.8 LST

The LST can be retrieved from the remotely sensed dataset. Land surface temperature (LST) is sensitive to vegetation and soil moisture. Therefore, it can be used to monitor the drought situation in the study area. The LST fraction images produced according to the equations supplied by YCEO (2010).

15.3.9 NDWI

The NDWI was used to oversee the conditions of the soil moisture and status of the water-surfaces in the studied area. The ratio between TM/ETM+ Band 3 (R: Red) and Band 5 (SWIR: short-wave IR): (1.55–1.75 μm), can highlight water bodies as brighter pixels (CPM 2003).

$$NDWI = \frac{(NIR - SWIR)}{(NIR + SWIR)}$$

15.4 Results and Discussions

15.4.1 NDVI

In consideration of the NDVI results and rainfall amounts in the study area in the years 1990, 2007, and 2008 it was obvious that there was a positive correlation between the

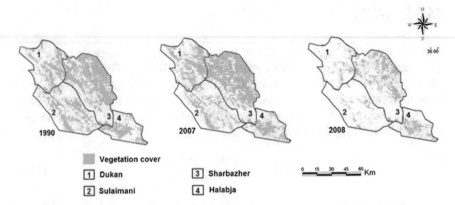

Fig. 15.3 The spatial distribution of the NDVI based vegetative cover in the study area for the years 1990, 2007, and 2008

Table 15.1 The total vegetative cover in the four studied districts for the years 1990, 2007, and 2008

District name	District area	Total veg. cover 1990		Total veg. cover 2007		Total veg. cover 2008	
	km^2	km^2	%	km^2	%	km^2	%
Sulaimaniyah	2,018.3	661	32.7	602.4	29.8	130.4	6.4
Sharbazher	2,387.8	1,304.1	54.6	1,444.9	60.5	716.4	30
Halabja	920.3	334.7	36.3	281.1	30.5	108.9	11.8
Dukan	1,537.9	574.1	37.3	772	50.2	184.4	11.9
Total	6,864.3	2,873.9	41.7	3,100.4	45.2	1,140.1	16.6

rainfall and the vegetation cover. These results extracted from the NDVI results and the rainfall amounts during the study period. The vegetation indices analyses were computed on a DN scale range between 0 and 255. The results (Fig. 15.3, Tables 15.1 and 15.2) showed that the total vegetation coverage area was 2,873.9 km^2 (41.7% of the total area) in 1990, while it increased by 3,100.4 km^2 (45.2%) in 2007. Conversely, the vegetation cover decreased to 1,140.1 km^2 (16.6%) in 2008.

The NDVI results showed that in Dukan district was the biggest decline by 38.2% in the vegetation cover area, which has occurred in the period 2007–2008. Similarly, the highest vegetation cover decrease happened in the same district from 1990 to 2007. That vegetation cover decrease has been accompanied by a significant shrinkage in the water surface area in 2007 and 2008. Since the annual rainfall in 2007 was the closest to the overall mean annual rainfall of the study area during the period 1971–2007, therefore the vegetation cover in Dukan and Sharbazher districts increased between 1990 and 2007 that can be considered the highest among the other studied districts. The increases were by 5.9 and 12.9% of Dukan and Sharbazher districts area, respectively. Conversely, there was a remarkable dwindling in

Table 15.2 The areas and rates of the vegetation cover changes in the studied districts during the study periods

District name	2008–2007	Change rate	2008–1990	Change rate	2007–1990	Change rate
	%	Km^2 year^{-1}	%	Km^2 year^{-1}	%	Km^2 year^{-1}
Sulaimaniyah	−23.3	−472	−26.3	−29.4	−2.9	−3.4
Sharbazher	−30.5	−728.4	−24.6	−32.6	5.9	8.2
Halabja	−18.7	−172.2	−24.5	−12.5	−5.8	−3.1
Dukan	−38.2	−587.6	−25.3	−21.6	12.9	11.6
Total	−28.6		−25.3		3.3	

the vegetative cover in Sulaimaniyah district during the span of 18 years from 1990 to 2008. The vegetation cover decline can be attributed to the lack of irrigation water and water resources during the drought periods, as well as to the huge urban sprawl in Sulaimaniyah city and its surrounding counties.

The vegetation cover decrease in the studied area was due to the small rainfall amounts during the study period particularly in the hydrological year 2007/2008. Thereupon, that rainfall amounts decrease led to a significant reduction in the surface and subsurface water resources of the study area such as the rivers, lakes, groundwater, and consequently soil moisture content.

15.4.2 NDWI

Water is a substantial factor for plant growth due to it helps in seed germination, helps in the process of photosynthesis by which plants make their food, and helps in the transport of nutrients and minerals from the soil to the plants. The results revealed that there was a significant shrinkage in the water surfaces areas in the study region, such as the lakes and rivers of the region in 2007 and 2008. Under those circumstances, the results showed that the entire water surface area was 187.4 km^2 in 1990, while it shrank to 153.8, 125.6 km^2 in 2007, and 2009, respectively. It was clear from the study results (Fig. 15.4) that Dukan and Darbandikhan lakes have shrunk significantly in size over the eighteen years of the study period. Namely, Dukan Lake was decreased from 161.2 km^2 in 1990 to 134.6 and 108.7 km^2 in 2007 and 2008, respectively. In short, it lost 16.5 and 32.6% of its total area in 2007 and 2008, respectively. On the other side, the results showed that the Darbandikhan Lake area increased to 11.69 km^2 in 2007, while it was 6.52 km^2 in 1990. However, it shrunk to be a small stream in 2008 because of the harsh effects of the drought that hit the Iraqi Kurdistan region in the aforementioned year.

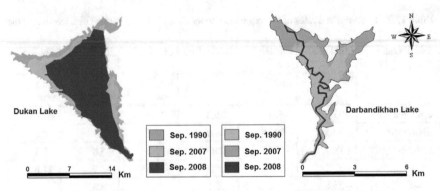

Fig. 15.4 The surface size of Dukan and Darbandikhan lakes in 1990, 2007, and 2008

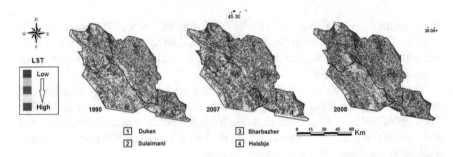

Fig. 15.5 The spatial distribution of the LST values in the study area in Sep. of 1990, 2007, and 2008

15.4.3 LST

Generally, LST in the study area tends to decrease in the orientation SW to NE associated with increasing altitude. Consequently, the results (Fig. 15.5) showed large differences in land surface temperature values between the hottest and the coldest sites in the study area. It was found that the bare land or the poor vegetative cover had high LST values coupled with low soil/vegetation moisture content. The statistical analysis results revealed that there was a significant negative correlation between the LST and the soil/vegetation moisture (-0.453), (-0.551), and (-0.233) for 1990, 2007, and 2008, respectively. Decreasing the correlation value in 2008 indicates the declining role of water in the soil wetting during the drought year of 2008.

15.4.4 SPI

The results of SPI values and its interpolated maps for the hydrological years 2006/2007 and 2007/2008 were presented in Figs. 15.6, and 15.7.

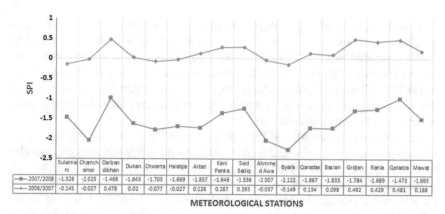

	Sulaima ni	Chamch amal	Darban dikhan	Dukan	Chwarta	Halabja	Arbat	Kani Panka	Said Sadiq	Ahmme d Awa	Byara	Qaradax	Bazian	Grdjan	Rania	Qaladza	Mawax
2007/2008	-1.326	-2.025	-1.468	-1.643	-1.703	-1.669	-1.857	-1.646	-1.536	-2.007	-2.122	-1.867	-1.833	-1.784	-1.689	-1.473	-1.693
2006/2007	-0.145	-0.027	0.478	0.02	-0.077	-0.027	0.126	0.287	0.293	-0.037	-0.149	0.134	0.098	0.492	0.429	0.481	0.188

METEOROLOGICAL STATIONS

Fig. 15.6 The SPI values of the 17 meteorological stations in the study area for the hydrological years of 2006/2007 and 2007/2008

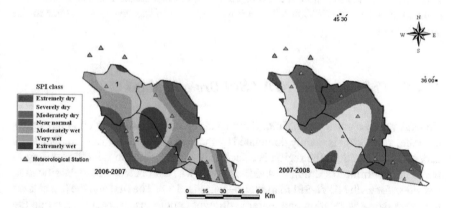

Fig. 15.7 The interpolated SPI categories maps of the study area for the hydrological years of 2006/2007 and 2007/2008

The SPI interpolated maps revealed that most of the studied districts were affected by drought during the hydrological year 2007/2008. Specifically, Chamchamal, Ahmed Awa, and Byara sites suffered the most severe drought that recorded at their meteorological stations. Figure 15.7 shows SPI drought categories maps for the hydrological years 2006/2007 and 2007/2008. The results showed a reduction in the rainfall amounts at these stations during the period 2007–2008. Consequently, that reduction led to a decrease in the vegetation cover areas and the soil/vegetation moisture in the studied sites; hence, the bare soil area was increased. These results also showed that the rainfall factor represented by SPI was more influential on the vegetative growth in 2007 than 2008 due to the significant correlation, which found between SPI and the NDVI. Furthermore, the drought severity was much higher in 2008 than 2007.

In terms of the declining percentage of the vegetation cover, Dukan was the most attacked district by the drought, which its vegetation cover in 2008 decreased by 38.3% compared to 2007. While Sharbazher and Sulaimaniyah districts were at the second and third order due to the reduction of vegetation cover were 30.5 and 23.4%, respectively. However, looking into the vegetation cover remaining over the whole districts, Sulaimaniyah district had only 130.4 km^2 accounting 6.4% of vegetation cover across the whole district area.

The difference between the SPI results of the years 2007 and 2008 was completely obvious as a result of the severe drought in 2008. Near Normal SPI class was recorded for all the districts in 2007 due to the normal rainy season of 2006/2007 in the area. However, significant rainfall decreasing in 2008 caused negative values of SPI across the area under severely dry class. Among all districts, Sulaimaniyah had the lowest SPI value in 2008, which was (-1.782), while it was (0.037) in 2007. Similarly, with a slightly different, Halabja district had the second order level with (-1.670) in 2008. Briefly, the reduction of the rainfall amounts in all studied districts during the hydrological year 2007/2008 led to a huge decrease in the vegetation cover in the studied areas.

15.4.5   The Combined NDVI-SPI Drought Maps

Generally, in the north part of Iraq, (Kurdistan region) including Sulaimaniyah governorate, vegetation is mainly dominated by the availability of the rainfall. Due to its geographical location, the region is affected by irregularities in rainfall resulting in frequent occurrence of drought. A GIS analysis was conducted, and reclassification was done for each NDVI-SPI map as shown in Fig. 15.8. The first and the last classes represent dense vegetation and extreme drought conditions, respectively, while the other classes represent the rest classes.

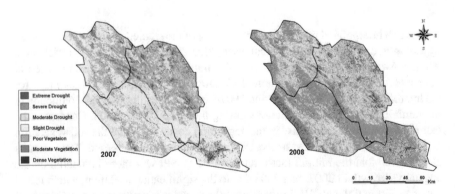

Fig. 15.8 The combined NDVI-SPI index based drought maps of the study area for the years 2007 and 2008

The ArcGIS ver. 10.3 was employed to generate the combination of NDVI-SPI layers and mapped for the entire study area in 2007 and 2008. Due to unavailability of the rainfall data of 1990, the combined NDVI-SPI map could not created. As the region is mostly relied on the rainfall, therefore, a positive correlation between vegetation cover and rainfall might exist. Based on this, the combined NDVI-SPI layers were generated in order to provide better understanding of the impact of the drought through overlay method. It is apparent from Fig. 15.8 that there was a positive correlation between rainfall rates and vegetation cover over the region. In other words, an increase in the rainfall amounts led to accelerating vegetation growth and vice versa.

As the study area is situated in semi-arid region and the region is mostly rain-fed area, therefore, vegetation growth is mostly based on rainfall. The region has a very high seasonal rainfall pattern and its geographical location made the region vulnerable to sustain vegetation cover. Traditionally, ether vegetation indices or meteorological indices have been used to assess and map drought events. In order to understand the correlation between SPI and NDVI, and the impact of the drought as a conclusion of both vegetation and meteorological indices, a combined NDVI-SPI drought map was developed. As Fig. 15.8 shows, generally, there is a clear difference between the results of 2007 and 2008, as 2008 was experienced a severe drought during the last decade in the region.

During 2007, the Sulaimaniyah shows normal vegetation and rainfall. However, the vegetation experienced stress and loss of vegetation health in 2008 mainly due to lack of rainfall. It can be seen from Fig. 15.8 that there was a large increase in the areas that were classified under the first three classes of drought (extreme, severe and moderate) from 2007 to 2008. For instance, the areas of these three drought classes were increased by around 5,575.7 km^2 from 2007 to 2008. Likewise, a considerable decline of the vegetation cover area was recorded by around 1,676.1 km^2 from 2007 to 2008.

Looking into the individual classes in the combined NDVI-SPI index, also confirmed that drought in 2008 had a huge impact on the vegetation cover all over the region. In this regard, subtracting classes between 2007 and 2008 was conducted in order to observe the impact of the drought at the class level. From (Fig. 15.8), we can see there was a large increase in the percentage of drought classes, particularly the extreme, severe and moderate classes over the region. On the other side, large decline of vegetation classes was also observed including dense, moderate and poor from 2007 to 2008. Interestingly, the current study showed that extreme and moderate drought classes did not exist in 2007, except for a few areas in Sulaimaniyah district. However, the highest percentage of the extreme drought area over the region was recorded for Sulaimaniyah district (9.7%) in 2008, that the area was increased by around 187 km^2 from 2007 to 2008. Therefore, the combined NDVI-SPI index behavior has been assessed for drought events that happened in 2008, showing its capability to distinguish between areas affected by drought.

15.4.6 The Statistical Analysis

Linear regression analysis was applied to investigate correlations among the various variables, which have been employed in this study. What is interesting in the results, that is a positive correlation between NDVI and SPI was observed. Increasing rainfall and moisture are helping to vegetation growth over the region as the region is the mostly rain-fed area. Additionally, a positive correlation was found between NDVI and the Landsat image band 4 (Near Infrared NIR), as in the NIR the reflectance is much higher than that in the visible bands because of the cellular structure in leaves. However, the correlations were negative between NDVI and the bands 1, 2, 3, 5, and 7. Regarding to the soil characteristics, positive correlations were found between NDVI and EC, O.M. and $CaCO_3$, while those correlations were negative with pH and CEC. This result could be explained by the fact that some soil chemical properties might help to improve the growth of vegetation and this might be different form an area to other based on the environment and soil conditions. The results also revealed that there was a positive correlation between SPI and each of soil moisture content, and CEC, in contrast, the correlation was negative with pH, silt percentage and LST. In addition, comparison among the statistical results of the year 2007 and 2008 was also conducted, and the results showed that the correlation between SPI and NDVI was much higher in 2007 than 2008. This result also indicated that a shortage of the rainfall over the region had a considerable negative impact on vegetation cover.

15.5 Conclusions

Developing an integrated index for quantifying drought severity is a challenge for decision makers. The drought indices, which have been used for this study, were: (1) NDVI, as a satellite-based vegetation cover, and (2) SPI, as a meteorological index. Due to restrictions of these indices in drought assessing, the combined NDVI-SPI drought index has been developed using the NDVI and SPI indices. Sulaimaniyah governorate due to its geographical location, the region is affected by irregularities in rainfall resulting frequent occurrence of drought events. Rainfall data, which are collected from 19 meteorological stations, were analysed for the hydrological years 2006/2007 and 2007/2008. The result showed that there was a large drop in rainfall in 2008 than 2007. Generally, spatial distribution of the rainfall is highly varied with respect to altitudinal variation, for instance, the amount of rainfall in mountain areas is higher than in the flat and plain areas. This could be the cause that flat and plain areas are the most affected areas by drought in the current study compare to the high altitude areas. This study found that the combined NDVI-SPI index was more accurate and promising in depicting drought severity than the use of NDVI or SPI individually.

References

Almamalachy YS, Al-Quraishi AMF, Moradkhani H (2019) Agricultural drought monitoring over Iraq utilizing MODIS products. In: Al-Quraishi AMF, Negm AM (eds), Environmental Remote Sensing and GIS in Iraq, Springer Water

Al-Quraishi AMF (2004) Design a dynamic monitoring system of land degradation using Geoinformation technology for the northern part of Shaanxi Province, China. J Appl Sci 4(4):669–674. https://doi.org/10.3923/jas.2004.669.674

Beg AAF, Al-Sulttani AH (2019) Spatial Assessment of drought conditions over Iraq using the Standardized Precipitation Index (SPI) and GIS Techniques. In: Al-Quraishi AMF, Negm AM (eds), Environmental Remote Sensing and GIS in Iraq, Springer Water

Bowers SA, Hanks AJ (1965) Reflection of radiant energy from soil. Soil Sci 100:130–138

CPM (2003) Processing technique for marsh surface condition index. University of Marryland, global land cover facility. Coastal Marsh Project

Dutta D, Kundu A, Patel NR, Saha SK, Siddiqui AR (2015) Assessment of agricultural drought in Rajasthan (India) using remote sensing derived Vegetation Condition Index (VCI) and Standardized Precipitation Index (SPI). Egyptian J Remote Sens Space Sci 18(1):53–63

Edwards DC, McKee TB (2006) Characteristics of 20th century drought in the United States at multiple time scales. Climatology Report No. 97–2, Department of Atmospheric Science, Colorado State University, Fort Collins, CO 80523-1371

Eklund L, Thompson D (2017) Differences in resource management affects drought vulnerability across the borders between Iraq, Syria, and Turkey. Ecol Soc 22(4):9. https://doi.org/10.5751/ES-09179-220409

Fadhil AM (2006) Environmental change monitoring by Geoinformation technology for Baghdad and its neighboring areas. In: Proceeding of international scientific conference of map Asia 2006: the 5th Asian conference in GIS, GPS, aerial photography and remote sensing. Bangkok, Thailand, 28 Aug–1 Sep., 2006. http://www.gisdevelopment.net/application/environment/conservation/ma06-103.htm

Fadhil AM (2009) Land degradation detection using geo-information technology for some sites in Iraq. J Al-Nahrain Univ Sci 12(3):94–108. https://doi.org/10.22401/jnus.12.3.13

Fadhil AM (2011) Drought mapping using Geoinformation technology for some sites in the Iraqi Kurdistan region. Int J Dig Earth 4(3):239–257. https://doi.org/10.1080/17538947.2010.489971

Fadhil AM (2013) Sand dunes monitoring using remote sensing and GIS techniques for some sites in Iraq. In: Proceedings SPIE 8762, PIAGENG 2013: intelligent information, control, and communication technology for agricultural engineering, p 876206. http://dx.doi.org/10.1117/12.2019735

Fadhil AM, Hu GD, Chen JG (2004) Land degradation detection, mapping, and monitoring in the northwestern part of Hebei Province, China, using RS and GIS technologies. In: Proceeding of map Asia 2004: the 3rd Asian conference in GIS, GPS, aerial photography and remote sensing. Beijing, China, 26–29 Aug 2004. https://www.geospatialworld.net/article/land-degradation-detection-mapping-and-monitoring-in-the-northwestern-part-of-hebei-province-china-using-rs-and-gis-technologies/

Giddings L, Soto M, Rutherford BM, Maarouf A (2005) Standardized precipitation index zones for Mexico. Atmosfera 18:33–56

Hameed M, Ahmadalipour A, Moradkhani H (2018) Apprehensive drought characteristics over Iraq: results of a multidecadal spatiotemporal assessment. Geosciences 8(2):58. https://doi.org/10.3390/geosciences8020058

Hammouri N, El-Naqa A (2007) Hydrological modeling of ungauged wadis in arid environments using GIS: a case study of Wadi Madoneh in Jordan. Revista Mexicana de Ciencias Geológicas 24(2):185–196

Hayes JM (1999) Drought indices. National drought mitigation centre. www.civil.utah.edu/~cv5450/swsi/indices.htm. Accessed on 20 Apr 2011

Hayes M (2012) The drought risk management paradigm in the context of climate change. In: Pryor SC (ed) Climate change in the midwest: impacts, risks, vulnerability, and adaptation, Chap 13, Indiana University Press, Indiana

Ionita M, Scholz P, Chelcea S (2016) Assessment of droughts in Romania using the standardized precipitation index. Nat Hazards 81:1483. https://doi.org/10.1007/s11069-015-2141-8

Khalili D, Famound T, Jamshidi H, Kamgar-Haghighi A, Zand-Parsa S (2011) Comparability analyses of the SPI and RDI meteorological drought indices in different climatic zones. Water Resour Manag 25:1737–1757

Kogan F (2008) Monitoring drought and impact on vegetation from space. In: NIDIS knowledge assessment workshop, contributions of satellite remote sensing to drought monitoring, Boulder, CO, USA. URL http://www.drought.gov/workshops/remotesensing/presentations/NIDIS_kogan_presen.pdf. Accessed on 14 May 2009

Kriegler FJ, Malila WA, Nalepka RF, Richardson W (1969) Preprocessing transformations and their effects on multispectral recognition. In: Sixth international symposium on remote sensing of environment, University of Michigan, Ann Arbor, MI, pp 97–131

Lunetta RS, Elvidge CD (1998) Remote sensing change detection: environmental monitoring methods and applications. Ann Arbor Press, Michigan, U.S.A

McKee TB, Doesken NJ, Kleist J (1993) The relationship of drought frequency and duration to time scales. In: Preprints, 8th conference on applied climatology, vol 17 no 22, Anaheim, CA, pp 179–184

Mohammad AH, Jung HC, Odeh T (2018) Understanding the impact of droughts in the Yarmouk Basin, Jordan: monitoring droughts through meteorological and hydrological drought indices. Arab J Geosci 11:103. https://doi.org/10.1007/s12517-018-3433-6

Moradi M, Safari Y, Biglari H, Ghayebzadeh M, Darvishmotevalli M, Fallah M, Nesari S, Sharafi H (2015) Multi-year assessment of drought changes in the Kermanshah city by standardized precipitation index. Int J Pharm Technol 8(3):17975–17987

Murad H, Saiful Islam AKM (2011) Drought assessment using remote sensing and GIS in northwest region of Bangladesh. In: 3rd International Conference on Water & Flood Management (ICWFM-2011)

Notaro M, Yu Y, Kalashnikova OV (2015) Regime shift in Arabian dust activity, triggered by persistent fertile crescent drought. J Geophys Res Atmos 120:10229–10249. https://doi.org/10.1002/2015JD023855

Portela MM, Silva AT, Santos JF, Zeleňáková M, Hlavatá H (2017) Assessing the use of SPI in detecting agricultural and hydrological droughts and their temporal cyclicity: some Slovakian case studies. Eur Water 60:233–239

Rasul G, Ibrahim F (2017) Urban land use land cover changes and their effect on land surface temperature: case study using Dohuk City in the Kurdistan Region of Iraq. Climate 5:13. https://doi.org/10.3390/cli5010013

Richards JA, Jia X (1999) Remote sensing digital image analysis—an introduction, 3rd edn. Springer, New York

Rouse JW, Haas RH, Schell JA, Deering DW (1973) Monitoring vegetation systems in the great plains with ERTS. In: Third ERTS symposium, NASA SP-351 I, pp 309–317

Stow DA (1999) Reducing mis-registration effects for pixel-level analysis of land-cover change. Int J Remote Sens 20:2477–2483

Thenkabail PS, Rhee J (2017) GIScience and remote sensing (TGRS) special issue on advances in remote sensing and GIS-based drought monitoring. GISci Remote Sens 54(2):141–143. https://doi.org/10.1080/15481603.2017.1296219

Tucker CJ, Choudhury BJ (1987) Satellite remote sensing of drought conditions. Remot Sens Environ 23:243–251

Tucker CJ (1979) Red and photographic infrared linear combinations for monitoring vegetation. Remote Sens Environ 1979(8):127–150

Vicente-Serrano SM, López-Moreno JI (2005) Hydrological response to different time scales of climatological drought: an evaluation of the standardized precipitation index in a mountainous mediterranean basin. Hydrol Earth Sys Sci Discuss 2 (4):1221–1246

Wilhite DA (2005) Drought and water crises: science, technology, and management issues. Taylor and Francis, New Work, p 432

Wilhite DA, Glantz MH (1985) Understanding the drought phenomenon: the role of definitions. Water Int 10:111–120

Yagoub YE, Li Z, Musa OS, Anjum MN, Wang F, Bo Z (2017) Detection of drought cycles pattern in two countries (Sudan and South Sudan) by using standardized precipitation index SPI. Am J Environ Eng 7(4):93–105. https://doi.org/10.5923/j.ajee.20170704.03

YCEO (2010) Converting landsat TM and ETM+ thermal bands to temperature. Yale center for earth observation. URL http://www.yale.edu/ceo/Documentation/Landsat_DN_to_Kelvin.pdf

Ye X, Zhang Q, Liu J, Li X, Xu C (2018) Distinguishing the relative impacts of climate change and human activities on variation of stream flow in the Poyang Lake catchment. China J Hydrol 494:83–95

Part VI
RS and GIS for Natural Resources

Chapter 16
Geo-Morphometric Analysis and Flood Simulation of the Tigris River Due to a Predicted Failure of the Mosul Dam, Mosul, Iraq

Younis Saida Saeedrashed and Ali C. Benim

Abstract Floodplain due to a possible failure of Mosul dam is delineated based on the worst scenario corresponding to a dam initial state with a water surface level at 330 m above the sea level. The results show that the flood wave will reach Mosul city within 2 h approximately and attain the maximum wave height of 24.1 m in 8 h after the dam break. The average flood velocity in the city will be 3.9 m/s. The stream power of the Tigris river during the flooding event will be in the range between 3.3–12.39 MW, which is extremely high compared to the range of 1.49–54.9 kW corresponding to normal conditions. The present results are compared with those of the previous studies, discussing the similarities and differences. Finally, the basic design parameters of the Badush dam are predicted, which is envisaged as a repulse dam to protect Mosul city and the other settlement areas downstream. For this dam, the calculations suggest a height of 92 m corresponding to 312 masl leading to the water storage of 9.8×10^{-9} m^3.

Keywords Digital elevation model (DEM) · River analysis system (RAS) · Morphometric analysis · Land use/land cover (LULC) · Flood risk analysis · Dam failure · Mosul dam · Iraq

16.1 Introduction

Mosul is one of the most important cities in Iraq because of its historical background and strategic location. Besides, economically, it is a very important city and has a high population of about 2 millions. Mosul city is located in the North West of Iraq, its geographic coordinates are Latitude: 36° 20′ 06″ (36.3350000°) N, Longitude: 43° 07′ 08″ (43.1188900°) E, and elevation above sea level: 228 m,

Y. S. Saeedrashed (✉) · A. C. Benim
Faculty of Mechanical and Process Engineering, Center of Flow Simulation (CFS),
Düsseldorf University of Applied Sciences, 40476 Düsseldorf, Germany
e-mail: younis.saeedrashed@hs-duesseldorf.de

© Springer Nature Switzerland AG 2020
A. M. F. Al-Quraishi and A. M. Negm (eds.),
Environmental Remote Sensing and GIS in Iraq, Springer Water,
https://doi.org/10.1007/978-3-030-21344-2_16

while dam location is Latitude: 36° 37′ 49″ (36.630178°) N, Longitude: 42° 49′ 23″ (42.823056°) E.

Between 1981 and 1983, Mosul Dam, which is a large multi-purpose dam located at about 60 km northwest of the Mosul city was constructed. Mosul dam is approximately 3.4 km long and consists of an earth-fill embankment dam and a concrete dam structure and in between several units including powerhouse, bottom outlet, concrete-lined spillway, and fuse-plugged secondary spillway. The height of the Mosul Dam is 113 m and the length is 3400 m. The top level, crest level, and bed level are 343.04, 341 and 236 masl, respectively. The maximum flood level is 338.45 masl, while the design operation level (at spillway invert) is 330 masl, which is reduced now to 319 masl due to the fragility of the current situation. The total storage at maximum operation pool is 11.1 billion m^3 and the maximum discharge of Spillway at the maximum pool is 13,000 m^3/s. The embankment is a 113 m high-zoned earth embankment. It is constructed on a foundation of soluble soils that are continuously dissolving, resulting in the formation of underground cavities and voids that place the dam under some continuing risk and thus requires a continuous grouting program to mitigate. The most significant feature regarding the dam foundation is the presence of anhydrite and gypsum bedding and the perpetual grouting program undertaken. Since 1989, Mosul Dam is considered as a floating dam because of the problems at its foundation. So, the dam is continuously under intensive monitoring by the local and international expertise. It is obvious that a failure of the Mosul Dam would have severe consequences. The present study aims to provide a numerical prediction for a failure of the Mosul Dam.

Several computational investigations were performed previously for predicting the hydrologic scenarios in case of a failure of Mosul Dam. In this chapter, three of the most well-known efforts which are SWISS Consultants (1984), Al-Taiee and Rasheed (2009), and Alessandro and Probst (2016) have been chosen in order to be discussed and compared with the present study. Some of the previous investigations are based on the Digital Elevation Model (DEM) and Geographic Information System (GIS) techniques, like the present investigation does. Indeed, the improved computational capabilities based on DEM and GIS techniques, give, in general, new possibilities for hydrologic research in the understanding of the fundamental physical processes underlying the hydrologic cycle and of the solution of the mathematical equations representing those processes.

There are eight factors which influencing on flood condition such as elevation, rainfall, geomorphology, LULC, soil type, drainage, slope (degree), and topographical wetness index (TWI). Elevation is one of the most important criteria for flood hazard assessment. The elevation map of the study area could be obtained from resampled SRTM-DEM data up to 20 m pixel size (Samanta et al. 2018).

Investigation of the geomorphological characteristics of the region under study need to be performed in order to determine active and inactive areas. Landsat aerial photographs are quite essential to determine active or inactive areas and evaluation can be performed according to the surface colour and vegetation detection in photos. A dark surface colour indicated an inactive zone, while lighter surfaces represented more active areas. Besides, field trips need to be carried out to assess some geomorphological features for delineation of active and inactive areas. These information included slop, drainage patterns, topographic contour, superficial characteristics, desert pavement, desert varnish, colour, and distinctive vegetation. Thus, flood hazard map could be delineated based on estimation of flood criteria which are velocity (m/s), depth (m), and geomorphological characteristics which range between very low inactive area up to very high active area (Mollaei et al. 2018).

A diffusive wave approach was proposed for GIS based flow modelling by Liu et al. (2003) for the determination of rainfall runoff and flood routing, which was successfully applied to the Attert catchment in Luxembourg. A further study based on GIS techniques was presented by Vaidja and Chauhan (2013) who analysed the morphometric activities of the lower Satluj river catchment in Himachal Pradesh, India. Youssef et al. (2011) used GIS based morphometry and satellite imagery for flood risk estimation along the St. Katherine road, southern Sinai, Egypt. In a further GIS based study, possible flood damage was estimated by Patel and Srivastava (2013) as a function of maximum water level, the velocity of water flow and the amount of time flood remained in a given land in Surat district, Gujarat, India. Zerger and Wealands (2004) suggested a procedure for risk reduction by providing disaster managers with access to model results in a structured and flexible framework that allows consequences of different hazard scenarios to be assessed and mapped, whereas GIS formed the basis of a common interface with seamless data exchange. A case study based in Cairns, in far-north coastal Australia was presented to illustrate how the system was developed. Liu and Smedt (2005) presented a spatially distributed hydrological model based on GIS and tested it on a small catchment in Belgium. Gallegos et al. (2009) used GIS techniques for modelling dam-break flooding in the urban residential area, for the Baldwin Hills Reservoir in Los Angeles County, USA. In the present work, DEM and GIS based procedures are applied to predict the Mosul Dam failure.

In the previous investigations on the Mosul Dam failure different modelling approaches were used. Principally, for such a complex flow, any numerical prediction procedure must rely on a number of assumptions (Wainwright and Muligan 2004), leading to modelling and numerical errors. Thus, predictions are afflicted with uncertainties. The purpose of the present numerical study is to provide further analysis to the problem using the recent software and paying special attention to achieving high numerical accuracy, as a complementary investigation to the previous studies, for achieving a more reliable assessment of the potentially existing flood risk due to Mosul dam brake. The present solution is compared with the previous ones, outlining the similarities and differences. Compared to the previous studies, a larger amount of solution parameters are presented that offers a broader base for the assessment of the predicted scenario. Finally, basic design parameters are predicted for the Badush dam, which is intended to be built as a repulse dam to protect the Mosul city and the other settlement areas on its downstream.

16.2   Methodology

Below sketched workflow indicates the methodology that enables researchers and engineers to perform geomorphological calculations and investigate inundation zone due to the dam failure using GIS and hydraulic modelling system. The required data such as DEM, Satellite Images and related tabular data need to be inventoried and indexed prior to start the simulation process. Bearing in mind, that the accuracy of hydraulic model being designed and obtained results strongly depend of the accuracy of the hydrologic model and geomorphology of the river-basin being investigated. In this study, hydraulic model is designed based on HEC-RAS program. Besides, GIS software such as ArcGIS and QGIS which they play vital role in pre-processing data that have been used through post-processing stage which include simulation, verification, validation, and calibration consecutively. Figure 16.1 shows the schematic workflow of the hydraulic model.

16.2.1   Software and Geometric Data

Remote sensing and Geographic Information Systems (GIS) are effectively being used in recent times as important tools to determine the quantitative description of morphometry of a basin (Vaidya and Chauhan 2013). GIS provides an integrated platform for using Digital Elevation Model (DEM) data, which can be used for watershed and stream delineation and computation of watershed hydrologic parameters (Shamsi 2005).

In the present work, ArcGIS 10.5 (www.esri.com/argis) and an open source QGIS 3.2 are used as GIS software tool in order to: process 10 m resolution of DEM for extracting related topography of the region under study; process 0.60 m resolution of satellite image to digitalize the area of Mosul city; process 15 m resolution of Landsat satellite image to obtain land use/land cover which is required for determining Manning's values (Dorn et al. 2014) for feature classes: water body, vegetation, soil, built-up areas; prepare hydrograph data, which is tabulated based on recorded daily and monthly discharge. Unsteady, two-dimensional flow simulations are performed using the hydrologic flow prediction software HEC-RAS 5.0.3 (www.hec.usace.army.mil/software/hec-ras/).

16.2.2   Morphometric Analysis

The process of morphometric analysis and design of hydrologic model system require the use of geometric data with sufficient quality from both categories, i.e. raster data and vector data. In this study, DEM with 10 m resolution and topographic sheets with scales; 1:50,000 were used. Besides, related satellite images with 0.6 m resolution

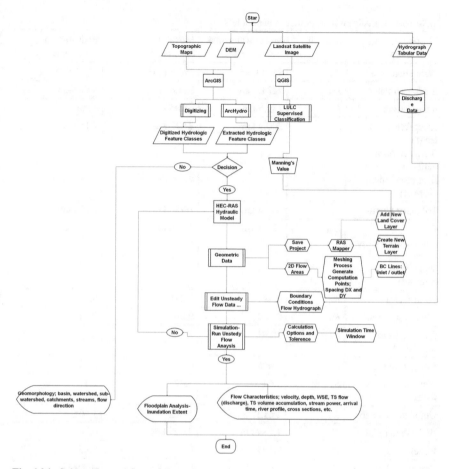

Fig. 16.1 Schematic workflow of the hydraulic model

were used for checking up the compatibility between the extracted feature classes from the DEM such as main river-channels, tributaries and watershed boundaries with the digitized feature classes from the hydro-topographic maps. It is essential that the process of extracting and digitizing such feature classes are done accurately, since the accuracy of morphometric calculations mainly depends on the quality of these feature classes being extracted from the DEM or being digitized from the topographic sheets. The extracted feature classes are rivers, tributaries, watershed, and catchments from the digital elevation model. These feature classes have been examined to remove topological errors. The centroid of the watershed and the longest flow path of the watershed were determined. These feature classes were delineated for morphometric calculations. Figure 16.2 shows the centroid of the watershed and the flow path of the main channel. The accordingly calculated parameters are tabulated in Table 16.1.

Table 16.1 Calculated morphometric parameters

Morphometric parameters	Symbol	Mosul watershed	Mosul dam catchment	Badush dam catchment
Watershed area	A_w	8512 km^2	295.8 km^2	31.41 km^2
Basin perimeter	L_p	7124.3 km	134.6 km	39.79 km
Basin length	L_b	149.5 km	32.04 km	9 km
River (channel) length* (between basin outlet and end of river along main channel)	L_c	200.9 km	27.22 km	7.159 km
Tributary length	L_t	500.4 km	100.8 km	14.26 km
L_{ca}* (length from basin outlet to a point adjacent to centroid)	L_{ca}	177.5 km	19.18 km	4.762 km
Form factor	R_{ff}	0.38	0.288	0.387
Circularity ratio	R_c	0.209	0.205	0.249
Elongation of watershed	L_l	15.98	5.165	2.323
Elongation ratio	R_e	0.696	0.605	0.702
Watershed shape factor	S_b	2.62	3.47	2.57
Unity shape factor	R_u	1.612	1.862	1.605
Watershed relief	H	1.239 km	0.429 km	0.18
Relief ratio	R_h	0.174	0.318	0.452
Relative relief	R_p	0.00174	0.00318	0.00452
Drainage density	D	0.082	0.4328	0.677
Ruggedness number	R_n	0.1019	0.1856	0.1218
Constant of channel maintenance	C	12.147	2.31	1.475
Fineness ratio	R_f	0.282	0.202	0.179
Stream frequency	C_f	0.0823 km/km^2	0.432	0.677
Average length of contour	L_{ac}	3.68	3.93	2.69
Watershed slope %	L_s	0.76	0.9	1.61
Main "River" (channel) slope*	C	0.00203	0.00565	0.00517
Watershed (Basin) Slope*	S	0.00828	0.013	0.02

(continued)

Table 16.1 (continued)

Morphometric parameters	Symbol	Mosul watershed	Mosul dam catchment	Badush dam catchment
Mean slope of overall basin %		0.69	0.65	0.78
Gray's factor	G	3939.58	255.09	66.2
Murphy		0.0003078	0.0117	0.082
Lag time*	t_p	48.2 h	8.85 h	2.89 h

Some of the parameters shown in Table 16.1 are used as input parameters for HEC-RAS. These parameters are indicated by an asterisk "*". The remaining parameters are displayed for having provided a more complete data for researchers that may be interested in the problem, as such data was normally not presented in the previous studies.

The "lag time" (t_P) is defined as the time elapsed until the flood reaches its maximum level after it reached the corresponding destiny. Based on the basin slope (S), the river channel length (L_c), the watershed shape parameter (L_{ca}) (Table 16.1), it can be estimated from the following empirical relationship (Ghanshyam 2010).

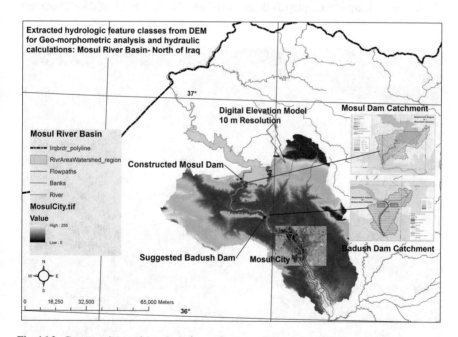

Fig. 16.2 Geo-morphometric analysis for catchments of Mosul dam and Badush dam

$$t_p = 0.72 C_l \left(\frac{L_c L_{ca}}{\sqrt{S}} \right)^{0.38} \tag{16.1}$$

where the empirical coefficient C_l takes the value $C_l = 1.715$ for the mountainous region, $C_l = 0.5$ for valleys and $C_l = 1.03$ for foothills. The estimated lag times according to Eq. (16.1) care also displayed in the table.

16.2.3 Hydraulic Modeling

Land use/land cover (LULC) classification is one of the most important elements in the hydraulic calculations, since the local Manning's roughness coefficients (Dorn et al. 2014), which represent the roughness of the river-basin, and, thus, have a decisive influence on the flow prediction are strongly dependent on this classification. In this study, Landsat satellite image with 15 m resolution is used for performing classification of the region under study. Generally, four different land covers were classified which are water body, vegetation area, urban area, and the soil texture. Figure 16.3 shows the classified LULC using image analysis process.

For the Manning friction coefficients of the different classes, the values are used, which were commonly employed in the literature (Bedient et al. 2008). These are:

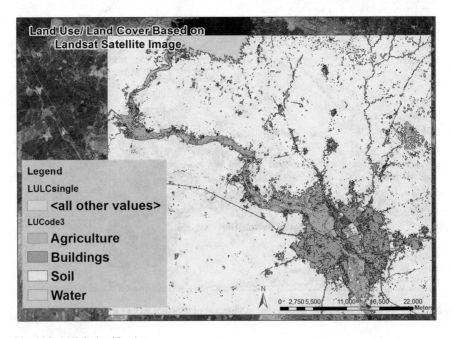

Fig. 16.3 LULC classification

0.03 for Tigris river channel (water body), 0.04 for vegetation and river banks, 0.01 for Mosul city (urban area) and 0.05 soil texture. In this model, the Manning roughness coefficient for the river channel was calculated and calibrated using Manning equation for a real cross section near Mosul hydrological station for many measured discharges that had been recorded in the river (Al-Taiee and Rasheed 2009).

The inlet boundary at upstream is defined to be at the location of the reservoir storage and Mosul Dam. Monthly hydrograph for the inlet is determined as an upstream boundary condition with energy grade slope equal to 0.05. The normal depth is determined for the downstream as outlet boundary condition.

In the current study, instead of depending on different scenarios of possible overtopping processes, it assumed that the Mosul Dam would fail to discharge water at a required rate via spillways and the bottom outlets on time. It should be noted that the problems at the foundation implicitly increases the likeliness of such a scenario, since the dam top level is continuously decreasing, due to the gradual sinking of the dam structure, despite grouting. Thus, the water flow will exceed the capacity of the spillways and the process of the flooding will start at elevation 330 masl as the worst scenario. The distance between the Mosul Dam to the Mosul city is considered to be 69,000 m. The approximate total area of Mosul city is 209.5 km^2, the elevation of Mosul city is about 228 masl and elevation of the Tigris river bed is 210 masl.

Generally, HEC-RAS uses two basic methods for simulating water: Either the full momentum equations are used (Saint Venant Equations), or a diffusion wave approximation is applied. The present calculations are performed using the diffusion wave approach, since otherwise it was not possible to obtain a stable solution.

16.2.4 Meshing and Grid Independence Study

The two-dimensional domain in the x–y plane was meshed by an equidistant spacing in both spatial directions. The cell sizes in both directions were also equal to each other ($\Delta x = \Delta y$) so that square shaped cells result. In mesh-based numerical simulation, a grid independence study should always be performed for ensuring that the discretization errors are sufficiently small and do not falsify the prediction. For this purpose, three solutions were obtained on three grids with different resolutions, and the results were compared, as shown in Table 16.2.

Table 16.2 Grid-independence study and obtained results from the simulation process

Predicted flow characteristics	Grid spacing and number of grid nodes		
	$\Delta x = \Delta y = 50$ m 894,196 Cells	$\Delta x = \Delta y = 100$ m 223,375 Cells	$\Delta x = \Delta y = 150$ m 99,141 Cells
Max. flood depth (m)	44.46	44.67	44.71
Max. water surface elevation (masl)	247	247.69	249.72

The results obtained by the three different grids are quite similar and indicate that grid resolution effects are rather small within the considered range (Table 16.2). The medium size grid with $\Delta x = \Delta y = 100$ m and 223,375 cells is used as the basic grid.

16.3 Results

16.3.1 Predicted Field Distributions

A detailed plot of the calculated 2D flow field at an instant of time is presented in Fig. 16.4, where the meandering shape of the Tigris River and the resulting flow pattern can be observed.

The flood front arriving at the urban areas with an indication of the velocity distribution is depicted in Fig. 16.5a, b.

16.3.2 Outline of the Previous Investigations

A very detailed study of the Mosul dam failure was presented by SWISS Consultants (1984). In the following, this investigation, will be referred to as the "SWISS" study.

Fig. 16.4 Detail view of predicted the flow field

A further study was provided by (Al-Taiee and Rasheed 2009). In the following, this investigation will be referred to as the "IWTC" study. The most recent numerical investigation of Mosul dam failure was published by Alessandro and Probst (2016). In the following, this investigation will be referred to as the "JRC" study. The present study will be referred to as the "CFS" study in the following.

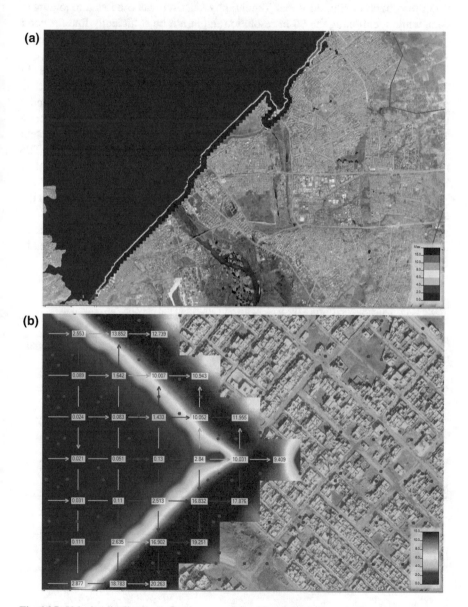

Fig. 16.5 Velocity distribution: **a** flood wave arriving Mosul city, **b** the wave front in larger detail

In the present study a high resolution DEM with 10 m resolution was used, while in the SWISS study, traditional surveying method was used for collecting topographic data. The IWTC study was simply based on two elevations, namely the water level of the reservoir, and the bottom outlet elevation of the breach, without considering any topologic detail data of the related terrain surfaces. In the JRC study, resampled 90 m resolution of Shuttle Radar Topography Mission was used as a topographic data, while mentioning the 90 m resolution might not be sufficiently fine for good accuracy. In all three previous studies (SWISS, IWTC, and JRC) no grid dependence study was performed.

In all studies, different software were used for simulating the flood wave. In the SWISS study, the program (FLORIS) which is based on a one-dimensional mathematical model was used. In the IWTC study, the simplified dam break model (SMPDBK) was used. The most recent study JRC used HyFlux2 computer code which was an in-house developed code to analyse dam break problem. In the present study, the flood wave is simulated by the two-dimensional HEC-RAS code. The main differences between the previous and present investigations are outlined in Table 16.3.

There are also some deviations in the initial and boundary conditions between the studies. These are summarized in Table 16.4.

Note that the distance corresponds to the city centre. The city boarder has a smaller distance to the dam, which is 54 km in the present study. This was not indicated in the previous studies.

Table 16.3 Differences between the present investigation and the previous investigation

	SWISS	IWTC	JRC	CFS
Software	FLORIS (1D)	SMPDBK (2D)	HyFlux2 (2D)	HEC-RAS (2D)
Topographic data	Surveying	Maps/S. Image	DEM-90 m	DEM-10 m
Grid-resolution	No information provided	No information provided	180 m/cell	50 m/cell
Grid-independence study	No information provided	No information provided	No information provided	Performed

Table 16.4 Differences in the boundary and initial conditions of different studies

	SWISS	IWTC	JRC	CFS
River bed at Mosul city (masl)	208.2	210	209	210
River water level at Mosul city (masl)	218.4	222	217	221
Distance (Mosul city dam) (km)	69.2	67.4	69	69

16.3.3 Results in Comparison with Previous Investigations

Figure 16.6 shows the comparison between inundated areas that obtained from the mentioned studies. It should be taken into consideration that the Mosul city has expanded in the area during the last decades.

One can see that the results are qualitatively similar. A quantitative comparison will be discussed below, based on Table 16.5, where the predicted major parameters by all three studies are presented.

The maximum discharge levels and initial speeds are similar between the SWISS, JRC and present studies, whereas IWTC source provided no data for these quantities.

For the wave arrival time, the present study predicts, with 1.9 h, the longest time. With 1.4 h, the shortest wave arrival time was predicted by JRC. Wave arrival times predicted by SWISS and IWTC predict are 1.5 and 1.7 h, respectively, which remain within the range given by JRC and CFS. The range of deviation of the predictions (from 1.9 to 1.4 h) is about 30% with respect to an arithmetic average of the four predicted values.

For the time required for the maximum wave height to be achieved in the Mosul city, the minimum time is, again predicted by JRC study, as 6.2 h. The IWTC study, which predicted an intermediate value for the wave arrival time, predicts, with 9.0 h, the maximum time for the achievement of the maximum wave height. SWISS and CFS studies predict very close values with 8.0 and 8.5 h, respectively. For the prediction of the time of maximum wave height, a deviation (in a sense described above) of

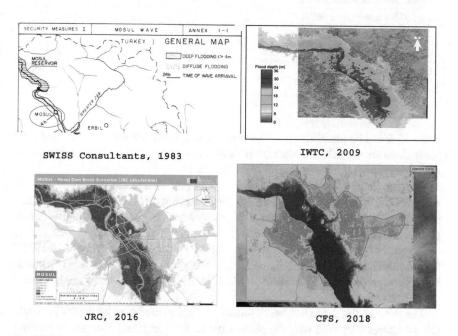

Fig. 16.6 Inundated areas inside Mosul city

Table 16.5 Comparison between the recent study and the previous studies

Hydraulic parameters	Studies			
	SWISS, 1984	IWTC, 2009	JRC, 2016	CFS, 2017
Maximum flood discharge (10^3 m^3)	402.000	–	–	404.875
Speed of flood at the beginning of dam failure (km/h)	40	–	40	38
Time of wave arrival to the Mosul city (h)	1.5	1.7	1.4	1.9
Time to maximum wave height in Mosul city (h)	8.5	9	6.2	8
Average flood velocity in Mosul city (m/s)	4.4	3.5	–	3.9
Max. flood level in Mosul city (WSE) (masl)	242.7	235.2	236.2	232.3
Maximum flood depth in Mosul city (m)	–	36	–	36.7
Maximum flood depth in Tigris River (m)	46.3	–	–	42.64
Maximum wave height in Mosul city (m)	24.3	25.2	26.3	24.01
Inundated area within the city (km^2)	74	–	–	121.6
% Inundated area within the city	–	54	–	58
Flood width around river bed (km)	3–10	–	–	4.5–11
Advisable evacuation distance (km)	–	–	4–5	5
Shear Stress in river bed (N/m^2)	–	–	–	1.5–42.6
Stream power of river (kW)	–	–	–	1.49–54.9
Stream power of flood (MW)	–	–	–	3.3–12.39

about 35% can be identified, whereas one can also see that SWISS, IWTC and CFS predictions are very close to each other (with a deviation of about 10%), whereas the JRC study shows a large deviation from the others.

Average flood velocity in Mosul city was not provided in the JRC study. SWISS and IWTC studies predict 4.4 and 3.5 m/s, respectively, whereas the CFS study predicts a value, i.e. 3.9 m/s, which is nearly equal to the arithmetic average of both values. Here, a deviation of the available results within a range of 28% can be observed. For the average flood level in the Mosul city, the predictions of IWTC (235.2 masl) and JRC (236.2 masl) studies are very close to each other, whereas the values predicted by the SWISS (242.7 masl) and CFS (245.39 masl) show a similarity.

The average flood depth value in Mosul city which predicted by the IWTC study (36 m) agrees very well with the present prediction (36.12 m). For this parameter, no values were presented by the SWISS and JRC studies. For the maximum flood depth in Mosul city, a value of 46.3 m was provided by the SWISS study, whereas no information was given by the IWTC and JRC sources. The presently predicted value of 44.67 m/s for the maximum flood depth in Mosul city agrees rather well by the prediction of the SWISS study (with a deviation of about 3.5%).

For the maximum wave height in Mosul city, the four predictions show a comparably good agreement with each other. The minimum value, 24.3 m, is predicted by the SWISS study, which is very close to the present prediction of 24.39 m. IWTC and JRC studies predict slightly higher values, with 25.20 and 26.30 m, respectively. In general, the deviations occur within a range of about 8%.

The inundated areas predicted by the SWISS ($74 km^2$) and present ($121.6 km^2$) studies differ mainly because of the time lag between both studies (33 years from 1984 to 2017) during which the city experienced a considerable expansion. For the percentage inundated area the value provided by IWTC (54%) show a rather good agreement with the presently predicted value (58%). The flood with around the river bed was given within the range 3–11 km by the SWISS study. These numbers show a fair agreement with the presently obtained values of 4.5–11.0 km. Some additional parameters such as Shear stress in the river bed, stream powers of river and flood, as predicted by the present study, are provided in Table 16.5, for information (for these parameters, no information was provided in the previous studies).

16.3.4 Required Height of the Repulse Badush Dam

For the case of a break of the Mosul dam, it is of utmost importance to have a repulse dam between the Mosul dam and Mosul city that can absorb the flooding water and protect the city. Such a dam should have the capability of storing a huge amount of water, as the water volume kept by the Mosul dam is about 11.1 billion m^3. As already identified by the previous studies, the already existing Badush dam, which is situated 40 km downstream the Mosul dam has the potential of being reconstructed for this purpose. The Badush dam currently has a design flood level of 250 masl. corresponding to a storage capacity of $0.39 \times 10^9 m^3$, which is too small to act as a repulse dam. With the help of the digital elevation model, we have determined that a reconstruction of the dam to a flood level up to 307 masl enables a quite large reservoir of $9.8 \times 10^9 m^3$, which is sufficiently close to the required capacity. The calculated numbers for the proposed reconstruction of the Badush dam are summarized in Table 16.6.

The reservoir shapes corresponding to the two different flood levels are shown in Fig. 16.7.

Table 16.6 Badush dam date reservoir for two cases

Dam	Water level (masl)	Dam height (m)	Reservoir volume (m³)	Reservoir surface area (m²)
Badush (existing)	260	35	0.39×10^9	40.1×10^6
Badush (proposed)	312	92	9.8×10^9	357.69×10^6

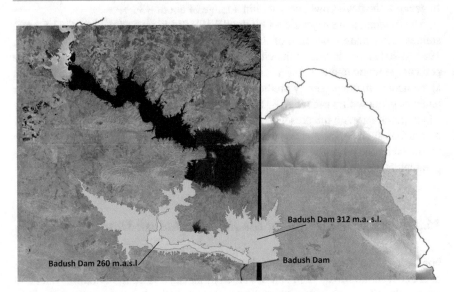

Fig. 16.7 Badush dam as a repulse dam (for two different flood levels)

16.4 Summary

A geo-morphometric analysis and flood simulation of the Tigris River due to a predicted failure of the Mosul Dam is presented. The results are compared with those of other authors. The present predictions are observed to be in good agreement with those of authors, specifically with the results of the SWISS study. Given the uncertainties in many model parameters and modelling assumptions, this can be seen as a reciprocal validation of the models that increases the confidence in the numerical results. In the present work, a larger amount of predicted data is presented in comparison to previous studies, which may be useful for the analysis of the case. The present predictions indicate that the initial flood wave will reach the Mosul city in around 2 h and the height of the flood wave will reach approx. 24 m within 8 h, while the average flood velocity is predicted to be 3.9 m/s. The stream power of the Tigris River which ranges between 1.49 and 54.9 kW under normal conditions, is predicted to attain an enormous increase during the flooding event to reach the range 3.3–12.39 MW. The sufficient height of the repulse Dam, which is envisaged to be a reconstruction of the already existing Badush Dam, is predicted to be 92 m, which corresponding to a water level of 312 masl. Dam that is capable of storing 9.8×10^9 m^3 of water and protect the Mosul city and the other settlement areas downstream.

Acknowledgements Authors would like to express their great thankful to both editorial board and reviewers for their constructive comments and suggestions to improve the manuscript. We are thankful also to the centre of flow simulation at Dusseldorf University of Applied Sciences for providing required support.

References

ArcGIS http://www.esri.com/arcgis

Alessandro A, Probst P (2016) Impact of flood by a possible failure of the Mosul dam, version 2. European Commission

Al-Taiee TM, Rasheed AM (2009) Simulation Tigris river flood wave in Mosul city due to a hypothetical Mosul dam break. In: Proceedings 13th international water technology conference, IWTC 13, Egypt, pp 283–299

Bedient PB, Huber WC, Vieux BE (2008) Hydrology and floodplain analysis. Pearson Hall

Dorn H, Vetter M, Hofle B (2014) GIS-based roughness derivation for flood simulation. Remote Sens 6:1739–1759

Ghanshyam D (2010) Hydrology and soil conservation engineering-watershed management. PHI Press

Gallegos HA, Schubert JE, Sanders BF (2009) Two dimensional, high resolution modelling of urban dam-break flooding: a case study of Baldwin Hills, California. Adv Water Resour 32:1323–1335

HEC-RAS http://www.hec.usace.army.mil/software/hec-ras/

Liu YB, De Smedt F (2005) Flood modeling for complex terrain using GIS and remote sensed information. Water Resour Manage 19:605–624

Liu YB, Gebremeskel S, De Smedt F, Hoffmann L, Pfister L (2003) A diffusive transport approach for flow routing in GIS-based flood modeling. J Hydrol 283:91–106

Mollaei Z, Davary K, Hasheminia SM, Faridhosseini A, Pourmohamad Y (2018) Enhancing flood hazard estimation methods on alluvial fans using an integrated hydraulic, geological and geomorphological approach. Nat Hazards Earth Syst Sci 18:1159–1171. https://doi.org/10.5194/nhess-18-1159-2018

Patel DP, Srivastava PK (2013) Flood hazards mitigation analysis using remote sensing and GIS: correspondence with town planning scheme. Water Resour Manage 27:2353–2368

Samanta RK, Bhunia GS, Shit PK, Pourghasemi HR (2018) Flood susceptibility mapping using geospatial frequency ratio technique: a case study of Subarnarekha River Basin, India. Model Earth Syst Environ 10(1007):s 40808-018-0427-z. https://doi.org/10.1007/s40808-018-0427-z

Shamsi UM (2005) GIS applications for water, wastewater, and stormwater system. CRC Press

SWISS Consultants—Consortium for Consulting Engineering Services (1984) Mosul flood wave, vols 1, 2, and 3. Ministry of Irrigation, Iraq

Vaidya K, Chauhan R (2013) Morphometric analysis using GIS for sustainable development of hydropower project in the lower Satluj river catchment in Himachal Pradesh. Int J Geomatics Geosci 3:464–473

Wainwright J, Mulligan M (2004) Environmental modelling-finding simplicity in complexity. Wiley Ltd.

Youssef AM, Pradhan B, Hassan AM (2011) Flash flood risk estimation along the St. Katherine road, Southern Sini, Egypt using GIS based morphometry and satellite imagery. Environ Earth Sci 62:611–623

Zerger A, Wealands S (2004) Beyond modelling: linking models with GIS for flood risk management. Nat Hazards 33:191–208

Chapter 17
Hydrologic and Hydraulic Modelling of the Greater Zab River-Basin for an Effective Management of Water Resources in the Kurdistan Region of Iraq Using DEM and Raster Images

Younis Saida Saeedrashed

Abstract The Greater Zab river-basin is one of the most important catchment areas in the Kurdistan region amongst the other four catchments which are; Lesser Zab, Sirwan, Khabur, and Uzem. Thus, this research paper seeks to enhance hydrologic information systems in the region under study via advanced technologies such as geographical information systems and hydrologic and hydraulic modelling system using remotely sensed data such as digital elevation models and satellite images. These techniques provide powerful and cost-effective tools for managing water resources and enable hydrologists and researchers to handle up-to-date hydrologic and hydraulic data and information. In this study, geomorphological analysis of the Greater Zab River-Basin has been performed by using digital elevation model and the obtained results are compared with that already obtained from the digitizing process. Computational hydrologic and hydraulic modelling systems are designed using interface method which links GIS with the modelling systems (HEC-HMS and HEC-RAS). Floodplain analysis of the Greater Zab River has been presented. Hydraulic model has been validated based on the obtained hydrograph from the Khabat flow gauge station which is located at the lower reach of the Greater Zab River. The comparison between the gauge station measurements (average discharge 750 m^3/s and pick discharge 1,500 m^3/s) and the results that obtained from the hydraulic model (707.92 m^3/s and pick discharge 1415.18 m^3/s) shows good agreement between both results.

Keywords Digital elevation model (DEM) · Hydrologic modelling system (HMS) · River analysis system (RAS) · Hydrologic and hydraulic (H&H) · Land use/land cover (LULC)

Y. S. Saeedrashed (✉)
Faculty of Mechanical and Process Engineering, Center of Flow Simulation (CFS),
Düsseldorf University of Applied Sciences, 40476 Düsseldorf, Germany
e-mail: younis.saeedrashed@hs-duesseldorf.de

© Springer Nature Switzerland AG 2020
A. M. F. Al-Quraishi and A. M. Negm (eds.),
Environmental Remote Sensing and GIS in Iraq, Springer Water,
https://doi.org/10.1007/978-3-030-21344-2_17

17.1 Introduction

During the last decades, global climate change has caused water shortages and droughts in many different areas all over the world. Consequently, some regional conflicts have raised between some neighbour countries about water. On the contrary, also global climate change has caused flooding in some other parts all over the world. Thus, water resources are paid intensive attention by the local authorities and international community as one of the most important ingredients of the environment that need to be protected and managed correctly.

Droughts can be defined as a deficit in effective rainfall in a period compared to previous periods, or is a reduction of agricultural products and loss of plants (Karamouz et al. 2003). While, flood is defined as extremely high flows or levels of rivers, lakes, ponds, reservoirs, and any other water bodies, whereby water inundates outside of the water body area (Marfai 2003). During the last decades, many studies and much research have been undertaken and published about the use of GIS applications for water resources. Thus, the use of GIS in the field of water resources is becoming more and more common (Brown and Dee 2000). Consequently, GIS applications have played a vital role in developing water resources management (Rivas and Lizama 2008). GIS applications can be defined as computer based-technology for handling geographical data in digital form. It is designed to capture, store, manipulate, analyze, and display diverse sets of spatial or geo-referenced data (Singh and Fiorentino 1996). However, remote sensing can be defined as the science and technology of extracting information about an object, area, or phenomenon through the process of image analysis of data that is captured in the distance (Lillesand et al. 2008).

During the 1990s, GIS was used as an effective tool for hydrologic modelling using digital elevation models (DEMs) for watershed and stream network delineation (Maidment 2002). The development of DEM processing algorithms as well as relevant software to extract hydrological information from DEM is increasing, making its application wide (Nguyen et al. 2003). Singh and Fiorentino (1996) pointed out that DTMs and GIS have been used by Djokic and Maidment for urban storm water modelling. Lanza et al. developed a set of automatic procedures for flash flood forecasting using a DTM and hydrologically oriented GIS. Cupertino et al. used DTM and GIS for calculating geomorphologic parameters which is useful in modelling flood forecasting. Hydrological information system (HIS) is a combination of hydrologic data, tools, and simulation models that supports hydrologic science, education and practice (Maidment 2005).

Understanding hydrologic and geomorphologic characteristics of the region under study and preparing the required hydrologic and climatic datasets is a very important step in watershed planning and management. Watershed characteristics: area, length, slope, shape, surface roughness, soil type, vegetation, and land cover need to be calculated precisely before undertaking the watershed planning process. In addition to watershed characteristics, river geomorphology; channel length, channel slope, channel roughness, channel form, and cross section information is usually necessary for hydrologic analysis, planning, and management (Karamouz et al. 2003).

Investigation of the geomorphological characteristics of the region under study need to be performed in order to determine active and inactive areas. Landsat aerial photographs are quite essential to determine active or inactive areas and evaluation can be performed according to the surface colour and vegetation detection in photos. A dark surface colour indicated an inactive zone, while lighter surfaces represented more active areas. Besides, field trips need to be carried out to assess some geomorphological features for delineation of active and inactive areas. These information included slop, drainage patterns, topographic contour, superficial characteristics, desert pavement, desert varnish, colour, and distinctive vegetation. Thus, flood hazard map could be delineated based on estimation of flood criteria which are velocity (m/s), depth (m), and geomorphological characteristics which range between very low inactive area up to very high active area (Mollaei et al. 2018).

GIS has the capability to integrate with the other applications such as Hydrologic and Hydraulic (H&H). This potential allows the user to devote more time to solving problems and less time for inputting data and interpreting reams of model output (Shamsi 2005).

An integrated and numerically efficient hydrological-hydraulic model for practical simulation of surface water dynamics could be developed. Hydrologic and hydraulic modelling could be developed for flood inundation mapping. Flood risk is quite high in areas where the catchment is large with defined flow route that inundates vast downstream flood plain region during high flood level (Khadka and Bhaukajee 2018). There are eight factors which influencing on flood condition such as elevation, rainfall, geomorphology, LULC, soil type, drainage, slope (degree), and topographical wetness index (TWI). Elevation is one of the most important criteria for flood hazard assessment. The elevation map of the study area could be obtained from resampled SRTM-DEM data up to 20 m pixel size (Samanta et al. 2018).

GIS can be an important tool in modelling many different variables that vary in space and time. Generally, using GIS for analysis or environmental modelling can be employed either internally within the GIS itself or externally by linking the GIS to a computer model. In the first case, where the modelling is not too complex, the GIS can be used for all three main stages: data preparation, modelling, and result presentation. While in the second case, the GIS can be used for data preparation and result presentation, but the more complex modelling can be implemented externally by linking the GIS to a model (Joao and Fonseca 1996). Perhaps the greatest application of GIS in hydrology has been in the area of watershed modelling. This is understandable because the computer models of watershed hydrology are highly data intensive and GIS is a natural technology for processing volumes of data (Singh and Fiorentino 1996).

Watershed models can be classified into two main types according to the method of treating spatial component of watershed hydrology which are the Lumped parameter model and distributed models. Lumped-parameter models lump the input parameters of a study area over polygons and use vector GIS applications. While, distributed models distribute the input parameters of a studied area over grid cells and use raster GIS applications (Shamsi 2005). Over the past decade, numerous significant advances have been made in the linkage of GIS and various research and application models (Miller et al. 2007). According to the three methods of GIS linkage to computer models that have been developed by Shamsi, there are three methods of using GIS applications for modelling watershed: Interchange Method, Interface Method, and Integration Method. In the first case, there is no direct link between a GIS and the model, both can be run separately and independently. While in the second case, there is a direct link for automate transfer of data and information between the GIS and computer model. This can be undertaken by adding new menus or buttons to a GIS. As opposed to both, an integration method merges between a model and GIS. This program combination offers both GIS and modelling functions. This approach is called GIS-based integration (Shamsi 2005). GIS-based model/ model-based GIS have been widely released and developed. For example, EPA's BASINS Program has been developed by the U.S. Environmental Protection Agency and Watershed Modelling System (WMS) by Environmental Modelling Research Laboratory, Brigham Young University (An 2007).

GIS-based tools, such as Automated Geospatial Watershed Assessment (AGWA) and Soil and Water Assessment Tool (SWAT), can be used for supporting watershed analysis (Rivas and Lizama 2008). ArcGIS/Arc Hydro is an example of GIS-based tools that enhance hydrologic information systems (HIS) and synthesize geospatial and temporal water resources data to support hydrologic analysis and modelling (Maidment 2002). Arc Hydro is a powerful framework for building a water resources information system (Fox et al. 2006). However, it differs from the simulation model (e.g. HEC-RAS, HEC-HMS, and SWAT), because it does not simulate or model any physical or environmental processes (Merwade 2008).

17.2   Objectives and Scope

The main objectives of this chapter are creating both hydrologic model and hydraulic model for the Greater Zab River-Basin by using GIS and remotely sensed data and raster images such as DEM, Satellite Images, and topographic maps with different scales which consist an effective platform for calculating hydrologic and hydraulic parameters accurately. While the scope of this research paper includes; geomorphologic calculations of the Greater Zab watershed, flood mapping of the Greater Zab River, and validation of the hydraulic model.

17.3   Physical Characteristics of the Study Area

Administratively, the Kurdistan region of Iraq includes three governorates which are; Erbil, Duhok, and Sulaimaniya. The total area of each governorate is 15,370, 10,200, and 19,340 km^2 respectively, with the total population approximately about

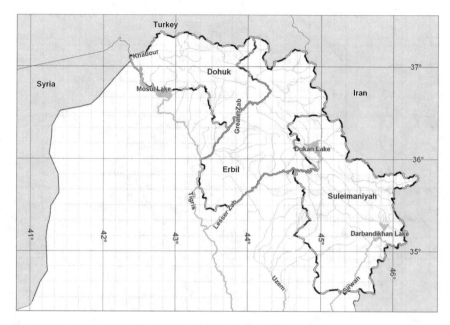

Fig. 17.1 Latitude and longitude of the Kurdistan region of Iraq

4,382,291.[1] The Kurdistan region of Iraq is located approximately between the SN Latitude 34° 30′ and 37° 20′ and the EW Longitude 42° 20′ and 46° 20′, bordered from the north by Turkey and from the east by Iran, as shown in the Fig. 17.1.

The Greater Zab watershed is the largest watershed in the Kurdistan region. It is one of the five major catchments. The other four catchments are: Khabour, Lesser Zab, Uzem, and Sirwan (Fig. 17.2). The Greater Zab watershed covers an area of 16,690 km², with the total perimeter 1,053 km. However, the total area of this watershed is 20,513.53 km² with total perimeter 1,141.47 km, since 81% of the watershed area is located within the international border of Iraq, and 19% inside Turkey. The Greater Zab watershed is located approximately between SN Latitude 35° 40′ to 37° 20′ and EW Longitude 43° 05′ to 45° 05′ and it is within the administrative boundaries of both of the governorates; Erbil and Duhok as shown in (Fig. 17.3). The elevations of this watershed range approximately from 200 to 3300 masl.

17.3.1 Hydrologic Variables and Parameters

The major hydrologic variable and parameters are; precipitation (P), infiltration (I), evaporation (E) which includes evapotranspiration from plants (T_s) and evaporation

---

[1] According to the last population published by the committee of elections in the Kurdistan Region (May 2009).

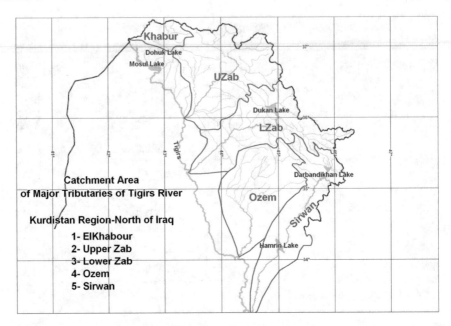

Fig. 17.2 Catchment area of major tributaries of Tigris river

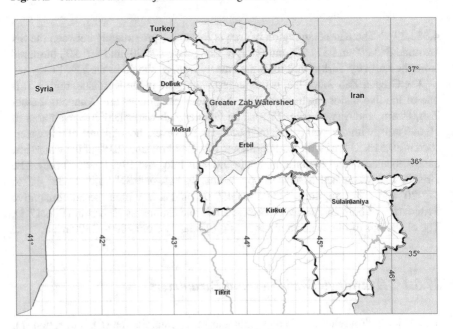

Fig. 17.3 Physical boundary of the Greater Zab watershed

from soil moisture $\left(E_g\right)$. Additionally, surface Runoff (R), ground water appearing as surface runoff; springs and wells $\left(R_g\right)$, subsurface flow (R_s), and finally groundwater flow (G).

These hydrologic parameters and variables are included within the main four hydrologic continuous elements which are; basin, watersheds, sub-watersheds, catchments, and streams. Figure 17.4 shows these five elements which extracted from the related DEM to the Greater Zab River-Basin.

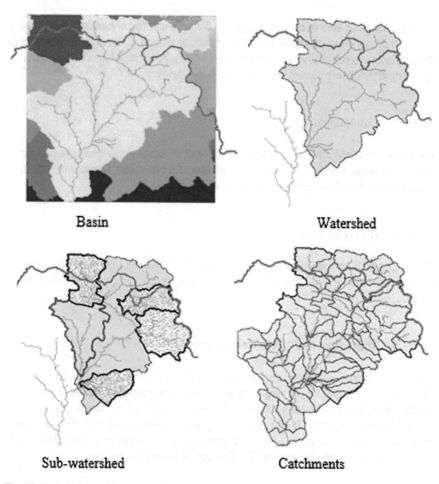

Basin   Watershed

Sub-watershed   Catchments

Fig. 17.4 Hydrologic elements of the Greater Zab river-basin

17.3.1.1 Basins

Basins can be defined as main huge drainage areas which include all types of water bodies such as rivers, streams, lakes, and ponds. Topographically, each basin is separated form an adjacent basin by a divide such as ridge, hill, or mountain.

17.3.1.2 Watershed and Sub-watersheds

A watershed is an area of land that captures water in any form, such as rain, snow, or dew, and drains it to a common water body, such as stream, river, or lake. Watersheds are a tessellation or subdivision of a basin into drainage areas selected for a particular hydrologic purpose which is so-called sub-watersheds.

17.3.1.3 Catchments

Each sub watershed includes a specific number of drainage areas which can be called catchment areas. Catchments are a subdivision of a basin into elementary drainage areas defined by a consistent set of physical rules.

17.3.1.4 Rivers and Streams

Rivers and Streams can be defined as a network of an open channel drainage lines consists of main channel and tributaries that determined by upper reach and lower reach.

17.3.1.5 Watershed and River Geomorphology

Watershed geomorphology can be represented mathematically by calculating its drainage area, length, slope, shape. While, river geomorphology can be represented mathematically by calculating its length, slope, cross-section, roughness, channel form. Channel roughness is an important input because it can be used for estimation of water elevation, velocity of stream flow, and travel time of runoff. Manning's roughness coefficient can be used for calculating these parameters.

17.3.1.6 Thalweg

It can be defined as the lowest point of the main discharge–carrying portion of a cross section.

17.3.1.7   Cross Section and Triangulated Irregular Network (TIN)

Tin data could be obtained either from the process of converting DEM or from converting other formats of grid raster data by using 3D analyst tools of ArcGIS. Accordingly, cross sections could be delineated perpendicularly throughout the river channel and tributaries in order to store distributed elevations within the area of interest. Besides, distributed roughness values which are already extracted from LULC could be stored by each cross section as well. Thus, the profile of the river channel in both ways longitudinal and crosswise could be investigated.

17.3.1.8   Land Use/ Land Cover (LU/LC)

The LULC is an essential data that need to be calculated precisely in order to perform the flood plain analysis adequately. Since flood plain analysis strongly correlated to the Manning's values (surface roughness) which can be obtained from the surface analysis of both sides of the main river channel. In other words, distributed roughness from the left bank and right bank to opposite perpendicular directions within the urban area and other surrounding areas.

17.3.1.9   Hydrograph

Generally, hydrograph represent the flow discharge of a river as a function of time. This parameter effectively plays the role of reach boundary conditions. Besides, hydrograph is strongly correlated to the other factors such as rainfall intensity, rainfall duration, watershed size, watershed slope, watershed shape, watershed storage, watershed morphology, channel type, LULC, soil type, and percent impervious. Thus, in order to model hydrologic and hydraulic systems of any river-basin adequately, all above mentioned data need to be pre-processed. Figure 17.5 shows used data in modelling Greater Zab River-Basin.

17.4   Methodology

As shown in the schematic workflow of the hydrologic and hydraulic modelling system for the Greater Zab river-basin, the methodology includes both pre-processing and post processing of the GIS data and information. These two processes interact with each other in rational method which is so-called Interface method. In which GIS could be linked with the simulation process.

In the first stage, three extension tools which are Arc Hydro, HEC-GeoRAS, and HEC-GeoHMS have been used for preparing required geo-database that can be used later for executing hydrologic and hydraulic model using HEC-RAS and HEC-HMS. Figure 17.6 shows sketched schematic of H&H modelling system.

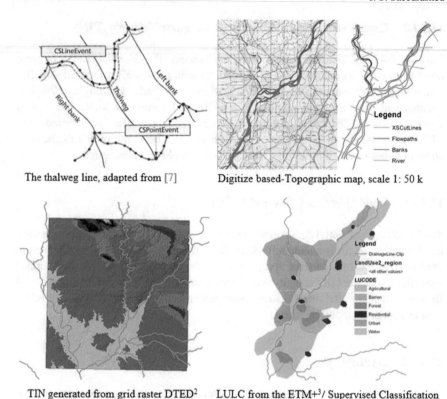

The thalweg line, adapted from [7]   Digitize based-Topographic map, scale 1: 50 k

TIN generated from grid raster DTED[2]   LULC from the ETM+[3]/ Supervised Classification

Fig. 17.5 Required data and information for H&H modelling

17.4.1 DEM-Based Hydrological Feature Classes

Hydrological feature classes such as watershed, sub-watershed, catchments and main channel and streams are extracted from the DEM by using Arc Hydro tool. This tool provide us an opportunity to pre-process the digital elevation model in a way that assure viability of the digital elevation model for further analysis avoiding any possibility of errors inherent might existed in the internal structure of the DEM. Errors such as pits and peaks can be removed from the DEM prior extracting feature classes. Any sinks might be existed in the cells can be filled automatically as the fill sinks function interpolate the elevation value to eliminate these problems. Additionally, DEM with low resolution in which the cell size is too large can be treated via reconditioning process. Meanwhile the technique of D-8 model determine the flow direction. Delineated catchments can be converted automatically from the grid raster to the polygon vector. Also, stream segment can be converted automatically from the stream segment link grid to the drainage polyline. The centroid of watershed and sub-watersheds which represents the centre of gravity can be determined and the

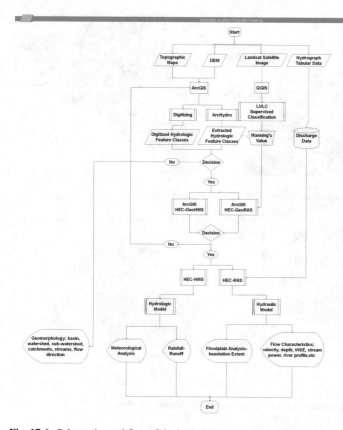

Fig. 17.6 Schematic workflow of the hydrologic and hydraulic modelling system

longest flow path within the watershed can be measured and identified. Figure 17.7 shows extracted feature classes in 3D and 2D.

17.4.1.1   Geometric Network and Flow Direction

This process creates a systematic relationship between catchment areas and associated stream network. The geometric network tool connects the stream segments, keeping track of points–where they start, end, and intersect each other, and places each stream segment within a catchment. Creating a geometric network will be followed by performing extra calculations that generate new fields and organize feature class attributes accordingly. For example, created field NextDownID will populate with the HydroID of the next down feature. Length downstream for both of the Edges and junctions can be calculated. Next downstream junction can be calculated. Outlet junction can be assigned. Layers that have a relationship can be consolidated; for example, Hydro Junction as a target layer will consolidate with the catchment layer

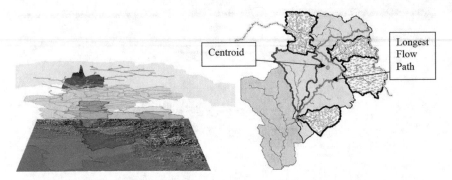

Fig. 17.7 Longest flow path and centroid of the Greater Zab watershed

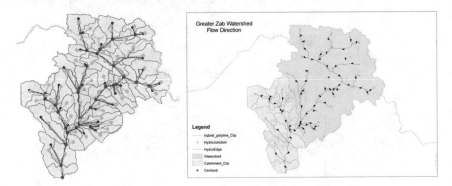

Fig. 17.8 Geometric network and flow directions of the Greater Zab river

as a source layer. ArcMap includes the utility network analysis toolbar that can be used to display arrows and other symbols representing of flow direction. Figure 17.8 shows geometric network and flow directions of the Greater Zab River.

17.4.2 HEC-GeoHMS Based Hydrologic Parameters

In this process, the required data are: raster dataset which includes; DEM for the study area, flow direction grid, flow accumulation grid, stream grid, stream link grid, catchment grid, curve number grid, slope grid and vector dataset which includes; stream network, catchment polygons, and adjoint catchment polygons. Both datasets have already been created by using Arc Hydro tool. HEC-GeoHMS has tools for extracting required information from these datasets. The extracting approach involves specifying control points at the basin outlet, which defines the tributary of the HMS basin. This requires creating outlet point (Watershed Outlet) as shown in Fig. 17.9.

The hydrologic parameters menu in HEC-GeoHMS provides tools to estimate and assign a number of watershed and stream parameters for use in HMS. These parameters include SCS curve number, time of concentration, channel routing coefficient, etc. The HMS process function can be used for transforming rainfall to runoff and routing channel routing; where, SCS-loss method is for getting excess rainfall from total rainfall, SCS-Transform method is for converting excess rainfall to direct runoff, and Muskingum is for Route Method. Depending on the method HMS process, each sub-basin must have parameters such as SCS curve number for SCS method and initial loss constant, etc. These parameters can be assigned using the sub basin parameters option. This function overlays sub basins over grids and compute average value for each basin. SCS curve number is extracted using a grid, but parameters can also be extracted by using a feature class and its intersection with sub-basins by using the sub-basin parameters from the features option. CN Lag Method: this function computes basin lag in hour (weighted time of concentration) or time from the centre of mass of excess rainfall hyetograph to the peak of runoff hydrograph. Figure 17.10 shows geometric network of the sub-basins (catchments).

17.4.2.1   River Profile

The river profile tool ⌂ allows the profile of selected river reaches to be displayed. If the river slope changes significantly over the reach length, it may be useful to split the river/ sub basin at such a slope change. The river segment can be split at the selected point by just making a right click inside dockable window. While, the sub basin can be split at the point displayed in the map by using the sub basin divide tool A⁺ , and making a click at the point displayed on the map. Figure 17.11 shows the river profile of the selected tributary.

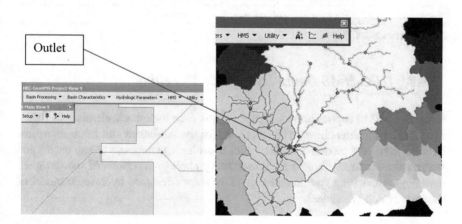

Fig. 17.9 Assigning outlet of the Greater Zab watershed

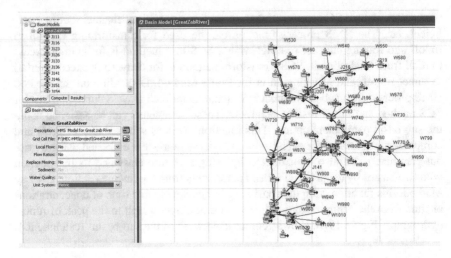

Fig. 17.10 Geometric network of the sub-basins (catchments), HEC-GeoHMS

Table 17.1 Hydrologic elements

HMS symbology	Function
Diversion	To model abstraction of flow from the main channel
Junction	To combine flows from upstream reaches and sub-basins
Reservoir	To model the detention and attenuation of a hydrograph caused by a reservoir or detention pond
Sink	To represent the outlet of the physical watershed. Sink has no outflow
Source	To introduce flow into the basin model (from a stream crossing the boundary of the modelled region). Source has no outflow
Sub basin	For rainfall-runoff computation on a watershed

Source Merwade(2007)

17.4.3 HEC-HMS Based Hydrologic Modelling

Since HEC-HMS applies lumped models within each hydrologic element, therefore hydrologic parameters have to be calculated for the sub-basins and reach segments. In order to run the model, computation of hydrologic parameters of sub basins and reaches are required. Table 17.1 illustrates symbology functions of the designed hydrologic modelling system and Fig. 17.12 shows designed hydrologic model of the Greater Zab Watershed.

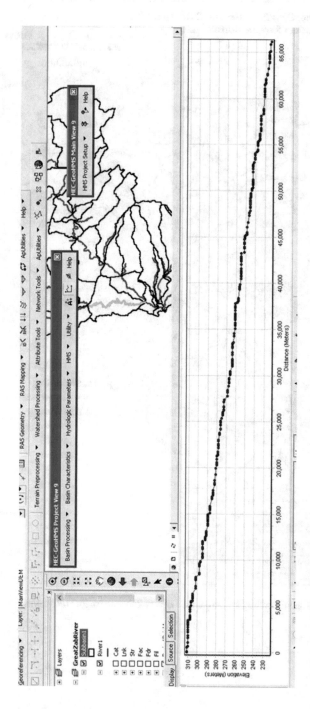

Fig. 17.11 Profile of the selected tributary

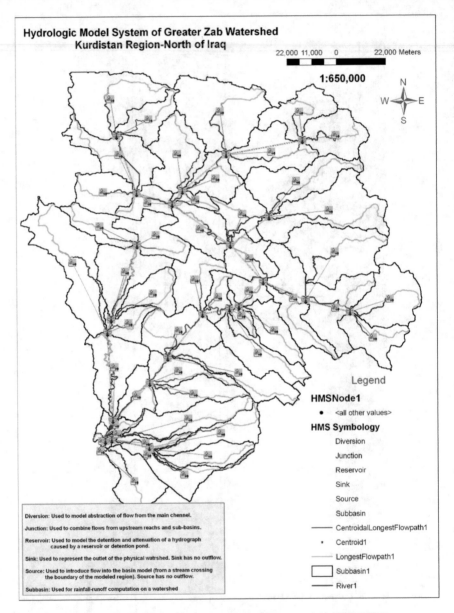

Fig. 17.12 Hydrologic modelling system of the Greater Zab watershed, HEC-HMS

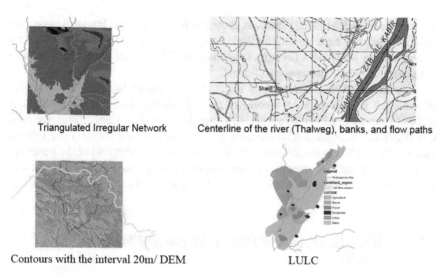

Triangulated Irregular Network   Centerline of the river (Thalweg), banks, and flow paths

Contours with the interval 20m/ DEM   LULC

Fig. 17.13 TIN, River geomorphology, contour lines, and LULC

17.4.4 HEC-GeoRAS Based Hydraulic Parameters

HEC-GeoRAS provide the functions that enable us to delineate feature classes such as river centreline, bank lines, flow path centreline, and cross-section lines. Additionally, other available features in the area under study such as bridges, culverts, levee, ineffective flow area,[2] and structures[3] which surround the river or the stream can be entered into the calculations. Created river centreline can be used as a flow path centreline, if it lied approximately in the centre of the main channel. HEC-GeoRAS tool is used for floodplain delineation and creating required geo-database that can be used later with the other parameters such as land use information in the process of the simulation model using HEC-RAS program. This process includes two main tasks: RAS Geometry and RAS Mapping. RAS Geometry includes the following steps: Creating TIN, Creating RAS Layers, and Extracting LULC Information from the satellite image, and assigning Manning's (n) value to cross-sections. Topographic maps (NIMA) in scale 1:50,000 are used as a base map for delineating river centreline, banks, flow paths, and cross-sections. Furthermore, Contours that have 20 m interval are extracted from the DEM. Figure 17.13 shows required data for pre-processing the hydraulic modelling system of the Greater Zab River-Basin.

Perpendicular delineated cross sections throughout river reaches intersect with the land use polygons and Manning's (n) are extracted for each cross section, and

---

[2]It is the areas with water but no flow/zero velocity of the floodplain. For example, areas behind bridge abutments representing contraction and expansion zones can be considered as ineffective flow areas.

[3]It is the areas with no water and no flow. For Example, buildings in the floodplain, it can be considered with the levees as obstructions.

reported in the XS Manning Table, as shown in Fig. 17.14. The land use table must have a descriptive field identifying land use type, which is LU Code, and a field for corresponding Manning's (n) values. In addition, HEC-GeoRAS requires the land use polygons to be non-multi-part features.[4]

Created layers were exported by using the function of Export GIS Data. This was the last task in the RAS Geometry step. It was checked that the right layers had been exported using the function of Layer Setup. This process creates two files: GIS2RAS.xml and GIS2RASImport.sdf. The next step is to import the created sdf file into a HEC-RAS model. After running the model successfully, one of the generated profiles which had the maximum flow was exported as an sdf file. Then, it was converted to an xml file so as to be used in the process of flood inundation mapping.

17.4.5 HEC-GeoRAS and HEC-RAS Based Floodplain Analysis

This step is considered as a HEC-GeoRAS post processing function. In this stage, the terrain TIN was converted to a dtmgrid file. In addition, bounding polygon was created, which basically defines the analysis extent for inundation mapping, by connecting the endpoints of XS cut lines. This operation was followed by another, in which the water surface TIN (tP003) was converted to a GRID. The generated dtmgrid must be subtracted from the water surface grid. The created Inundation polygon must be checked for its quality. It must be corrected by looking at the inundation map and the underlying terrain. The terrain might have errors which need to be resolved in the HEC-RAS geometry file. The refinement of flood inundation results to create hydraulically correct output requires several iterations between HEC-GeoRAS and HEC-RAS. The ability to judge the quality of terrain and flood inundation polygon comes with the knowledge of the study area and experience. Figure 17.15 shows generated Inundation polygon via HEC-GeoRAS tool and geometric data editor of HEC-RAS.

17.4.6 HEC-RAS Based Hydraulic Modelling

Generally, there are two methods of interacting between GIS and H&H modelling systems which are integration method and interface method. The Interface method has been concentrated on because it includes much effective potential, especially pertaining cross sections and 3D representation of river profiles. Additionally, the interface method provides a direct link to transfer information between the GIS and a model. The interface method consists of: a pre-processor that analyzes and exports the GIS data to create model input files and a postprocessor that imports the model

---

[4]A multi-part feature has multiple geometries in the same feature.

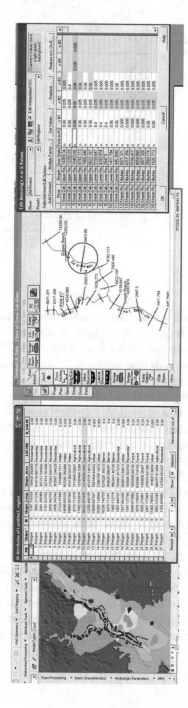

Fig. 17.14 Corresponding Manning's value to the reach stations

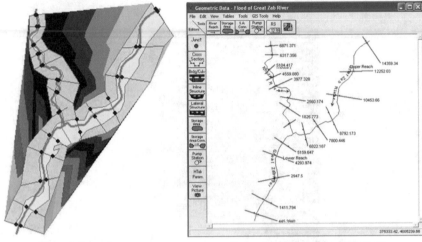

Created Inundation Polygon   Geometric Data Editor, HEC-RAS

Fig. 17.15 Floodplain analysis

output and displays it as a GIS theme. The model is executed independently of the GIS; however, the input file is created in the GIS. In the interface method, options for data editing and launching the model from within the GIS software are not available. An interface simply adds new menus or buttons to a GIS to automate the transfer of data between a computer model and a GIS. In order to set a homogeneous methodology for hydraulic design in the whole basin, the designed peak discharges must be consistent with the ones derived by statistical regionalisation of discharge data available. Designed discharges often obtained from rainfall data through simple conceptual rainfall-runoff models (rational method and SCS-Curve Number method), especially where direct discharge measurements are not available.

This process starts by creating a new project and importing created geometry data during the implementation of HEC-GeoRAS. Geometric data editor enables us to perform a quality check on the data to make sure that no erroneous information is imported from GIS. It can be used to move bank stations, change the distribution of Manning's n, add/move/delete ground points, edit structure, and cross-sections. Flows are typically defined at the most upstream location of each river/tributary, and at junctions. However, there are situations where it is necessary to define flows at additional locations. Each flow that needs to be simulated is called a profile in HEC-RAS. After exporting created SDF file, we will return to ArcMap to create a flood inundation map. This is a final step of simulation process. If we get an error, we will need to modify geometry or flow data based on error messages to run the simulation successfully.

Table 17.2 Differences between extracted and digitized Greater Zab watershed

Watershed	Digitized from raster image (km^2)	Extracted from DEM (km^2)	Difference (km^2)
Greater Zab	17,180	16,690	490

17.4.7 Validation of the Hydraulic Model

Generally, any hydrologic and hydraulic model need to pass through three main processes of evaluation in order to be viable and reliable for further tasks and analysis. Firstly, the model need to be verified which means that the model is capable of solving inherent related equations correctly. Secondly, the model need to be validated via comparing the obtained results with that have been collected from the hydrological, meteorological, and flow gauge stations. Finally, the model need to be calibrated in order to bring the obtained results as much as possible to that have been already collected experimentally or locally from the gauge stations. In our case study, validation has been performed via discharge parameter based on the recorded flow discharge at the Khabat gauge station.

17.5 Results and Discussion

17.5.1 Geo-Morphometric Calculations of the Greater Zab River Basin (Large Scale Watershed)

17.5.1.1 Accuracy of the Delineated Watersheds and Streams

Comparing to the manual digitizing method for delineation of watersheds and streams depending on the topographic maps and the other raster images as a base map, the use of digital elevation models and an automated extracting such hydrological feature classes from the DEM provide more accurate results of watershed delineation, discretization, and parameterization than that method of digitizing. Thus, more accurate geo-morphometric calculations for the Greater Zab watershed can be expected. Figure 17.16 shows the difference between both methods and Table 17.2 shows the difference in area between both methods which is (490 km^2).

17.5.1.2 Drainage Density Calculation

Drainage density and watershed area are the most important factors that affect flooding because they affect the time of concentration and the total volume of stream flow.

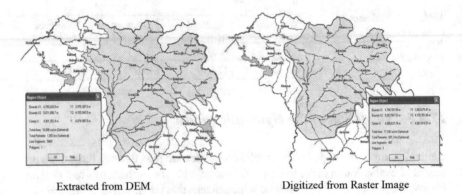

| Extracted from DEM | Digitized from Raster Image |

Fig. 17.16 Greater Zab watershed based on DEM and manual digitizing process

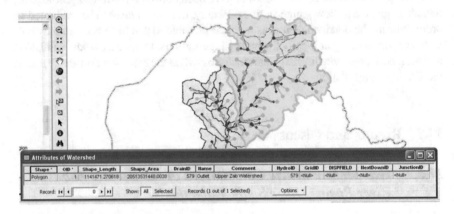

Fig. 17.17 Attributes of Greater Zab watershed

Drainage density is high in regions that have high rainfall intensity which means greater flood risk. Figures 17.17 and 17.18 illustrate the method of calculation.

Area of the Greater Zab watershed $= 20,513.53$ km^2; Perimeter $= 1141.471$ km
Total length of drainage lines $= 1,091.676$ km
Drainage Density $=$ Total Stream Length/Watershed Area
Drainage Density $=$ Length/Area $= 1,091.676/20,513.53 = 0.053$.

17.5.1.3 Longest Flow Path

Figure 17.19 illustrate calculation of LFP

Longest flow path $= 352.0171$ km.

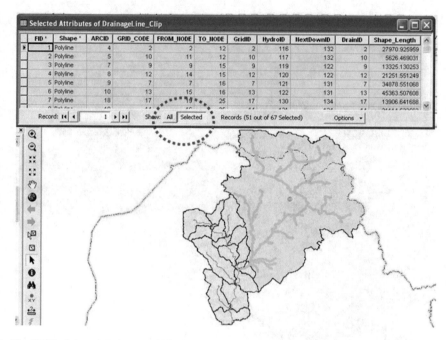

Fig. 17.18 Selected drainages within the watershed and statistics of drainages

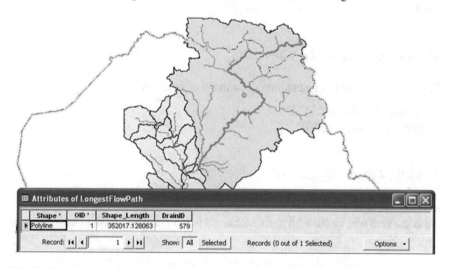

Fig. 17.19 Attributes of longest flow path

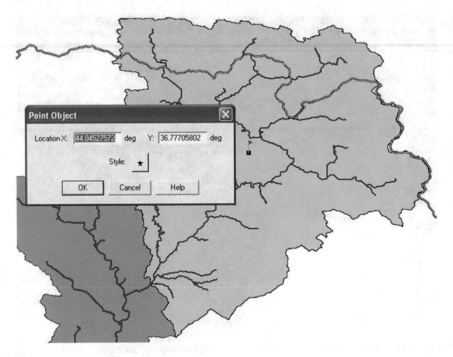

Fig. 17.20 Long/Lat of watershed's centroid

17.5.1.4 Centroid of the Watershed

Figure 17.20 illustrates calculation of centroid of watershed

Longitude X = 44.04527572°
Latitude Y = 36.77705802°.

17.5.1.5 Circularity Ratio (R_c)

A is Area of watershed = 20,513.53 km^2
A_0 is the area of a circle with perimeter equal to the perimeter of the watershed

$$p = 2 \times \pi \times R \quad \Rightarrow \quad 1141.471 = 2 \times 3.14 \times R$$
$$R = 181.76 \text{ km} \quad \Rightarrow \quad A_0 = 3.14 \times (181.76)2 = 103{,}735.23 \text{ km}^2$$
$$R_c = \frac{A}{A_0} = 20{,}513.53/103{,}735.23 = 0.197.$$

(a)

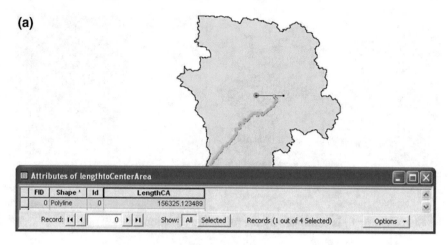

(b)

Fig. 17.21 Measuring **a** L_{ca}. **b** L

17.5.1.6 Watershed Elongation and Watershed Length

Figure 17.21a, b illustrate calculation of elongation and length of watershed

Watershed Length: $L = 184$ km

Fig. 17.22 Upper elevation and lower elevation of Greater Zab river

Shape factor L_l is estimated by the following equation:

- $L_l = (L \cdot L_{ca}/2.58)^{0.3}$

where L_l shows the elongation of a watershed.

L is the watershed length in kilometres, it can be defined as a distance measured from the watershed outlet to the farthest point on the basin divide. It is usually a measure of the travel time of water through the watershed.

L_{ca} (Length to the centre of area) is the distance along the main channel between the basin outlet and the point on the main channel opposite the centre of area (Measured in km).

$$\therefore \; L_l = (184 \times 156.3251/2.58)^{03} = 16.37 \text{ km.}$$

17.5.1.7 River Length and River Slope

The river (channel) length is the distance between the river basin outlet and the end of the river along the main channel L_c.

The river (channel) slope is the difference in elevation between the upper and lower elevation of river ΔE_c over the length L_c. Figure 17.22 shows inlet and outlet elevations of the Greater Zab River.

- $S_c = \frac{\Delta E_c}{L_c} \Rightarrow (2640 - 220)/352.0171 = 6.87$

Table 17.3, shows the geo-morphometric calculations of the Greater Zab River Basin.

Table 17.3 Geomorphological results of the Greater Zab watershed

Geo-morphometric parameters	Results
Area of watershed	20,513.53 km^2
Circularity ratio	0.197
Drainage density	0.053
Watershed length L	184 km
Length to the centre of area L_{ca}	156.3251 km
Shape Factor (Watershed elongation) L_l	16.37 km
River (Channel) length or longest flow path L_c	352.0171 km
River (Channel) slope	6.87

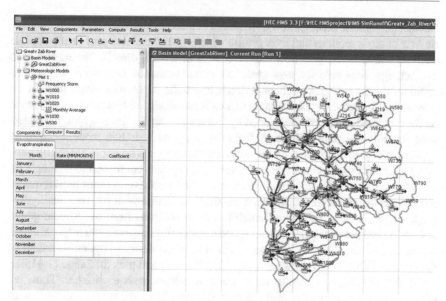

Fig. 17.23 Meteorological model system (MMS) of the Greater Zab watershed

17.5.2 Hydrologic and Meteorological Modelling Using HEC-GeoHMS and HEC-HMS

Hydrologic and climatic parameters can be prepared in advance as tabular data so as to be used for calculating geomorphological characteristics of each catchment area within the Greater Zab Watershed. Figure 17.23 shows the hydrologic model which enable us to perform monthly and yearly meteorological calculations.

17.5.3   Hydraulic Modelling Using HEC-GeoRAS and HEC-RAS

Designed hydraulic model with HEC-RAS enables researchers and engineers to deal with each specific location along the river reaches through the delineated cross section in order to investigate hydraulic parameters and perform required flood plain analysis. Figure 17.24 shows cross section output.

17.5.4   Validation of the Hydrologic Model

Designed hydraulic model, allow us to obtain detailed information about the geometric data of cross sections and river profile which is very useful to perform floodplain analysis and to investigate flow characteristics as well. Hydraulic parameters such as discharge, river velocity, flow area, and wetted perimeter have been obtained. In addition, there are details of channel depth, hydraulic depth, loss, slope, and stream power. Additionally, we can validate and calibrate designed hydraulic model based on the collected data from the existed hydrological gauge stations within the watershed. Usually, the obtained results from the designed hydraulic model need to be verified, if any errors exist, then the correction process can be implemented by verifying input data such as cross-sections, surface roughness coefficient (Manning's n), and LULC values. The simulation process can be rerun several times until required reliable results are achieved. In our case study, in comparison between the hydrograph of the Greater Zab River which has been obtained from the Khabat gauge station located lower reach which recorded average discharge of ($750 \text{ m}^3/\text{s}$) and pick discharge ($1,500 \text{ m}^3/\text{s}$) with that we obtained from the designed hydraulic model of the Greater Zab River-Basin which was ($707.92 \text{ m}^3/\text{s}$) and pick discharge ($1,415.18 \text{ m}^3/\text{s}$) that the obtained results are in good agreement with each other. Thus, we conclude that the calibration of the designed hydraulic model is quite possible with additional cross sections and rerunning the simulation process several times until we get much closer to that recorded at the Khabat flow gauge station. Figure 17.25 and Table 17.4 illustrates both results of flow gauge station and the hydraulic model being designed.

17.6   Summary

In conclusion, the geo-morphological calculations of the Greater Zab River-Basin are quite reliable with an automated extraction of the hydrological feature classes from the digital elevation model rather than the manual digitize method. The more accu-

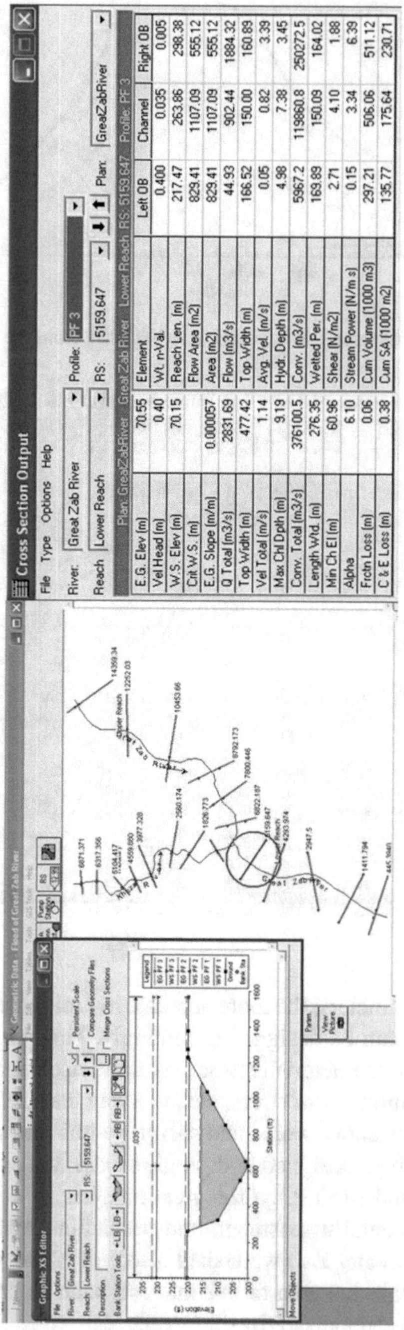

Fig. 17.24 XS-1 of the Greater Zab River-lower reach

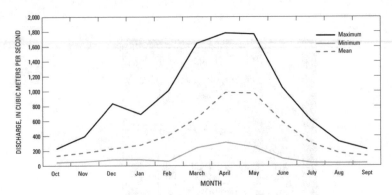

Fig. 17.25 Hydrograph of the Greater Zab River (Q in m³/s) taken from the Khabat flow station

Table 17.4 Profile output table-standard Table 1 in (meter)—lower reach (PF1, PF2, and PF3)

Reach	River Sta	Profile	Q Total (m3/s)	Min Ch El (m)	W.S. Elev (m)	Crit W.S. (m)	E.G. Elev (m)	E.G. Slope (m/m)	Vel Chnl (m/s)	Flow Area (m2)	Top Width (m)	Froude # Chl
Lower Reach	5159.647	PF 1	707.92	60.96	65.69		65.81	0.000687	1.53	645.14	246.23	0.29
Lower Reach	5159.647	PF 2	1415.84	60.96	66.93		67.25	0.000611	1.83	977.80	291.75	0.29
Lower Reach	5159.647	PF 3	2831.69	60.96	70.15		70.55	0.000057	0.82	2491.61	477.42	0.10
Lower Reach	4293.974	PF 1	707.92	60.96	64.40	64.40	65.28	0.010286	4.15	170.48	99.14	1.01
Lower Reach	4293.974	PF 2	1415.84	60.96	65.32	65.32	66.72	0.010497	5.13	273.86	125.64	1.07
Lower Reach	4293.974	PF 3	2831.69	60.96	65.96	65.96	70.12	0.017605	7.47	360.59	144.17	1.43
Lower Reach	2947.5	PF 1	707.92	60.96	62.00	62.00	62.51	0.000258	0.44	279.66	282.66	0.14
Lower Reach	2947.5	PF 2	1415.84	60.96	62.61	62.61	63.38	0.000221	0.54	455.41	299.42	0.14
Lower Reach	2947.5	PF 3	2831.69	60.96	63.54	63.54	64.72	0.000191	0.68	747.42	325.35	0.14
Lower Reach	1411.794	PF 1	707.92	60.96	61.75		62.33	0.000486	0.54	412.12	532.12	0.19
Lower Reach	1411.794	PF 2	1415.84	60.96	62.38	62.38	63.05	0.000265	0.59	747.37	541.30	0.16
Lower Reach	1411.794	PF 3	2831.69	60.96	63.17	63.17	64.24	0.000227	0.73	1177.67	543.82	0.16
Lower Reach	445.3940	PF 1	707.92	60.96	61.49	61.49	61.76	0.032076	3.35	332.69	626.68	1.47
Lower Reach	445.3940	PF 2	1415.84	60.96	61.80	61.80	62.23	0.027517	4.23	528.18	626.68	1.47
Lower Reach	445.3940	PF 3	2831.69	60.96	62.30	62.30	62.98	0.023343	5.31	841.57	626.68	1.46

rate geomorphological parameters, the more accurate hydrologic model. Thus, more accurate hydraulic model can be designed accordingly. This will enable engineers to get more accurate flow characteristics such as flow discharge, velocity, stream power, water surface elevation, shear stress, etc., in a way that facilitate the design of hydraulic structures such as dams, weirs, and culverts within the selected catchment area economically and safely. Designed hydrologic model will enable us to perform meteorological analysis and further hydrological analysis in a way that facilitate drought studies, environmental assessment, and prediction. Meanwhile, designed hydraulic model for the Greater Zab watershed also will enable us to perform lots of hydraulic activities within the area of our interests such as flood plain analysis, culverts and artificial open channel nearby the Greater Zab River. Designed hydraulic model for the Greater Zab River-Basin has been validated according to the existed flow gauge station at the lower reach of the Greater Zab River which is so-called Khabat flow gauge station. This means that we can go further with additional hydraulic calculations with different scenarios.

Acknowledgements Authors would like to express their great thankful to both editorial board and reviewers for their constructive comments and suggestions to improve the manuscript. We are thankful also to the centre of Flow Simulation at Dusseldorf University of Applied Sciences for providing required support.

References

An W (2007) The study of GIS-based hydrological model in highway environmental assessment, Ph.D., thesis, University of Pittsburgh, School of Engineering, USA, [Internet]. Available from: http://etd.library.pitt.edu/ETD/available/etd-03212007-140530/unrestricted/Weizhe_An_2007.pdf

Brown S, Dee DD (2000) Using a GIS-based solution for watershed analysis and automation: a case study. In: ASCE conference proceeding paper, section 63, Ch. 1, 30 July–2 Aug 2000, Minnesota, USA, [Internet]. Available from: http://www.pbworld.com/library/technical_papers/pdf/67_Using_a_GIS_Based_Solution.pdf

Fox S, Clapp D, Mundy C (2006) The NHD: QAQC-editing and arc hydro application in Northeast Florida. In: Conference on geographic information systems and water resources IV, AWRA, 8–10 May, USA [Internet]. Available from: http://sjr.state.fl.us/archydro/pdfs/AWRA_GIS_final.pdf

Joao E, Fonseca A (1996) The role of GIS in improving environmental assessment effectiveness. Theory Pract 14:371–387

Karamouz M, Szidarovszky F, Zahraie B (2003) Water resources systems analysis. CRC Press, USA

Khadka J, Bhaujee J (2018) Rainfall-Raunoff simulation and modelling using HEC-HMS and HEC-RAS models: case studies from Nepal and Sweden. Division of Water Resources Engineering, Department of Building and Environmental Technology, Master thesis, TVVR 18/5009 Lund University [Internet]. Available from: http://lup.lub.lu.se/luur/download?func=downloadFile&recordOId=8956602&fileOId=8956603

Lillesand TM, Kiefer RW, Chipman JW (2008) Remote sensing and image interpretation, 6th edn. Wiley Inc.

Marfai MA (2003) GIS modeling of river and tidal flood hazards in a Waterfront City, M.Sc. thesis, International Institute For Geo-Information Science and Earth Observation Enscheda, The Netherlands, [Internet]. Available from: http://www.itc.nl/library/Papers_2003/msc/ereg/marfai.pdf

Maidment DR (2002) Chapter 1, Arc hydro: GIS for water resources. ESRI Press, USA

Maidment DR (2005) Hydrologic information system: status report. J Water Resour Plan Manag 134(2):95–96. (CUAHSI HIS, USA)

Merwade V (2008) Building arc hydro using national hydrography dataset, vol 7, no 7. Purdue University, School of Civil Engineering, USA, USGS-NHD Newsletter, [Internet]. Available from: http://web.ics.purdue.edu/~vmerwade/education/archydro.pdf

Merwade V (2007) Hydrologic modelling using HEC-HMS. School of Civil Engineering, Purdue University, [Internet]. Available from: https://web.ics.purdue.edu/~vmerwade/education/hechms.pdf

Miller SN, Semmens DJ, Goodrich DC, Hernandez M, Miller RC, Kepner WG, Cuertin DP (2007) The automated geospatial watershed assessment tool. J Environ Model Softw 22(3):365–377, [Internet]. Available from: http://www.epa.gov/esd//land-sci/agwa/pdf/agwa_tool_web.pdf

Mollaei Z, Davary K, Hasheminia SM, Faridhosseini A, Pourmohamad Y (2018) Enhancing flood hazard estimation methods on alluvial fans using an integrated hydraulic, geological and geomorphological approach. Nat Hazards Earth Syst Sci 18:1159–1171. https://doi.org/10.5194/nhess-18-1159-2018

Nguyen HQ, Rientjes T, Maathuis B (2003) Geo-informatics in the context of watershed modeling: a case study of rainfall-runoff modeling. J Photogram Remote Sens 57(4):241, [Internet]. Available from: http://www.water.utwente.nl/publs/downloads/Nguyen-et-al-2006.pdf

Rivas BL, Lizama IK (2008) GIS technology in watershed analysis. In: Conference of BALWOIS, 27–30 May 2008. Republic of Macedonia, [Internet]. Available from: http://balwois.com/balwois/administration/full_paper/ffp-1076.pdf

Samanta RK, Bhunia GS, Shit PK, Pourghasemi HR (2018) Flood susceptibility mapping using geospatial frequency ratio technique: a case study of Subarnarekha River Basin, India. Model Earth Syst Environ 10(1007):s 40808-018-0427-z. https://doi.org/10.1007/s40808-018-0427-z

Singh VP, Fiorentino M (1996) Geographic information systems in hydrology. Kluwer Academic Publishers, Netherlands

Shamsi UM (2005) GIS applications for water, wastewater, and stormwater systems. CRC Press, USA

Chapter 18
Spatial Assessment of Drought Conditions Over Iraq Using the Standardized Precipitation Index (SPI) and GIS Techniques

Ayad Ali Faris Beg and Ahmed Hashem Al-Sulttani

Abstract Iraq is located between latitudes 29° 00′ to 37° 15′N and longitudes 38° 45′ to 48° 25′E and covering an area of 438,320 km². The topography of Iraq consists of alluvial plain in the middle and south parts and desert plateau in the west as well as rising terrain to the mountain ranges in north and northeast parts of Iraq. The geographical location and topographic diversity are induced a climate variation over Iraq. The current study aims to use the standardized precipitation index (SPI) in the evaluation of drought conditions based on long-term monthly precipitation data during the period between 1980 and 2010, collected from 18 weather stations. SPI values at five accumulated periods, i.e., 3, 6, 9, 12, and 24-month are calculated using SPI software developed by National Drought Mitigation Center, University of Nebraska-Lincoln. Average SPI values and SPI frequencies for wet and dry conditions are calculated with Microsoft Excel software and analyzed using GIS techniques. Results of spatial interpolation of SPI values are presented with maps and time series graphs. Analysis of the results shows the wet and drought conditions are periodic and alternative at most of the Iraqi regions. The SPI values show wide ranges of variation in their durations and magnitudes at all weather stations. Frequency percentage of SPI values show high variations have reached up to 20% for wet conditions, 22% of dry conditions and the rest is near normal conditions. Time series analyses for the graphs of SPI at different stations insured the trends toward increasing drought and repeated periodic dry and wet conditions over Iraq. That means the agricultural, hydrological and socioeconomic situations in Iraq will face many crises in future.

Keywords Drought · SPI · GIS · Hydrology · Precipitation · Iraq

A. A. F. Beg
College of Education, Department of Geography, Mustansiriyah University, Baghdad, Iraq
e-mail: aafbeg64@uomustansiriyah.edu.iq

A. H. Al-Sulttani (✉)
Faculty of Physical Planning, Department of Environmental Planning, University of Kufa, Najaf, Iraq
e-mail: ahmedh.alsulttani@uokufa.edu.iq

© Springer Nature Switzerland AG 2020
A. M. F. Al-Quraishi and A. M. Negm (eds.),
Environmental Remote Sensing and GIS in Iraq, Springer Water,
https://doi.org/10.1007/978-3-030-21344-2_18

18.1   Introduction

Understanding of drought attracts the scientists' attention in geology, hydrology, meteorology, agriculture, and play an important factor in water resources planning and management (Mishra and Singh 2010). In recent years abrupt events represented by drought and flood have occurred frequently and widely, because of many reasons like regional climate change, geographical location, terrain conditions and other reasons (Yan et al. 2013).

Gillette (1950), Wilhite and Vanyarkho (2000) describe drought as "creeping phenomenon". Drought is not always related to the areas with low rainfall rates, but also the areas of high rainfall rates as well, because drought is relative to long-term average conditions of balance between rainfall and evapotranspiration in one place, meanwhile, the mentioned four types of droughts, i.e., meteorological, agricultural, hydrologic, and socioeconomic. Meteorological drought is related to site-specific and to short-term periods of dry (Wilhite and Glantz 1985). Agricultural drought is a connection of various characteristics of meteorological drought to agricultural impacts with focusing on precipitation shortages. The hydrologic drought has defined a drought in a year and is related to water supply, surface runoff, ground water, and precipitation. Socioeconomic drought is an integrated impact of drought on supplying economic good and relatively related to a longer period (Wilhite and Glantz 1985; Wilhite and Vanyarkho 2000; Yan et al. 2013).

Standardized Precipitation Index (SPI) is one of the methods used in the assessment of drought conditions, which is normalized spatially and temporally, therefore it becomes widely accepted and used in many countries for operational surveying of drought (Du et al. 2012; Karavitis et al. 2011; Khan and Gadiwala 2013; Łabędzki 2007; McKee et al. 1993; Al-Quraishi et al. 2019; Paulo et al. 2012; Radzka 2015; Almamalachy et al. 2019; Yan et al. 2013).

Evaluation of drought conditions using Standardized precipitation index (SPI) was mentioned by Edwards and Makee (1997) and Svoboda et al. (2012) as a long-term precipitation records fitted using a probability distribution function and then transformed by a normal distribution function, so the mean of SPI value is 0 for the given location and required interval.

Assessment of drought conditions in Iraq is a challenge and rigorous task for the researchers, due to the shortage of climatic data. In Iraq, drought conditions have a negative impact on agriculture, hydrology and socioeconomic by deteriorating the life quality and reducing the yields, and most of the rural citizens and nomadic peoples depend on precipitation and groundwater in irrigation and managing their life, domestic and livestock needs. Although there are many rivers that flow into Iraqi territory, with several lakes, still Iraq suffers from the drought and desertification growing. Topographic diversity and geographical location lead to spatial and temporal variations in drought conditions and the expansion of desertification phenomena.

The study aims to: (a). Analysis of the drought conditions abroad Iraqi regions using standardized precipitation index (SPI), (b) Interpolating their spatial variation using GIS techniques, (c). Assessment of the time series trend of drought based on monthly precipitation records from 18 weather stations for the period from 1980 to 2010, (d). Quantify the wet and dry frequencies, durations and magnitudes of drought.

18.2   Materials and Methods

To achieve the aims of current study represented by spatial assessment of Drought condition over Iraqi territories; monthly rainfall data are collected from meteorological organization and seismology in Baghdad at 18 weather stations during the period between 1980 and 2010. Standardized precipitation index (SPI) was computed using SPI Generator software developed by National Drought Mitigation Center (NDMC), University of Nebraska-Lincoln (NDMC 2018). Statistical data were calculated using Excel sheet i.e., averages, counts, minimum, maximum, and frequency values of dry and wet conditions at each station. Whereas, statistical data are spatially distributed on maps using ArcGIS software, while assessment of drought trends at the stations are computed and plotted using GraphPad Prism 5 software (Fig. 18.1).

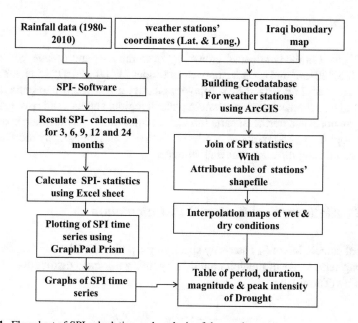

Fig. 18.1 Flowchart of SPI calculation and analysis of the results

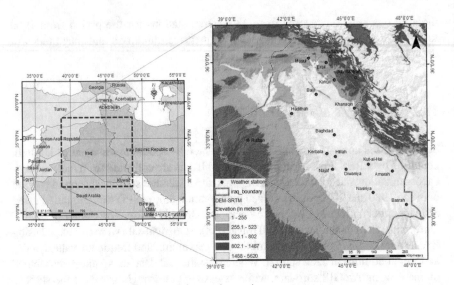

Fig. 18.2 Locations of study area and the weather stations

18.2.1   Study Area

Iraq is located in the continental area away from the seas and oceans, covering an area of 438,320 km^2 and bounded between latitudes 29° 00′ and 37° 15′N, longitudes 38° 45′ and 48° 25′E (Fig. 18.2). Topographically; it is divided into plain areas in the central and southern region, western plateau region, and the land rises gradually toward the north and northeast parts forming undulate land and mountain series with height more than 3550 m above sea level. Location and topography are the major factors controlling the dominant Iraqi climate.

18.2.2   Mathematical Basis of SPI Calculation

A typical calculation of SPI starts by specifying a probability density function (pdf) for a long-term series of precipitation data by fitting a gamma distribution, which is defined by (Thom 1958).

$$g(x) = \frac{1}{\beta^{\alpha}\Gamma(\alpha)}x^{\alpha-1}e^{-x/\beta} \quad for \; x > 0 \tag{1}$$

where x is precipitation, $\alpha > 0$ is shape e parameter, and $\beta > 0$ is the scale parameter. The parameters of the gamma probability density function are estimated at the data location for each time scale, and each observation. The maximum likelihood solutions used to estimate α and β parameters (Edwards and Makee 1997; Thom 1966)

$$\tilde{\alpha} = \frac{1}{4A}\left(1 + \sqrt{1 + \frac{4A}{3}}\right), \tilde{\beta} = \frac{\bar{x}}{\tilde{\alpha}} \tag{2}$$

$$A = \ln(\bar{x}) - \frac{\Sigma \ln(x)}{n} \tag{3}$$

where n is the number of precipitation data observations. The output parameters are then used to find the cumulative probability of an observation event for the given month and time scale for the station. The cumulative probability is given by:

$$G(x) = \int_0^x g(x)dx = \frac{1}{\Gamma(\alpha)} \int_0^x t^{\tilde{\alpha}-1} e^{-t} dt \tag{4}$$

The cumulative probability, letting $t = x/\beta\tilde{\,}$, this equation becomes the incomplete gamma function (Edwards and Makee 1997), because of gamma function is undefined for x= 0 and may precipitation data distribution contains zero, so the cumulative probability becomes (Edwards and Makee 1997; Wu et al. 2007):

$$H(x) = q + (1 - q)G(x) \tag{5}$$

where q represents the probability of zeros in precipitation data. Thom (1966) used q as the mean of numbers of zeros in the precipitation data. The SPI can be defined as the cumulative probability $H(x)$ is transformed to the standard normal random variable Z with mean zero and variance of one. The SPI value is the final step of this equiprobability transformation, which is based on an approach proposed by Abramowitz and Stegun (1964) as follow:

$$Z = SPI = -\left(t - \frac{c_0 + c_1 t + c_2 t^2}{1 + d_1 t + d_2 t^2 + d_3 t^3}\right) \quad for\ 0 < H(x) \le 0.5 \tag{6}$$

$$Z = SPI = +\left(t - \frac{c_0 + c_1 t + c_2 t^2}{1 + d_1 t + d_2 t^2 + d_3 t^3}\right) \quad for\ 0.5 < H(x) < 1 \tag{7}$$

where

$$t = \sqrt{\ln\left(\frac{1}{(H(x))^2}\right)} \quad for\ 0 < H(x) \le 0.5 \tag{8}$$

$$t = \sqrt{\ln\left(\frac{1}{(1.0 - H(x))^2}\right)} \quad for\ 0.5 < H(x) < 1.0 \tag{9}$$

where $c_0, c_1, c_2, d_1, d_2, d_e$ are constants as follow:

$$c_0 = 2.515517 \quad d_1 = 1.432788$$
$$c_1 = 0.802853 \quad d_2 = 0.189269$$
$$c_2 = 0.010328 \quad d_3 = 0.001308$$

18.2.3   Materials and Analysis Methods

Standardized precipitation index (SPI) values are calculated based on monthly precipitation data collected from 18 weather stations abroad Iraqi regions during the period from 1980 to 2010. SPI values at five accumulated intervals, i.e., 3, 6, 9, 12 and 24 months are carried out using the SPI Generator software (McKee et al. 1993).

Average SPI values for wet (SPI > 0.99), and dry (SPI < −0.99) conditions are calculated using Microsoft Excel software, which are used to prepare the spatial maps of average drought conditions using ArcGIS v. 10.3 software and inverse distance weighting (IDW) interpolation method with processing parameters i.e., distance power equal to 3, variable search radius including 4 stations. Time series graphs are prepared with GraphPad prism 5, and the graphs are assigned with drought categories intervals to follow the trend and durations of dry or wet conditions.

The drought is assigned when the value of the SPI is equal or less than −1, and the duration is the number of months until the SPI value increased above −1, while the drought magnitude is the accumulation SPI values for particular drought duration (Zin et al. 2012). McKee et al. (1993) Explained drought in terms of duration, intensity, and magnitude of drought; duration is the period from beginning to ending of drought event, intensity is the peak value of SPI, and the drought magnitude (DM) is the summation of SPI values during specified drought duration as given by the following equation:

$$DM = -\left(\sum_{j=1}^{x} SPI_{ij} \right) \tag{11}$$

where j is start month to end of drought (x), for any i time scales.

Calculated SPI values are used to classify the drought conditions into many categories based on the classes mentioned in McKee et al. (1993), Lloyd-Hughes and Saunders (2002), Du et al. (2012), Zeng et al. (2012) and Blain (2014) given in (Table 18.1).

18.3   Results

Standardized precipitation index (SPI) values at five accumulated periods are carried out from precipitation data collected at 18 weather stations during the period from 1980 to 2010. In spite the distribution of stations is not well covering Iraqi territories,

Table 18.1 Categories of drought based on SPI values

S. No.	Intensity	SPI value
1	Extremely wet	=>2
2	Very wet	1.50 to 1.99
3	Moderately wet	1.00 to1.49
4	Mildly wet (near normal)	0.0 to 0.99
5	Mild drought (near normal)	0.0 to −0.99
6	Moderately drought	−1.00 to −1.49
7	Severely drought	−1.5 to −2.00
8	Extremely drought	=<−2

due to the lack of weather stations in southwestern and western regions, so the results are valid for areas surrounding the weather stations.

Average SPI values of wet and dry conditions and the percentage of the frequency of dry months with SPI value <−0.99 and wet months with SPI value >+0.99 are illustrated in spatial distribution maps shown in (Figs. 18.3 and 18.4) using ArcGIS v.10.5. The classes of average SPI maps have been done according to the McKee et al. (1993) (Table 18.1). Time series graphs are used for analyzing the magnitude, duration, and trend of drought conditions (Fig. 18.5).

18.4 Discussions

From maps given in (Fig. 18.3), the accumulated 3-month SPI average values show two classes; moderate wet conditions covering the areas of Hadithah, Baghdad, Sulaymaniyah and Kirkuk stations, the other stations are situated in very wet conditions. Frequency percentage map for the same period are ranging from 2 to 39% and the stations of Hadithah, Khanaqin, Baghdad, Kerbala, Kut-al-Hai, Amarah and Nasiriyah have shown more frequently of wet conditions ranged from 26 to 39% of the study period.

Moderate wet class of average SPI for a 6-month are extended to cover the regions of Sallahadeen, Erbil, Sulaymaniyah, Kirkuk, Khanaqin, Baghdad, Kerbala, Hillah, Kut-al-Hai, Diwaniya and Nasiriya stations, with frequency percentage, ranged from 11 to 20% increased from central toward eastern regions of Iraq. The central and eastern regions of Iraq in accumulated intervals of 9 and 12-month SPI values are covered by moderate wet class, other stations have shown very wet conditions, while frequency percentage of wet conditions have ranged from 6 to 12% and 3 to 22%, respectively.

Through accumulative period 24-month SPI average values, the results show most of the Iraqi regions are located in moderate wet conditions, with frequency percentage ranged from 10–27% of the study period. All analyses of the wet conditions are

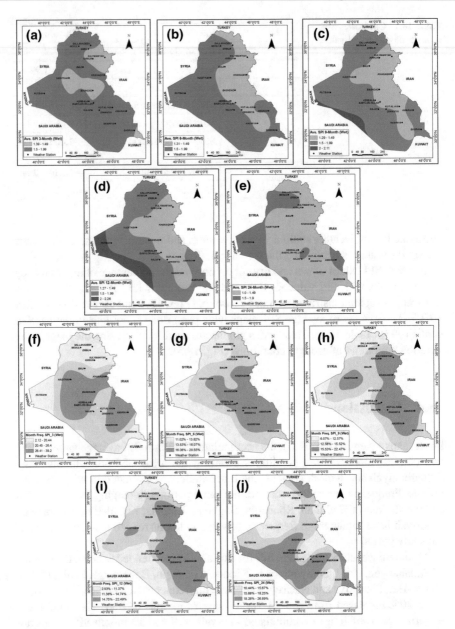

Fig. 18.3 Distribution of average SPI values: **a** 3-month, **b** 6-month, **c** 9-month, **d** 12-month, and **e** 24-month; and frequency percentage: **f** 3-month, **g** 6-month, **h** 9-month, **i** 12-month, and **j** 24-month during wet conditions

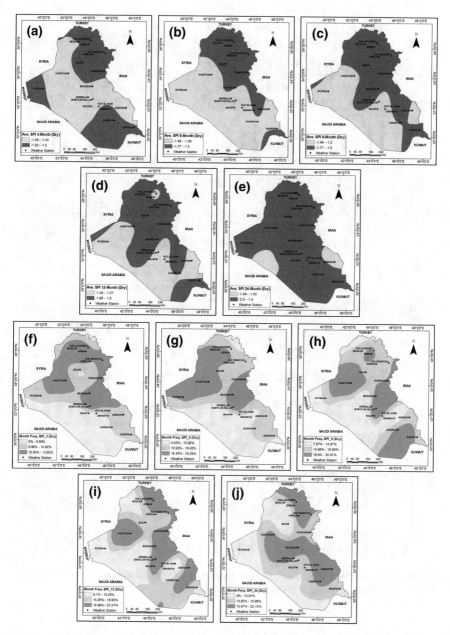

Fig. 18.4 Distribution of average SPI values: **a** 3-month, **b** 6-month, **c** 9-month, **d** 12-month, and **e** 24-month; and frequency percentage: **f** 3-month, **g** 6-month, **h** 9-month, **i** 12-month, and **j** 24-month during dry conditions

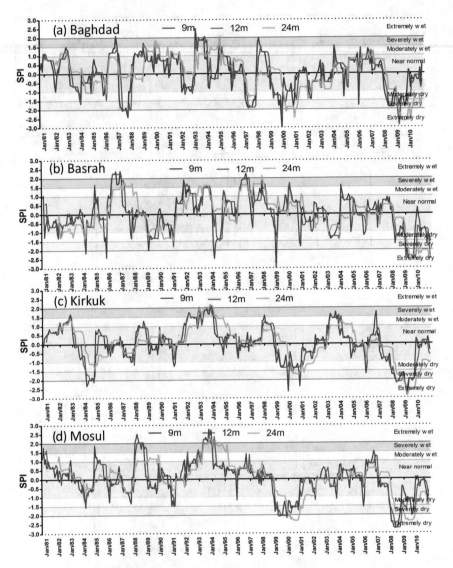

Fig. 18.5 The SPI Time series for 9, 12 and 24-month in the selected weather-stations: **a** Baghdad, **b** Basrah, **c** Kirkuk, **d** Mosul, **e** Rutba and **f** Sulymaniyah

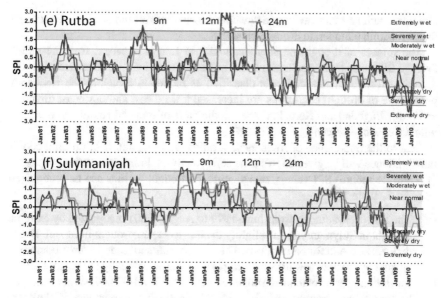

Fig. 18.5 (continued)

pointing out that Iraqi regions are subjected to very short periods of wet conditions. Indeed, this will have an impact on the socioeconomic of Iraqi rural people.

The spatial distribution of the high average value of wet SPI substituted by the less percentage of frequency and vice versa (Fig. 18.3). This due to the short intense storms in the western parts of Iraq which have increased the value of average wet SPI while in the eastern and northern east. This proves that the frequency of represents the actual spatial distribution of wet SPI than average values.

To explain the spatial distribution of dry conditions over Iraq. Accumulated 3-month average SPI values show the average SPI values are ranged from -1.34 to -1.83, and the stations of Basrah and Nasiriya in the south of Iraq and Erbil, Sallahadeen, Sulymaniyah, Kirkuk, Baiji and Khanaqin in the north and northeast parts and Rutbah in the west of Iraq are under severe drought conditions (Fig. 18.4). The frequency percentages have ranged from zero to 14%, and the frequency is increased toward the stations of Hillah, Kerbala, Baghdad, Khanaqin, Sulymaniyah, Kirkuk, Erbil, Sallahadeen and Hadithah with values ranging from 11 to 14%. The same pattern of distribution of SPI values appears with a 6-month interval, except Nasiriya and Rutbah stations show moderate drought conditions, and their frequency is ranging from 4.5 to 19%.

To highlight the hydrological impact of drought on Iraq 9 and 12-month SPI average values are used. Severe drought conditions with 9-month continued with the same stations and extended to cover other situations like Kerbala, Nasiriya, Basrah with frequency percentage between 7 and 21% with extending toward the northeast to cover Hadithah station in the west of Iraq. The same pattern of drought conditions continues with 12-month average SPI values except for Erbil and Sallahadeen stations that are within a moderate drought class.

The impact of drought on socioeconomic in Iraq is expected from 24-month average SPI values, the results show that most of the stations are classified as severe drought except Amarah station is classified as moderate drought. The frequency percentage of drought conditions are increased toward east and northeast direction and cover Baghdad and Kerbala in the middle parts of Iraq and Haditha in the west, with a frequency percentage of 16–22%.

Due to the importance of frequency, duration, and magnitude of drought conditions; reliance will be given to SPI time series analysis. From time-series graphs for (12- and 24-month) SPI values, two significant drought periods can be recognized (Fig. 18.5, Table 18.2):

(a) The first period started from June-1999 to August-2001 recognized by extreme drought in Sulymaniyah and slightly decline to severe drought in Mosul, Baghdad, Kirkuk, and Rutbah stations. The drought starts to recede from moderate to normal by shifting toward southern parts of Iraq specifically at Basrah station.
(b) The second period extended from mid of 2003 up to end of 2007 has been marked by long severe drought periods in Mosul, Baghdad, Sulymaniyah, Kirkuk stations and in a later time of the same period, the drought was shifted to extreme conditions at Basrah station.

The values of duration, magnitude and peak intensity of drought in 12-month and 24-month accumulated periods are identified and summarized in (Table 18.2) based on time series graphs.

In reverse to the wet SPI, average and percentage frequency of dry SPI show so far same spatial distribution. This due to the essential differences between dry and wet conditions and its climatic elements.

18.5 Conclusions

Analysis of the spatial distribution maps and graphs of SPI values indicates there are three main drought periods that have happened in Iraq. The first one started in 1984, and it was in a limited area, short (one year) and shallow. The second period starts with 1999 and third period in 2008. Second and third drought period have lasted for more than two years in some places, and extremely dry, especially the third period which covers most of Iraq area.

The interval time between drought periods is decreased in the recent years. It ranges from 13 to 15 years between first and second periods, while between second and third periods ranges from 6 to 9 years. The severity and frequency of drought are important to indicate their impact on hydrological settings, agriculture, and food production in Iraq.

Table 18.2 Details of drought parameters for selected weather stations

Weather station	Drought (12-month SPI)				Drought (24-month SPI)				
	Period	Duration (months)	Magnitude	Peak intensity	Period	Duration (months)	Magnitude	Peak intensity	
Baghdad	Nov–Dec 83	2	3.09	−1.67	**May 84–Nov 84**	7	**13.5**	**−2.65**	
	Apr–Dec 87	**9**	**17.57**	**−2.03**	**Feb–Oct 97**	**9**	**10.8**	**−1.29**	
	Jan–Oct 97	**10**	**17.34**	**−1.93**	**Jan 00–Mar 01**	**15**	**27.4**	**−2.46**	
	Mar 99–Sep 00	**19**	**24.51**	**−2.42**	Dec 01–Mar 02	4	5.1	−1.43	
	Dec 00–Jan 01	2	2.68	−1.48	**Dec 08–Dec 10**	**25**	**48.2**	**−2.65**	
	Mar 08–Jan 10	**23**	**37.47**	**−2.89**					
Basrah	Feb–Mar 82	2	2.32	−1.25	Feb–Apr 82	3	3.5	−1.3	
	Jan–Feb 84	2	2.36	−1.25	Feb–Mar 83	2	2.3	−1.27	
	Jan–Oct 89	**10**	**13.07**	**−1.37**	Jan–Feb 84	2	2.6	−1.31	
	Dec 98–Jan 99	2	2.5	−1.4	Dec–Jan 89	2	2.2	−1.16	
	Feb–Nov 03	**10**	**12.7**	**−1.41**	**Jan–Dec 90**	**12**	**15.3**	**−1.45**	
	Dec 08–Nov 09	**12**	**27.05**	**−2.49**	**Jan 09–Dec 10**	**24**	**50.5**	**−2.55**	
	Nov–Dec 10	2	4.02	−2.49					
Kirkuk	**Nov 83–Oct 84**	**12**	**20.55**	**−1.99**	**May–Dec 84**	**8**	**9.1**	**−1.43**	
	Apr–Nov 99	8	10.34	−1.33	**Jan 00–Nov 01**	**23**	**31.5**	**−1.7**	
	Feb–Nov 000	**10**	**13.67**	**−1.59**	**Feb 08–Apr 10**	**27**	**55.0**	**−2.82**	
	Jan–Feb 01	2	2.23	−1.22	Oct–Dec 10	3	3.1	−1.1	
									(continued)

Table 18.2 (continued)

Weather station	Drought (12-month SPI)				Drought (24-month SPI)			
	Period	Duration (months)	Magnitude	Peak intensity	Period	Duration (months)	Magnitude	Peak intensity
Mosul	**Oct 07–Dec 09**	**27**	**52.12**	**−2.88**	**Oct 99–Apr 01**	**19**	**35.7**	**−2.32**
	Jan–Feb 84	2	2.58	−1.39	**Feb 08–Apr 10**	**27**	**48.9**	**−2.36**
	Jan–Feb 91	2	2.63	−1.49	Nov–Dec 10	2	2.9	−1.48
	Dec 98–Nov 00	**24**	**41.66**	**−2.06**				
	Nov 07–Dec 09	**26**	**46.01**	**−2.74**				
Rutba	**Mar–Nov 99**	**9**	**14.46**	**−2.03**	**Nov 99–Nov 00**	**13**	**21.9**	**−2.00**
	Jan–Apr 00	4	5.17	−1.44	Dec 05–Jan 06	2	2.2	−1.13
	Feb–Sep 02	**8**	**12.13**	**−1.7**	**Oct 08–Feb 10**	**17**	**28.6**	**−2.03**
	Nov 04–Jan 05	3	3.48	−1.25	Nov–Dec 10	2	2.3	−1.23
	Apr–Oct 008	**8**	**11.81**	**−1.61**				
	Jan–Mar 09	3	3.44	−1.22				
	Sep 09–Feb 10	**6**	**10.01**	**−2.52**				
Sulymaniyah	**Nov 83–Sep 84**	**11**	**14.82**	**−2.05**	Dec–Jan 91	2	2.41	−1.33
	Jan 99–Feb 01	**26**	**50.39**	**−2.89**	**Nov 99–Feb 02**	**28**	**55.5**	**−2.77**
	Apr 08–Oct 09	**19**	**29.8**	**−2.18**	**Apr 08–Mar 10**	**24**	**37.1**	**−2.00**
	Nov–Dec 10	2	2.24	−1.23	Oct–Dec 10	3	3.4	−1.16

Using the frequencies of wet and dry SPI values in the GIS mapping displayed real spatial distribution and variation of wet and dry SPI rather than the average values of SPI epically for wet SPI in our case.

Spatial interpolation of average wet and dry SPI values and graphs of SPI time series insuring the fact of Iraqi territories that drought and wet conditions are subjected to periodic, alternative and repeated. Despite the drought conditions are not long in most of the time, but their impact on Iraqi agriculture activities and hydrological balance in surface and groundwater will be continuously under severe conditions. Drought conditions will affect socioeconomic conditions of Iraqi people, especially in the area where they depend on agriculture as the main income source.

18.6 Recommendations

Based on the results of the spatial analysis of the drought in Iraq for the period 1980–2010, the study recommends the following:

- Continuous monitoring the drought cycles over Iraq by adopting new drought indices to determine which ones are more suitable for the Iraqi climatic conditions.
- Implement agricultural plans to overcome the drought crises in the areas most vulnerable to drought conditions.
- Construct dams in large valleys basins to harvesting rainfall water during rainy years, and exploitation of groundwater in the area prone to periodic drought conditions.

Acknowledgements At the end of this work, the authors would like to present their sincere thanks and appreciation to the Mustansiriyah University and the University of Kufa for their support in collecting the data and completing of this work. Special thanks also to the staff of the Iraqi meteorological organisation and seismology in Baghdad for supplying the required rainfall data.

References

Abramowitz M, Stegun IA (1964) Handbook of mathematical functions: with formulas, graphs, and mathematical tables. Courier Corporation

Almamalachy YS, Al-Quraishi AMF, Moradkhani H (2019) Agricultural drought monitoring over Iraq utilizing MODIS products. In: Al-Quraishi AMF, Negm AM (eds), Environmental Remote Sensing and GIS in Iraq, Springer Water

Al-Quraishi AM, Qader SH, Wu W (2019) Drought Monitoring using Spectral and Meteorological based Indices Combination: A Case Study in Sulaimaniyah, Kurdistan Region of Iraq. In: Al-Quraishi AMF, Negm AM (eds), Environmental Remote Sensing and GIS in Iraq, Springer Water

Blain GC (2014) Extreme value theory applied to the standardized precipitation index. Acta Sci Technol 36(1):147–155

Du J, Fang J, Xu W, Shi P (2012) Analysis of dry/wet conditions using the standardized precipitation index and its potential usefulness for drought/flood monitoring in Hunan Province. China Stoch Env Res Risk Assess 27(2):377–387

Edwards DC, Makee TB (1997) Characteristics of 20th century drought in the United States at multiple time scales. DTIC Document

Gillette H (1950) A creeping drought under way. Water Sew Works 104:e105

Karavitis CA, Alexandris S, Tsesmelis DE, Athanasopoulos G (2011) Application of the standardized precipitation index (SPI) in Greece. Water 3(4):787–805

Khan M, Gadiwala M (2013) A study of drought over Sindh (Pakistan) using standardized precipitation index (SPI) 1951 to 2010. Pak J Meteorol 9(18)

Łabędzki L (2007) Estimation of local drought frequency in central Poland using the standardized precipitation index SPI. Irrig Drain 56(1):67–77

Lloyd Hughes B, Saunders MA (2002) A drought climatology for Europe. Int J Climatol 22(13):1571–1592

McKee TB, Doesken NJ, Kleist J (1993) The relationship of drought frequency and duration to time scales. In: Proceedings of the 8th conference on applied climatology, vol 17. American Meteorological Society Boston, MA, USA, pp 179–183

Mishra AK, Singh VP (2010) A review of drought concepts. J Hydrol 391(1):202–216

NDMC (2018) SPIGenerator software link: https://drought.unl.edu/droughtmonitoring/SPI/SPIProgram.aspx. University of Nebraska-Lincoln, USA

Paulo A, Rosa R, Pereira L (2012) Climate trends and behaviour of drought indices based on precipitation and evapotranspiration in Portugal. Nat Hazard Earth Syst Sci 12:1481–1491

Radzka E (2015) The assessment of atmospheric drought during vegetation season (according to standardized precipitation index SPI) in central-eastern Poland. J Ecol Eng 16(1)

Svoboda M, Hayes M, Wood D (2012) Standardized precipitation index user guide. World Meteorological Organization Geneva, Switzerland

Thom HC (1958) A note on the gamma distribution. Mon Weather Rev 86(4):117–122

Thom HCS (1966) Some methods of climatological analysis

Wilhite DA, Glantz MH (1985) Understanding: the drought phenomenon: the role of definitions. Water Int 10(3):111–120

Wilhite DA, Vanyarkho OV (2000) Drought: pervasive impacts of a creeping phenomenon. In: Wilhite DA (ed) Drought: a global assessment, vol I. Routledge, London, pp. 245–255

Wu H, Svoboda MD, Hayes MJ, Wilhite DA, Wen F (2007) Appropriate application of the standardized precipitation index in arid locations and dry seasons. Int J Climatol 27(1):65–79

Yan D, Wu D, Huang R, Wang L, Yang G (2013) Drought evolution characteristics and precipitation intensity changes during alternating dry–wet changes in the Huang–Huai–Hai River basin. Hydrol Earth Syst Sci 17(7):2859–2871

Zeng H, Li L, Li J (2012) The evaluation of TRMM multisatellite precipitation analysis (TMPA) in drought monitoring in the Lancang River Basin. J Geograph Sci 22(2):273–282

Zin WZW, Jemain AA, Ibrahim K (2012) Analysis of drought condition and risk in Peninsular Malaysia using standardized precipitation index. Theoret Appl Climatol 111(3–4):559–568

Chapter 19
Assessing the Impacts of Climate Change on Natural Resources in Erbil Area, the Iraqi Kurdistan Using Geo-Information and Landsat Data

Huner Abdulla Kak Ahmed Khayyat, Azad Jalal Mohammed Sharif
and Mattia Crespi

Abstract Kurdistan Region of Iraq faces a large-scale semi-aridization of the climate, revealed by the rise of temperatures and the decline of the amount of precipitations, with negative effects visible in the desiccation of vegetation cover and surface water. To quantitatively study this situation, proper methods have been set up, and a retrospective analysis was performed about climate changes and LULCC that occurred in more than two decades (1992–2014) in Erbil Area. Particular attention was devoted to analyze the role of climate change and urban and built-up areas expansion on the degradation of vegetation cover and surface water in this area. The investigation was based both on climate data (temperature and rainfall) acquired from three climate stations available for the whole considered period and on the only two available Landsat satellite images, respectively taken on 28 May 1992 by TM sensor on board Landsat-4 and on 25 May 2014 by ETM+ sensor on board Landsat-7. As regards climate data, spring and summer seasons were mainly affected by temperature increase (20 out 23 years with significant increase with respect to the World Meteorological Organization Standard Normal Period 1961–1990), and spring seasons were mainly affect by rainfall decrease (16 out 23 years with significant decrease with respect to the World Meteorological Organization Standard Normal Period 1961–1990). The vegetation cover was lost for more than 50%, mainly for both these climate change effects (>94%) and for the small remaining part (<6%) for urban and built-up areas expansion mostly concentrated (>50%) in Erbil district. Similarly, the surface water resources also suffered a strong reduction (>41%) due to the increase of temperatures and decrease of rainfall. Even in a not optimal situation

H. A. K. A. Khayyat (✉) · A. J. M. Sharif
Department of Geography, Salahaddin University, Erbil, Kurdistan Region, Iraq
e-mail: hunerak@gmail.com; huner.khayyat@su.edu.krd

A. J. M. Sharif
e-mail: azad.j.m.sherif@gmail.com

M. Crespi
Geodesy and Geomatics Division, Department of Civil, Constructional and Environmental
Engineering, University of Rome La Sapienza, Rome, Italy
e-mail: mattia.crespi@uniroma1.it

© Springer Nature Switzerland AG 2020
A. M. F. Al-Quraishi and A. M. Negm (eds.),
Environmental Remote Sensing and GIS in Iraq, Springer Water,
https://doi.org/10.1007/978-3-030-21344-2_19

463

about data availability (only three climatological stations, only two Landsat images), it was more than clear the benefit to combine climate and remote sensing data for monitoring and understanding LULC changes.

19.1 Introduction

Climate change is one of the most important environmental issues that threaten the entire planet. It does not only affect the environment, but it also has severe economic, social and political impacts (IPCC 2014).

Climate change in IPCC (Intergovernmental Panel on Climate Change—https://www.ipcc.ch/) usage refers to a change in the state of the climate that can be identified (e.g. using statistical tests) by changes in average conditions and changes in variability (including, for example, extreme events) and that persists for a decade or longer. It refers to any change in climate over time, whether due to natural variability or as a result of human activity, related to precipitation, temperature, air composition, atmospheric circulations, weather extremes, and solar radiation. The main causes of climate changes are the natural process; however, human's activities can also cause changes to the climate. The Earth's climate has never been completely static, and in the past, the planet climate has changed due to natural causes. However, the natural causes cannot be causing current global warming, and the climate changes seen today are being caused by the increase of carbon dioxide (CO_2) and methane and other greenhouse gas emissions by humans. Increase in CO_2 concentration and other greenhouse gases have raised concerns about global warming and climatic changes. According to IPCC Report, CO_2 in the atmosphere is increasing by 1.4 ppm per year, and this will contribute to the increase in temperature by 1.5–4 °C by the end of this century.

The effects of climate change can be investigated monitoring land use and land cover change (LULCC), to understand how land surfaces are impacted by human decisions.

Nowadays, Kurdistan Region of Iraq faces a large-scale semi-aridization of the climate, revealed by the rise of temperatures and the decline of the amount of precipitations, with negative effects clearly visible, among others, in the desiccation of vegetation cover and surface water.

This critical situation was the main reason to develop this work, whose aim was to set up proper methods and perform a retrospective analysis about climate changes and LULCC that occurred in more than two decades (1992–2014) in Erbil Area. Attention was devoted to analyze the role of climate change and urban and built-up areas expansion on the degradation of vegetation cover and surface water in this area.

To capture the evolution in time of climate aridization, a first step consisted in collecting climate data (temperature and rainfall), acquired from the three climate stations available for the whole considered period, from the responsible authorities (General Directorate of Seismology and Meteorology of both Republic of Iraq and Kurdistan Regional Government). These data were analyzed both individually and compared to the Standard Normal Period (SNP) of climate (more recent 30 years period, in this case 1961–1990) indicated by the World Meteorological Organization (WMO), which is commonly used in climate maps and climate statistics and which is the base period for most climate change studies.

In order to quantify spatial and temporal dynamics of LULCC, in particular the changes in vegetation, surface water and urban and built-up areas, unfortunately, only two Landsat satellite images were available: a scene, (path/row 169/35) taken on 28 May 1992 by TM sensor on board Landsat-4; a second scene (same path/row) taken on 25 May 2014 by ETM+ sensor on board Landsat-7.

The analysis was based on the Modified Normalized Difference Water Index (MNDWI) to extract the water surface, and on the Modifies Soil Adjusted Vegetation Index2 (MSAVI2) to extract the areas of vegetation cover, then measuring the positive and negative and null changes between the two considered epochs; a segmentation classification was also used for extracting urban and built-up areas expansion and for quantifying their impacts on other land covers.

Overall, the main objectives of the study were (Fig. 19.1): (1) to investigate and determine the trends in temperatures and rainfall in annual and seasonal scale in Erbil Area for the period 1992–2014 and assess the evidence of climate change in the mentioned period; (2) to monitor the spatial and temporal changes of vegetation cover and surface water in Erbil Area by analyzing remote sensing data; (3) to identify the main drivers of vegetation cover and surface water changes and to identify the impacts of climatic change in relation to these vegetation cover and surface water dynamics; (4) to identify and quantify the impact of urban and built-up expansion on vegetation cover.

This chapter is structured in three sections, besides this Introduction (Sect. 19.1). Section 19.2 gives an overview on the main physical, climatological and human features of the study area. Section 19.3 presents temperature and rainfall data for the considered period and their analysis, in the light of commonly used WMO SNP 1961–1990. Section 19.4 illustrates the Landsat image processing and analysis, in order to temporally and spatially quantify the changes in the three-main land cover and land use categories (vegetation cover, surface water and urban and built-up areas), and also discusses the comparison between the climatological and remote sensing results to identify the responsibilities of vegetation and surface water degradation. Conclusions and prospects (Sect. 19.5) are finally outlined.

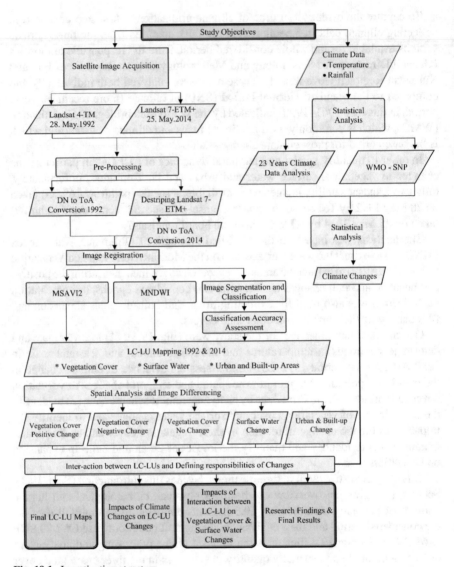

Fig. 19.1 Investigation structure

19.2 Main Features of the Study Area

In this section, the authors is going to explain the main physical and human characteristics of the study area, started with identifying the geographical and astronomical location of the study area, the explaining the main natural characteristics of the study area like (Tectonic and Geological setting of the study area, topography, water resources: surface water and underground water resources, climatical characteristics of the study area, and also the main human characteristics of the study area have been explained like the administrative divisions, border with adjacent administrative units and population and population growth of the study area.

19.2.1   Location and Boundaries

The study area occupies the central and southern parts of Erbil Governorate and covers 6,116.81 km^2, constituting about 41.1% of the total area of this Governorate (Figs. 19.2 and 19.3), and comprising five districts (Table 19.1).

The study area has an irregular shape and the total length of borders is 577.03 km: Kore and Tauska valleys are defining its northern border; the valleys of Gomaspan, Bastora and Sharabot heights at the contour line (600 m) and the small valley of little Zab are defining its eastern border; the Awana Mountain series and the administrative borders of Erbil are its southern and south-western borders; its western border is the Great Zab river. The maximum length of the study area between the south-east and north-west is about 140 km. The maximum width is about 82 km from east to west.

19.2.2   Geology and Topography

The study area is a part of the Arabian plate. Regional geology of northern Iraq is controlled by the tectonic evolution of this plate. The study area is located within the Low Folded Zone (Foot Hill Zone) and High Folded Zone of (Buday and Jassim 1987) tectonic divisions. It is covered with different formations of Tertiary Period and Quaternary sediments of variable thickness. In general, Erbil plain which forms the biggest part of the study area is a part of Low Folded Zone of Alpine orogeny in the north and northeast of Iraq. It is bounded by two main anticlines; Pirmam anticline to the north direction and Kirkuk group anticlines to the south direction. According to Buday and Jassim (1987), these two anticlines are separated by a large sub-basin syncline which represents the middle part of Erbil plain. The Quaternary sediments accumulated under the effect of weathering and erosion of surrounding elevated area; particularly the North and South parts of this plain are covered by thick succession of limestone, sandstones, marlstone and clay stone. These rocks have been divided into eight mapable formations (from oldest to youngest 1-Aqra-Bekhme formation, 2-Shiranish formation, 3-Tanjero formation, 4-Kolosh formation, 5-Khurmala formation, 6-Gercus formation, 7-Pila Spi formation, 8-Fatha (Lower

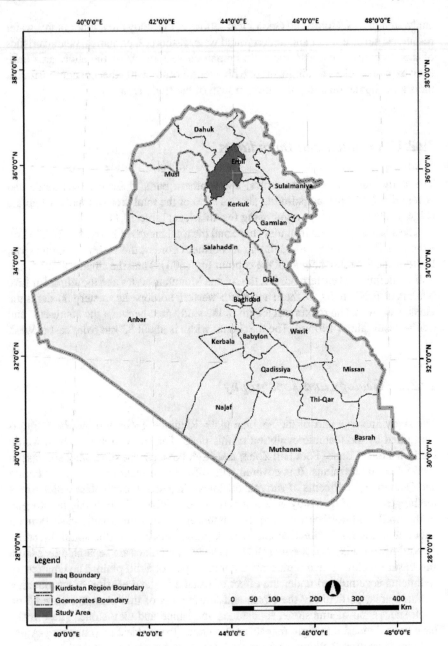

Fig. 19.2 Location of the study area in Iraq. *Source* Kurdistan Regional Government, Ministry of Planning, KRSO, Administrative Map of Erbil Governorate

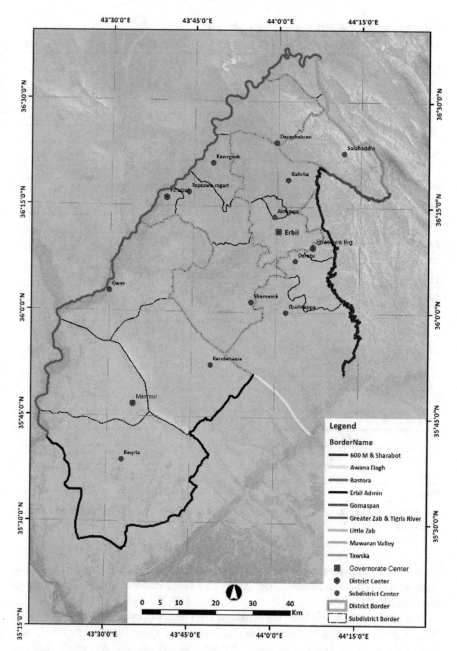

Fig. 19.3 The study area and its surrounding borders

Table 19.1 Administrative divisions of the study area

ID	Sub districts	District	Area Km^2	% of Erbil Area
1	Qaraj	Makhmur	1,058.7	17.3
2	Kandenawa		503.1	8.2
3	Gwer		513.0	8.4
4	Malaqara		240.6	3.9
5	Makhmur		530.1	8.7
Total			2845.5	46.5
6	Darashkran	Khabat	296.8	4.9
7	Rizgari		253.0	4.1
8	Kawr Gosk		111.6	1.8
9	Khabat		43.2	0.7
Total			704.5	11.5
10	Qushtapa	Dashti Hawler	786.3	12.9
11	Daratu		151.4	2.5
12	Kasnazan		24.9	0.4
13	Bnaslawa		57.8	0.9
Total			1,020.3	16.7
14	Bahrka	Erbil	370.9	6.1
15	Erbil Center		152.8	2.5
16	Shamamk		558.7	9.1
17	Ainkawa		36.3	0.6
Total			1,118.7	18.3
18	Salahaddin	Shaqlawa	334.9	5.5
19	Harir		92.9	1.5
Total			427.8	7.0
Total			6,116.8	100.0

Reference: Kurdistan Regional Government, Ministry of Planning, KRSO, Administrative Map of Erbil Governorate

Fars) formation) belonging to Pre-Quaternary sediments, ranging in age from Upper Cretaceous to Pliocene, covered by Quaternary sediments Pleistocene-Holocene in age (Fig. 19.4).

The study area is located within three different topographic units: (1) Wide Plain Region, including Flat Plains (Erbil, Kandenawa, Qaraj Plains), Highlands and curved plains (Demir Highlands) and Low Contortion Mountains (Awana, Qarachugh mountains); (2) Mountain Plains Region, including Kore—Kudarian Plain; (3) Low Folded Mountains, including Pirmam Series—Bnabawe. For more details related to topography, refer to Kahraman (2004).

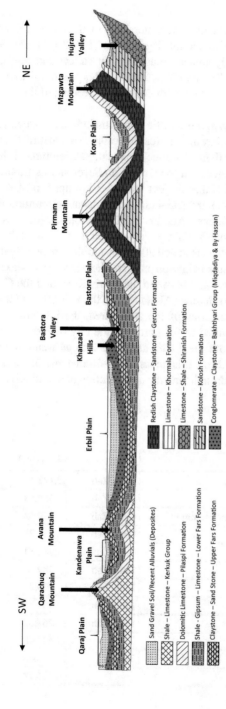

Fig. 19.4 The geological section of the study area (Kahraman 2004, p. 48)

19.2.3 Surface and Underground Waters

The study area is located within the watersheds of the Little Zab river (23.40%), the Great Zab river (61.78%) and the Tigris River (14.72%) (Table 19.2). The main water resources in the study area are seasonal river valleys and wadis, such as Wadi Muawaran, Bastoura and Kudara (Shiwasur), Qashqqa and Shirawa. Water is only used in rainy seasons due to the lack of continuous feeding of these valleys, limited to rainwater only.

Although the Great Zab and the Little Zab constitute the boundaries of the study area and do not penetrate, they are of great importance in meeting the water needs of the residents of Erbil and their activities (especially agriculture). It has established a number of irrigation projects on the Great Zab River and its tributaries, the most important being the Aski Kalak project located 3 km north of Aski Kalak bridge in Khabat district, which can irrigate 41,000 dunums of agricultural land within the study area, benefiting 16 villages. Another irrigation project is located east of the Aski Kalk project and the Shamamk Plain project, which irrigates 47.2 thousand dunums of land in the Shamamak Plain (Ismaeil 2006). In addition to the irrigation projects of Aski Kalak and the Shamamk irrigation projects, there is also the Makhmur irrigation project, which includes the land between the Tigris River and the Great Zab River. It irrigates about 356,000 dunums (Al-Haddad 2000, p. 132). As for the Little Zab, which is part of the boundaries of the study area, the investment of agriculture and household is very limited to the part near the city of Prde. There is no doubt that the importance of both great Zab and little Zab as water sources which the study area strongly depends on, this importance will increase in the future due to the repeated phenomenon of drought and the deterioration of quantitative and qualitative groundwater.

Table 19.2 River watersheds in the study area

Name	River	Area Km^2	%
Erbil valleys basin (Kurdara)	Great Zab	2,122.17	34.69
Shamamik valley basin	Great Zab	769.52	12.58
Kore river Basin and Harir and Hujran valleys	Great Zab	629.42	10.29
Bastora river basin	Great Zab	263.73	4.31
Total		3,784.84	61.88
Bash Tapa valley basin	Little Zab	309.55	5.06
Khazne valley basin	Little Zab	670.28	10.96
Dibaga valley basin	Little Zab	184.33	3.01
Shalgha river basin	Little Zab	267.31	4.37
Total		1,431.48	23.40
Tigris river basin (Ashur Valley)	Tigris	900.48	14.72
Total areas		6,116.79	100.00

The study area comprises six main groundwater basins (1. The Kapran basin, 2. The Erbil central basin, 3. The Bashbatba basin, 4. The mountain basin, the Dibaga basin, the Makhmur plain basin) (Fig. 19.5). Underground water is the main source of water on which the residents of Erbil depend on for their daily water needs as well as their economic activities (particularly agriculture). The sources of groundwater in the study area consist of the rainwater falling in the ponds of the study area during the rainy periods as well as the winter rain falling on the ponds of the area and adjacent to it, where a large part of them flows into the ground. The main means of using ground water in the study area at present is wells (shallow wells and artesian). The total number of wells in the study area is 4488, of which 1659 are for drinking water, 2363 are for irrigation and 466 are for special purposes (KRG, Ministry of Agriculture 2013).

19.2.4   Climate

The study area is divided into two climatic zones according to the Koppen classification:

- the Mediterranean climate zone (Csa), with hot and dry summers and cold and rainy winters, with an annual rainfall of 600–800 mm; this region is located in the areas between the Mohawaran valley and the southern slopes of Pirmam mountain (Naqshbandi 1997, p. 114; Naqshbandi 2008, p. 44)
- the semi-arid climate zone (BSh), which includes all the other parts of the study area, with an annual rainfall rate below 500 mm (Hassn 2006, p. 144).

Considering the lack of detailed climatic data of the monthly temperature and rainfall rate of the stations of the study area for historical periods, due to the political situation experienced by the region over the last three decades, it was possible to use the climate data (temperature, rainfall) collected by three climatological stations only for the period 1992–2014 (Table 19.3).

The spatial difference between the two climatological zones is clear considering the monthly and annual mean both for temperatures and rainfalls.

The two stations (Erbil and Makhmur—semi-arid climate zone [BSh]) with the monthly and annual average temperatures higher than the global average are those located in the central and south-western parts of the study area,

While the third station (Salahaddin—Mediterranean climate zone [Csa]), which recorded monthly and annual average lower than the global average, is in the northern part of the study area (Table 19.3, Fig. 19.6).

Correspondingly, the monthly and annual average rainfall at Salahaddin station is more than double of those at the southern station Makhmur, while the monthly and annual average rainfall at Erbil station is in between the stations of Salahaddin and Makhmur, noting that the amounts of rainfall are increasing from south-west to north-east (Table 19.4, Fig. 19.7).

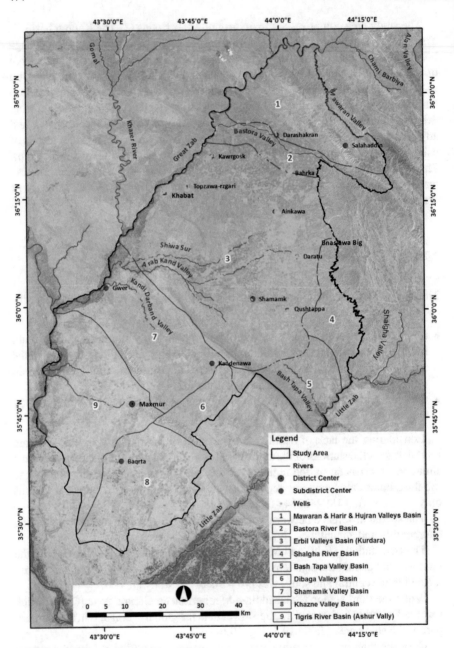

Fig. 19.5 Distribution of wells and river watersheds in the study area

Table 19.3 Mean monthly and annual temperatures for the period 1992–2014

1992–2014	Salahaddin	Erbil	Makhmur	Erbil Area average
January	5.37	8.53	9.47	7.79
February	6.27	9.88	11.32	9.16
March	10.28	13.71	15.34	13.11
April	15.57	18.72	20.83	18.38
May	21.46	26.27	27.42	25.05
June	27.08	31.13	33.05	30.42
July	30.98	34.40	36.11	33.83
August	30.82	34.05	35.89	33.59
September	26.23	29.36	31.02	28.87
October	20.27	23.60	25.24	23.04
November	12.28	15.48	16.74	14.83
December	7.89	10.22	11.07	9.73
Annual average	17.88	21.28	22.79	20.65

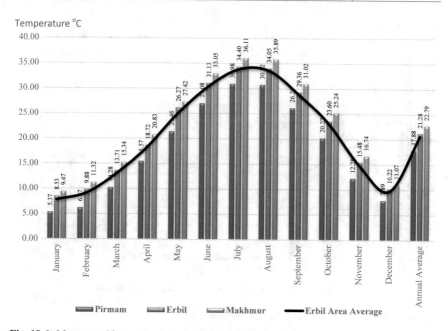

Fig. 19.6 Mean monthly temperatures for the period 1992–2014

Table 19.4 Mean monthly rainfall for the period 1992–2014

1992–2014	Salahaddin (mm)	% to annual	Erbil (mm)	% to annual	Makhmur (mm)	% to annual	Erbil Area (mm)	%
Jan	121.01	20.33	82.63	20.36	52.43	19.62	85.36	20.19
Feb	105.63	17.75	65.42	16.12	40.74	15.24	70.60	16.70
Mar	87.09	14.63	62.70	15.45	33.18	12.42	60.99	14.42
Apr	60.82	10.22	45.61	11.24	35.45	13.27	47.30	11.19
May	24.43	4.11	16.53	4.07	11.70	4.38	17.56	4.15
Jun	2.18	0.37	1.73	0.43	0.92	0.34	1.61	0.38
Jul	0.76	0.13	0.28	0.07	0.17	0.06	0.40	0.09
Aug	0.18	0.03	0.22	0.05	0.01	0.00	0.14	0.03
Sep	3.38	0.57	2.55	0.63	1.62	0.61	2.52	0.60
Oct	38.65	6.49	22.44	5.53	15.65	5.86	25.58	6.05
Nov	63.14	10.61	42.30	10.42	31.51	11.79	45.65	10.80
Dec	87.96	14.78	63.53	15.65	43.85	16.41	65.11	15.40
Total	595.22	100	405.96	100	267.24	100	422.81	100

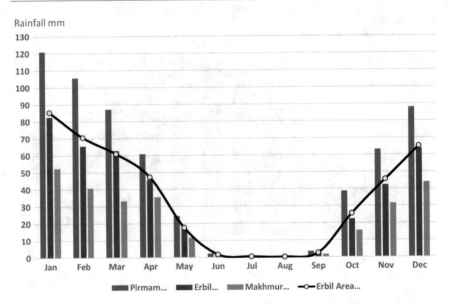

Fig. 19.7 Mean monthly rainfall for the period 1992–2014

19.2.5 Population

The population of Erbil Area in 2014 was estimated to be 1,541,196 persons; Table 19.5 and Fig. 19.8 show a steady growth of the population between 1977 and 2010, which led to a significant increase in population density. The large demographic growth of the study area, beside the natural population growth, is due primarily to the forced relocation of the Kurdish villages during the 1970s and 1980s by the Iraqi central government. Also, the growth is due to the continuous migration of the inhabitants of the villages in the study area towards the cities of the study area (especially the governorate center and the capital city of Erbil), and finally to the movement of the population from the cities of Kirkuk and Khanaqin and Mandali to Erbil, because of the policy of displacement and Arabization of the previous Iraqi government.

Table 19.5 Population and population densities in Erbil Area for the period 1977–2014

Year	Population	Population density (people/km^2)
1977	357,322	58
1987	621,708	102
1992	714,964	117
2004	1,128,376	184
2010	1,376,068	225
2014	1,541,196	252

References: the statistics of this table has been prepared from the data got from
1. Republic of Iraq, Ministry of planning, Central Statistics Organization (CSO), summery of population census results for the rural and urban areas of Erbil governorate for 1977–1987
2. Republic of Iraq, Kurdistan regional government, Directorate of Statistics in Erbil, summery results of Inventory and numbering in the urban and rural areas for Erbil governorate population for the years 2004–2009

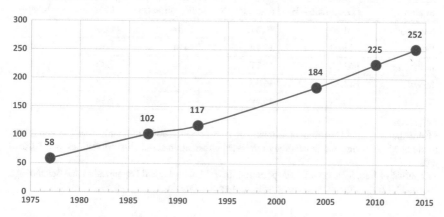

Fig. 19.8 Population density (people/Km^2) for the period 1977–2014

19.3   Evidence of Climate Changes in the Erbil Area

19.3.1   Data Collection and Analysis Method

As explained previously, the temperature and rainfall data of three main climatological stations, namely Erbil, Makhmur, Salahaddin were used, both to find out the main features of the climate of the study area and to evaluate their changes over the period 1992–2014.

In addition, the historical data of Mosul station (82 km West of Erbil Station and 77 km North West of Makhmur Station) and Kerkuk station (88 km South East of Erbil Station and 81 km South East of Makhmur Station) were used to extract the available historical data (1961–1992) (Table 19.6).

The World Meteorological Organization (WMO) method of Standard Normal Period (SNP) compared to the Current Climate Situation has been used to detect significant changes in both temperature and rainfall, through comparing the data of all the years of the study period with WMO-SNP values. The widely used WMO 1961–1990 SNP was considered as baseline period in this study, and it has been compared to other consecutive 23-years periods of this study in detail to extract the positive and negative percentages of changes in the annual, monthly and seasonal averages of temperature and rainfall between both periods (Sulaiman 1992–2014). In details, the monthly averages of minimum and maximum temperature and the monthly total of rainfall has been used to extract the annual and seasons—annual averages of both elements (temperature and rainfall) of each year of the consecutive 23-years of the study period, then compare all these to the corresponding temperatures and rainfall of the WMO SNP 1961–1990.

Table 19.6 Climatological stations used in the study

Station	Longitude—E	Latitude—N	Elevation (m)	Climate class (Koppen)
Salahaddin	44.20	36.37	1088	Csa
Erbil	44.03	36.19	414	BSh
Makhmur	43.58	35.77	270	BSh
Mosul	43.14	36.37	223	BSh
Kerkuk	44.34	35.49	331	BSh

Reference
• Kurdistan Regional Government, Ministry of Transportation and Communication, General Directorate of Meteorology and Seismology, statistical information about climatic stations, unpublished data
• Republic of Iraq, Ministry of Transportation, Iraqi Meteorological Organization and Seismology, statistical information about climatic stations, unpublished data

19.3.2   Annual and Seasonal Temperature Analysis

The comparison between the annual average temperatures for the study period 1992–2014 and the WMO SNP 1961–1990 clearly shows that all the three climatological stations in the study area recorded a positive change, whose average percentage (for Erbil Area) is equal to 4.3% and average increase is equal to 0.8 °C; among the three stations, Salahaddin station recorded the highest positive change in the annual temperature, with a percentage of change 5.1% and an increase of 0.9 °C (Table 19.7).

As regards the analysis of seasonal average temperatures, it is known that the lengths of the four seasons in the study area are not equal and vary from one season to another, and from one place to another in the region. Anyway, in this study, the following division of four seasons in equal length was adopted:

- Winter: December–January–February
- Spring: March–April–May
- Summer: June–July–August
- Fall: September–October–November

Also, in this case, it is clear that the three stations recorded positive changes, higher in percentage for winter (7.2%, for Erbil Area), and higher in absolute values for spring and summer seasons (1.0 °C, for Erbil Area) (Table 19.8).

A significant index confirming the positive trend of the temperature in the Erbil Area is also the number of the years of the study period 1992–2014 with negative and positive changes compared with WMO SNP 1961–1990, both at annual and seasonal level. The number of positive changes is by far dominant at annual level (3 negative, 20 positives, for Erbil Area), being especially driven by the positive changes during springs and summers (Table 19.9).

19.3.3   Annual and Seasonal Rainfall Analysis

Rainfall is the most important factor for aquifer recharge in the region; the rainfall season starts from September till end of May, but most of the rainfall occurs in the region from October through end of May; during summer no rainfall is usually recorded.

Table 19.7 Comparison between the annual temperature averages for the study period 1992–2014 and the WMO SNP 1961–1990 (in blue negative changes, in red positive changes)

Annual Temperature Average °C

	Salahaddin				Erbil				Makhmur				Erbil Area			
Year	Annual	change %	New-Base	Year	Annual	change %	New-Base	Year	Annual	change %	New-Base	Year	Annual	change %	New-Base
61-90	17.0			61-90	20.5			61-90	22.0			61-90	19.8		
1992	15.1	-11.1	-1.9	1992	19.5	-5.0	-1.0	1992	20.9	-5.2	-1.1	1992	18.48	-6.6	-1.3
1993	16.9	-0.6	-0.1	1993	19.7	-3.8	-0.8	1993	21.5	-2.2	-0.5	1993	19.37	-2.2	-0.4
1994	18.0	5.8	1.0	1994	20.2	-1.3	-0.3	1994	22.7	3.4	0.7	1994	20.32	2.6	0.5
1995	17.7	4.0	0.7	1995	20.2	-1.7	-0.4	1995	22.2	0.9	0.2	1995	20.01	1.1	0.2
1996	17.7	3.9	0.7	1996	21.1	3.1	0.6	1996	23.0	4.7	1.0	1996	20.61	4.1	0.8
1997	16.7	-1.5	-0.3	1997	19.8	-3.2	-0.7	1997	22.1	0.4	0.1	1997	19.56	-1.2	-0.2
1998	18.7	10.1	1.7	1998	21.7	5.9	1.2	1998	23.7	7.9	1.7	1998	21.39	8.0	1.6
1999	18.9	11.4	1.9	1999	21.6	5.2	1.1	1999	23.6	7.2	1.6	1999	21.36	7.9	1.6
2000	18.2	6.8	1.2	2000	21.3	3.7	0.8	2000	22.9	4.2	0.9	2000	20.78	4.9	1.0
2001	18.6	9.6	1.6	2001	21.8	6.1	1.3	2001	23.0	4.3	1.0	2001	21.11	6.6	1.3
2002	17.8	4.6	0.8	2002	21.3	3.7	0.8	2002	22.8	3.7	0.8	2002	20.62	4.1	0.8
2003	17.7	4.0	0.7	2003	20.9	2.0	0.4	2003	22.6	2.9	0.6	2003	20.41	3.1	0.6
2004	17.8	4.6	0.8	2004	21.1	3.1	0.6	2004	22.5	2.5	0.5	2004	20.49	3.5	0.7
2005	18.3	7.5	1.3	2005	21.7	6.0	1.2	2005	22.5	2.3	0.5	2005	20.84	5.2	1.0
2006	17.8	4.7	0.8	2006	22.1	7.8	1.6	2006	23.2	5.2	1.2	2006	21.01	6.1	1.2
2007	17.9	5.3	0.9	2007	21.5	4.9	1.0	2007	23.0	4.7	1.0	2007	20.81	5.1	1.0
2008	18.2	7.0	1.2	2008	22.0	7.4	1.5	2008	23.0	4.5	1.0	2008	21.07	6.4	1.3
2009	17.7	4.1	0.7	2009	21.5	4.7	1.0	2009	22.5	2.1	0.5	2009	20.54	3.7	0.7
2010	20.0	17.6	3.0	2010	23.3	13.6	2.8	2010	24.2	10.1	2.2	2010	22.51	13.7	2.7
2011	17.7	4.1	0.7	2011	21.2	3.3	0.7	2011	23.0	4.4	1.0	2011	20.62	4.1	0.8
2012	17.5	3.0	0.5	2012	22.1	7.6	1.6	2012	23.1	5.0	1.1	2012	20.89	5.5	1.1
2013	17.8	4.7	0.8	2013	21.7	5.8	1.2	2013	22.7	3.4	0.7	2013	20.75	4.8	0.9
2014	18.4	8.5	1.4	2014	22.2	8.4	1.7	2014	23.5	6.7	1.5	2014	21.38	8.0	1.6
Average	17.9	5.1	0.9		21.3	3.8	0.8		22.80	3.6	0.8		20.65	4.3	0.8
sd		5.3				4.4				3.2				4.0	

Table 19.8 Comparison between the seasonal temperature averages for the study period 1992–2014 and the WMO SNP 1961–1990

Season	Salahaddin			Erbil			Makhmur			Erbil area		
	Average	Change (%)	New-base	Average	Change (%)	New-base	Average	Change (%)	New-base	Average	Change (%)	New-base
Winter	6.5	16.3	0.9	9.5	6.0	0.5	10.6	4.3	0.4	8.9	7.2	0.6
Spring	15.8	8.1	1.2	19.6	4.7	0.9	21.2	4.5	0.9	18.9	5.3	1.0
Summer	29.6	4.3	1.2	33.2	2.5	0.8	35.0	3.3	1.1	32.6	3.2	1.0
Fall	19.7	1.5	0.3	22.8	3.7	0.8	24.3	2.2	0.5	22.3	2.7	0.6

Table 19.9 Annual and seasonal negative (−) and positive (+) changes of the temperature averages for the study period 1992–2014 with respect to the WMO SNP 1961–1990

Station	Annual (−) change	Annual (+) change	Winter (−) change	Winter (+) change	Spring (−) change	Spring (+) change	Summer (−) change	Summer (+) change	Autumn (−) change	Autumn (+) change
Salahaddin	3	20	5	18	3	20	1	22	10	13
Erbil	5	18	9	14	6	17	5	18	6	17
Makhmur	2	21	9	14	3	20	1	22	7	16
Erbil Area	3	20	7	16	3	20	3	20	6	17

The rainfall varies in study area substantially year to year and sharply decrease from North to South (ExxonMobil Kurdistan Region of Iraq 2012a, b). The rainfall regime in Kurdistan region is highly affected by the orographical features: the rainfall amount is increasing as moving from South-West to the North-East part of the region (Table 19.10): at the lowest station Makhmur (elevation of 270 m above the sea level) the mean annual rainfall is 263.5 mm, which is the lowest in the study area, and at the highest station Salahaddin (elevation of 1088 m above the sea level) the mean annual rainfall is 587.4 mm (Abbas 2008).

On the basis of the annual mean distribution of rainfall in Kurdistan region, the Ministry of Agriculture of Kurdistan region classifies zones as guaranteed, semi-guaranteed and non-guaranteed (Abbas 2008; Ministry of agriculture/Kurdistan Region-Iraq 2007):

- guaranteed rainfall zone, where annual rainfall exceeds 500 mm
- semi-guarantied rainfall zone, where annual rainfall is in the range 300–500 mm
- non-guaranteed rainfall zone, where annual rainfall is below 300 mm

All the three stations in the study area recorded a negative change in the annual rainfall averages, and among them Makhmur station recorded the highest negative change of −102.3 mm, in percentage equal to −28.0% compared to the average of the WMO SNP 1961–1990; Salahaddin station also recorded a negative change of −77.3 mm, in percentage equal to −11.6%, and similarly Erbil station recorded a negative change of −48.7 mm, in percentage equal to −10.8%. Overall, in the average for the Erbil Area, the negative change in the annual rainfall averages was −76.1 mm, in a percentage equal to −15.4% (Table 19.10). In the average, Salahaddin station zone is still a guaranteed rainfall one, Erbil station zone is still a semi-guarantied rainfall one, but Makhmur became a non-guaranteed rainfall zone.

As regards the analysis of seasonal rainfall averages, the three stations recorded negative changes, higher both in absolute values and in percentage for spring (−39.9 mm, −24.1%, for Erbil Area), whereas the negative change for fall is limited (−5.9 mm, −7.9%, for Erbil Area) (Table 19.11).

A significant index confirming the negative trend of the rainfall in the Erbil Area is also the number of the years of the study period 1992–2014 with negative and positive changes compared with WMO SNP 1961–1990, both at the annual and seasonal level. The number of negative changes is by far dominant at annual level (16 negative, 7 positive, for Erbil Area), being especially driven by the negative changes during springs and winters (Table 19.12).

Table 19.10 Comparison between the annual rainfall averages for the study period 1992–2014 and the WMO SNP 1961–1990 (in blue negative changes, in red positive changes)

Annual Rainfall mm

	Salahaddin				Erbil				Makhmur				Erbil Area		
Year	Annual	change %	Difference	Year	Annual	change	Difference	Year	Annual	change	Difference	Year	Annual	change	Difference
60 - 90	664.7			60 - 90	451.9	0		60 - 90	365.8	0		60 - 90	494.1	0	
1992	686.5	3.3	21.8	1992	522.9	15.7	71.0	1992	203.8	-44.3	-162.0	1992	471.07	-4.7	-23.0
1993	778.8	17.2	114.1	1993	743.5	64.5	291.6	1993	671.2	83.5	305.4	1993	731.17	48.0	237.1
1994	820.1	23.4	155.4	1994	505.7	11.9	53.8	1994	319.9	-12.5	-45.9	1994	548.57	11.0	54.5
1995	745.0	12.1	80.3	1995	652.4	44.4	200.5	1995	450.3	23.1	84.5	1995	615.90	24.6	121.8
1996	523.3	-21.3	-141.4	1996	359.5	-20.4	-92.4	1996	322.2	-11.9	-43.6	1996	401.67	-18.7	-92.4
1997	464.4	-30.1	-200.3	1997	395.1	-12.6	-56.8	1997	279.0	-23.7	-86.8	1997	379.50	-23.2	-114.6
1998	794.0	19.5	129.3	1998	461.6	2.1	9.7	1998	296.6	-18.9	-69.2	1998	517.40	4.7	23.3
1999	358.3	-46.1	-306.4	1999	188.1	-58.4	-263.8	1999	98.1	-73.2	-267.7	1999	214.83	-56.5	-279.3
2000	381.1	-42.7	-283.6	2000	267.0	-40.9	-184.9	2000	134.6	-63.2	-231.2	2000	260.90	-47.2	-233.2
2001	430.9	-35.2	-233.8	2001	346.7	-23.3	-105.2	2001	230.8	-36.9	-135.0	2001	336.13	-32.0	-158.0
2002	660.8	-0.6	-3.9	2002	356.1	-21.2	-95.8	2002	249.5	-31.8	-116.3	2002	422.13	-14.6	-72.0
2003	748.2	12.6	83.5	2003	528.2	16.9	76.3	2003	306.6	-16.2	-59.2	2003	527.67	6.8	33.6
2004	738.3	11.1	73.6	2004	496.5	9.9	44.6	2004	344.6	-5.8	-21.2	2004	526.47	6.5	32.4
2005	652.2	-1.9	-12.5	2005	412.2	-8.8	-39.7	2005	260.2	-28.9	-105.6	2005	441.53	-10.6	-52.6
2006	708.5	6.6	43.8	2006	431.6	-4.5	-20.3	2006	339.2	-7.3	-26.6	2006	493.10	-0.2	-1.0
2007	616.3	-7.3	-48.4	2007	386.0	-14.6	-65.9	2007	275.5	-24.7	-90.3	2007	425.93	-13.8	-68.2
2008	291.2	-56.2	-373.5	2008	177.8	-60.7	-274.1	2008	125.7	-65.6	-240.1	2008	198.23	-59.9	-295.9
2009	404.9	-39.1	-259.8	2009	287.6	-36.4	-164.3	2009	144.6	-60.5	-221.2	2009	279.03	-43.5	-215.1
2010	634.7	-4.5	-30.0	2010	369.6	-18.2	-82.3	2010	208.3	-43.1	-157.5	2010	404.20	-18.2	-89.9
2011	461.6	-30.6	-203.1	2011	350.9	-22.4	-101.0	2011	194.4	-46.9	-171.4	2011	335.63	-32.1	-158.5
2012	380.6	-42.7	-284.1	2012	242.9	-46.2	-209.0	2012	110.5	-69.8	-255.3	2012	244.67	-50.5	-249.4
2013	771.6	16.1	106.9	2013	491.3	8.7	39.4	2013	303.7	-17.0	-62.1	2013	522.20	5.7	28.1
2014	458.9	-31.0	-205.8	2014	300.7	-33.5	-151.2	2014	192.0	-47.5	-173.8	2014	317.20	-35.8	-176.9
Average	587.4	-11.6	-77.3		403.2	-10.8	-48.7		263.54	-28.0	-102.3		418.05	-15.4	-76.1
sd		25.1				30.4				34.1				27.2	

Table 19.11 Comparison between the seasonal rainfall averages for the study period 1992–2014 and the WMO SNP 1961–1990

Season	Salahaddin			Erbil			Makhmur			Erbil area		
	Average	Change (%)	New-base (mm)	Average	Change (%)	New-base (mm)	Average	Change (%)	New-base (mm)	Average	Change (%)	New-Base (mm)
Winter	315.5	−4.5	−14.7	211.9	−12.0	−28.8	136.8	−27.4	−51.7	221.4	−12.5	−31.7
Spring	172.3	−20.0	−43.2	124.8	−20.5	−32.3	80.3	−35.4	−44.1	125.8	−24.1	−39.9
Summer	No Rainfall			No Rainfall			No Rainfall			No Rainfall		
Fall	96.5	−17.6	−20.6	64.2	19.2	10.3	45.3	−14.4	−7.6	68.7	−7.9	−5.9

Table 19.12 Annual and seasonal negative (−) and positive (+) changes of the rainfall averages for the study period 1992–2014 with respect to the WMO SNP 1961–1990

Station	Annual (−) change	Annual (+) change	Winter (−) change	Winter(+) change	Spring (−) change	Spring (+) change	Autumn (−) change	Autumn (+) change
Salahaddin	14	9	17	6	19	4	18	5
Erbil	15	8	15	8	17	6	13	10
Makhmur	21	2	17	6	21	2	15	8
Erbil Area	16	7	15	8	19	4	14	9

19.4 Land Use/Land Cover Change in Erbil Area from Remote Sensing Data

19.4.1 Landsat Image

One of the most critical issues facing Iraq and Kurdistan Region is the threat of continued drought desertification resulting from climatic factors and human activities. Geographic Information System and remote sensing play a key role in developing a global and local operational capability for monitoring desertification in dry lands (Ziboon 2015).

Gather information about Land Use/Land Cover changes (LULCC) is fundamental for better understanding the relationships and interactions between humans and the natural environment. Remote sensing (RS) data have been one of the most important data sources for studies of LULC spatial and temporal changes.

LC/LU are two separate terminologies which are often used interchangeably. Land cover (LC) refers to the physical characteristics of Earth's surface, captured in the distribution of vegetation, water, soil and other physical features of the land, including those created solely by human activities, e.g., settlements. Land use (LU) refers to the way in which land has been used by humans and their habitat, usually with accent on the functional role of land for economic activities. The land use/cover pattern of a region is an outcome of natural and socio-economic factors and their utilization by man in time and space. Information on land use/cover and possibilities for their optimal use are essential for the selection, planning and implementation of land use schemes to meet the increasing demands for basic human needs and welfare (Rawat et al. 2015).

This paragraph illustrates the approaches adopted to evaluate LU/LC with respect to vegetation, water surface and built-up areas, and then to evaluate their changes in Erbil area between 1992 and 2014 using Landsat satellite images. In details, spatial and temporal dynamics of LU/LC were quantified using two Landsat scenes (path/row 169/35), respectively taken in 28 May 1992 by TM sensor on board Landsat-4 and in 25 May 2014 by ETM+ sensor on board Landsat-7. The TM and ETM+ sensors have a spatial resolution of 30 meters for bands (1–5, 7), a spatial resolution of 120 m for band 6 in TM and 60 m in ETM+.

19.4.2 Image Processing for Land Use/Land Cover Evaluation

The Landsat scenes have been processed as outlined hereafter in order to detect changes related to vegetation, water surface and built-up areas.

19.4.2.1   Vegetation Cover

It is well known that information contained in a single spectral band is usually insufficient to characterize vegetation status, so that vegetation indices are usually developed by combining two or more spectral bands (Qi et al. 1994); the most used vegetation index is NDVI. Anyway, in areas with little vegetation cover, the influence of soil noise is naturally more significant and NDVI is strongly affected by these soil properties (Pretorius 2015). Variations in texture, color, composition and moisture content of soils, will influence its reflectance spectra. Therefore, the Modified Second Adjusted Vegetation Indices 1 and 2 (MSAVI1, MSAVI2) were developed (Qi et al. 1994); here, the more refined and simply to use MSAVI2 was used to detect vegetation cover in Erbil area:

$$MSAVI_2 = \frac{2NIR + 1 - \sqrt{(2NIR + 1)^2 - 8(NIR - R)}}{2}$$

MSAVI2 has been proposed as good predictors of arid and semiarid grassland vegetation cover (Purevdorj 1998; Ji 2007; Mundava 2014; Wiesmair 2015).

19.4.2.2   Surface Water

Accurate and frequent update of surface water has been made possible by remote sensing technologies. Normally, to detect surface water changes, water features are extracted individually using multitemporal satellite data, and then, those data are analyzed and compared to detect their changes (Rokni 2014).

The most common index to extract water from remote sensing imagery is the Normalized Difference Water Index (NDWI) (Acharya 2017), which that has been developed to delineate open water features and enhance their presence in remotely-sensed digital imagery. The NDWI makes use of reflected near-infrared radiation and visible green light to enhance the presence of such features while eliminating the presence of soil and terrestrial vegetation features; the NDWI was derived using principles similar to those that were used to derive the NDVI. Due to the limitations of NDWI, Xu (2006) proposed the Modified Normalized Difference Water Index MNDWI, which was found to be efficient in distinguishing water and urban areas: MNDWI can enhance open water features while efficiently suppressing and even removing built-up land noise as well as vegetation and soil noise. The NDWI is modified by substituting the MIR band for the NIR band so that MNDWI can be expressed as follows (Xu 2006; Du 2016):

$$MNDWI = \frac{Green - MIR}{Green + MIR}$$

In this study, the Modified Normalized Difference Water Index (MNDWI) was used to extract and quantify the spatial distribution of surface water in Erbil Area.

19.4.2.3   Built-Up Areas

Various approaches have been employed to extract built-up areas from Landsat imagery, like pixel/object-based classification, spectral mixture analysis (SMA), regression model, segmentation classification (Sun 2017). In this study, object-oriented classification method has been used to delimit the urban area, performing a segmentation followed by classification of these segments using their spectral characteristics, textural, morphological and context (De Maeyer 2010; Benz 2004).

The relative importance of each spectral band has to be specified for the segmentation step. Several segmentation tests have been made, and the best result includes the green, red and mid infra-red bands with respective weights equals to 1, 1 and 2 (De Maeyer 2010; Almeida 2007).

The extraction of urban and built-up areas has been quantified using ArcGIS Pro software through the segmentation and classification tools, which provide an approach to extracting features from imagery based on objects. These objects are created via an image segmentation process where pixels in close proximity and having similar spectral characteristics are grouped together into a segment. Segments are exhibiting certain shapes, spectral, and spatial characteristics can be further grouped into objects. The objects can then be grouped into classes that represent real-world features on the ground. The image segmentation is here based on the Mean Shift approach. This technique uses a moving window that calculates an average pixel value to determine which pixels should be included in each segment.

It is important to note that it is crucial to perform a proper collection of training samples to define the classes, in order to guarantee as much as possible the classification correctness after segmentation. In the current study, the Maximum Likelihood method based on Bayes' theorem has been used to classify the segmented image. It assumes samples in each class follow the normal distribution and calculates probabilities of all classes for each sample; then it assigns the class with the highest probability to that sample.

The segmented Landsat 4TM scene for 28 May 1992 used in this study (Fig. 19.9) shows similar areas grouped together into objects without many speckles. It generalizes the area to keep all the features as a larger continuous area.

19.4.3   Land Use/Land Cover (LU/LC) in Erbil Area in 1992 and 2014 and Their Changes

The results of spatial and quantitative distribution of vegetation cover, surface water and urban and built-up areas in Erbil Area, as derived from the results of MSAVI2 and MNDWI indices and from the segmentation/classification, are reported in Table 19.13 and Fig. 19.10 for the Landsat 4 image taken in 28 May 1992 by TM sensor and in Table 19.14 and Fig. 19.11 for Landsat-7 image taken in the 25 May 2014 by ETM+ sensor.

Table 19.13 Total areas and percentages of MSAVI2, MNDWI, and classified segmented based class of Erbil Area in 1992

District		Shaqlawa	Khabat	Erbil	Dashti Hawler	Makhmur	Total
Area hectare		42,782	70,449	111,869	102,028	284,554	611,681
% of Erbil Area		6.99	11.52	18.29	16.68	46.52	100
1992	1992 vegetation cover area size/hectare	37,805	38,622	65,212	40,662	20,052	202,353
	Vegetation cover % from Erbil Area	6.18	6.31	10.66	6.65	3.28	33.08
	1992 urban and built-up area size/hectare	390	2,391	7,101	1,101	1,851	12,834
	1992 surface water area size/hectare	7,911.5					7,911.5

Table 19.14 Total areas and percentages of MSAVI2, MNDWI, and classified segmented based class of Erbil Area in 2014

District		Shaqlawa	Khabat	Erbil	Dashti Hawler	Makhmur	Total
Area Hectare		427,82	70,449	111,869	102,028	284,554	611,681
% of Erbil Area		6.99	11.52	18.29	16.68	46.52	100
2014	2014 vegetation cover area size/hectare	22,548	10,663	26,859	11,669	18,957	90,696
	Vegetation cover % from Erbil Area	3.69	1.74	4.39	1.91	3.10	14.83
	2014 urban and built-up area size/hectare	1152	3859	13,686	3375	3783	25,855
	2014 surface water area size/hectare	4,605.9					4,605.9

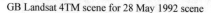

GB Landsat 4TM scene for 28 May 1992 scene | Segmented image based on
Landsat 4TM scene for 28 May 1992 scene

Fig. 19.9 Sample of the segmented landsat 4TM image for extraction of urban and built-up areas

The main results for 1992 are hereafter summarized:

- the total vegetation cover was 202,353 ha, which constitutes a percentage of 33.08% of the total study area
- the total urban and built-up areas were 12,834 ha, which constitutes a percentage of only 2.10% of the total study area
- the total surface water area was 7,911.5 ha, which constitutes a percentage of only 1.29% of the total study area Correspondingly, the main results for 2014 are hereafter summarized.
- the total vegetation cover was 90,696 ha, which constitutes a percentage of 14.83% of the total study area
- the total urban and built-up areas were 25,855 ha, which constitutes a percentage of only 4.23% of the total study area
- the total surface water area was 4,605.9 ha, which constitutes a percentage of only 0.75% of the total study area

The carried-out images' analysis were able to detect the positive, negative and net changes in vegetation, surface water and urban and built-up areas in Erbil Area (Table 19.15). Significant degradation in the total area of vegetation cover and surface water cover is evident, as well as a remarkable growth of urban and built-up areas expansion.

The relevant results are summarized hereafter:

1. As regards the overall net changes, the vegetation cover has been decreased from 202,353 ha in 1992 into 90,696 ha in 2014, with a vegetation loss of 111,657 ha (−55.18%), and the surface water also has been decreased from 7,911.5 ha in 1992 into 4,605.9 ha in 2014, with a surface area loss of 3,306 ha (−41.8%). On the contrary, the urban and built-up areas expanded from 12,834 ha in 1992 into 25,855 ha in 2014 (+101.5%).
2. Spatial analysis was performed to quantify the impact of urban and built-up areas expansion on vegetation cover loss. The spatial analysis was then able to

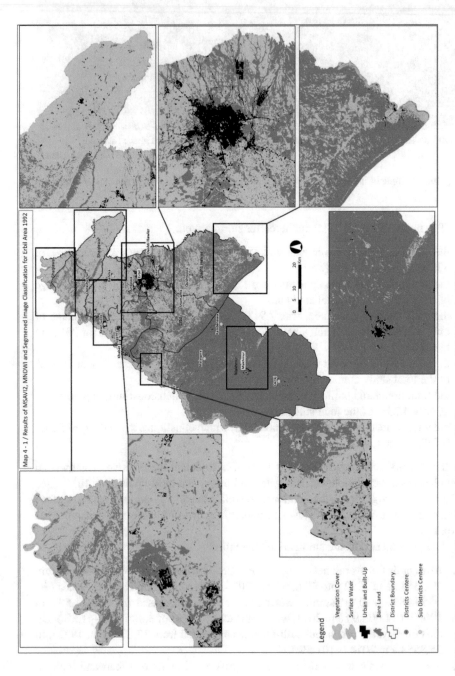

Fig. 19.10 Vegetation cover, surface water and urban and built-up areas in Erbil Area in 1992

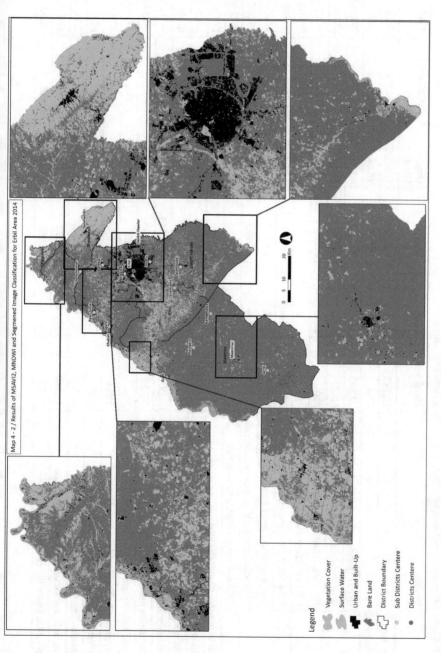

Fig. 19.11 Vegetation cover, surface water and urban and built-up areas in Erbil Area in 2014

Table 19.15 LULC positive, negative, no, net changes in Erbil Area 1992–2014

District		Shaqlawa	Khabat	Erbil	Dashti Hawler	Makhmur	Total
Area hectare		42,782	70,449	111,869	102,028	284,554	611,681
% of Erbil Area		6.99	11.52	18.29	16.68	46.52	100
1992	1992 vegetation cover area size/hectare	37,805	38,622	65,212	40,662	20,052	202,353
	% of vegetation cover from Erbil	6.18	6.31	10.66	6.65	3.28	33.08
	1992 urban and built-up area size/hectare	390	2391	7101	1101	1851	12,834
	1992 surface water/hectare					7911.5	7911.5
2014	2014 vegetation cover area size/hectare	22,548	10,663	26,859	11,669	18,957	90,696
	% of vegetation cover from Erbil Area	3.69	1.74	4.39	1.91	3.10	14.83
	2014 urban and built-up area size/hectare	1152	3859	13,686	3375	3783	25,855
	2014 surface water area size/hectare					4605.9	4605.9
Vegetation positive change/hectare		329	1983	9,598	5,055	11,132	28,097
Vegetation negative change/hectare		−15,599	−29,920	−47,941	−34,052	−12,242	−139,754
Vegetation no change/hectare						62,599	62,599
Vegetation net change/hectare		−15,269	−27,937	−38,343	−28,997	−1111	−111,657
% of vegetarian losses from net loses		−7.55	−13.81	−18.95	−14.33	−0.55	−55.18
Urban and built-up area expansion hectare		762	1467	6585	2275	1932	13,021
% of urban and built-up area expansion		195.4	61.4	92.7	206.6	104.4	101.5
Vegetation losses for urban expansion hectare		702	734	3670	879	349	6335
% vegetation losses for urban expansion		0.63	0.66	3.29	0.79	0.31	5.67
Vegetation losses for climate change/hectare		**14,567**	**27,203**	**34,673**	**28,118**	**761**	**105,322**
% vegetation losses for climate changes		**13.05**	**24.36**	**31.05**	**25.18**	**0.68**	**94.33**

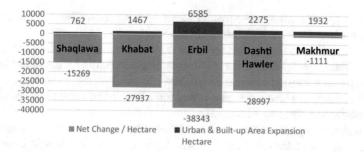

Fig. 19.12 Net vegetation cover changes versus urban and built-up areas expansion/hectare in Erbil Area 1992–2014

highlight that the area with positive, negative and no changes for vegetation. Its results show that urban expansion was responsible for 6,335 ha of vegetation cover loss (5.67% only) (Fig. 19.12), while climate change (here represented and quantified through annual averages of temperature and rainfall) was responsible on the remaining much wider loss of 105,322 ha (94.33%).

3. The LULCC results are in good agreement with the results about climate change which was explained in Tables 19.7 and 19.10, which show an increase in the annual temperature averages (0.8 °C, +4.3%) and a decrease in annual rainfall averages (−76.1 mm, −15,4%) compared to the WMO SNP 1961–1990 reference.

4. All districts of the study area recorded different values of net vegetation cover loss, due to the big impact of climate change, but, among them, Erbil and Dashti Hawler districts had the highest net vegetation loss respectively. Beside the impact of climate change, the main reason for the high vegetation loss in the two mentioned districts was their high urban expansion (6,585 ha in Erbil and 2,275 ha in Dashti Hawler district).

19.5   Conclusions

The large-scale degradation of vegetation cover and surface water in Kurdistan Region of Iraq was here studied, focusing on climate changes and LULCC that occurred in more than two decades (1992–2014) in Erbil Area. The role of climate change and urban and built-up areas expansion was investigated, on the basis of both climate data (temperature and rainfall) and the only two available Landsat satellite images, respectively taken on 28 May 1992 by TM sensor on board Landsat-4 and on 25 May 2014 by ETM+ sensor on board Landsat-7.

As regards climate data, spring and summer seasons were mainly affected by temperature increase (20 out 23 years with significant increase with respect to the

SNP), and spring seasons were mainly affect by rainfall decrease (16 out 23 years with significant decrease with respect to the SNP).

The vegetation cover was lost for more than 50%, mainly for both these climate change effects (>94%) and for the small remaining part (<6%) for urban and built-up areas expansion mostly concentrated (>50%) in Erbil district. Similarly, the surface water resources also suffered a strong reduction (>41%) due to the increase in temperatures and decrease of rainfall.

Even in a not optimal situation about data availability (only three climatological stations, only two Landsat images), it was more than clear the benefit to combine climate and remote sensing data for monitoring and understanding LULC changes.

19.6 Recommendations

1. For the future, at first it is important to continue and improve such monitoring, possibly making it continuous using the new free remote sensing data supplied by EU Copernicus Programme, especially Sentinel-3 imagery.
2. In addition, on the political side, it is necessary to establish laws to ensure the proper management and use of land and natural resources; on the political and economical side, it is necessary the involvement of stakeholders and decision makers at the local, regional level in natural resource management; on the educational side, environmental awareness programs are necessary to enable rural people to have a better understanding of and ability to manage their land and natural environment.
3. Implementing more research projects with the application of multi-temporal satellite imagery (with high resolution) and GIS tools for studying LULC change in the area.
4. Improvement of agricultural activities via the introduction of technologies and modifying cropping patterns.
5. Conducting environmental awareness programs to enable rural people to have a better understanding of and ability to manage their land and natural environment.
6. Implementation of rangeland management by improving grazing land, grazing patterns, and by protecting shrubs and trees.
7. Involvement of stakeholders, decision makers at the local, regional level in natural resource management.
8. Requirement for extensive research to develop methodologies that combines remotely sensed and physical and socio-economic data to evaluate the impact of physical factors and human-driven forces on climate changes an LULC change and its impact on livelihood, which requires increasing the area to obtain representative cover over all climatic zones in the Southern Darfur region.

References

Abbas KA (2008) Investigation and analysis of precipitation variability and drought conditions in the federal region of Kurdistan. Ph. D. thesis, Iraq Salahaddin University, pp 56–58

Acharya TD et al (2017) Combining water indices for water and background threshold in landsat image. In: Presented at the 4th international electronic conference on sensors and applications (ECSA 2017), 15–30 November 2017, pp 2–5

Al-Haddad HY (2000) Natural resources atlas of Erbil governorate and land management for agricultural purposes. A master thesis submitted to the council of college of Arts, University of Salahaddin—Erbil, pp 74–78, 132

Almeida CM et al (2007) Multilevel object-oriented classification of quickbird images for urban population estimates. In: GIS'07 proceedings of the 15th annual ACM international symposium on advances in geographic information systems Article No 12, pp 1–8

Buday T, Jassim SZ (1987) The regional geology of Iraq. Vol. 2. Tectonism, magmatism and metamorphism. In: Abbas MJ, Kassab II (eds) GEOSURV. Baghdad, p 246

Benz UC et al (2004) Multi-resolution, object-oriented fuzzy analysis of remote sensing data for GIS-ready information, ISPRS J Photogramm Remote Sens 58(3–4):239–248

De Maeyer M et al (2010) Comparison of standardized methods (object-oriented vs. per pixel) to extract the urban built-up area: example of Lubumbashi (DRC). Int Arch Photogramm Remote Sens Spat Inf Sci 38(4):89

Dewan AM et al (2009) Using remote sensing and GIS to detect and monitor land use and land cover change in Dhaka metropolitan of Bangladesh during 1960–2005. Environ Monit Assess 2009:237–245

Directorate of Statistics—Ministry of agriculture/Kurdistan Region-Iraq, Ministry of agriculture profile 2007" profile (2007)

Du Y (2016) Water bodies' mapping from sentinel-2 imagery with modified normalized difference water index at 10-m spatial resolution produced by sharpening the SWIR band. Remote Sens 8(354):2–8

ExxonMobil Kurdistan Region of Iraq Limited (2012) Environmental baseline study report for onshore seismic, drilling and well testing Salahaddin block. Kurdistan Region of Northern Iraq, pp 5–6

ExxonMobil Kurdistan Region of Iraq Limited (2012) Environmental impact study report for onshore seismic, drilling and well testing Salahaddin block. Kurdistan Region of Northern Iraq, pp 2–3

Hassan TK (2006) The geographical analysis of temperature characteristics in Iraqi Kurdistan Region. A thesis submitted to the council of the College of Arts, University of Salahaddin—Erbil, in partial fulfilment of the requirements for the Degree of Master in Geography, p 46, 144

http://cdiac.esd.ornl.gov/pns/current/ghg/htmll; Date: 23. January. 2015

http://epa.gov/ghginfo/topics/table1-2.html; Date: 23 January 2015

IPCC (2014) Climate change 2014: synthesis report. contribution of working groups I, II and III to the fifth assessment report of the intergovernmental panel on climate change [Core writing team, Pachauri RK, Meyer LA (eds)]. IPCC, Geneva, Switzerland, 151 pp. https://www.ipcc.ch/site/assets/uploads/2018/02/SYR_AR5_FINAL_full.pdf

Ismaeel SA (2014) The change in trends of index temperature in Sulaimania city between 1992–2014, p 3

Ismaeil AI (2006) Characteristics of water discharge of great Zab River in Iraq I Kurdistan region, the geographical analysis of temperature characteristics in Iraqi Kurdistan Region. A thesis submitted to the council of the College of Arts, University of Salahaddin—Erbil, in partial fulfillment of the requirements for the Degree of Master in Geography, pp 131–132

Ji L et al (2007) Performance evaluation of spectral vegetation indices using a statistical sensitivity function. Remote Sens Environ 106(2007):59–65

Kahraman LM (2004) Geographical analysis of soil characteristics & problem in Erbil governorate and its land capability. A thesis submitted to the Council of College of Arts–Salahaddin University—Erbil, in Partial fulfilment for the degree of doctor philosophy in Geography, pp 19–31

Kurdistan Regional Government, Ministry of Agriculture and Water Resources, Directorate of Underground Water, non-published wells statistics, 2013

Mundava C et al (2014) Evaluation of vegetation indices for rangeland biomass estimation in the Kimberley area of western Australia. ISPRS Ann Photogramm Remote Sens Spat Inf Sci II(7):47–54

Naqshbandi AMA (1997) Climate of Kurdistan region—Iraq. Matin magazine, issue 63. Khabat Press, Duhok, p 101, 114

Naqshbandi AMA (2008) Impacts of global warming on Kurdistan region, first edn. Ministry of Culture Press, Erbil, p 44

Pretorius E (2015) Improving the potential of pixel-based supervised classification in the absence of quality ground truth data, S Afr J Geomat 4(3):250–258

Purevdorj T (1998) Relationships between percent vegetation cover and vegetation indices. Int J Remote Sens 19:3519–3525

Qi J et al (1994) A modified soil adjusted vegetation index. Remote Sens Environ 48:119–126

Rawat JS et al (2015) Monitoring land use/cover change using remote sensing and GIS techniques: a case study of Hawalbagh block, district Almora, Uttarakhand, India. Egypt J Remote Sens Space Sci 18:77–84

Rokni K (2014) A new approach for detection of surface water changes based on principal component analysis of multitemporal normalized difference water index. J Coast Res 32(2):443–451

Sun Z et al (2017) A modified normalized difference impervious surface index (MNDISI) for automatic urban mapping from landsat imagery. Remote Sens 9(942):2–14

Wiesmair M et al (2015) Estimating vegetation cover from high-resolution satellite data to assess grassland degradation in the Georgian Caucasus, Mt Res Dev (MRD), 56–63

Xu H (2006) Modification of normalized difference water index (NDWI) to enhance open water features in remotely sensed imagery. Int J Remote Sens 27(14):3025–3033. https://doi.org/10.1080/01431160600589179

Ziboon ART et al (2015) Study and analysis of desertification phenomenon in Karbala governorate by remote sensing data and GIS. Iraqi Bull Geol Min 11(1):143–156

Chapter 20
Mapping Forest-Fire Potentiality Using Remote Sensing and GIS, Case Study: Kurdistan Region-Iraq

Iraj Rahimi, Salim N. Azeez and Imran H. Ahmed

Abstract During recent years a large number of wildfires have been reported among Kurdistan region forests and rangelands. Forest fires are a source of concern for environmental, economy, society, human safety and population in many parts of the world. From an ecological point of view, fire is an important factor that plays a basic role to determine vegetation diversity and dynamics. Since Kurdistan's region, north of Iraq, is almost the only area in Iraq where forests are still remaining so they have been playing a vital role in the region's ecosystem. Regarding these facts, it is highly important to develop rapid, accurate and reliable maps, as this study is aiming to, that show fire potentiality to take precautionary action beforehand. Remote Sensing (RS) data and techniques are, today, one of the most reliable tools that provides temporal and spatial coverage of biomass burning, defining Vegetation Indices (VI), without costly and expensive fieldwork. Normalized Difference Vegetation Index (NDVI), among them, have been strongly proposed to be used as a useful tool in order to estimate the proneness of vegetation to fire. Accordingly, this study has tried to develop a fire potential map by integration of satellite and field data, for Kurdistan region, in the North of Iraq. MODIS times series with 250 m of spatial resolution, taken in 2010, used to prepare NDVI layer. The developed map revealed a high match between the potential map, develop based on 2010 image, and fired location from 2014 to 2015. The output map reveals that, Rs and GIS are a priceless tool to manage and monitor natural hazardous phenomenon, like wildfires.

Keywords Remote sensing (RS) · Vegetation index (VI) · NDVI · Forest fire · Wildfire · MODIS · Kurdistan Regional Government (KRG)

I. Rahimi (✉) · S. N. Azeez · I. H. Ahmed
Department of Surveying, Darbandikhan Technical Institute,
Sulaimani Polytechnic University, Darbandikhan, Sulaimani, Iraq
e-mail: iraj.amin@spu.edu.iq

© Springer Nature Switzerland AG 2020
A. M. F. Al-Quraishi and A. M. Negm (eds.),
Environmental Remote Sensing and GIS in Iraq, Springer Water,
https://doi.org/10.1007/978-3-030-21344-2_20

20.1 Introduction

A forest fire is considered as an important issue for the environmental, economy, population, and human safety in many forested areas in the world. The scale of this phenomenon may range from small fires with little impacts to very large fires having serious large-scale impacts. Forest fires are a source of concern for environmental, economy, society, human safety and population in many parts of the world (Bajocco et al. 2009). From an ecological point of view, fire is an important factor that plays a basic role to determine vegetation diversity and dynamics in time and space (Bajocco et al. 2009). In addition, forest fires release a remarkable amount of greenhouse gasses, particulates and aerosol into the atmosphere, which significantly increases the anthropogenic CO_2 emissions (Levine 1999).

Kurdistan's region, north of Iraq, is almost the only area in Iraq where forests are remaining (Chapman 1950). They have been playing a vital role in the region's ecosystem and are considered as an important source of income among rural communities (Warveen 2016). Furthermore, these forests have served, in the past, as the residence for a variety of local species (Guest et al. 1966). Kurdish natural resources, also, were destructed widely during the political conflict (Black 1993). Those incidents resulted in burning forest areas and agriculture lands, as well as forcing millions of Rural residents to leave rural areas (Black 1993). Many attempts including reforestation campaign, wildfires control have been carried out by Kurdistan Regional Government (KRG) that were unfortunately unsuccessful due to poor planning and lack of expertise (Warveen 2016).

Forest fires, sometimes, can become very disastrous after 15–20 min. Therefore, it is really important to develop rapid, accurate potential maps, and the maps that show fire probability. A number of techniques are being used for fire detection, like airborne fire detection where people observe large areas during the flight and record the location information using Global Positioning System (GPS). Another method is the watch tower where large areas can be viewed. However, the most efficient, easy and cost-effective is nowadays suggested by **R**emote **S**ensing (**RS**). Satellite remote sensing is today regarded as the main source of data for mapping fire risk, assessing forest fuel, monitoring forest fires, as well as, estimating post-fire damages (Santi et al. 2017; Meng et al. 2017; Frazier et al. 2018; McCarley et al. 2017; Verger et al. 2016). As proved in many researches, remote sensing data are one of the most reliable tools that are supposed viable to this purpose. The use of remote sensing data provides temporal and spatial coverage of biomass burning without costly and expensive fieldwork (Chuvieco 1996). The resulting information is suitable for its integration into a Geographic Information System (GIS) which allows the storage and processing of large volumes of spatial data (Chuvieco 1996), as well as the production of spatial analysis (Sunar and Ozkan 2001). Related to forest fires, remote sensing data provide information on environmental conditions before, during and after a fire occurs (Roy et al. 1999). At regional to global scales, the detection of burned areas using satellite data has been traditionally carried out by the Advanced Very High-Resolution Radiometer (AVHRR) because of its high temporal resolution (Kaufman et al. 1990).

Table 20.1 RS indices

RS indices	References
Land surface temperature (**LST**)	Mao et al. (2018), Peng et al. (2007)
Fuel moisture content (**FMC**)	Chuvieco et al. (2004), Yebra et al. (2008)
Normalized differential vegetation index (**NDVI**)	Jasinski (1990), Alonso et al. (1997), Chuvieco et al. (2004), Yankovich et al. (2019)
Enhanced vegetation index (**EVI**)	Huete et al. (2002)
Grassland fire danger index (**GFDI**)	Wang et al. (2004)
Normalized burn ratio (**NBR**)	Key and Benson (2005)

However, the MODerate resolution Imaging Spectrometer (MODIS) sensor is opening a new era in the remote sensing of burned areas (Kaufman et al. 1990). MODIS is a sensor housed on Terra and Aqua NASA (National Aeronautics and Space Administration) satellites with more than 30 narrow bands at wavelengths from the visible to the thermal infrared and at variable spatial resolutions (250–1000 m) (Lentile 2006). Different studying in the field of remotely based forest fire investigation in different study areas, have resulted in different indices. Table 20.1 shows some of the remotely sensed indices developed to study various aspect of wildfire phenomenon.

Although satellite imagery has been widely used for the assessment of fire related studies at local (Jia et al. 2006; López García and Caselles 1991), regional (Collins et al. 2007; Díaz-Delgado and Pons 2001) and global scales (Grégoire et al. 2003; Justice et al. 2002), but there are rare number of remotely sensed environmental researches about Kurdistan region. This research, getting aid from the priceless ability of RS and GIS techniques, is aiming to provide a map to show the potentiality of fires among Kurdistan forest and rangeland areas.

The factors, which could generally cause the forest and rangeland fires are identified as the followings:

- Fires among dry plants caused by lightning.
- Fires due to the existence of objects such as glass bottom, broken glass or bottle, since these bodies sometimes act as magnifier and light plants.
- Plants self-burning due to severe heating and dryness among forage rangelands.
- Human activities.

The last factor, i.e. human activities, is the most frequent causes of ignition in Kurdistan's forest areas. This issue seems more serious when the frequency of fire occurrences is more than what naturally happens, exactly like what happened in this study area, where more than 100 fires occurred in less than 4 months in a recent decade. Mateescu (2006) classify Forest Fire Management into three phases. The pre-fires phase which involves prevention and risk assessment. The second phase is Effect fire phase which includes response and mitigation, and the last phase is the Post fire phase, which is engaged with damage assessment and rehabilitation policy.

This study is not seeking the first phase, mapping areas of the most potentiality to face forest fires. Since most of the fires in the study area, occur during the hot and dry months, this study has been temporally limited to the period of June to September. During this period, rangeland and grass mass gradually get dry and trees water content decrease as well. As a result, there will be a great mass of flammable fuel among forest areas. The major matter, at the first step, is temporal and spatial propagation of fuel. That is why this the research mainly focuses on vegetation cover, as a start step, among different dynamic and static factors which are proved to be effective in fire occurrence and propagation (Lozano et al. 2007).

20.2 Study Area

As reported by Kurdistan Regional Government (KRG), Kurdistan Region is located at the north of Iraq (Fig. 20.1). It borders Syria to the west, Iran to the east, and Turkey to the north, respectively, and lying in fertile plains meet the Zagros mountains (KRG 2003). It is traversed by the Sirwan river and the Tigris and its tributaries, the Great Zab and the Little Zab (KRG 2003). This region covers an area which estimate about 40,643 km (KRG 2003), resided by almost 3,800,000 residents (Oil-for-Food Distribution Plan 2002), and the annual rainfall of 375–724 mm (FAO 2000). It is also reported that, KR, the climate of the Kurdistan Region is semi-arid continental. It means it is very hot and dry in summer, and cold and wet in winter. Especially, the summer months from June to September are very hot and dry. In July and August, the hottest months, mean highs are 39–43°, and often reach nearly 50° (KRG 2003). This condition can result in wildfires. besides, the mountains of the Kurdistan Region have an average height of about 2,400 m, rising to 3,000–3,300 m in places. The highest peak, Halgurd, is near the border with Iran and measures 3,660 m (UNEP 2003).

20.2.1 Forests of Kurdistan

Historical evidence supports the argument that most of Kurdistan mountains have been covered with well-stocked and healthy forests compared to the currently degraded forests conditions (Chapman 1950; Nasser 1984). Nasser mentioned that the forest cover in Kurdistan starts in the northwestern part of the country extending from Zakho area near Turkish borders between (37° 08′ 22″N 42° 40′ 30″E) and (34° 40′ 50″N 45° 30′ 45″E) ending on the Iranian border in an area called Horin Shirin (Nasser 1984). Natural forests were confined to the mountainous areas only, covering 4% of the total area about (438,466 km^2), and about 60% of the total mountainous region- approximately about 30,000 km^2 (Warveen 2016). Based on previous studies, broad-leaved oak forests comprise the largest percentage of the total forested area about 90%, with the remaining 10% of other forest types such as pine, riverine, plantation, and others (Nasser 1984; Chapman 1984; Şefik 1981). Broad-leaved

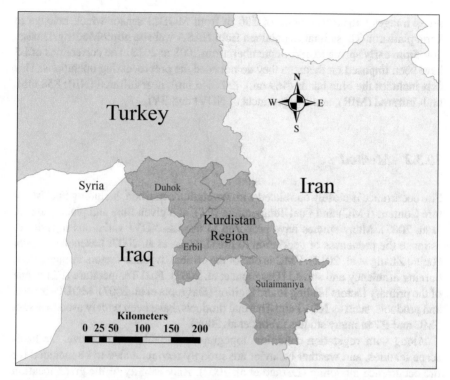

Fig. 20.1 Study area, Kurdistan Region, located in the North of Iraq. It is created in Arc GIS. The shapefiles of Administrative Boundaries were downloaded at Esri open data: https://hub.arcgis. com/pages/open-data

forests are known to be resistance to grazing and the region's harsh weather condition especially during hot and dry summer season. However, these forests have suffered various man-made and natural disturbances are especially shifting cultivation and fires (wild and man-made), tree cutting and others (Nasser 1984; Chapman 1984; Şefik 1981).

20.3   Methods and Materials

20.3.1   Used Data

Two main data sets were used in this study. Field data, at the first place, including the date and location of Forest fires among Sulaimani province boundary, prepared by Department of Agriculture and Forest of Sulaimani. The Satellite image, MODIS products, at the other place, are the other used data sets which include 16-day time-series images with a resolution of 250 m from MODIS sensor which belongs to Terra platform. These images obtained from NASA website http://Modis.gsfs.nasa. gov/ from early spring to mid-September, from 2008 to 2012. The corrections of L3 have been imposed on them, so they do not need the preprocessing operations. This data includes the blue bands (469 nm), red (645 nm), near-infrared (NIR: 858 nm), mid- infrared (MIR), and two products of NDVI and EVI.

20.3.2 Method

Fire occurrence is closely correlated with vegetation condition, including Fuel Moisture Content (FMC) and Fuel Temperature (FT), at a given time and place (Lozano et al. 2007). Many studies have proposed to use the NDVI variations in order to estimate the proneness of vegetation to fire (Lozano et al. 2007; Lasaponara 2005; Haijun Zhang et al. 2012). FMC is one of the critical dynamic factors initiating fire, burning efficiency and spread (Dasgupta et al. 2007). Fuel Temperature (FT) is one of the primary factors leading to fire ignition (Dasgupta et al. 2007). MODIS images and products, such as NDVI and Thermal products, have been widely used to assess FMC and FT in many studies (Yebra et al. 2008).

Along with vegetation condition, topographic variables characterize, the landscape features, and weather behavior are strongly recommended to be included for fire occurrence modeling (Lozano et al. 2007). Accessibility of the given location may be an influencing factor which usually determines where and when will burn, especially for fires caused by human, especially in the touristic areas. Lozano et al. (2008) have mentioned dynamic and static predictors used to model fire occurrences as seen in Table 20.2.

RS is considered as a very strong tool to estimate all almost all factors mentioned in Table 20.2 and preparing thematic maps to show each. The ability to locate the places where a fire trends to occur is important for the fire management. Vegetation cover is considered as the most important factor in forest fires, because it shows the presence and content of fuels in forest and rangeland areas (Roy 2003). In this study, the main focus is on using MODIS-based NDVI (Eq. 20.1) as a reliable index to shows and model vegetation cover condition and how it would help us to see where might have more potential to experience forest fire.

$$NDVI = \frac{NIR - RED}{NIR + RED} \tag{20.1}$$

NDVI values vary from -1 to $+1$. Because of the high reflectance of healthy vegetation in the NIR part of the electromagnetic spectrum, they are represented by high NDVI values between 0.05 and 1. In contrast, non-vegetated surfaces such as water bodies have negative values of NDVI. Bare soil NDVI values are close to 0 due to high reflectance in both the visible and NIR portions of the electromagnetic spectrum (Lillesand and Keifer 1994). NDVI values for each pixel of satellite images

Table 20.2 Environmental predictors used to model fire occurrence

Static landscape variables at the considered temporal scale (yearly)
• Elevation (m)
• Slope (degrees)
• Inner pixel standard deviation of the elevation derived from a 5 m DEM
• Inner pixel variation coefficient of the elevation derived from a 5 m DEM
• Annual solar radiation (MJ/(cm^2 year))
• Heat load index (no unit)
• Distance to the nearest village (m)
• Distance to the nearest path (m)
• Distance to the nearest isolated building (m)
• Frequency (0–1) of heathland (dominated by Erica spp.) in a 7 × 7 kernel
Dynamic landscape variables at the considered temporal scale (yearly)
• **NBR** (normalized burned ratio) index values of the four previous years
• **NDMI** (normalized difference moisture index) index value of the four previous years
• **NDVI** (normalized difference vegetation index) index value of the four previous years
• **TCW** (tasseled cap wetness) index value of the four previous years
• **CG** (tasseled cap greenness) index value of the four previous years

related to a specific area with specific vegetation, depends on a number of parameters, such as (I) the type and density of plants, (II) plant health and (III) volume of water in plant tissue. Therefore, interpretation and comparisons of these circumstances must be considered. High NDVI values indicate high leafy biomass, canopy closure, leaf area. It can be used to assess vegetation proneness to fire and to access post-hurricane and fire forest damage (Chuvieco et al. 2004).

The Global M1ODIS vegetation indices are designed to provide consistent spatial and temporal comparisons of vegetation conditions. For example, the MODIS Normalized Difference Vegetation Index (NDVI) products provide continuity for time series historical applications.

Integrating Field and satellite data, the following diagram (Fig. 20.2) can simply shows the steps to prepare the output of this study.

Field data, includes fired forests recorded in 2014 and 2015, divided into two main groups. The first group, including 20 diffused locations, which was used to train the model. And other det of field records, which used for verifying the output map to see how accurate it would be. On the other hands, it is satellite data. The study area was extracted from the satellite image using ENVI software. The extracted average value of NDVI for training locations, form late Winter to early Autumn, (Fig. 20.3), reveals that NDVI meets its highest value in the April and May, which the area is covered with fresh grassland and healthy and green tree covers, and then gradually decreased to its lower level in about August which the only rest green cover is trees. It is tried to pick the training data from different locations of different fire- strength

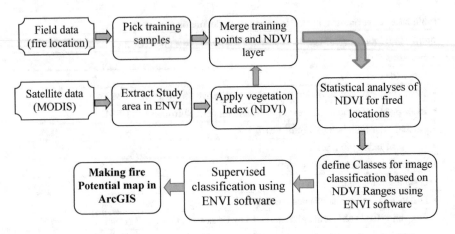

Fig. 20.2 Sequential steps from input to output

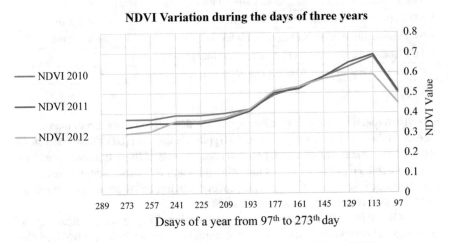

Fig. 20.3 Average NDVI Value for training locations during 2010, 2011, and 2012

to see how their NDVI value. The extracted NDVI values from satellite-based data for these location shows the more the NDVI value during the April and the May, the higher the degree of destruction and wider scar of fire during summer.

The inclination of fired forest and rangeland areas to follow NDVI value, for three years, was used as an index to show vegetation distribution and FMC seemed quite helpful to classify study area into some nominal classes based on NDVI different ranges to show the level of firing potentiality. Four classes defined as showed in Table 20.3.

These ranges are extracted based on NDVI trends (Fig. 20.3). These values do not necessarily match other years NDVI behavior, considering different trends of vegetation condition in different years regarding precipitation and weather condition. ENVI 4.8 IDL software was used to analyze and process the satellite image.

Table 20.3 NDVI ranges for defining classes

Classes	NDVI value
Very hight potential	NDVI > 0.6
High potential	0.45 < NDVI < 0.6
Low potential	0.3 < NDVI < 0.45
No potential	NDVI < 0.3

The supervised classification, maximum likehood, used to classify the NDVI image considering the classes defined in Table 20.3. The classified image then uploaded into Arc GIS to apply more analyzes including displaying, defining scale, and adding other layers. The satellite image and other information layers were not in the same coordinate system, so, they would be transferred into the same coordinate system, geographical coordinate system. The layers of residency, fired places' 1000 buffer, districts and subdistricts then added to the produced potential map.

20.4   Results

Figure 20.4 shows the final map which is the result of applying supervised classification on NDVI raster layer to produce a final classified map. The classes included NDVI ranges (assigned in Table 20.3). The used satellite image was taken in 2010, four years before recorded field data about wildfire locations. As seen on the map, black circles, which are the locations of fired forest from 2014 to 2015, closely match the Red and Yellow areas on the map which are classified as Very high potential and High potential, respectively. Based on the developed map Penjwen, located in the west of the map, had the most potential of forest fire considering the distribution of Red and Yellow classes on the map. Soran, Amedi, Zakho, Choman, Pishdar, Gharadakh, and Dukan are other areas which considered as high potential areas to face forest fires. Surprisingly, the data reported officially for Dukan, Gharadakh, Pishdar, and Halabja also match the prediction made by developed map, nevertheless, lacking data for the northern and central part of study area would not let us say a valid report about these areas. Green and White areas on the map, which are show mid and low potential classes, also are quite consistent with field recorded statistics. The study, also, shows that less than 15% of Kurdistan region have a serious potential of facing forest fire, so, considering the importance of forest in Iraq's environment, preventive actions should be considered seriously. Figure 20.5 shows the fact in a closer view. As seen, black circles are almost lie in the areas which are predicted as **High** and **Very High Potential** areas. Especially, the western part, Penjwen and Sharbazher. The same also is observable for Pishdar and Dukan subdistricts. The map shows that, the wider and denser level of Red and Yellow area among the study area, the higher number of fire occurrences have been recorded in later years. To have an even better view, Fig. 20.6 has also been developed which especially has

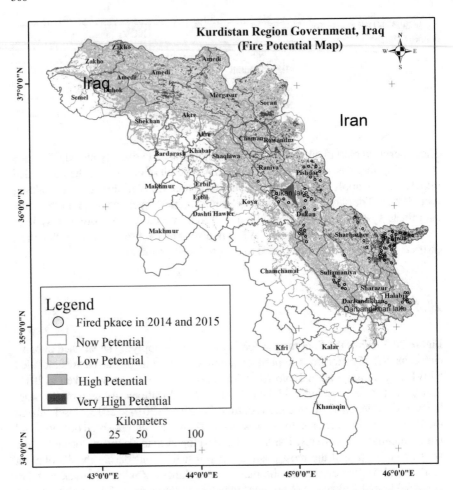

Fig. 20.4 Result of Fire potential classification based only on applying on NDVI. The used satellite image belongs to 2010. The Black circles are <u>1000</u> m buffers of fired places among Sulaimani province's forests and rangeland areas. The shapefiles of Administrative Boundaries were downloaded at Esri open data: https://hub.arcgis.com/pages/open-data

focused on only fired areas. In Fig. 20.6 the occupied areas inside the circles simply reveals the high-level presence of Red and Yellow pixels which represent Very High and High Potential areas, respectively. So, the developed map based on an image was taken in 2010, could maps the fire-potentiality of forest and rangeland areas of the study area.

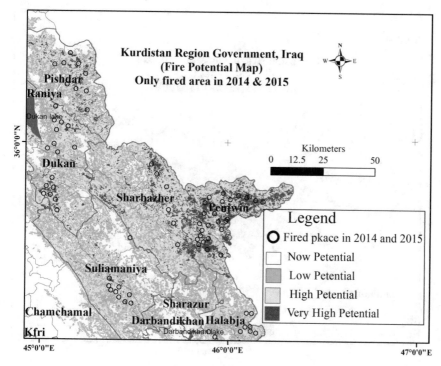

Fig. 20.5 The western part of the map in more details

20.5   Discussion

There are many researches at local, regional and global scales focusing on the application of remote sensing in environmental issues including forest and rangeland fires. There are also some very systematic models to monitor and predict forest fires using RS. Among them, The McArthur Forest Fire Danger Rating System and the McArthur Grassland Fire Danger Rating System is widely used in Australia. In addition, The North American models National Fire Danger Rating System (NFDRS), Fire behavior (BEHAVE), Fire Area simulator (FARSITE), and National Fire Management Analysis System (NFMAS) (Deeming et al. 1978; Lundgren et al. 1995). Furthermore, The Canadian Forest Fire Danger Rating System which is mostly used in Canada (Arroyo et al. 2008; Van Wagner 1987). The Russian, also, developed The Forest Fire Satellite Monitoring Information System of Russian Federal Forestry Agency (SMIS-Rosleshoz) (Arroyo et al. 2008; Lasaponara and Lanorte 2007). Despite all priceless research done in different part of the world in the field of forest fire monitoring using RS, in Iraq, especially in Kurdistan region, the use of remote sensing in environmental investigation is almost ignored. This study, has not aimed to develop an innovative technique and method, but it mainly focused on

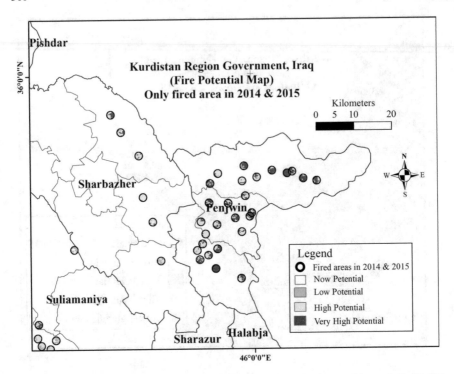

Fig. 20.6 The western part of the study area, focusing only on fired areas. The map shows that most of the fired area in 2014 and 2015 have been predicted as High potential areas before, based on 2010 developed map

using existing possibilities for developing a Fire-potential maps for the North of Iraq, which it can be regarded as a new research in this study area.

20.6 Conclusion and Recommendation

This study reveals that, Rs and GIS can play a vital role in managing natural hazardous phenomenon, like wildfires, as wee as, monitoring natural resources. RS helps us produce variety of thematic maps in different scales. Fire occurrence is closely correlated with vegetation condition, including Fuel Moisture Content (FMC) and Fuel Temperature (FT), at a given time and place. Many studies have proposed to use the NDVI variations in order to estimate the proneness of vegetation to fire. Since rangeland and forages (grass species) have less endurance against heat so they have the most potential to be fired. NDVI index can play a major role in studying of these kinds of fires because NDVI index can give a reliable and attributable view about the health and density of vegetation cover, and also its growth for this region. Penjwen, located in the west of the map, had the most potential of forest fire considering the distribution of Red and Yellow classes on the map. Dukan, Gharadakh, Pishdar, and Halabja, respectively, were the other areas in Sulaimani province which the highest number of fires have been recorded for them. The Potential map (Fig. 20.6) show a high consistency with the field observations. There was no access, during the research, to get any records from Soran, Amedi, Zakho, Choman, Pishdar, Gharadakh, and Dukan subdistricts. However, they are also classified as high potential areas to face forest fires. It is recommended to add their data to the map to get a better assessment. Furthermore, access to more detailed field data as well as using other multispectral and hyperspectral satellite image, accompanied by other satellite-based indices might result in more reliable and precise maps. Conducting some research on studying Water Stress (WS) among Kurdistan forest, as well as, FMC and FT among different species during different seasons are strongly recommended. A precise research on mapping vegetation type in the study area will be pretty helpful for spatial to estimate fire risk and fire potentiality.

Acknowledgements The authors would like acknowledge the Department of Forest and Rangeland management of Sulaimani Province for their cooperation providing the data for this project.

References

Alonso FG, Cuevas JM, Casanova JL, Calle A, Illera P (1997) A forest fire risk assessment using NOAA AVHRR images in the Valencia area, eastern Spain. Int J Remote Sens 18(10):2201–2207

Arroyo LA, Pascual C, Manzanares JA (2008) Fire models and methods to map fuel types: the role of remote sensing. For Ecol Manag 256(6):1239–1252

Bajocco S, Rosati L, Ricotta C (2009) Knowing fire incidence through fuel phenology: a remotely sensed approach. Ecol Model 221(1):59–66

Black G (1993) Genocide in Iraq: the Anfal campaign against the Kurds. Human Rights Watch

Chapman GW (1950) Notes on forestry in Iraq. Emp For Rev 132–135

Chapman GW (1984) Forestry in Iraq. Unasylva 2(5):251–253

Chuvieco E (1996) Fundamentos de teledetección Especial. Rialp Press, Madrid, Espana

Chuvieco E, Cocero D, Riano D (2004) Combining NDVI and surface temperature for the estimation of live fuel moisture content in forest fire danger rating. Remote Sens Environ 92:322–331

Collins BM, Kelly M, van Wagendon JW, Stephens SL (2007) Spatial patterns of large natural fires in Sierra Nevada wilderness areas. Landsc Ecol 22:545–557

Dasgupta S et al (2007) Evaluating remotely sensed live fuel moisture estimations for fire behavior predictions in Georgia, USA. Remote Sens Environ 108(2):138–150

Deeming JE, Burgan RE, Cohen JD (1978) The national fire-danger rating system 1978. General technical report INT-39. Intermountain Forest and Range Experiment Station. USDA Forest Service, Ogden, Utah

Díaz-Delgado R, Pons X (2001) Spatial patterns of forest fires in Catalonia (NE Spain) along the period 1975–1995. Analysis of vegetation recovery after fire. For Ecol Manage 147:67–74

FAO (2000) Derived from the global agro-ecological zones study, Food and Agriculture Organization of the United Nations (FAO), Land and Water Development Division (AGL), with the collaboration of the International Institute for Applied Systems Analysis (IIASA), 2000. Data averaged over a period of 37 years. Raster data-set has been exported as ASCII raster file type

Frazier RJ, Coops NC, Wulder MA, Hermosilla T, White JC (2018) Analyzing spatial and temporally variability in short-term rates of post-fire vegetation return from Landsat timeseries. Remote Sens Environ 205:32–45

Grégoire JM, Tansey K, Silva JMN (2003) The GBA2000 initiative: developing a global burned area database from SPOT-VEGETATION imagery. Int J Remote Sens 24:1369–1376

Guest E, Townsend CC (1966) Flora of Iraq

Huete A, Didan K, Miura T, Rodriguez EP, Gao X, Ferreira LG (2002) Overview of the radiometric and biophysical performance of the MODIS vegetation indices. Remote Sens Environ 83:195–213

Jasinski FM (1990) Sensitivity of the normalized difference vegetation index to subpixel canopy cover, soil albedo, and pixel scale. Remote Sens Environ 32(2–3):169–187

Jia GJ, Burke IC, Goetz AF, Kaufmann MR, Kindel BC (2006) Assessing spatial patterns of forest fuel using AVIRIS data. Remote Sens Environ 102:318–327

Justice CO, Townshend JRG, Vermote EF, Masuoka E, Wolfe RE, Saleous N et al (2002) The MODIS fire products. Remote Sens Environ 83:244–262

Kaufman Y, Tucker C, Fung I (1990) Remote sensing of biomass burning in the tropics. J Geophys 95:9927–9939

Key CH, Benson NC (2005) Landscape assessment: remote sensing of severity, the normalized burn ratio; and ground measure of severity, the composite burn index. FIREMON: Fire Effects Monitoring and Inventory System

KRG (2003) KRG administered territory. Compiled by the Food and Agricultural Organization (FAO) from various national and regional sources: International Boundaries from National Imagery and Mapping Agency (NIMA) Digital Chart of the World (DCW). www.cabinet.gov. krda

Lasaponara R (2005) Inter-comparison of AHVRR-based fire susceptibility indicators for the Mediterranean ecosystems of Southern Italy. Int J Remote Sens 26(5):853–870

Lasaponara R, Lanorte A (2007) On the capability of satellite VHR Quick Bird data for fuel type characterization in fragmented landscape. Ecol Model 204(1–2):79–84

Lentile L (2006) Remote sensing techniques to assess active fire characteristics and post-fire effects. Wildland Fire 319–345

Levine J (1999) Introduction. In global biomass burning: atmospheric, climatic and biospheric implications. USA: MIT Press, Cambridge

Lillesand TM, Keifer W (1994) Remote sensing and image interpretation. Wiley, New York

López García MJ, Caselles V (1991) Mapping burns and natural reforestation using thematic mapper data. Geocarto Int 1:31–37

Lozano FJ, Suárez-Seoane S, de Luis E (2007) Assessment of several spectral indices derived from multi-temporal Landsat data for fire occurrence probability modeling. Remote Sens Environ 107(4):533–544

Lozano FJ, Suarez-seoane S, Kelly M, Calabuig EL (2008) A multi-scale approach for modeling fire occurrence probability using satellite data and classification trees: a case study in a mountainous Mediterranean region. Remote Sens Environ 112(3):708–719

Lundgren S, Mitchell W, Wallace M (1995) Status report on NFMAS—an inter-agency system update project. Fire Manag Notes 55:11–12

Mao K, Zuo Z, Shen X, Xu T (2018) Retrieval of land-surface temperature from AMSR2 data using a deep dynamic learning neural network. Chin Geogra Sci 28(1):1–11

Mateescu M (2006) Burnt area statistics 3D GIS tool for post-burn assessment. Geogr Tech 2(2):56–65

McCarley TR, Kolden CA, Vaillant NM et al (2017) Multi-temporal LiDAR and landsat quantification of fire-induced changes to forests structure. Remote Sens Environ 191:419–432

Meng R, Wu J, Schwager KL et al (2017) Using high spatial resolution satellite imagery to map forest burn severity across spatial scales in a Pine Barrens ecosystem. Remote Sens Environ 191:95–109

Nasser MH (1984) Forests and forestry in Iraq: prospects and limitations. The Commonw For Rev 299–304

Oil-for-Food Distribution Plan (2002) Approved by the UN, December

Peng GX, Li J, Chen YH, Abdul-patah N (2007) A forest fire risk assessment using ASTER images in Peninsular Malaysia. J China Univ Min Technol 17(2):232–237

Roy PS (2003) Forest fire and degradation assessment using satellite remote sensing and geographic information system. Available at: http://www.wamis.org/agm/pubs/agm8/Paper-18.pdf

Roy PS, Giglio L, Kendall JD, Justice CO (1999) Multi-temporal active-fire based burn scar detection algorithm. Int J Remote Sens 20:1031–1038

Santi E, Paloscia S, Pettinato S et al (2017) The potential of multifrequency SAR images for estimating forest biomass in Mediterranean areas. Remote Sens Environ 200:63–73

Şefik Y (1981) Forests of Iraq. J Fac For Istanbul Univ (JFFIU) 31(1)

Sunar F, Ozkan C (2001) Forest fire analysis with remote sensing data. Int J Remote Sens 22(12):2265–2277

UNEP (2003) United Nations Environment Programme (UNEP). http://sea.unepwcmc.org/latenews/Iraq_2003/facts.htm

Van Wagner CE (1987) Development and structure of the Canadian forest fire weather index system. Forest Technology Report 35, Canadian Forestry Service, Ottawa, Canada

Verger A, Filella I, Baret F, Peñuelas J (2016) Vegetation base-line phenology from kilometric global LAI satellite products. Remote Sens Environ 178:1–14

Wang L, Zhou Y, Wang S, Chen S (2004) Monitoring for grassland and forest fire danger using remote sensing data. In: Proceedings IGARSS'04, IEEE International geoscience and remote sensing symposium, pp 2095–2098

Warveen LM (2016) Forest cover change and migration in Iraqi Kurdistan: a case study from Zawita Sub-district. dissertation in Michigan State University

Yankovich KS, Yankovich EP, Baranovskiy NV (2019) Classification of vegetation to estimate forest fire danger using landsat 8 images: case study. Math Prob Eng 4:1–14

Yebra M, Chuvieco E, Rian D (2008) Estimation of live fuel moisture content MODIS images for fire risk assessment. Agric For Meteorol 148:523–536

Zhang H, Han X, Dai S (2012) Fire occurrence probability mapping of Northeast China with binary logistic regression model. IEEE J Sel Top Appl Earth Obs Remote Sens 6(1):121–127

Part VII
Conclusions

Chapter 21
Updates, Conclusions, and Recommendations for Environmental Remote Sensing and GIS in Iraq

Ayad M. Fadhil Al-Quraishi and Abdelazim M. Negm

Abstract This chapter aims to highlight the main conclusions and recommendations that can be extracted from the chapters contributed to this book. It includes a brief information on some distinguished studies that used remote sensing (RS) and Geographical Information Systems (GIS) techniques for environmental applications in some sites in Iraq. It also throws light on the role of remotely sensed dataset, remote sensing and GIS technologies for modeling, mapping, characterizing of soil properties, proximal soil sensing, land cover/land use change monitoring, land degradation, drought, sand dunes, dust storms, as well as to hydraulic modeling, climate change, and forest fires. The current chapter includes set of recommendations for the future works that aim to get benefits from the facilities and the capabilities of remote sensing and GIS technologies toward the environmental sectors in Iraq, which are the most vital issues in the present and the future of Iraq.

Keywords Monitoring · Modeling · Environment · Natural resources · Remote sensing · GIS · Proximal · Mapping · Drought · Flood · Soil · Forest · Climate change · Iraq

21.1 Update

In the following paragraphs, the national studies concerning the environmental applications of remote sensing and GIS in the Republic of Iraq are presented. In addition,

A. M. F. Al-Quraishi (✉)
Environmental Engineering Department, College of Engineering, Knowledge University, Erbil 44001, Kurdistan Region, Iraq
e-mail: ayad.alquraishi@gmail.com; ayad.alquraishi@knowledge.edu.krd

A. M. Negm
Water and Water Structures Engineering Department, Faculty of Engineering, Zagazig University, Zagazig 44519, Egypt
e-mail: Amnegm@zu.edu.eg

some studies related the hydraulic and the precipitation measurement are mentioned. The brief results of the studies are introduced in this chapter.

Soil salinity is considered as one of the most active land degradation phenomena in the middle and south of Iraq, whereas it is estimated that approximately 60% of the cultivated lands has been seriously affected by salinity (Wu et al. 2014). However, the outdating and low resolution of the available maps and the satellite images cannot meet the requirement of land management and salinity control at the farm/local scale in the region. Therefore, the intention was to incorporate radar data, taking their advantages of independence of weather condition and penetration to subsoil, with optical imagery to utilizing them for salinity assessment in the middle and south of Iraq. The remotely sensed dataset used to salinity assessment were ALOS (Advanced Land Observing Satellite) PALSAR (Phased Array L-band Synthetic Aperture Radar) data and Landsat 5 TM (Thematic Mapper) imagery acquired at almost the same time were employed for that purpose.

The soil is one of the most important components of agricultural production and can have a dominant effect on crop yields and quality. Infield soil information has been used for centuries by farmers to make decisions concerning crop manage-ment practices. Quantitative information and spatial distribution of soil properties are among the main prerequisites for achieving sustainable land management. The Landsat 8 OLI image was used in a study to predict some soil properties such as soil organic carbon and nitrogen forms, soil salinity, and soil water in Al-Kufa, and Musayb in the middle of Iraq.

The reflected radiation depends on the soil's physical and chemical properties, such as texture, structure, moisture, soil minerals, organic matter, gypsum, and car-bonate, which considered an essential feature in recording the spectral signature of soil. Therefore, different types of soils can be characterized and separated (AL-Rajehy 2002). In the study that employed the Landsat satellite images to classify the Soil Map Units in Bahr Al-Najaf, Iraq, the results revealed that all soils are located belong to the Entisol order, Fluvents and Psamments suborder.

The need for soil information is higher now than ever before. Agriculture and how we use and manage our soils are being changed with the concerns over food security and global climate change. Mainly, soil data is necessary to be used in soil and natural resource management. Proximal Soil Sensing (PSS), has become a multidisciplinary area, aims to develop field-based techniques for acquiring information on the soil from close by, or within the soil. It can be used to monitor soil both surface and subsurface spatial and temporal information rapidly, cheaply and with less labor (Viscarra Rossel and McBratney 1998). A study on employing soil spectroscopy in Sulaimani, Iraqi Kurdistan Region was presented in this book. The spectroscopy technique was used for monitoring and mapping total Fe, and Fe_2O_3 rich soils of some sites in Sulaimani governorate, Iraqi Kurdistan Region as any such a research had not been conducted in the region, and there was no up-to-date map that present and show soil Fe and Fe_2O_3 in Sulaimani governorate.

Proximal soil sensing and Geostatistics along with its applications represent poten-tial analytical techniques for studying the soil properties. It is well know that most of the soil properties are both spatially and temporally variable over a short distance

(Harris et al. 1996). Geostatistics is another approach of describing soil variability based on the regionalized variable theory. GIS-Kriging as one of Geostatistics method makes it possible to determine values at the un-sampled location by using the spatial correlation between estimated points to predict un-estimated samples. The study results of proximal soil sensing for Samawa and Rumetha soils indicated that NIR Spectroscopy is an effective tool for a rapid assessment of soil information under field conditions, through which decision on fertilizer requirements can be based (under the conditions of this study).

Land use/cover (LULC) affects local, global environment, climate and land degradation that reduces ecosystem services and functions (Tolessa et al. 2017). Monitoring of LULC is important to assess the change and manage the environment (Jawarneh and Biradar 2017). Historical and current status of the land is essential for efficient environmental management. This can especially be noticed in regions that are vitally affected by climate variability and human activities such as Zakho district, Kurdistan Region-Iraq. The information and status of land use/cover (LULC) help to design an efficient and sustainable environmental management program. The present study illustrates the spatiotemporal dynamics of LULC in Zakho district, Iraq.

Land cover is the physical condition and biotic component of the earth's surface, whereas land use is the modification of the human activities and climate change on the land cover (Friedl et al. 2010). Determination of modification in land cover over a certain period is called change detection. The change in the land cover is the most important aspect of global environmental change. MODIS-NDVI scenes were used to detect the relationship between the NDVI values and both of the annual precipitation and elevation. An inverse relationship was observed for the NDVI values of the grasslands class and the elevation values.

Evaluating the effect of land cover and land use changes on hydrologic conditions is vital for catchment area management and development (Woldesenbet et al. 2017). The change in land cover has a considerable impact on the nature of runoff. Land cover in some areas remains unchanged for long periods, while in other regions land cover sees drastic changes more frequently. Economic and social activities such as urban growth and agricultural development have a direct effect on land cover changes (Letha et al. 2011). Geographic Information Systems (GIS) was used to prepare different layers belonging to rainfall spatial distribution using Inverse Distance Weighting (IDW) tools, and various land covers were determined from remotely sensed data in Duhok watershed.

A catchment area is an area from which runoff is resulting from precipitation flows. Land cover changes are the most significant factors that directly impact the runoff process (Woldesenbet et al. 2017). Much research on runoff response has focused on projected climate variation, while the endemic catchment area is directly affected by urban growth. Due to urbanization, the continuous growth in urban areas has led to a significant transformation of land cover pattern in built-up areas which has significantly affected the surface runoff behavior in the urban realm.

Land use and land cover were integrated with a hydrological model SCS-CN to compute the runoff volume from the catchment area. Geographic Information Systems (GIS) was used to prepare different layers belonging to rainfall spatial distribution using Inverse Distance Weighting (IDW) tools, and various land covers were determined from remotely sensed data.

Remote sensing may offer possibilities for extending existing soil survey data sets. The data it provides can be used in various ways. Firstly, it may help in segmenting the landscape into internally more or less homogeneous soil–landscape units for which soil composition could be assessed by sampling using classical or more advanced methods. Secondly, remotely sensed data could be analysed using physically-based or empirical methods to derive soil properties. Moreover, remotely sensed imagery could be used as a data source supporting digital soil mapping (Ben-Dor 2008; Slaymaker 2001).

Four processes of land degradation are usually recognized in Iraq including salinization, erosion, sand dunes, and urbanization. Many studies have shown that salinity is one of the most serious degradation processes in the central and southern Iraq lands.

Iraq is a country that was well known for its agricultural production and fertile soil. However, agricultural drought overshadows the vegetative cover in general and cropland specifically in Iraq as it represents a creeping disaster. The arable lands in Iraq experienced increasing drought events that led to land degradation, desertification, economic losses, food insecurity, and deteriorating environment, particularly in recent years. UNESCO (2014) highlighted that Iraq has suffered from several drought events in the period of 2003–2012, where different factors contributed in the occurrence of such events including low rainfall rates, higher temperatures, lower water income from upstream countries, and low efficiency in water utilization. These factors caused a multidimensional effect on the region such as the lower discharge of Tigris and Euphrates, less available and more saline groundwater, population migration, and agricultural degradation (UNESCO 2014).

Remote sensing dataset and techniques were employed in this chapter. Four different spectral indices; Vegetation Health Index (VHI), Vegetation Drought Index (VDI), Visible and Shortwave infrared Drought Index (VSDI), Temperature–Vegetation Dryness Index (TVDI) were utilized, each of them is derived from MODIS dataset of Terra satellite for monitoring the agricultural drought in Iraq.

Desertification process is the persistent degradation of land in arid and semi-arid environments due to the drought (particularly the global warming) and overexploitation of soil through human intervention (e.g., overgrazing, over-cultivation) (Lam et al. 2011). It can be considered as the major challenge in the arid and semi-arid regions, particularly in the last decades. Aeolian sediments arise as one of the main factors of desertification in term of extent and movement in the Iraqi territory. Aeolian sand dunes are one of the most amazing natural features on Earth. Understanding how aeolian sediments (i.e., sand dunes) form and move has long been a research topic in Earth surface processes. This chapter describes a remote sensing approach utilized to monitor temporal and spatial changes of aeolian sand dunes in Hor Al-Dalmaj area, which is classified according to climatology as an arid area. The aeolian sand dunes in Hor Al-Dalmaj area characterized by NW-SE direction make them parallel to the fold axes extend.

In the study, it was utilized two-primary types of sensors: passive and active to detect the aeolian sand dunes changes using Landsat imagery for entire Iraq. Moreover, the study selected a specific sample area in the central part of the Mesopotamia to determine the activity of the aeolian sand dunes movement by applying the DIn-SAR technique and Sentinel 1A images.

Drought is classified as an environmental hazard and natural disaster that depreciates the sustainable development of society. Its long-lasting impacts badly have increased its extent on agricultural production, livestock, physical environment, and the overall economy. Over the last three decades, many world regions have suffered from water crises, and drought caused serious impacts on local economies. Iraq, Syria, Turkey, and Iran, have been dealing with decreased rainfall affected the agricultural sector, livelihood system, employment and water allowable quantity and quality (UNDP 2010) negatively.

This study emphasized the use of Remote Sensing and GIS in the field of drought risk evaluation. The results showed that the NDVI is an efficient way to monitor changes in vegetation conditions (weekly or daily) during the growing season, and can be used as simple and cost-efficient drought index to monitor agricultural drought at a small or large scale.

Dust storms have been the focal point of a vast body of projects and studies during the last decades. Almost all instances of such studies explored the identification of the sources, causes, emission mechanisms, health and socioeconomic impacts, and warning/monitoring as their main objectives. Because dust storms emerge as the final result of numerous composite factors, various methods, data and disciplines in Earth-sciences have been proposed and implemented in this regard. In addition to environmental factors such as land surface vegetation cover, soil, and water resources, dust storms and high-speed winds may also originate from special atmospheric conditions. One of the main objectives of this chapter, in aiming to study the dominant factors causing dust storms is to scrutinize RS and GIS, at which point, due to the multi-disciplinary, multi-factor, multi-cause and transnational nature of dust storms, it is necessary to consider several scientific disciplines, simultaneously.

Therefore, other fields of study, such as soil sciences and meteorology are also accounted for. Furthermore, this chapter will discuss the potentials of RS and GIS in providing applicable solutions for planning to combat against dust storms. In this regard, the RS and GIS-based problem-solving approaches are discussed.

Iraq has suffered in the last fifteen years an environmentally difficult period, where the serious decline in water discharges of its two main Tigris and Euphrates and their tributaries, as well as the significant drop in the precipitation averages throughout the country. That reduction has disastrously affected the agricultural areas, the quantities of available water for drinking, industrial and agricultural uses. The situation of drought has been reported recently in Iraq since the annual rainfall has dramatically decreased in the past few years.

Severe drought has affected the Kurdistan region as well as the other parts of Iraq throughout the last years which was characterized by a significant drop in the rainfall amounts (Fadhil 2011). Particularly, the Kurdistan region was considered the Iraqi's historical breadbasket, where rain-fed wheat is grown. The study aims s to investigate the role of the integration of NDVI as a satellite-based vegetation index and SPI as a meteorological-based index for drought monitoring in the Kurdistan region of Iraq during eighteen years.

Since GIS, RS and hydrologic and hydraulic modelling systems are capable of integrating each other to perform required hydrologic and hydraulic investigation and analysis precisely, therefore in this chapter through a specific schematic workflow the methodology of this research is applied in an efficient and reliable way. In this study, remotely sensed images such as DEM, topographic maps, Landsat satellite image, and tabulated hydrograph data are used. Thus, obtained geo-morphometric parameters for the Mosul Dam River-Basin are tabulated. Also, the discharge flow based on unsteady flow analysis is simulated based on a computational-2D program which is so-called HEC-RAS in order to calculate related hydraulic parameters such as time of arrival, flood depth, water surface elevation, and stream power and to delineate flooded zone inside Mosul city as well. The obtained result from this study compared to the previous studies SWISS 1984, IWTC 2009 and JRC 2016 show good agreement with those of authors, specifically with the results of the SWISS study which is more detailed comparatively.

Digital elevation model (DEM) and raster images are used as essential raw data to build reliable modelling systems. The schematic workflow of this chapter shows the methodology that is applied in order to create computational hydrologic and hydraulic modelling systems. This methodology includes two main stages of processing; pre-processing of data and post-processing. In the first stage, both geo-morphological and hydrological feature classes are obtained and calculated by using ArcGIS tools such as Arc Hydro, HEC-GeoHMS, and HEC-GeoRAS. While, in the second stage, 2D-simulation programs such as HEC-HMS and HEC-RAS are used for creating both hydrologic and hydraulic modeling systems respectively. This chapter shows the results that have been obtained. First, the main hydrologic feature classes such as (basin, watersheds, sub-watersheds, catchments, streams, and rivers) are extracted from the digital elevation model. Thus, the main geo-morphometric parameters for the Greater Zab watershed are calculated. Second, the hydrologic model is designed for calculating Rainfall-Runoff and performing floodplain analysis.

The study presents the computational design of the hydrologic and hydraulic modelling system for the Greater Zab River-Basin in the Kurdistan Region of Iraq.

Assessment of drought conditions in Iraq is a very important task because of its location in the arid and semi-arid region. To assess and monitor drought conditions in Iraq, monthly complete precipitation data from 18 weather stations for the period (1980–2010) have been used to calculate Standardized Precipitation Index (SPI) using SPI Generator software. Statistical analysis has been done for SPI data to calculate averages, counts, minimum, maximum, and frequency values of dry and wet conditions at each station. Statistical data have been joined with the location of weather stations in ArcGIS for mapping SPI. IDW interpolation is used with specific parameters for mapping SPI average and SPI wet and dry frequencies data. The maps show that SPI frequency for wet and dry conditions give better and real spatial distribution and variation rather than average of SPI. Time series figures and summarize table for selected weather stations in Iraq were used to analyses the trend, duration, and frequency of wet and dry periods.

Assessing, monitoring, and mapping of long-term drought and wet condition by using Standardized Precipitation Index (SPI) is the best way to put efficient water policy and management in Iraq.

Climate change in IPCC (Intergovernmental Panel on Climate Change—https://www.ipcc.ch/) usage refers to a change in the state of the climate that can be identified (e.g., using statistical tests) by changes in average conditions and changes in variability (including, for example, extreme events) and that persists for a decade or longer. It refers to any change in climate over time, whether due to natural variability or as a result of human activity, related to precipitation, temperature, air composition, atmospheric circulations, weather extremes, and solar radiation. The main causes of climate changes are the natural process; however, human's activities can also cause changes to the climate. The Earth's climate has never been completely static, and in the past, the planet climate has changed due to natural causes. However, the natural causes cannot be causing current global warming, and the climate changes seen today are being caused by the increase of carbon dioxide (CO_2) and methane and other greenhouse gas emissions by humans. Increase in CO_2 concentration and other greenhouse gases have raised concerns about global warming and climatic changes. According to IPCC Report, CO_2 in the atmosphere is increasing by 1.4 ppm per year, and this will contribute to the increase in temperature by 1.5–4 °C by the end of this century.

In order to quantify spatial and temporal dynamics of LULCC, in particular the changes in vegetation, surface water and urban and built-up areas, unfortunately, only two Landsat satellite images were available: a scene, (path/row 169/35) taken on 28 May 1992 by TM sensor on board Landsat-4; a second scene (same path/row) taken on 25 May 2014 by ETM+ sensor on board Landsat-7.

North of Iraq is almost the only area in Iraq where forests are remaining. A great number of forest fires have been recorded in this region. This study is aiming to map the areas of most potential of firing to help the manager to follow preventive actions.

A forest fire is considered as an important issue for the environmental, economy, population, and human safety in many forested areas in the world. Satellite Remote Sensing (RS) is today regarded as the main source of data for mapping fire risk, assessing forest fuel, monitoring forest fires, as well as, estimating post-fire damages. This research, getting aid from the priceless ability of RS and GIS techniques, is aiming to provide a map to show the potentiality of fires among the Iraqi Kurdistan region forest and rangeland areas. Two sets of data, fields data (the location and date of fires in 2014 and 2015), and satellite data (MODIS NDVI-product time-series) were used. It has been proposed in many studies to use the NDVI variations in order to estimate the proneness of vegetation to fire. By classifying the NDVI image by a supervised classification (the maximum likelihood method), Using ENVI software, a classified image produced based on the satellite images which was taken in August 2010. This classified image was regarded as the Fire Potential map. Then, the location of the fires from 2014 to 2015 were added to the map. The result showed there is a high level of overlap between the fired locations recorded in 2014 and 2015, and the areas, which named as Very High and High Fire-potential areas on the developed map.

21.2 Conclusions

The soil, water, and vegetation are the most important elements of the environmental systems in our globe, and play vital roles in human life. In the Republic of Iraq, several environmental issues have occurred during the last decades, as a result of natural and man-induced reasons. The quantitative characteristics of those reasons on the environment in Iraq were studied, and in the next sections, some of the conclusions and recommendations of the chapters in this book of the Environmental Remote Sensing and GIS in Iraq are presented.

Chapter 2 "Using Radar and Optical Data for Soil Salinity Modeling and Mapping in Central Iraq" has concluded that the removal of vegetation cover impacts can greatly improve the salinity prediction accuracy and reliability by radar data, but radar-optical combined dataset can deliver better soil salinity prediction and mapping results.

Chapter 3 "Using Remote Sensing to Predict Soil Properties in Iraq" indicates that remotely sensed multi-temporal satellite dataset has proven to be a vital tool to predict some physicochemical soil properties on a large scale using different Landsat images-based indices, such as; NDVI, SAVI, and GDVI.

Chapter 4 entitled "Characterization and Classification of Soil Map Units by Using Remote Sensing and GIS in Bahar Al-Najaf, Iraq" showed that the remote sensing and GIS techniques contributed effectively in the identification and separation of soil units. The overall accuracy of the matrix error that obtained from the Landsat 8 image supervised classification technique was 90% for Kappa statistical coefficient, which means a very good acceptance for the thematic mapping of the classified soils.

Chapter 5 "Proximal Soil Sensing for Soil Monitoring" has concluded that the accuracy and distribution maps derived from the laboratory spectroscopy measurements proved the capability of laboratory proximal level prediction for assessing the soil Fe and Fe_2O_3 in the study area in Sulaimani, Iraqi Kurdistan region.

Chapter 6 "Proximal soil sensing Applications in soil fertility", whereas this chapter states several important conclusions, such as; there are several bands have been identified with high correlation and low RMSE. The obtained prediction model quality parameter values were at best successfully model to predict soil total N and available P, and well suited for a large variety of low to high concentrations under the condition of this study.

Chapter 7 "Multi-Temporal Satellite Data for Land Use/Cover (LULC) Change Detection in Zakho, Kurdistan Region-Iraq". The overall pattern of LULC change in the Zakho district over the past 28 years was one of the sprawl of crops and built-up lands. Also, a substantial reduction of forest and grassland indicates an acceleration stage of agriculture and urbanization. Despite the pressing land requirements for urbanization, land development, and consolidation in grass and forest areas, and the adjustment of the agricultural structure, the foundation was put for the transition to intensively use the land in the Zakho district.

Chapter 8 "Monitoring of Land Cover Changes in Iraq". One of the major causes of eco-environmental degradation in Iraq is ineffective and wasteful utilization of land cover, particularly the vegetation cover that, to a large extent, contributes to climate change. Four main factors have resulted in the development of modern land cover in Iraq; these are climate, surface and subsurface water, lithology, and relief. Hence, a strong correlation was found between these factors, and the spatial distribution of the land cover classes is strong.

Chapter 9 "Effects of Land Cover Change on Surface Runoff Using GIS and Remote Sensing: a Case Study Duhok Sub-Basin". Duhok sub-basin, Duhok, Kurdistan region of Iraq was subjected to significant land use changes in the period from 1990 to 2016. The study in this chapter indicates that the urban growth of the watershed increased from 10% in 1990 to 70% in 2016. Surface runoff volume increased from 12% in 1990 to 36% in 2016, while the vegetation land decreased from 47 to 14% in the same period.

Chapter 10 "Monitoring and Mapping of Land Threats in Iraq using Remote Sensing". This study extracted that remote sensing and GIS techniques are very useful for monitoring and mapping the main type of Iraq land degradation processes as reflected by high salt accumulation and sand dunes movements.

Chapter 11 entitled "Agricultural Drought Monitoring over Iraq utilizing MODIS Products". This chapter revealed that the year 2008 was found to be the most severe drought year during the study period (2003–2015) dominated by around 37% of severe drought. While 2009, 2011, and 2012 were the less-severe drought years dominated by mild or moderate drought with an areal coverage of 44, 50, and 48.5% respectively.

Chapter 12 "The Aeolian Sand Dunes in Iraq: A New Insight". Based on remote sensing studies, the chapter concludes that the aeolian sediments in Iraq represent an active system whose on-going migration results in loss of agricultural land, highway obstruction, etc. Moreover, the desertification is increased in the Mesopotamian Plain for recent decades.

Chapter 13 entitled "Drought Monitoring for Northern part of Iraq using Temporal NDVI and Rainfall Indices". The conclusion of this chapter finish with some remarks. First, the advantage of using SPI index is that it detects the onset of drought events and the short -time scales could be closely related to soil moisture. Second, the NDVI index is more sensitive to detect drought events and seasonal vegetation changes across all seasons. Third, drought risk-prone areas are well defined by combination of both index SPI and NDVI.

Chapter 14 "Remote Sensing and GIS for Dust Storm Studies in Iraq". The study in this chapter indicated the sand/dust storms (SDS) studies requires a comprehensive framework for implementing remote sensing and GIS in combination with other disciplines of the sciences and technologies.

Chapter 15 entitled "Drought Monitoring using a Combination of Spectral and Meteorological Indices: A Case Study in Sulaimaniyah, Kurdistan Region, Iraq". The study concludes that the use of a combination of NDVI-SPI indices provides more reliable results for drought monitoring than any single index in the study area, as well as the year 2008 was the driest one that affected Kurdistan region of Iraq.

Chapter 16 "Geo-morphometric analysis and flood simulation of the Tigris River due to a predicted failure of the Mosul Dam". It could be concluded from this chapter that the geo-morphometric analysis is quite essential for designing reliable computational hydrologic and hydraulic modelling system. Besides, the present predictions indicate that the initial flood wave will reach the Mosul city in around 2 h and the height of the flood wave will reach approx. 24 m within 8 h, while the average flood velocity is predicted to be 3.9 m/s.

Chapter 17 entitled "Hydrologic and hydraulic modelling of the Greater Zab River-Basin for an effective management of water resources in the Kurdistan region of Iraq using DEM and Raster Images". This study indicates that from this chapter that the accuracy of the computational hydrologic and hydraulic modelling systems depends on the accuracy of the extracted hydrological feature classes from the digital elevation model and calculated geo-morphological parameters.

Chapter 18 "Spatial Assessment of drought conditions over Iraq using the Standardized Precipitation Index (SPI) and GIS Techniques". The SPI values indicate that there are three main drought periods that have happened in Iraq in (1984, 1999, and 2008). The interval time between drought periods is decreased in the recent years. Using the frequency of wet and dry SPI is better for mapping than average SPI.

Chapter 19 entitled "Chapter 19 Assessing the Impacts of Climate Change on Natural Resources in Erbil Area, the Iraqi Kurdistan using Geo-Information and Landsat Data". Even in a shortage of the available data, it was more than clear the benefit to combine climate and remote sensing data for monitoring and understanding LULC changes.

Chapter 20 "Mapping Forest-Fire Potentiality using Remote Sensing and GIS, Case study: Kurdistan Region-Iraq". The results of this study for the study area showed that the remote sensing (RS) and GIS could be used to map the potential of fire. Therefore, RS and GIS are a very strong and cost-effective tools to estimate forest fire potentiality, as well as, a reliable tool for managers to take preventive actions.

21.3   Recommendations

The materials presented in the chapters of "Environmental Remote Sensing and GIS in Iraq" emphasises the recommendations:

1. Radar data have great potential for soil salinity mapping. Development of radar-based approach for minimization of vegetation cover impacts on the backscattering coefficients without any dependence on the optical data is recommended in future work.
2. The Landsat images and remote sensing techniques are very accurate, useful and helpful to predict the properties of the surface soil as they can consume efforts, time and money.

3. Emphasize the importance of using RS and GIS techniques in soil survey and classification studies because of the speed and accuracy of the work, preparation of soil survey and classification maps, Add to reducing the effort and cost compared to traditional field surveys.

4. There is still adherence to a consistent and standard protocol in Proximal Soil Sensing (PSS) is needed. More importantly, development of Soil Spectral Libraries (SSLs) is required. Also, detail chemical and physical data need to be recorded.

5. Finally, chapter puts its recommendations to pay attention toward strategic of the Iraqi Soil Spectral Library at the country to directly be used in soil monitoring and fertilization management.

6. Work on forming institutions that are concerned with environmental affairs at the regional, local and national levels, and provide them with accurate applied research to show other factors that have a significant impact on land cover and land use. Reducing urban expansion in agricultural areas, especially in the Sindi plain. Moreover, work on the vertical expansion of residential areas to accommodate population increment.

7. We recommend formulating science-based policies for setting up a program to monitor the changes in the land cover of the Iraqi territory. Also, efforts to educate farmers and rural communities on the importance of using latest technology of land management and conservation. Finally, it is important to reduce soil salinity and control sabkha development by preventing use of Al-Tharthar Lake water for irrigation as it contributes to washing out of salts that ultimately enters the Euphrates River.

8. The study recommend that the future studies should use high-resolution satellite images and different techniques to clarify impact of climate change and urban growth on the Run-off. The new studies should be adopted object-based analysis to classify imagery. It is called a segment that is a cluster of the pixel with comparable spectral, spatial and texture attributes for urban detection.

9. There is great need to monitor the natural resources and their properties in practical place and time. The application of remote sensing data have proven useful technology to monitor and mapping the natural resources and their properties and improving the accuracy and consumed time, effort and coast.

10. Given the latest advances made in hydrologic/land surface modeling, availability of other satellite data including those with soil moisture which can characterize the antecedent soil moisture condition, and also state of the art data assimilation, Iraq can benefit from new approaches for drought monitoring and forecasting and recovery.

11. We recommend formulating science-based policies for mitigating impacts of the hazardous aeolian processes, particularly, in fertile agricultural regions. Also, current soil-eroding surface irrigation method should be replaced by water-conserving modern irrigation methods, such as drip irrigation or subsurface irrigation. Furthermore, mobile sand dunes should be stabilized through the development of natural herbaceous cover, using special perennial species that can tolerate harsh climatic condition and salty soil. Finally, strict regulations

should be enforced to protect date palm orchards and other agriculture lands from pressures of urban expansion, particularly around main cities.

12. The chapter recommended to build up future studies on the specific region information to get early warning information to reduce the impact of drought. More attention should be focused toward archiving and reporting data and information to simulate Iraq drought trends and monitor future drought events.

13. The most important recommendations are to invest in data collection, data sharing, and data analysis to support the remote sensing and GIS applications in SDS studies.

14. The using of time series of satellite images for monitoring drought corresponding with satellite-based meteorological dataset could lead to promising studies for monitoring and forecasting the droughts events is recommended in future work.

15. The using of time series of satellite images for monitoring drought corresponding with satellite-based meteorological dataset could lead to promising studies for monitoring and forecasting the droughts events is recommended in future work.

16. This chapter includes the most important recommendation which suggest that the required sufficient height of the repulse dam downstream of the Mosul Dam should be no less than 92 m high in order to protect the Mosul city from the flooding disaster in case of the collapse of the Mosul Dam.

17. This chapter includes the most important recommendation which suggest the use of digital elevation model and the other raster images through most developed GIS and RS software in order to obtain more reliable hydrologic hand hydraulic results. Thus, the existed water resources in the region under study could be managed more effectively and accurately comparing to the other traditional methods.

18. Continuous monitoring the drought cycles over Iraq. Implement agricultural plans to overcome the drought crises. Construct dams in large valleys basins to harvesting rainfall water during rainy years, and exploitation of groundwater in the area prone to periodic drought conditions.

19. For the future, at first, it is important to continue and improve such monitoring of climate change and LULCC. Additionally, on the political side, it is necessary to establish laws to ensure the proper management and use of land and natural resources. From political and economical side, it is necessary to involve stakeholders and decision makers at the local and regional level when thinking in natural resource management. Moreover, on the educational side, environmental awareness programs are necessary to enable rural people to have a better understanding of and ability to manage their land and natural environment.

20. To get fire statistics also from Soran, Amedi, Zakho, Choman, Pishdar, Gharadakh, and Dukan subdistricts to make better assessments. Furthermore, using other multispectral and hyperspectral satellite image, accompanied by other satellite-based indices might result in more reliable and precise maps.

References

AL-Rajehy AM (2002) Relationships between soil reflectance and soil physical and chemical properties, M.Sc. Thesis, Mississippi State University, Mississippi, USA, pp 75

Ben-Dor E (2008) Imaging spectrometry for soil applications. Adv Agron 97:321–392

Fadhil AM (2011) Drought mapping using geoinformation technology for some sites in the Iraqi Kurdistan region. Int J Digit Earth 4(3):239–257

Friedl MA, Sulla-Menashe D, Tan B, Schneider A, Ramankutty N, Sibley A, Huang X (2010) MODIS Collection 5 global land cover: algorithm refinements and characterization of new datasets. Remote Sens Environ 114:168–182. https://doi.org/10.1016/j.rse.2009.08.016

Harris RF, Karlen DL, Mulla DJ (1996) A conceptual framework for assessment and management of soil quality and health. In: Jones AJ, Doran JW (eds) Methods for assessing soil quality, vol 49. SSSA Special Publication, Madison, WI, pp 61–82

Jawarneh RN, Biradar CN (2017) Decadal national land cover database for jordan at 30 m resolution. Arab J Geosci 10(22)

Lam DK, Remmel TK, Drezner TD (2011) Tracking desertification in California using remote sensing: a sand dune encroachment approach. Remote Sens. 3:1–13. https://doi.org/10.3390/rs3010001

Letha J, Thulasidharan Nair B, Amruth Chand B (2011) Effect of land use/land cover changes on runoff in a river basin: a case study. WIT Trans Ecol Environ 145:139–149

Slaymaker O (2001) The role of remote sensing in geomorphology and terrain analysis in the Canadian Cordilleran. J Appl Earth Obs Geoinf 3(1):7

Tolessa T, Senbeta F, Kidane M (2017) The impact of land use/land cover change on ecosystem services in the central highlands of Ethiopia. Ecosyst Serv 23(Supplement C):47–54. https://doi.org/10.1016/j.ecoser.2016.11.010

UNESCO (2014) Integrated drought risk management, DRM: national framework for Iraq, an analysis report (Tech. Rep.). Amman: UNESCO Office Iraq (Jordan). Retrieved from http://www.unesco.org/new/fileadmin/MULTIMEDIA/FIELD/Iraq/pdf/Publications/DRM.pdf

United Nations Development Programme UNDP (2010) Drought impact assessment, recovery and mitigation framework and regional project design in Kurdistan region (KR). December 2010

Viscarra Rossel RA, McBratney AB (1998) Laboratory evaluation of a proximal sensing technique for simultaneous measurement of clay and water content. Geoderma 85:19–39

Woldesenbet TA, Elagib NA, Ribbe L, Heinrich J (2017) Hydrological responses to land use/cover changes in the source region of the Upper Blue Nile Basin, Ethiopia. Sci Total Environ 575:724–741

Wu W, Al-Shafie WM, Muhaimeed AS, Ziadat F, Nangia V, Payne W (2014) Soil salinity mapping by multiscale remote sensing in Mesopotamia, Iraq. IEEE J Sel Top Appl Earth Obs Remote Sens 7(11):4442–4452. https://doi.org/10.1109/jstars.2014.2360411